ISBN 978-0-282-65742-0
PIBN 10086217

# 1 MONTH OF
# FREE
# READING

## at
## www.ForgottenBooks.com

By purchasing this book you are eligible for one month membership to ForgottenBooks.com, giving you unlimited access to our entire collection of over 1,000,000 titles via our web site and mobile apps.

To claim your free month visit:
www.forgottenbooks.com/free86217

English
Français
Deutsche
Italiano
Español
Português

# www.forgottenbooks.com

**Mythology** Photography **Fiction**
Fishing Christianity **Art** Cooking
Essays Buddhism Freemasonry
Medicine **Biology** Music **Ancient
Egypt** Evolution Carpentry Physics
Dance Geology **Mathematics** Fitness
Shakespeare **Folklore** Yoga Marketing
**Confidence** Immortality Biographies
Poetry **Psychology** Witchcraft
Electronics Chemistry History **Law**
Accounting **Philosophy** Anthropology
Alchemy Drama Quantum Mechanics
Atheism Sexual Health **Ancient History**
**Entrepreneurship** Languages Sport
Paleontology Needlework Islam
**Metaphysics** Investment Archaeology
Parenting Statistics Criminology
**Motivational**

THE

# SOUTH AMERICA PILOT.

## PART II.

### FROM THE RIO DE LA PLATA TO THE BAY OF PANAMA, INCLUDING MAGELLAN STRAIT, THE FALKLAND, AND GALAPAGOS ISLANDS.

By Captains PHILLIP PARKER KING and ROBERT FITZROY,

ROYAL NAVY.

*FIFTH EDITION.*

PUBLISHED BY ORDER OF THE LORDS COMMISSIONERS OF THE ADMIRALTY.

LONDON:
PRINTED FOR THE HYDROGRAPHIC OFFICE, ADMIRALTY;

AND SOLD BY

J. D. POTTER, *Agent for the Admiralty Charts,*
31 POULTRY, AND 11 KING STREET, TOWER HILL.
1860.

*Price 5s.*

# ADVERTISEMENT

## TO THE FIFTH EDITION.

THE Fifth Edition of the South America Pilot contains Sailing Directions for the East and West Coasts of that Continent, from the Rio de la Plata round Cape Horn to Guayaquil and the Bay of Panama, including Magellan Strait and the Falkland and Galapagos islands.

The former editions were compiled by Captain Robert Fitz Roy, from surveys in H.M.S. *Beagle*, and those of Captain P. P. King, in H.M.S. *Adventure*, made between the years 1826 and 1834; the directions for the Coasts of Patagonia and Magellan Strait being by the latter officer. A further examination of the interior of the Falkland Islands and their numerous harbours was made by Commander W. Robinson, and subsequently by Captain B. F. Sulivan, C.B., during the years 1838–45; and from the surveys of these officers the third and fourth chapters have been compiled.

The whole has been revised by Mr. Thomas A. Hull, Master R.N., who has added directions for the West Coast from Guayaquil to the Bay of Panama, derived chiefly from the surveys of Captain H. Kellett, C.B., and Commander James Wood, R.N., in the years 1845–8, and from the Remark books of officers in Her Majesty's ships on that coast up to the present period.

I. W.

Hydrographic Office, Admiralty,
    1st June, 1860.

# CONTENTS.

## CHAPTER I.

### RIO DE LA PLATA TO THE RIO NEGRO.

| | Page |
|---|---|
| Banks and currents between Piedras point and Cape Corrientes - - | 1–5 |
| Mogotes point.  Sierra Ventana.  Ascuncion point.  El Rincon or Bahia Blanco; directions - - - - - - - | 5–10 |
| Port Belgrano; dangers at entrance; and directions - - - | 11–14 |
| False Bay.  Horn spit, Labyrinth shoals, and Brightman inlet - - | 15, 16 |
| Rio Colorado.  Union bay.  Anegada bay - - - - | 18–20 |
| San Blas; directions.  San Blas banks - - - - - | 21–24 |
| Rio Negro; tides, and directions - - - - - | 25–30 |

## CHAPTER II.

### RIO NEGRO TO THE STRAIT OF MAGELLAN.

| | Page |
|---|---|
| Cape Bermejo - - - - - - - - | 31 |
| Port San Antonio.  Port San Josef - - - - - | 33–36 |
| Valdes creek.  Neuvo gulf - - - - - - | 37–39 |
| Castro point to Port Santa Elena - - - - - | 41–43 |
| Leones isle.  Port Melo.  Gulf of St. George - - - - | 44–48 |
| Cape Blanco.  Port Desire.  Sea Bear bay - - - - | 49–53 |
| Spiring bay.  Cape Watchman.  Bellaco rock - - - - | 53–55 |
| Port San Julian.  River Santa Cruz - - - - - | 56, 57 |
| Cape Fairweather.  Port Gallegos - - - - - | 61, 62 |
| Winds, weather, tides, and currents - - - - - | 63 |

## CHAPTER III.

### THE FALKLAND ISLANDS.

| | Page |
|---|---|
| East Falkland; general description - - - - - | 68–74 |
| Making the land.  Cape Dolphin.  Port Salvador - - - | 75–77 |
| Uranie rock.  Berkeley sound - - - - - | 78 |
| Port William.  Stanley harbour; lights; tides - - - - | 80, 81 |
| Port Harriet.  Port Fitzroy - - - - - | 84–86 |
| Port Pleasant.  Pleasant road.  Choiseul sound - - - | 88–90 |
| Lively sound.  Shag rock.  Beauchêne island - - - | 92–94 |
| Adventure sound.  Bay of harbours; Bull road - - - | 94–96 |
| Eagle passage - - - - - - - | 97 |

# CHAPTER IV.

## FALKLAND SOUND AND WEST FALKLAND.

|  | Page |
|---|---|
| Port San Carlos.  Port Sussex.  Newhaven | 100 |
| Cygnet, King, and Findlay harbours | 101 |
| White Rock bay.  Manybranch harbour.  Fox bay | 102 |
| Tides and directions for Falkland sound - | 103 |
| Tamar harbour.  Pebble sound.  Kepple sound.  Port Egmont | 105–107 |
| Race rocks.  Brett harbour.  Hope harbour | 107, 108 |
| Tides on the North coast.  Jason islands | 108–110 |
| King George bay.  Queen Charlotte bay.  Port Philomel | 111–113 |
| New island, Grey channel.  Weddell island, Smylie channel | 115–118 |
| Port Stephens.  Port Albemarle - | 118–120 |
| Port Edgar | 120 |

# CHAPTER V.

## STATEN' ISLAND, AND THE OUTER OR SEA COAST OF TIERRA DEL FUEGO:

| | |
|---|---|
| New Year islands.  St. John harbour | 123 |
| Port Cook.  Port Basil Hall.  Port Parry | 124, 125 |
| Port Hoppner.  Port Vancouver.  Back harbour | 127, 128 |
| Natural history; geology; vegetation | 128 |
| Coast of Tierra del Fuego; aspect; soundings | 129–131 |
| Catherine point to Cape San Diego | 132–134 |
| Good Success bay.  Strait Le Maire | 134, 135 |
| Beagle channel.  Goree road.  Nassau bay | 136 |
| Wollaston islands.  Hermite islands.  St. Martin cove.  Cape Horn | 137–140 |
| Hardy Peninsula.  Diego Ramirez.  Ildefonsos.  New Year sound | 141, 142 |
| Christmas sound.  Londonderry islands - | 143, 144 |
| Camden islands.  Magill isles - | 145, 146 |
| Barbara channel.  Noir island.  Grafton islands | 147–149 |
| Landfall islands.  Week islands.  Dislocation harbour.  Cape Pillar | 150–152 |
| General observations; Cape Horn currents | 153 |

# CHAPTER VI.

## STRAIT OF MAGELLAN; CAPE VIRGINS TO THE BARBARA CHANNEL.

| | |
|---|---|
| Cape Virgins.  Dungeness point.  Sarmiento bank.  Virgins reef | 157, 158 |
| Wallis shoal.  Cape Possession.  Orange bank - | 159, 160 |
| First Narrows.  Barranca ledge.  Triton bank - | 161 |
| Second Narrows.  Elizabeth island.  Royal road | 162, 163 |

Page

Santa Magdalena bay. Laredo bay. Sandy point. Freshwater bay.
Port Famine - . . . . . . - 164–167
Cape San Isidro to Glascott point. Cape Froward. Useless bay - 169–172
Dawson island. Gabriel channel. Mount Sarmiento. Admiralty sound 173–176
Magdalen sound. Cockburn channel - . . . - 176–178
Barbara channel - . . . . . . - 179


## CHAPTER VII.

### STRAIT OF MAGELLAN; BARBARA CHANNEL TO CAPE PILLAR.

Clarence island. Fortescue bay, Port Gallant - . . - 183–187
Jerome channel, Otway water - . . . . - 188, 189
Charles islands. Crooked reach, Borja bay, El Morion - . - - 190–194
Long reach. Swallow bay. Playa Parda - . . - 194–197
Gulf of Xaultegua. Sea reach. Cape Tamar - . . - 200–202
Port Tamar. Sholl bay. Smyth channel - . . - 203, 204
Valentine harbour. Port Mercy. Cape Pillar - . . - 206, 207
Sir John Narborough islands. Los Evangelistas . . . 208


## CHAPTER VIII.

### PASSAGES ROUND CAPE HORN, AND THROUGH THE STRAIT OF MAGELLAN.

Strait le Maire. Weather off Cape Horn - . . - 209–212
Passage through the Strait of Magellan from the Atlantic to the Pacific.
First Narrows. Froward reach. Long reach - . . - 212–216
From the Pacific to the Atlantic. Cape Victory. Gregory bay - - 216–218
Remarks by Captain J. L. Stokes, R.N. - . . . . 219
Tides - . . . . . . . . 221


## CHAPTER IX.

### PATAGONIA, WEST COAST; INNER CHANNELS.

Smyth channel. Victory Pass. Interior sounds - . - 223, 228
Sarmiento channel. Estevan channel. Guia narrows - . - 228, 229
Concepcion channel. Wide channel. Indian reach. Eden harbour - 230, 231
English Narrows. Messier channel. Island harbour - . - 232–234
Cape Isabel. Madre Is. Port Henry. Gulf of Trinidad - - 234–236
Port Barbara. Fallos channel. Guaianeco Is. Kelly harbour - - 238–241
San Estevan gulf. Cape Tres Montes. Port Otway; tides. - - 242–245

# CHAPTER X.

## GULF OF PENAS TO CHILOE ISLAND.

Page

San Andres bay.   Port San Estevan   -   -   -   -   -   247
Anna Pink bay.   Vallenar road.   Haumblin island   -   -   - 248, 249
Guaytecas Is., Port Low -   -   -   -   -   -   - 250, 251
Chiloe I.   San Pedro harbour.   Port San Carlos;   light.   Chacao
    narrows   -   -   -   -   -   -   - 251–255
Ancud gulf.   Corcovado gulf   -   -   -   -   - 258–269
Port Montt.   Reloncavi sound; tides   -   -   -   - 270–272

# CHAPTER XI.

## CHILOE ISLAND TO COQUIMBO BAY.

Carelmapu Is.   Cape San Antonio   -   -   -   - 273, 274
Valdivia.   Mocha island -   -   -   -   -   - 274–276
Santa Maria island.   Arauco bay   -   -   -   - 278, 279
Concepcion.   Talcahuano.   River Maule   -   -   - 281–283
Valparaiso   -   -   -   -   -   -   -   286
Quintero and Horcon bays.   Port Papudo   -   -   - 289–290
Pichidanque.   Port Herradura.   Coquimbo bay.   La Serena   - 291–296

# CHAPTER XII.

## COQUIMBO BAY TO LAVATA BAY.

Pajaros islets.   Tortoralillo bay.   Toro reef   -   -   - 299, 300
Port Huasco.   Herradura de Carrisal   -   -   -   - 302–304
Pajonal cove.   Copiapo -   -   -   -   -   - 306–308
Port Yngles and Port Caldera.   Port Flamenco -   -   - 310–312
Lavata bay.   Hueso Parado   -   -   -   -   -   315
Tides and currents; winds; passages   -   -   -   - 316–318
Juan Fernandez, and Mas a fuera islands   -   -   - 318, 319
St. Ambrose, and St. Felix islands   -   -   -   -   320

# CHAPTER XIII.

## LAVATA BAY TO CALLAO ROAD.

Nuestra Señora bay.   Paposo   -   -   -   -   -   321
Mount Moreno.   Constitucion harbour -   -   -   -   322
Leading bluff.   Cobija bay.   Paquiqui -   -   -   - 323–325
Loa river.   Iquique.   River Pisagua   -   -   -   - 326–329
Port Arica.   Ylo road.   Islay bay   -   -   -   - 330–333
Chala point.   Port San Juan.   Port San Nicholas   -   - 337, 338
Independencia bay.   Boqueron de Pisco -   -   -   - 340, 341
Pisco bay.   Chincha islands   -   -   -   -   - 342, 343
Port Chilca.   Morro Solar.   Chorillos bay   -   -   - 345, 346

# CHAPTER XIV.

## CALLAO TO THE RIVER TUMBEZ.

| | Page |
|---|---|
| Callao, Lima, San Lorenzo island, El Boqueron. Hormigas de afuera | 347–351 |
| Haura islets. Huacho bay. Supé bay | 352–354 |
| Guarmey, Casma, and Samanco bays | 355–357 |
| Ferrol and Santa bays. Huanchaco road | 358–360 |
| Truxillo. Malabrigo road. Lambayeque road | 360–362 |
| Lobos de afuera, Lobos de tierra | 362, 363 |
| Aguja point. Port Payta | 363, 364 |
| Pariña point. Cape Blanco. River Tumbez | 365, 366 |
| Winds; weather; currents; passages | 367 |

# CHAPTER XV.

## RIVER TUMBEZ TO CAPE CORRIENTES, INCLUDING THE GALAPAGOS ISLANDS.

| | Page |
|---|---|
| Gulf of Guayaquil, Amortajada, Payana shoals | 371, 372 |
| Mala hill, Mala bank, Directions | 373–375 |
| River Guayaquil, Morro channel, Estero Salado | 376–379 |
| Santa Elena bay, Salango island, Port Manta | 380, 381 |
| Carracas river, Cape San Francisco, Atacames bay, Esmeraldas river | 382–385 |
| Posa harbour, Port Tumaco, Gorgona island, Buenaventura river | 386–389 |
| Magdalena bay, Point Chirambirá, Winds and Weather | 391–393 |
| Malpelo island, Rivadeneyra shoal | 393 |
| Galapagos islands | 394–400 |

# CHAPTER XVI.

## CAPE CORRIENTES TO PANAMA.

| | Page |
|---|---|
| Cape Corrientes, Cabita bay, Cupica bay | 401–403 |
| Piñas bay, Garachiné point, Bay of San Miguel | 404–407 |
| Darien harbour, Tuyra and Savannah rivers | 408, 409 |
| Pearl islands | 410–414 |
| Brava point, Trinidad and Chiman rivers, Pelada and Chepillo islands | 414, 415 |
| Panama road | 416–422 |
| Taboga island, Chamé bay, Parita bay | 423–425 |
| Cape Mala, Mariato point, Cocos island | 426, 427 |
| General observations, winds, currents, passages | 427–429 |
| Pacific Passage Table | 433 |
| Table of Positions | 437 |

IN THIS WORK THE BEARINGS ARE ALL MAGNETIC,
EXCEPT WHERE MARKED AS TRUE.

THE DISTANCES ARE EXPRESSED IN SEA MILES OF
60 TO A DEGREE OF LATITUDE.

A CABLE'S LENGTH IS ASSUMED TO BE EQUAL TO
100 FATHOMS.

THE

# SOUTH AMERICA PILOT.

## PART II.

### CHAPTER I.

RIO DE LA PLATA TO THE RIO NEGRO.

VARIATION in 1860, from 11° E. to 15° E.   Annual decrease about 4'.

PIEDRAS POINT, the south point of entrance of the Rio de la Plata, should be approached with more caution than any other land near the entrance of the river, excepting the north-east part of Cape San Antonio, which lies about 55 miles to the southward.   The point is low and ill defined.   A few trees of stunted growth show themselves at a small distance in shore; but as the land is almost flat, and not 20 feet above the level of the river, it is difficult for a stranger to recognize the spot, either by description or by a drawing.*

Very near this point, about half a mile in-shore, is a single tree ; 2 miles south-west of the point is another tree, rather larger.   Six miles to the south-west is a clump of trees ; 4 miles beyond which is another clump ; all easily distinguishable by those who are accustomed to low land, but appearing  like bushes to eyes familiar with such grand scenery as that of the coast of Brazil.

Northward, and to the north-west of Piedras point, there are no distinct trees ; although straggling bushes are sometimes altered in their appearance, by refraction, so much as to deceive a practised eye.

PIEDRAS BANK.—Towards the east and south of Piedras point extends a dangerous bank, to the distance of 7 miles in an easterly direction, and  15 miles towards the south.   The north part is hard clay, with many

* See Admiralty Chart :—East coast of South America, Rio de la Plata, No. 2,544 ; scale, m = 0·2 of an inch.

A

patches of *tosca* (clay hardened to the consistence of half-baked bricks), almost as injurious to a vessel as actual rock. Indeed, it is not improbable that greenstone similar to that of Monte Video and Maldonado, may reach the surface here, as well as at the Chico and the English bank.

Within the limits of the Piedras bank the bottom is uneven, and the lead cannot be trusted. Outside of the limits a vessel may go by the lead, with confidence, according to her draught of water. But in estimating distance from this low land by the eye, one may be very much deceived, so much is it at times either elevated or apparently depressed by refraction. In approaching the banks from the northward or eastward, the ground becomes harder, and the water decreases gradually in depth; from the southward the decrease of depth is somewhat quicker, though the bottom is not nearly so hard.

**SANBORONBON BAY.**—From Piedras point to the Salado river, a distance of about 20 miles to the southward, the coast forms a slight curve called Sanboronbon bay, from a small stream of that name which falls into it about 2 miles to the northward of the Salado. The shore is uniformly low and level. Besides the clumps already mentioned, there are but a few straggling trees. Having given a wide berth to the south-east part of the Piedras bank, and being to the southward of it, a vessel may close the land, from 5 to 8 miles, according to her draught, off the mouth of the Salado.

Northward the land is, as has been said, uniformly low, not exceeding 20 feet above the level of the water, in many places much lower; but to the southward of the river, distant 5 miles from its mouth, is a rising ground covered with trees, called Mount Rosas. The highest part of this mount may be 30 feet above the water. Being covered with trees, and higher than the adjoining land, it affords a good mark for the entrance of the Salado. Another remarkable object is a red brick-kiln upon the shore, 2½ miles S.S.W. of the mouth of the river. When within 4 miles distance from the beach, the entrance of the river is distinguishable.

**TUYU BANK.**—From Mount Rosas to Cape San Antonio the land is very low, and quite flat. In many places, especially near the little river Tuyu, by Cape San Antonio, it is a mere marsh. Trees occasionally show themselves as if to assure one that the dark line in the horizon is actually land, and not the shadow of a cloud. The great extent of the Tuyu bank, called also Arenas Gordas, which at Cape San Antonio extends nearly 10 miles from the shore, prevents any even the smallest vessel from approaching this half-drowned land. The ground near, and even on the bank, is extremely soft; the depth decreases gradually, and with the lead going there is no danger.

RIO SALADO.—The Salado is a very shallow bar river, unfit for anything but the smallest vessels. At times, when the Plata is high, there are 6, 8, or 10 feet water on the bar. But at other times the smallest boat cannot even approach the mouth of the river; and the mud is so soft that one cannot walk from the boat, lying aground, to the firm land. There are a few houses near the Salado, and on Mount Rosas.

CAPE SAN ANTONIO.—Rasa or Flat point, the northern extremity of the ill-defined Cape San Antonio, is a low sandy spit, extending to the northward, and under water towards some breakers, near the northern limit of the Tuyu bank. Close to the westward of Rasa point is the little river Tuyu, communicating with several lakes, having 2 fathoms water at its entrance. It forms a creek winding through the Tuyu bank, by which a very small vessel can approach and enter that river. The creek is difficult to find, and at the present day of no consequence; but in a few years its position may be changed, so soft and yielding is the ground through which it passes.

At Rasa point the shore assumes a different appearance. The almost united land and water of the coast of Sanboronbon bay is succeeded by a well defined, though low extent, of sand and shingle. The Tuyu bank gradually diminishes, and a few miles south of Rasa point vessels may approach the land as near as 2 or 3 miles. In the old charts, a space exceeding 30 miles of sea coast has been called Cape San Antonio; the part near Rasa point having been called the north end, and the other extreme the south end. Some confusion has thus arisen; and to avoid mistakes the northern part, near Rasa point, will here be considered, and called Cape San Antonio; the southern will be distinguished by the name Medano point, which is 42 miles from Rasa point.

The CURRENTS off Cape San Antonio set into or out of the Rio de la Plata, varying in their strength and duration as the winds vary, by which they are principally governed. Generally speaking, the current sets to the northward before and during the commencement of southerly winds, from 1 to 3 knots; and to the southward before and during the beginning of northerly winds, with about the same strength.

When there has been an unusual flood in the inland countries, and the sea is at low ebb, or when the sea spring tide is unusually high and the river is the reverse, the current may set round Cape San Antonio at least as strongly as it has been known to run past Lobos island, on the northern side of the entrance to the Plata, at the rate of 5 or 6 knots. These, however, are extreme cases, of rare occurrence.

ASPECT.—The coast from Cape San Antonio, southward, is of a light colour, low and sandy. Occasionally straggling bushes, or patches of

rough grass, are seen. Sand-hills between 20 and 40 feet in height begin
to show themselves 10 miles to the southward of Rasa point, gradually increasing in number and height as they approach Medano point, rising
near that point to 100 feet above the sea. Two of these sand-hills near
one another, in lat. 36° 27′ S., remind one of a Spanish saddle ; they are
rather higher than their neighbours. Two other rather marked sand-hills
have been used as fixed points in surveying the coast. One called Medano
Chato, or flat sand-hill, in lat. 36° 28′ S. ; the other Medano Alto, or high
sand-hill, in lat. 36° 46′ ; but whether a stranger would recognize them,
or whether the wind will leave them in their present form, is very
doubtful.*

Between 5 and 10 miles off shore E. S. E. from Rasa point, it was
found in the course of the survey that the quality of the bottom appeared
different, in two different years, although the depth remained the same.
Being satisfied of the vessel's position with respect to the land, this change
in the soundings occasioned a suspicion that soft oozy mud from the Tuyu
bank is at times carried round Rasa point, and deposited upon the sandy
bottom, usually found to the south-east of that point, until a strong current, or gale, from the sea washes it again into Sanboronbon bay.

MEDANO BANK is an extensive and dangerous shoal, stretching
6 miles to seaward from Medano point, and which at any time, even in fine
weather, must be given a wide berth. In crossing it, irregular and shoal
soundings were obtained ; and at 3 or 4 miles from the shore there are
places over which are not more than 2 fathoms water.

About Medano point the land is higher than to the northward or to the
southward. A range of hills between 100 and 200 feet in height, appears to stretch to the north-west, or more westerly. The Medano shoal
seems to be a submarine continuation of that range ; irregular soundings,
with many shoal spots, may therefore be expected.

From Medano point to the narrow isthmus, between the sea and the
lagoon called Mar Chiquito, the coast is lower than near Medano point ;
but it has a similar appearance, sand-hills, with a few patches of verdure,
being the only objects on which the eye can rest. These sand hills, and
the coast near them, have a whiter look than those to the northward of
the point.

In approaching this part, there is no danger while at a reasonable distance from the shore of from 1 to 3 miles, according to the weather ; but
as in some places, especially to the northward, near Medano point, the
soundings are irregular, shoaling suddenly, a fathom or two at a time, and
then deepening again, it is as well not to go nearer than three miles.

---

* See East Coast of America, Sheet IX., No. 1,324 ; scale, d = 4·0 inches.

**TIDES.**—It is high water, full and change, off Medano point at 11h., rise and fall being 6 feet.

**MAR CHIQUITO** is a lagoon of salt water (visible from the mast-head of a passing ship) into which flow the Tandil and other small rivers. It is 60 miles from Medano point, and at times overflows and runs into the sea, but generally there is a dry bank of shingle between the two. From the spot where the Mar Chiquito or Little sea overflows, the land rises and is no longer sandy. A low range of cliffs, from 20 to 30 feet in height, is surmounted by a rising ground, of which the highest part is about 80 feet above the sea. Pasture land now meets the eye. On that high ground near which is the Estancia de la Loberia chica (small Seal Farm), thousands of fine cattle may sometimes be seen feeding.

**CAPE CORRIENTES** is a high and rather a bold headland; the south-eastern extremity of a range of hills running nearly east and west. The Sierra Tandil and Sierra Vulcan form part of this range. In clear weather three ranges of the latter are visible from a vessel at sea, and have a singular wedge-like form, somewhat resembling the Bill of Portland. These ranges of high land (high compared with the pampa, or plain country) are like the downs on the English coast, but they do not end so abruptly, nor with such imposing cliffs. Near the sea they slope away gradually, and are ended by broken rocky shore.

Half a mile to the northward of the cape is a little bay, where a boat may land in fine weather. The southern side of the bay is bounded by the rising ground of the cape : the northern by the cliffy shore under the Estancia. A vessel may anchor in this bay during off-shore winds, in from 5 to 10 fathoms water, over a clean sandy bottom; but with easterly winds of any strength a heavy swell would set in, and render the anchorage unsafe.

There is a bare sandy place on the east side of Cape Corrientes, around the upper part of which the green turf has so regular an edge that it appears artificially cut; and all the higher part of the cape is covered by smooth green turf, without either trees or bushes. The highest part is 120 feet above the level of the sea.

**MOGOTES POINT,** lies 5 miles to the southward of Çape Corrientes ; it is high, bare, and sandy ; ending towards the sea in a low projecting spit, which requires a berth of 2 miles in rounding the point. The summit is 104 feet above the sea. When near the point several sand-hills may be distinguished, some of which are peaked and higher than others, whence the name Mogotes, signifying insulated rocks or pointed cornstacks, which these sand hills rather resemble.    Behind the sand-hills the down-like hills, described above, extend to the westward.

Off Mogotes point it has been said that there is a shoal ; but no less water than 7 fathoms near the shore, and 9 fathoms at 5 miles distance, was found in the course of the survey by H.M.S. *Beagle* in 1832, when seeking expressly for the reported shoal, excepting on the ridge which continues under water one mile from the spit off the point.

**ANDRES HEAD** is the south-west extremity of a range of high bold cliffs, which extend from the north-east half way between that head and Mogotes point.    From the place where they end the shore is low, sandy, and rocky.    The bay, or rather bight, between Mogotes point and Andres head should not be entered by ship or boat.    Many sunken rocks lie near the shore, causing blind breakers at sudden intervals. A boat and three of her crew were lost only a few years since in this place.    A wave rose under the boat without warning and filled her in an instant, washing the men overboard.    The highest part of the cliffs, near Andres head, is 70 feet above the sea.    A short distance in-shore of those cliffs is the Estancia de la Sociedad (Society farm), formerly the Estancia de la Loberia grande (Great Seal farm).

**TIDES and CURRENTS.**—It is high water, full and change, off Andres Head at 10h., rise and fall being 8 feet.    The currents off the projecting land (which extends from Andres head to Cape San Antonio) set strongly to the northward previous to and during southerly winds; and as strongly in the opposite direction under contrary circumstances.    From 1 to 3 knots are usually the limits of strength, although there are intervals when no current is perceptible, and times, although rare, when its strength may exceed that above mentioned.

**The COAST** from Andres head to Hermenegildo point is rugged, and from 30 to 80 feet in height.    There are a few detached irregular cliffs, and some gaps, or creeks, which might afford a landing-place for a boat in fine weather ; but there is neither shelter nor anchorage for a ship.    Close to Hermenegildo point is a little bight, into which runs a small stream of fresh water.

Very few bushes appear on this part of the shore, and scarcely a tree, excepting a few near the Estancia de la Sociedad.    Sometimes a considerable extent of grass land is seen, but in most places near the sea the ground appears sandy and barren, thinly covered here and there by coarse grass, or by low prickly shrubs.

From Hermenegildo point to Black point, and thence onwards to Asuncion point and Mount Hermoso, the coast has a similar appearance, and is equally unfit to approach. Occasionally the sand-hills rise higher, to 100 or 130 feet above the sea, and some are more than usually barren, or there are a few more bushes, and rather more grass, to vary the view; but there is no other variety in this monotonous coast.

GUEGUEN, or JOSEF RIVER, runs into the sea about 5 miles to the eastward of Black point. Its entrance is accessible to boats during moderate weather, when there is not much swell ; but it is inaccessible to small vessels. A heavy swell is generally rolling towards this shore, so fully exposed to the Southern Ocean.

CAUTION.—If a vessel should anchor near any part of this coast, it is probable that she will lose or break her anchor in endeavouring to weigh. Hard *tosca*, full of holes, receives the anchor in most places. In one week H.M.S. *Beagle* broke three anchors during rather fine weather. On the lead fine brown sand and broken shells always came up, indicating a clean bottom ; but the sand lay thinly over the treacherous tosca, as we afterwards found. Tosca is a sort of clay almost turned into stone. It is about as hard and as tough as a brick two-thirds baked. The constant action of the sea wears holes in the upper surface, and in those holes the flukes of the anchor become fixed. By heaving nearly up and down, and waiting several minutes before attempting to weigh the anchor, the tosca will probably crack or crumble, and give way before the steady strain of the anchor; but if a sudden, forcible strain is applied, the anchor or cable will give way sooner than the tosca.

There is sufficient depth of water for the largest ships near all the coast between Cape San Antonio and Asuncion point, excepting near the Medano bank and Mogotes spit. As the land is approached the soundings decrease gradually. They extend to a great distance. When 50 miles off shore the depth is between 30 and 60 fathoms. At 100 miles distance there are between 50 and 80 fathoms. The bank of soundings may be traced to an average distance of 150 miles from the land.

Ariel Rocks are said to have been discovered by the *Ariel*, of Whitehaven, Dixon master, and have been sought for repeatedly, without the least success. Their alleged situation is in the track of vessels trading between the rivers Plata and Negro, not one of which vessels have ever fallen in with them. On no part of the coast between those two rivers is there such deep water as 40 fathoms (the *Ariel's* soundings) within sight of land. It is most probable that they do not exist. It is to be regretted that masters of vessels, for general benefit, will not ascertain, as far as possible, the real nature of supposed dangers which had not been before discovered.

EL RINCON, or the Corner, is the deep bight formed by the sudden change in the direction of the coast on each side of Bahia Blanco. Generally speaking, it is shallow. Inside of an imaginary line drawn across this deep bight from the river Gueguen to the Colorado, not more than 20 fathoms water will be found. Between that and 10 fathoms will be the depth until either shore is approached near enough to see the highest parts in clear weather, with the eye 20 feet above the water; and throughout the space thus described there is anchorage during north or west winds. South or east winds send a swell into El Rincon which obliges vessels to keep under sail ; but north and west winds prevail during at least four days out of five.

SOUNDINGS.—In El Rincon, and along the coasts to the eastward and southward, the lead will invariably bring up sand, or sand mixed with broken shells, and perhaps some gravel; but the quality and colour of the sand is very different in different situations, and should be carefully noticed, whether with a view to anchoring in good ground, or avoiding any of the numerous and very dangerous sand-banks which throng the coasts between Bahia Blanco and the Rio Negro. On and near the banks the sand is always of a dark brown colour, very fine, and generally unmixed with other substances: sometimes bits of shell come up on the lead; seldom anything else. If an anchor is let go upon this sort of ground, its recovery is doubtful. There may be soft ground underneath, but rarely; most of the banks are formed of tosca, and this very fine dark brown sand is simply the tosca pulverized. In the offing, over soft ground, the sand is speckled, or black or white, rather fine generally; when coarse it is mixed with gravel: broken shells are frequent, though they do not occur so regularly as to assist in ascertaining a ship's place.

Having such soundings as those last described, the seaman may be certain that his ship is out of danger from a shoal; and that if necessary an anchor may be dropped with confidence. At night, if the weather is moderately fine and the wind off-shore, it is better to anchor than to keep under sail.

ASUNCION POINT is a projecting sand-hill 120 feet above the sea, lying 96 miles to the westward of the river Gueguen, difficult to distinguish with certainty, yet the most marked feature of this unvaried coast.

THE SIERRA VENTANA is a high mountain 3,500 feet above the sea, a considerable height anywhere; but in this low country extraordinary. It is situated 48 miles to the northward of Bahia Blanco, and the Gauchos (country people) call it Monte Hurtado (strayed or stolen), implying that it is out of place and belongs to some other country. It has also been called Monte Hermoso (beautiful) from its striking and fine appearance; but this latter name is now more properly given to the little mount (a molehill in comparison) at the entrance of Bahia Blanco. As a beacon that little mount well deserves the name of *hermoso*, but at a future day it will be more so, when distinguished by a lighthouse.

The name Ventana is said to be derived from an opening or cavity in the side of the mountain, resembling a large window. When seen from the south-east, the summit is peaked ; seen from the south, it appears rather square, with a notch in the middle.* A good bearing of this mountain (astronomical if possible) and the latitude of the ship will fix her position with certainty. In very clear weather the peaks may be visible when in lat. 39° 10′ S., bearing N.W. at 65 m. distance.

MONTE HERMOSO is a little round hill at the north entrance to Port Belgrano, higher than those around it (excepting one little hillock, which is rather more inland). It is close to the sea, and forms when seen from the eastward, a distinct finish to the sea coast ; westward of it the land is lower, and is not at first visible. It is 120 feet high, and upon it there was (in 1833) a pile of tosca, 10 feet square, raised by the *Beagle's* crew. It is to be hoped that this pile will be increased rather than suffered to diminish, for it is of much use to vessels entering the bay. Below the mount a low cliff of about 12 feet in height will be seen ; it is the only one hereabouts, and is called Parrot cliff. When seen from the southward, this little mount is confounded with the adjacent land, and by a stranger would hardly be made out, unless by seeing the *Beagle's* mark. If this mark were three times as large and of a whitish colour, it would be of great utility.

TIDES.—It is high water, full and change, in El Rincon at 5 h.; rise 6 to 8 feet. The tide-streams set strongly, the flood to the north, the ebb to the south, nearly 6 hours each way. They are much influenced by the winds, their strength varying from 1 to 4 knots when within 10 miles of the banks or land; and from half a knot to 2 knots when between 10 and 20 miles from the outer limit of the dangers.

DIRECTIONS.—When bound for El Rincon, if the weather threatens, or the wind is southerly or easterly, it is more prudent to stand directly off-shore during the greater part of the night. Heaving-to, or making

* *See* View of Sierra Ventana, on Sheet IX. No. 1,324.

free with the land, is not to be recommended on any coast, much less on this, which is considered by those who have frequented it during many years to be intricate and dangerous.

The land is extremely low, almost flat, in most places. The banks are very extensive, and suddenly steep. A vessel may shoalen her water from 10 to 2 fathoms in 2 cables' lengths, even while out of sight of land from the deck. To these inconveniences should also be added strong tides, and gales from the south-east, which bring thick weather and a heavy sea, overfalling and breaking as it approaches the banks.

When entering El Rincon, if the object be to anchor in or near Bahia Blanco, the northern shore between Black point and Asuncion point should be kept in sight from the main top; or the ship should be kept between the parallels of 39° 5′ and 39° 15′. Eastward of Asuncion point, the land may be approached as near as may be thought proper; but from that point to the westward more caution must be used, particularly when about 10 miles west of Asuncion point, as well as in the immediate vicinity of Bahia Blanco.

To the westward of Asuncion point, and thence along shore, the soundings are irregular, within 8 miles of the land. Ridges of tosca run out in a south-east direction, from 5 to 10 miles from the shore. These ridges are so frequent, and so regular, that crossing them from north-east to south-west gives one the idea of a vast land swell, the hollows of which are occasionally 6 fathoms lower than the risings, and about 2 cables' lengths from one hollow to another; the depths jump from 8 to 6, and at times from 10 to 4 fathoms; and the water deepens again as quickly as it shoals. At 6 miles from the land, and at 17 m. S.W. ¼ S. of Asuncion point, there is a spot with only 13 feet water; it is absolutely necessary, therefore, to keep fully 7 miles from the land.

When 30 miles to the westward of the meridian of Asuncion point, 60° 37′ W., and about the parallel of 39° 10′, the Sierra Ventana may be seen if the weather is very clear, bearing N. 45° W., distant 65 miles. Supposing that the weather is thick, or that the Sierra Ventana is not seen, the distance of the vessel from the north shore should first be ascertained, either by latitude or by steering due north, with attention to the lead and the look-out, until the water shoals to 8 or 7 fathoms, if the land is not seen previously to obtaining that depth. Reference to the chart and the ship's reckoning will show the position. By distance from the land, which runs nearly east and west from 10 miles west of Asuncion point, the latitude will be known.

Being about 10 miles west of Asuncion point, and 8 or 9 miles from the north shore, or in lat. 39° 10′ S., steer so as to make good a W. by S. ¼ S. course, and a distance (by ground log) of 30 miles, keeping the vessel as nearly as possible in the parallel of 39° 10′, not going to the north-ward of it nor to the southward of it more than can be avoided. Excepting the Sierra Ventana and the northern shore, no land will be seen. The northern shore will appear low, just topping above the horizon of an eye 20 feet above the water. The Sierra Ventana, if the weather be clear, may become visible when it bears N.W. ; it will then be distant 65 miles.

PORT BELGRANO, situated in the bight of El Rincon or Bahia Blanco, is the first harbour after quitting the river Plata. Vessels visiting this port should moor, because the tide runs strongly and rather irregularly, from 1 to 3 knots being the usual strength. South-east gales raise the water several feet ; those from the north-west have a contrary effect. Port Belgrano stretches inland for 25 miles, ending in a small creek. Throughout its entire extent the water is salt ; some small fresh-water rivulets run into it at and above the Guardia, which is situated in the creek. A boat can go from the anchorage off Anchorstock hill to the Guardia, a distance of 15 miles, in one tide, if the wind is not strong against her.* †

Supplies.—Plenty of excellent water may be obtained from wells be-tween Anchorstock hill and the beach. There is but little firewood, bushes, the stems and branches of which burn well, grow upon Zuraita island, and in hollows between the sand hills on the north shore. Deer, cavias, and ostriches are numerous. Fish swarm in the creeks and at the edges of the banks. Fresh beef may be procured at the settlement called Fuerte Argentino, at 5 miles to the north-west from the Guardia, and but little else. The country has been so much harassed by the Indians, as well as by the warfare carried on against them by Buenos Ayres, that the little colony has not had opportunity to improve its circumstances, in proportion to the time it has been established.‡

DANGERS at ENTRANCE.—The dangers to be guarded against in entering Port Belgrano are the two shoals, the Toro on the south-west, and the North bank to the north-east. The banks are all hard, of fine brown sand where they are steep and dangerous ; of coarser and lighter coloured sand where flatter and safer to approach. The Toro and the eastern extremity of the North bank are instances of the different qualities.

---

* Port Belgrano has often been confounded with Bahia Blanco, a name originally given to the outer bay, in compliment to General Blanco. *See* Voyages of H.M.S. *Adventure* and *Beagle*, vol. ii., page 101.

† *See* Admiralty Plan of Port Belgrano, No. 1,331 ; scale, m = 0·9 of an inch.

‡ While the river Plata was blockaded by the French and English squadrons, Rosas had his heavy guns landed at Port Belgrano, and transported across the country to Buenos Ayres, a distance of about 350 miles.

The fine brown sand, of a very dark colour, generally lies upon tosca. In the channels, between the banks, the bottom is everywhere soft; a dark, soft, sandy mud. On and near the banks it is everywhere hard. In the offing, when in the fair way, the ground will feel rather soft and sticky; still farther off it is somewhat hard, being clay covered by speckled sand, with broken shells.

**The Great North Bank,** projecting S.S.E. 10 miles from the shore, shoals very gradually, and may be approached by the lead in a large ship with safety. The Toro is the reverse. From the deepest water between the banks it shoals so suddenly that there is hardly time for the best leadsman to give warning. The Horn spit is particularly steep. From 7, 8, or, at high water, nearly 9 fathoms, over a soft muddy bottom, you will shoal in two casts of the lead, or less if going fast, to 2, 3, or 4 fathoms.

**Breakers** are sometimes found on the edges of these banks; at others, only ripplings. The Toro is generally more marked than the North bank. Sometimes there is not a mark on the water by which they can be distinguished by a stranger. Generally the water in the channels is less discoloured than that on the banks; but where all the water is extremely dirty and discoloured, a few shades are hardly noticed. From the mast-head, in fine settled weather, the water in the channels will appear bluish, while that on the banks is muddy; but in windy weather the colour is everywhere alike. Generally in fine weather the water is smoother on the banks than it is in the channels, where there is a slight tide ripple. This is also the case with a fresh breeze at high water; but at low water, particularly with a breeze, breakers show themselves upon the banks in all directions. With or after south-east winds, as long as the swell lasts, there are breakers on all the banks, the highest being on the Toro.

**TIDES.**—It is high water, full and change, in Port Belgrano at 6h. The tide sets along the north shore about one knot; the flood to the westward near that bay: but to the eastward, when off Asuncion point. In the entrance to Port Belgrano it runs between and parallel to the banks, nearly north-west and south-east, from 1 to 3 knots. The rise of tide in Port Belgrano is from 8 to 12, and at the extremity of the inlet from 10 to 14 feet is the usual difference between high and low water.

**DIRECTIONS.**—A good azimuth compass, a sextant, the plan of the bay, and good leadsmen should, wherever possible, be turned to account on entering Port Belgrano.

A vessel should not go to the westward of the meridian of Mount Hermoso before she sees the land, and sounds upon the edge of the North bank. When near that meridian, 61° 40′ W., it is best to steer due north until soundings of 6 and 7 fathoms water are obtained over a rather soft clayey bottom : Mount Hermoso should then bear about north-west. If the weather is thick, it is prudent to anchor or stand to the south-east, for it is out of the question attempting to enter until the vessel's position is correctly ascertained.

Having made out Mount Hermoso and sounded on the edge of the North bank, alter the course entirely, and steer between S.E. and S. by E. (according to the vessel's position) to cross or round the North bank. In doing this, if the water shoals, or requiring to get into a greater depth, south-easterly is the course upon which the soundings will increase fastest. All the banks and ridges lie north-west and south-east.

When the greater part of the land is sunk beneath the horizon of an eye 20 feet above the water, the depth will have been between 6 and 8 fathoms water (according to the time of tide), with a soft or sticky clay bottom. No land will be seen from the deck to the southward or west-ward, but from the mast-head some bushes will be seen nearly west, apparently rising out of the water, they are on a low, flat, marshy island, called Zuraita, which forms the southern shore of Port Belgrano. Those first seen are called the Laborde bushes ; afterwards the Ryan bushes, more to the westward, make their appearance.

It should be remembered that the tide is always setting strongly either in or out, excepting at the very few minutes of slack water ; that the fair way is narrow ; that the deepest water is close to the most dangerous bank, the Horn spit ; and that, if in doubt, an anchor should immediately be dropped.

Supposing that the vessel has rounded the north bank, and the depth is between 6 and 8 fathoms over a sticky bottom, steer N.W. by W. or from that to N.W. according to the wind and tide, until hard casts and shoaler soundings are again attained on the south side of the North bank. Do not go into a greater depth than 6 fathoms, or 7 fathoms at high water, while steering southerly, which will keep the vessel from crossing the fair-way, and getting too near the Horn spit. It is safer to borrow upon the North bank and not to go into the deepest water at all.

Near the Horn spit there are between 7 and 9 fathoms water over a soft bottom, but unless the edge of the Horn shows by rippling or breaking, it is rather too close to the deepest water. Having run along the south side of the North bank, and shoaled the water a little, one bearing of Mount Hermoso will show the vessel's position. Should it shoal too soon, keep more westerly, and deepen again. When Mount Hermoso bears N.N.E., steer one or two points more westerly to find the end of the East Gate Post.

 . **The Gateway.**—The tail of the East Gate Post shoaling very gradually, is an excellent guide to the Gateway channel, and a far better one than the Horn spit, which shoals so suddenly. When near the end of the Gate Post, Mount Hermoso, Texada point, (with its low, round, bare sand-hills) the Black spot or Medano mark, and the Laborde bushes, will be seen. Bearings of two, or angles between three of which will show the vessel's position.

   The Black spot, or Medano mark, 9 miles from Mount Hermoso, is a very singular hillock of sand close to the water, so covered with verdure as to look quite black, and well defined among the other bare sand-hills. Behind it the land is rather higher than on any other part of the shore. From the south-east end of the East Gate Post, Mount Hermoso bears N.E. by N.; Texada point, N. ¾ W.; Black spot, N.W. ½ N.; and Laborde bushes, W. ½ N.

   Having found this spot (which if the vessel anchors may be done by a boat), pass on the south side of it in 5, 6, or 7 fathoms water, but no more, and steer along the south-west side of the East Gate Post, until Mount Hermoso bears N.E.; then steer N.W. till Laborde bushes bear S.W. by W. ½ W., the highest part of the dry Toro sand, W. ¾ S.; Anchorstock hill, W. by N. ¾ N.; Black spot, N. ¼ W. Texada point, N.E. ½ E.; or Mount Hermoso, N.E. by E. ¾ E. When either of these bearings are on, the vessel will have passed the Gateway and be in Port Belgrano. Hence, W. by N. will lead up the harbour, and the plan will be a sufficient guide.

   The plan, the eye, and the lead ought to be trusted more than any written directions, for they cannot embrace all the contingencies of winds, weather, tides, and oversights. When in the Gateway the rippling on the East and West Gate Posts, or the smoothness of the water on them, and the rippling in the channels assist one much. At low water many parts of both banks show by breakers, and sometimes quite uncover.

   Anchorstock hill, 15 miles to the westward of Mount Hermoso, is the highest (57 feet) and most peaked of the hummocks seen to the north-westward; on it is a mark which looks at a distance like a mast. When in or near the Gateway, the Black spot is so useful and in so good a place, that one might easily believe it artificial.

   For passing the Gateway, the best time of tide is the last quarter ebb or the first quarter flood. The banks show more distinctly; and if the vessel should touch the ground the tide will soon lift her off. The dry part of the Toro sand is a good mark when near the north end of the Gateway. The breakers upon the Toro spit are also serviceable.

   H.M.S. *Beagle* worked in, under treble-reefed topsails and reefed courses, against a strong north-west wind. She went from the meridian of

Mount Hermoso to an anchorage off the Pareja creek, a distance of 24 miles, in 4 hours ; and she passed in and out half a dozen times without touching the ground ; therefore there can be no danger if care be taken. All the northern shore from Mount Hermoso westward is low, a succession of sand-hills, partly covered with shrubs and rough grass.

To go higher up the inlet, and anchor off the Pilot's house (or the Guardia), it is advisable to weigh at the last quarter ebb, and to anchor again before half-flood. When the banks are covered it is difficult to find the way, but easy when they show themselves at low water. At high tide an unbroken extent of water is seen ; when the tide is out, mud banks, with narrow channels winding between them, meet the eye in every direction. Above Anchorstock hill most of the banks are mud mixed with fine sand ; and in the creeks so soft in most places near the water that it will not support a man's weight.

FALSE BAY is an extensive and dreary waste. Sand-banks surround it, and neither land nor land-marks can be seen until a vessel is within the banks, and then they are not wanted. What it might become in the hands of an enterprising and seafaring people would be something widely different from its present state ; but now, without even a point of land in sight, the lead, the chart, and the latitude alone assist the seaman. The remarks on the precaution necessary on the nature of the banks, and on the passages between them, just given in the description of Bahia Blanco, apply equally to False bay, to Green bay, and to Brightman inlet.*

TIDES.—It is high water, full and change, in False bay at 5h. 30m.

HORN SPIT, TORO, and LOBOS BANKS lie between Bahia Blanco and False bay. The latter does not shoal so quickly as the others, particularly the Horn ; but it stretches out to the south-east as far as lat. 39° 21′ S. This and the other two banks should be carefully avoided, whether approaching False bay or Bahia Blanco. A large part of Lobos bank uncovers at half-ebb.

ARIADNE ISLAND and LABYRINTH SHOALS.—To the south and west of False bay the Labyrinth shoals extend to the south-east, from Ariadne island as far as the parallel of 39° 27′. They are nearly all under water. Here and there a patch of sand is uncovered temporarily, and affords a resting-place for seals.

---

* See Chart, Bahia Blanco to the Rio Negro, No. 1,358 ; scale, m = 0·33 of an inch.

**GREEN ISLAND** will be seen near these shoals, appearing rather high, because surrounded by a dead flat, although no more than 60 feet above the water. Off Labyrinth head, Green island spit extends from Green island to lat. 39° 29′ S.

**PAZ BANK**, at the south-east extremity of the Labyrinth shoals, is the most outlying, and dangerous. At low tide there are not 4 feet water on it. Its seaward side shoals gradually. As the land is in sight from the vicinity of this bank, it is not difficult to avoid it during daylight. From the Paz bank the heights upon Green island bear W. by N. ½ N. northerly. The nearest point of the island is distant 5 miles; the heights are distant 8 miles.

**LABYRINTH HEAD** bears W. by S. southerly from the tail of the Paz bank. It seems to be the termination of the shore on the south side of Brightman inlet, and makes as a perpendicular bluff 40 feet in height.

**GREEN BAY**, situated the southward of the Paz bank, is beset with shoals in the interior, though the entrance appears somewhat tempting. If it is necessary to enter, keep close to Green bank, carrying soundings from it, and anchor to the northward of the narrow passage (off the east end of Green island) with the peaked hillocks bearing W. by S. southerly. Remember that the banks on the east side of the entrance shoal suddenly. Low water should be chosen as the time to enter.

**BRIGHTMAN INLET** lies between Green island spit and the shore; a narrow bar harbour, at a distance it looks like a large river. On the south side of the entrance, the land is level and rather low, ending in a bluff, 40 feet above the water. A vessel approaching this inlet, with a view of anchoring, should not go to the northward of lat. 39° 30′ S. Farther south will be still safer, because the coast between Brightman inlet and the Rio Colorado is quite free from outlying dangers, while to the northward of that parallel the banks are extensive and dangerous.

To the northward upon Green island are some peaked hillocks, rising 60 feet above the sea. At the southern extremity of Green island is a single hillock, of use as a mark. At low water there are two fathoms on the bar, which is about 2 cables' lengths wide.

**Supplies.**—There is abundance of game on the main and on Green island. Good water may be obtained by digging wells, about 8 feet deep, on the island. Plenty of fuel may be cut on the main land.

It appears probable that a creek, fordable at low water, is the only separation between Green island and the main.. The tracks of cattle and horses are too numerous to suppose that there is not an easy communication. Perhaps there is no separation at all.

Green island has an excellent soil, and is capable of much improvement. At the north-east point, between the hillocks and Green bay, there is an eligible spot for a settlement.

Deep water close to the firm land, perfectly sheltered from all winds, and yet so near the sea; a fine tract of fertile land, in a fine climate, the possibility of insulation, if not already insulated, together with the extreme abundance of fish, and the increasing population of these shores, are reasons for looking at Green island, in its present state, with regret. What resources would not this broken and now avoided coast offer to an enterprising maritime nation; lighthouses, pilots, and docks would change its nature.

TIDES.—It is high water, full and change, in Brightman inlet at 5h. 10m., and springs rise 12 feet, neaps 8 feet. The flood-tide sets across the entrance, therefore a vessel should keep to the southward sufficiently to ensure a proper position. There is only half an hour's interval between high water in the harbour and in the offing. The strength of the tide stream is between 1 and 2 knots.

DIRECTIONS.—To pass over the bar, bring Labyrinth head to bear N.W. ¼ W., and when the east end of Green island bears N. by W. ⅓ W. the vessel will be close to or upon it. Keep Labyrinth head bearing N.W. ¼ W. until the water deepens to 3½ or 4 fathoms, with single hillock bearing N.N.W. ¼ W. Steer upon that line, keeping the above single hillock on the same bearing until the head bears W.S.W., then steer W. by N. ⅓ N., until the vessel is almost in a line between Single hillock and Labyrinth head, with about 4 fathoms water over a soft muddy bottom. There anchor.

The COAST from 3 miles south of Labyrinth head, or from lat. 39° 30' to the mouth of the Rio Colorado in lat. 39° 51½', runs directly south, and is quite free from obstructions or dangers of any description. The water shoals gradually, and regularly, therefore a vessel may go by her lead and close the land as much as she thinks proper. The land is low. A range of sand-hills, between 30 and 40 feet in height, extends parallel to the high-water mark. The beach is sandy, and in some places runs off more than a mile at low water. It has neither opening nor break of any kind.

Flat-top Hill is a sand-hill covered with verdure, lying three-quarters of a mile north of the Colorado. It is 40 feet in height, the highest about that part of the coast, and may easily be recognized by a stranger.

**THE RIO COLORADO** is accessible to vessels which do not draw more than 7 feet. A bar surrounds and defends the entrance. This bar, and the banks inside the river, are continually changing their position, caused by floods, high tides, and gales of wind.*

The entrance to the Colorado may be known by the abrupt ending of the sand-hills. South of the river the land is quite flat and low. When east of the river's mouth, trees will be seen growing on the banks, a short distance in-shore. They are a kind of willow, and the only trees on the coast. The entrance to the river is not more than half a cable's length in breadth, and has only 3 feet water when the tide is out. It is difficult to enter, even with the flood tide, unless the wind is fair, and not too strong. A strong south-east or easterly wind, throws so much sea upon the bar that it is then impossible to make the attempt.

The Cañada creek is an arm of the Colorado, less navigable than the latter. Between the Cañada and the main stream there are low marshy islets and winding streamlets, almost dry at low water. It is high water, full and change, on the bar at 4h. The tide rises from 6 to 9 feet. The stream of the river makes the flood weak, but the ebb runs very strong.

**DIRECTIONS.**—The best passage into the Colorado is from the northward; and when the western point is passed, keep over towards the banks on the east. All to the westward is flat and low. A stranger would perhaps suppose the river lay more to the westward, and would get rather hampered in consequence. This error must be guarded against by keeping over to the east side. *See* View on Sheet IX., No. 1,324.

The river does not show itself distinctly until the vessel is close to the green banks, which are nearly a mile from the bar. It is about a cable wide, with 2 fathoms at low water; the banks are steep-to, con-sisting of clayey earth. From the appearance of the banks, and the large trees that are scattered about the low land, it is evident that the river is subject to vast floods; in two months only the entrance had quite changed, what was the deepest part having become shallow, and the reverse.

No vessel, however small, should attempt to enter while there is a swell on the bar. It is also prudent to wait during one low water at the entrance, in order to see which is the best channel. It is the only safe way. Sometimes the water is fresh outside the bar.

---

* The Rio Colorado or Coluleuba of the Indians rises at the foot of the eastern slope of the Andes, and after a course of above 550 miles in a general south-east direction falls into the South Atlantic, 7 miles north of Union bay.

**UNION BAY**, situated 7 miles to the southward of the Colorado, is adapted to the use of vessels drawing less than 17 feet. In smooth water, with a fair wind, a line-of-battle ship might enter, as there are not less than 5 fathoms in the fairway at high water; but for general use, 15 or 16 feet is quite draught enough. Vessels intending to enter Union bay should make the land about the river Colorado, where the coast is clear, and the lead may be trusted.*

Indian Head is the north point of Union Bay, and appears as an island when seen from the northward, the land westward of it being very low. It is 45 feet high, bluff to the south-east, composed of sand hillocks, and partly covered with bushes. A little to the westward is a creek, through which boats may work their way to the Colorado at half-tide. It passes close to the Creek hills. Firewood is scarce about Indian head, although good water may be obtained there, by digging only a few feet into the sand. Union bay will be of great advantage when the banks of the Colorado are inhabited and cultivated. All intercourse with that river, and by that river with the interior, must be carried on through Union bay: so near the Colorado, both land carriage and (by the creeks) water carriage at command; so good a port close to the sea, yet entirely sheltered; landing-places so practicable, and the whole so accessible if a light stood upon Indian head; such advantages cannot much longer be allowed to remain neglected.

**Serpent Bank** is one of the chief dangers to be guarded against in approaching and entering Union bay. It extends in a long ridge 4 or 5 miles from the shore. As far out as 2 miles from the land it dries at low water. It shoals gradually on the north, but rather suddenly on the south side.

**Dog Bank** extends in a similar manner to the south-east, but it is of great extent to the southward. The water shoals upon its edge rather quickly. The quality of the bottom alters as the vessel approaches these banks, which is also the case in the other inlets and harbours of this coast. In the middle of the channels there is soft, dark-coloured sandy mud. Near and upon the banks there is hard, fine, brown sand.

**TIDES.**—It is high water, full and change, at 3h. 10m.; the rise of tide from 6 to 12 feet. The flood-tide sets to the northward right across the banks about 2 knots, seldom more, generally less. The ebb-tide sets right out at first, and then more to the southward as it clears the Dog bank.

---

* *See* Admiralty Plan of Union Bay, No. 1,329; scale, *m* = 1 inch.

**DIRECTIONS.**—In approaching Union bay, look out for Flat-top hill, 40 feet high, just north of the Colorado ; and when it bears W. by N., and is distant 5 miles, the depth will be 5 to 7 fathoms water over a dark sandy bottom. From thence steer S. by E. ¼ E., or rather steer so as to make good that course, until the water shoals upon the north side of the Serpent bank, or until Creek hill bears W. by S. Creek hill is the highest of three hillocks 2 miles N.W. of Indian head.

On Starve Island there is a peaked hillock (the middle one of three), which may be of use when crossing the tail of the Serpent. It ought to be kept to the westward of S.W. by W. while you are crossing, in order to insure having sufficient water. When Creek hill bears W. by S. and Indian head S.W. by W. ¼ W., steer S.E. 5 miles (by ground log), then S.SW., until Indian head bears W. by N. With Indian head bearing W. by N., steer for it (allowing for tide and keeping it on the same bearing) until the vessel is between the Serpent and the Dog banks, and between 2 and 3 miles from the head ; then steer west and anchor with Indian head bearing N.N.W. distant half a mile, in 4 fathoms at low water, over a soft muddy bottom.

This anchorage is preferable, because of ready communication with the shore. Mud flats extend so far from the land in all other parts that at low water a boat cannot land.

**ANEGADA BAY** is that portion of coast to the southward of Union bay lying between Indian and Rubio heads, a distance of 38 miles, and is appropriately named Anegada (lowland overflown) bay. Through it there are numerous creek communications, by which a boat may go to Union bay, and thence to the Colorado.

Extensive sand-banks, level with the water, stretch beyond eyesight soon after the first quarter ebb. At high water very little dry land can be distinguished, even from the mast-head. Every vessel, large or small, should give these shoals a wide berth ; more particularly when going northward. They extend to the south-east southerly, the flood tide sets up between them to the north-west northerly. In steering to the northward during the flood, a vessel will probably be set towards or amongst the banks ; and if so hampered she will have great difficulty in extricating herself, as with shoal water on each side, the tide setting 1, 2, or 3 knots towards the danger, and a southerly wind, there is no resource left but to anchor.

In this case it is probable that she will not be able to haul off to seaward, because banks extend in parallel ridges on each side. These ridges are very numerous, all lying north-west northerly and south-east southerly; upon them from 2 to 6, and between them from 8 to 12 fathoms water.

Between the parallels of 40° 5', and 40° 35', a vessel might run aground so far from the land that it could hardly be seen from the mast-head in the clearest weather. Between those parallels the banks extend from 10 to 15 miles to the eastward. Their length north-west and south-east is much greater. Some parts are more suddenly steep than others. The water over them is always more or less discoloured. When there is a swell the breakers are high, and the ground swell on the banks is sufficiently distinct to show their position. There are no rocks.

**The Viper Bank** and the tail of the Snake, with their accompaniment of parallel ridges, are extremely dangerous. In going towards the south there is less danger if the lead is carefully attended, because, the water shoaling upon the outside bank first, the vessel may at once haul off to seaward; besides the dangerous tide, that of flood will probably be a weather tide, while it is running. The quality of the soundings, as before mentioned, must be strictly watched, and the time of tide borne in mind.

Southerly winds, particularly if strong, raise the flood tide, causing it to run half an hour or an hour longer, and with more strength. Strong northerly winds have a diminishing effect upon the flood, but they make the ebb run stronger, and cause the water to fall unusually.

**Creek and Deer Islands** may be seen from the mast head if the weather is clear, but it is better to keep so far off as not to see land at all. On Deer Island there are wild dogs, the produce of those which have been left by vessels, either accidentally or on purpose. They are very savage.

If the land is seen between the parallels of 40° 5' and 40° 35' the vessel is in a more or less dangerous situation. There are always breakers near the edges of the banks even in a calm; but on the dangerous ridges, where are only 2 or 3 fathoms, there are no breakers in fine weather. A slight rippling, or an unusual smoothness, and some difference in the colour of the water, are, with the soundings, the only warnings of these dangers. When there is much swell the whole extent of the banks is shown, even where there are 4 or 5 fathoms water over them.

**North East Sand** is a large bank extending 20 miles to the south-eastward from the east coast of Deer island; its south-western point is called Hog island, on which a beacon was erected in 1833 as a leading mark into San Blas harbour.

**SAN BLAS HARBOUR**, lies to the southward of the low islands and extensive sand-banks of Anegada bay. Although opening into an extensive and well sheltered harbour, the entrance to San Blas is considerably obstructed by the small banks, narrow channels, and strong tides, between the North-east sand and Rubio head, its western point of entrance.*

Besides the injury to shipping, which is caused by the extreme dryness of this place, their chain cables suffer from some hidden cause. Whether there is copper or anything at the bottom of the harbour, which acts chemically upon the iron, is hard to say; but the fact is, that in a surprisingly short time chain cables are considerably corroded, if suffered to lie upon the ground. By lying moored in this harbour, during four months, an English ship, built of well-seasoned African oak, was rendered unseaworthy. Her chain cables were reduced one-third in size, and lost the greater number of the cross-bars. During most summer nights no dew falls. At no period of the year is there a rainy season. During winter there are occasional but not heavy rains.

**Supplies.** — There is a plentiful supply of good water in the wells near the beach in San Blas harbour. From one of these wells, formed by sinking two casks in the ground, one placed above the other, a ship obtained 5 tons of water in one day.

Fuel is scarce in the immediate neighbourhood, but can be procured in great quantity, and of excellent quality, by sending to the inhabitants, who bring it from the interior in carts. It is called Peccolini, and is the best 'fuel on this coast, and perhaps as good as any wood for burning. Iron may be brought by it to a welding heat. Fresh provisions may be procured in abundance from the neighbouring Estancia. Fish are plentiful, and there is frequent communication with San Carmen, the town on the banks of the river Negro.

**RASA POINT,** lying 17 miles to the southward of the entrance to San Blas, is low, but on it there is a bare round sand-hill, 30 feet in height. Seen from the southward, it appears like an island, the land to the westward being low. Five miles south-west of Rasa point are some sand-hills, rather more remarkable than their neighbours, being flat-topped, and partly covered by straggling bushes, which somewhat resemble a drove of cattle. Vessels wishing to enter San Blas harbour ought to make the land to the southward of Rasa point; on no account to the northward of it, unless absolutely certain of their latitude. Even then they should not close it to the northward of Second Barranca point unless the tide is ebbing.

**Second Barranca Point,** lying 7 miles to the northward of Rasa point, is a sand-hilly height, roughly covered with shrubs and grass. At its

---

* See Admiralty Plan of San Blas Harbour, No. 1,320; scale, m = 1 inch.

base near the sea is a low cliff, the only one hereabouts, and therefore remarkable. Rubio head and Rubia point, situated respectively 4½ and 11 miles to the northward of Second Barranca point, are in their appearance similar to it, excepting the cliff at the base. Rubia point is 40 feet, Rubio head 35 feet above the sea. Between them the coast is low, consisting of sand-hills more or less covered with verdure.

**SAN BLAS BANKS.** — Five miles E.N.E. of Rubia point is the southern extremity of the San Blas banks. Generally breakers are visible, though not always. Half way between the shore and this extremity of the banks is the proper distance to keep while steering for the entrance. There are three channels or passages lying close together, the Great Gat, the Little Gat, and the Ship Gat. These channels are formed by three banks and the shore. Between the shore and the Helgat bank lies the Little Gat, a narrow passage, fit only for the smallest coasting craft, or for boats.

Between Helgat bank and Middle bank, is the Great Gat, the least bad of the three; and between Middle bank and East bank is the Ship Gat, but unfit for any vessel.

Middle bank is a narrow ridge of shingle. Something was seen in the middle of it, which might have been a piece of wreck, but it had the appearance of a rock occasionally showing at low water. This bank generally shows at half tide.

**TIDES.**—At full and change it is high water at 12h. 30m. in the offing, at 1h. 30m. under Rubia point, and at 2h. in the harbour. The usual rise of the tide is between 8 and 12 feet; but it is affected here, as elsewhere on the coast, by the wind.

The tides run along this coast with dangerous strength, from 3 to 5 knots. The flood is the strongest by nearly a knot, and as it sets directly towards the outlying banks, a vessel must, when near the entrance, with the flood tide running, do one of three things; enter the channels leading to the harbour, stand to the southward, or anchor. By standing off and on, or to the eastward, she will be on the banks in a very short time. If unable to stem the tide by standing to the southward, and not choosing to enter the passage, the anchor must be dropped; and if it should come on to blow from the south-eastward, her situation will not be pleasant. With the flood tide, weighing would be out of the question; with the ebb she might get clear, but it must be remembered that a heavy sea tumbles in with a south-east gale.

With and after a gale from the south-eastward, both channels and banks are covered with heavy rollers and breakers. At such a time it would be highly improper attempting to enter. The last part of the flood tide comes from the northward, from Anegada bay, at which time the tide is beginning to ebb at the gats.

**DIRECTIONS.**—The most simple and the surest appear to be: wait for the last quarter-ebb, and enter as near low water as possible, as at any other time, more trouble will be required. Obtain soundings on the east side of Helgat bank, and go through the Great Gat by the lead, the eye, and the plan. When near Rubia point, bring it to bear West, distant about 3 miles; at which point Rubio head will bear N. by W. ½ W., and the Estancia (farm house) on the rising ground between Rubia point and Rubio head, N.W. Then steer N. ¼ W. until Rubio head bears N.N.W., and the Estancia W. ¾ N., when the vessel will be abreast of the south end of East bank in 4 fathoms at low water. From this spot N. by W. leads safely through the Great Gat into the broad channel. When past the Middle bank, the Estancia will bear S.W. ¾ W.

When in mid-channel, between Helgat bank and the south end of Middle bank, with the Estancia bearing W. by S. ½ S., and Rubia point S.W. by S., the deepest water will be found. From thence keep Hog Island beacon (*see* page 21) one degree open to the eastward of Rubio head, N. by W. ¼ W. which is the best course with a leading commanding breeze, as long as the beacon is visible. If it is not seen, it will be advisable to keep towards Helgat bank until Rubio head bears N.N.W. ¼ W., and then steer N. by W. ¼ W. The lead,—the breaking or rippling water upon the shoals, and the plan,—must assist in directing the vessel's track, and it must be remembered that the flood-tide sets strongly towards the south-east end of the Middle bank, the ebb towards the north-west extremity.

The dangerous part of this entrance is about a mile in length, after which Great Gat and Ship Gat join, making the channel comparatively wide and also deeper. While Hog island beacon is standing, the entrance will be less difficult; but it cannot be expected to stand long in the shifting sand.

Little Gat is used by small fishing vessels and boats. It is close to the shore, consequently a place of refuge is at hand, if a roller should fill or capsize them. Being within three banks, much sea is broken off before it reaches this passage. The eye and the lead are the only guide of those that enter by the Little Gat.

Ship Gat has a bar across its entrance, between Middle and East banks. It is probable that this bar and the Middle bank, both of which are chiefly composed of loose shingle, shift occasionally. From the north-west end of Middle bank the flood sets rather towards the north end of Helgat bank, therefore care must be taken in steering up N. by W. ¼ W.

Having entered by either of the two large Gats, and brought Rubio head to bear West, a W.N.W. course may be steered to the anchorage off the watering place. This part of the harbour is called the Broad channel; it is about a mile wide, and the tide sets directly through. It is best to keep on board the south shore, which is steep-to, and will afford some shelter from south-east winds if obliged to anchor sooner than intended.

In working up the Broad channel, do not stand far over to the north shore, as the bank shoals suddenly on that side. There are places where, in one cast of the lead, the water will change from 10 to 3 fathoms. The best anchorage is off the watering-place, within half a mile of the beach, secure from all winds, and well sheltered from the south-east. The best bower in-shore in 10 or 12, the small bower to the north-east in about 14 fathoms, and open hawse to the south.

The beach is steep-to, having 6 fathoms a few yards from it. The bottom is stiff sandy mud, covered with coarse gravel and shingle. Boats may land here in any weather.

**The COAST**, southward to the Rio Negro, is a line of sand-hills, few as high as Rubia point, here and there partly covered by rough grass, and by low prickly shrubs. This coast may be more closely approached. There are neither shoals nor rocks when one mile from the beach, and the lead may be trusted. On nearing the Rio Negro the sand-hills are lower; some hills of a different character just showing themselves inland.

Rasa point will be recognized by the trend of the coast, as well as by three or four flat-topped sand-hills. Some persons have mentioned these hills or rather hummocks as remarkable, and as sufficiently showing which is the point; but a stranger may find difficulty in recognizing them, particularly as their form may sometimes be changed by gales of wind.*

Around Main point, the eastern point of entrance to the Rio Negro, to the southward, lies the bar of the river, and beyond it is low land, which extends about 2 miles to the comparatively high and extensive range of cliffs called the South Barranca. *See* View on Sheet No. 1,324.

**TIDES** set strongly from 2 to 5 knots, following the coast line nearly 6 hours each way, so that a stranger, unaware of their existence, might run aground on the banks near San Blas, thinking himself to the southward of Rasa point.

---

* *See* Admiralty Chart :—East Coast of South America, Sheet X., from Rio Negro to Cape Three Points, No. 1,288 ; scale, *d* = 4 inches.

The **RIO NEGRO** separates the provinces of La Plata and Patagonia. In approaching this river it is best to keep to the southward, and to make the land about the False sisters, or Bermejo head (*see* page 31). Being certain of the latitude, and making proper allowance for the tide, may justify a direct course for the river, but on no account should a vessel incur the risk of being set to the northward of Rasa point.*

Northward of the Rio Negro the nearest land of any height is the North Barranca (Ravine), a range of sand-hills 160 feet high, without any cliff. From the north-eastward this ridge appears as a bluff headland, visible as soon as Rasa point is passed. All the coast northward and eastward of the Rio Negro is low, and generally shows sand hills slightly covered with vegetation, or entirely barren.

There is a bank on each side of the entrance of this river, the north-eastern being hard sand and shingle, the south-western chiefly a quick-sand. The sand composing the latter is very fine and dark-coloured; during one night, in the year 1827, a vessel of 60 tons burthen was buried in it, although the water was smooth, with very little swell.

**Supplies.**—Fresh provisions, in moderate quantity, some firewood, plenty of vegetables, and, during the season, abundance of excellent fruit, may be procured at San Carmen, a town situated 17 miles from the entrance, which by encouragement and protection would become a thriving settlement. The climate is delightful, the soil on the banks extremely rich, and watered by the periodical overflow of the river.

**Pilots.**—There is a pilot resident at the Boca (entrance), maintained by the Buenos Ayrean government. When vessels approach the bar, if the weather will allow of his going out in a whale boat, he readily embarks, and remains on board until an opportunity offers for their entering. Leaving the river is rather more difficult than entering, because of meeting instead of leaving the southerly swell. There are instances of vessels having been detained 40 days at the Boca, waiting for an opportunity to cross the bar. No vessel should make the trial without a commanding breeze, for light winds are treacherous, and they blow different ways on the river. North-east winds are the best, either for entering or leaving. Variation 15° East in 1860.

**Main, or Redonda Point,** is a low, rounded hummocky sand hill. From the eastward it appears to be formed by 3 hummocks, tolerably covered with verdure. On the highest, which is 40 feet above the sea, is a small half-formed battery, which cannot be distinguished outside

---

* *See* Admiralty Plan of Rio Negro, No. 1,310; scale, m = 1 inch.

the bar. At some distance in-shore two headlands will be observed.
one of which, Leading hill, bearing N.W. from Main point, is the mark
of most use in entering the river. It is the higher and the more eastern
of the two headlands, having a small round-topped hillock 180 feet high,
at the south-west extremity, which slopes suddenly to the south-west, and
may be seen from 3 to 4 miles outside the bar of the river.

**Medano Point**, on the opposite shore, is low, bare, and sandy. In high
tides it is overflowed, at which time the entrance of the river would appear
three times its usual breadth, and it would be difficult to hit the deep
water channel. South point is a cluster of sand hills to the north-west
of Medano point. Both Main point and Medano point are steep-to:
by closing them, the opposite banks, which are dangerous, are avoided.
Abreast of Main point the channel is only 300 yards in breadth.

**Flat Point**, on the eastern bank of the river, nearly a mile from Main
point, is sandy, 25 feet high, and covered with verdure. It shows a steep
side to the south-west, and is rather wedge-shaped; in 1833 this point
was distinguished by a mark, the bowsprit of a large ship, erected by the
*Beagle's* officers. This mark, if distinguishable, in one with the highest
part of Leading hill, is the best guide for crossing the bar in the deepest
water. In a line between Medano and Flat points there are from 3 to 4
fathoms at low water, with a soft, sticky, muddy bottom.

On the beach to the northward of the river are two wrecks (1833);
while they last they will be excellent marks. One, a brig, with her lower
masts standing, her hull half buried in the sand, lies rather more than 4
miles from Redonda or Main point. The other, without masts, is a mile
from Main point, and so far inside of high-water mark that one must
suppose the land is there gaining upon the water unusually fast. They
may soon decay, or be carried away for firewood, and yet they may be of
use longer than these remarks, for heavy gales and heavier seas assist
the torrent from the interior in altering and shifting the ground about the
entrance of the river. In 1827 there was a battery on a part of Main
point, where, in 1833, we found 2 fathoms water.

Again, within the river, the deep-water channel has changed sides.
Six years ago there were three fathoms where now are 6 feet. Yet the
principal entrance, the south-east, and the marks (Leading hill and the
summit of Flat point) have varied very little.

**South Barrancas.**—From Medano point, a low sandy-hilly shore extends 2 miles to the south-west. There the land changes its character,
suddenly rising into a range of perpendicular cliffs, from 150 to 200 feet
in height. These cliffs are called the South Barrancas. Above them no
land is seen; their upper outline is as horizontal as the surface of the sea
when they are viewed from the south-east in the offing.

**TIDES.**—It is high water, full and change, in the offing at 2h. 0m., 3 hours later than upon the bar and at the entrance of the river. Some say it is high water upon the bar at 11h. 15m.—others say 11h., full and change. It will vary much more than that quantity, even half an hour, or an hour, during gales of wind; but the usual time of high water upon the bar, during settled weather, is 11h., or a few minutes later, on the days of new and full moon. The tide stream runs parallel to the coast from 2 to 4 knots.

The ebb stream in the river runs at the rate of 2, 3, 4, 5, or 6 knots, according to the wind, the body of fresh water coming down from the interior, and the state of the offing tide. The flood-stream is less strong, seldom exceeding 2 or 3 knots. When an unusually great body of fresh water is brought down, owing to floods in the interior country, the ebb is at the strongest; and, on the contrary, the flood-tide is hardly to be noticed, though the water rises as usual. Like all large rivers having their sources in mountainous countries, it is subject to periodical inundations.

Directly after high water, the ebb tide begins to set strongly out of the river, and over the north-east bank. Care must be taken to avoid its effects by anchoring, if in sufficient water, or running out to sea again, if on the bank. There is no time to be lost; one or other alternative must be instantly adopted, if unable to obtain the proper anchorage before the tide makes out too strongly for the wind to enable the vessel to overcome it. Very few vessels have escaped that have once grounded on the banks or bar of this river. The tide reaches a few miles above the town of San Carmen, during the dry season. It is then high water at the town two hours later than at the Boca.

**DIRECTIONS.**—With Redonda or Main point bearing N.N.W., or more northerly, approach the bar until the water shoals to 10 fathoms. The lead lines should be carefully examined, marked to feet, and well used in both chains. The edge of the bar will always show either by breakers or by ripplings; rarely does a day occur on which there are no breakers; with any swell the breakers continue all round the bar, not excepting the channel.

Do not bring Main point to the westward of N.N.W. unless intending to pass the bar.

If circumstances do not admit of attempting the passage at once, anchor, or stand directly off shore again, because of the strong tide. Good anchoring-ground, and a convenient place for awaiting a proper time for entering, is to the southward of the bar, with Main point bearing north, and the North-east end of the cliffs (South Barrancas) bearing W.N.W., or from that to N.W. by W. The former bearing will give 8 fathoms, the latter 10, over a clean sandy bottom.

While the winds have northing and westing this is a good roadstead; but it must be quitted directly if the wind gets to the southward or eastward, threatening to blow strong. Heavy seas are sent in by southerly gales, and the strength of tide should be remembered.

The proper time for crossing the bar is during the last-quarter flood, before the time of high water by calculation. The water does not rise during the last-quarter flood, and it is of material consequence that a vessel should enter the river before the ebb-tide makes out with strength. At this time (last-quarter flood) the tide is setting strongly to the north-eastward, along shore over the bar. The difficulty is to avoid being set to the north-eastward in crossing the bank. After high water this difficulty is increased, because the powerful *ebb* out of the river meets the tide *flowing* to the north-eastward, along shore, and they together sweep over the eastern bank at the rate of 3, 4, or 5 knots.

The passage across the bar is very narrow, not more than a cable wide. At low water there are only 6 feet upon it. Great attention to the marks, a commanding breeze, and smooth water, are absolutely necessary to enable a vessel drawing 12 or 14 feet to enter the river. Fourteen feet is almost too heavy a draught. Not that vessels of 15 or even 16 feet might not occasionally pass the bar, but to do so they must have smooth water, a spring tide, and a fine breeze of wind. The least swell might be fatal; such smooth water is rare on this exposed coast, and generally speaking, 11 feet is the utmost draught that may enter the Rio Negro without incurring risk.

By keeping Redonda or Main point summit distinctly open to the eastward of Leading hill summit, so that a sail passing between them on the horizon would just fill the intermediate space ; or by keeping Flat point in one with the summit of Leading hill; or the extreme of Main point in one with the summit of Leading hill, about N.W. ¼ N., the bar may be crossed at the proper place. When on the shoalest part of the bar the north-east end of the South Barranca will bear W. by S. ⅓ S. At spring tides there is about 18 feet, at neaps about 14.

Keep the marks in one until the end of the South Barranca bears W.S.W. ; then steer one point more westerly until Flat point comes open of Main point, with the summit of the latter bearing N.N.W. From this steer direct for Flat point, passing close to Main point, having passed which, close Medano point, and anchor in a line between Medano and Flat points if wishing to proceed farther up the river. If intending to remain at the Boca (entrance), a better berth is rather more to the northward, out of the strength of the tide, in 3 fathoms (at low water) over a muddy bottom.

For boats there is a safer entrance than the south-east channel. They should pass the South Barrancas in 4 or 5 fathoms water, and steer about N. by E., till some bare sand-hills, rather inland, on the north-east side of the river, are distinguished. The second of these hills (South Channel hill) counting from the west, brought to bear N. by E. ¾ E., will carry a boat at three-quarters flood into the river. At high water a vessel drawing 7 feet may thus cross the south bank.

This is called the West Channel, though channel there is none ; but it is safer for boats to enter by this line, because they run less risk from blind breakers, which are common in the south-east channel. Also, if any accident does occur, the shore is close at hand, and the tide sets into the river.

**SAN CARMEN.**—From the mouth of the river Negro to the town of San Carmen, the plan and the lead will suffice. If a vessel should ground, with a flowing tide, smooth water, and soft banks, she will not be injured; but, both for the bar and for the river, a local pilot should always be procured if possible. The banks alter, more or less, every year, although the main channel is nearly stationary.

The town is on the side of a steep sandstone bank, on the northern or left bank of the river, at 17 miles from its entrance, in lat. 41° 4', long. 62° 50' W. The streets are irregular, and the houses chiefly huts of one story, built of mud and bricks. A ruinous mud fort exists, which was built by Francisco Viedma at the formation of the settlement in 1779. The soil in the neighbourhood of the town is fertile, and produces wheat, barley, indian corn, and various kinds of fruits and vegetables. Sheep, horses, cattle, and goats are also abundant. Pop. 1,230 in 1832.

**VARIATION.**—The variation at the beginning of the year 1860 off Piedras point was 11° E. ; off Cape Corrientes, 12° E. ; off Bahia Blanco, 14° E. ; and off the Rio Negro, 15° E. The curves of equal variation on this coast assume a S.E. by S. direction (true), and if crossed at right angles there would be a change in variation of a degree for each 90 miles of distance run. The variation is decreasing at the rate of about 4' annually.

## CHAPTER II.

### RIO NEGRO TO THE STRAIT OF MAGELLAN.

Variation in 1860, from 15° E. to 21½° East.   Annual decrease about 4′.   ·

---

**The Coast.**—The South Barranca range of hills extends from the river Negro to Belen bluff in the Gulf of San Matias, with only a slight interval, that of Rosas bay. Their utmost height (near Belen bluff) is 300 feet. At their beginning, near the river Negro, they are 200 feet above the sea. Thence, nearly to Bermejo (red) head, erroneously called by many Cape Two Sisters, the low land above and in-shore of the cliffs is so level as to appear parallel to the water line when seen from the south-eastward.* *See* View on Sheet X.

The bar of the Rio Negro is the only outlying danger (if it can be so called) on this part of the coast, or, it may be said, between San Blas bay and Port San Antonio. The banks off each of those places excepted, the intervening coast is quite clear.

**Cape Bermejo,** lying 10 miles westward of Medano point, forms the north point of entrance to the Gulf of San Matias; it is high and the water free from danger. If set out of her reckoning during the night or thick weather a vessel will not run any risk of getting ashore. Northward of Rasa point the case is reversed, as the flood tide will sweep her strongly towards the San Blas banks. Many vessels have been wrecked, owing to this cause. Bermejo head is an excellent point to make when approaching the Rio Negro, the Bay of San Blas, Port San Josef, or Port San Antonio. It will keep a vessel clear of numerous dangers which attend the approach to either of those places on a parallel, as is the frequent but, on this coast, dangerous practice. Near Bermejo head there are hummocks and irregular hills, nearly covered with rough verdure.

**Dos Hermanas.** At the north-eastern part of the Bermejo heights are two peaks, very small, yet showing distinctly when seen from the eastward ; these are called the False Sisters. These peaks stand nearly

---

* *See* Admiralty Chart :—East Coast of South America, Sheet X., No. 1,288 ; scale; *d* = 4 inches.

over two peculiar cliffs, which so resemble one another that they have obtained among the inhabitants of San Carmen and amongst the traders on the coast, the name of the Two Sisters (or Dos Hermanas), to the prejudice of the real Two Sisters, which remain neglected and unknown to those who have not the old Spanish chart. In the excellent original, executed by the officers of the *Atrevida* and *Descubierta*, the name Punta de las dos Hermanas is applied to the point at the west side of Rosas bay, which is so similar to the point or headland at the eastern side that the name appears very appropriate. *See* View on Chart No. 1,288.

As far as Belen bluff, which lies 30 miles to the westward of Bermejo head, soundings in from 10 to 20 or 25 fathoms will be found, while between one mile and 20 miles from the land. Everywhere there is a clean sandy or gravelly bottom. Thirty miles to the eastward of Bermejo head, there is only 20 fathoms water (a fathom more or less). Twenty miles to the southward, the same bank extends.

Westward and southward of Belen bluff, the *Beagle* had deep water excepting in one place, as shown upon the chart, where only 15 fathoms were found. It was dark; but two successive casts of that depth having been taken, the hands were immediately called, and sail shortened for anchoring, but the next cast gave no bottom; and as the shoal ledge could not be found again without a great loss of time, it was left unexamined. There certainly is, however, a ledge of rocks there, and perhaps with less than 13 fathoms over them. The water appeared light coloured, although the night was very dark, and the lead had been carefully attended. In all other parts of the Gulf of San Matias, there is deep water; from 30 fathoms when a mile or two from the shore, to 80 fathoms in the offing.

TIDES.—West of the meridian of Norte point, 63° 48' W., and northward of lat. 41° 50' S., but little stream of tide is felt; though the water rises 24 feet. Round Norte point and thence to the bay of San Josef, the tide sets strongly, with ripplings and races, dangerous for boats, or very small vessels. From Belen bluff it sets strongly to the eastward, along shore with the flood, and but faintly to the south, or south-eastward during the ebb. Westward of Belen bluff the flood-tide sets to the north-west, the ebb to the south-east.

When the swelling of the sea, or the tide wave, comes up the coast from the southward, it rushes round the projecting land of San Josef with much strength, causing violent and dangerous overfalls off Valdes creek, and off Norte point. Part of the body of water thus going northward, separates, and goes round Norte point. The main body continues its progress to the northward, inclining to the west, until near Belen bluff, when it divides ; one stream running to the north-west, the other to the eastward along the land. With the wind against the tide, there is a very cross short sea in the entrance of the gulf.

**TIDE CREEK** lies 10 miles westward of Belen bluff ; in it there are 3 or 4 fathoms at high water, but at low water it is dry. Westward of that creek, another range of cliffs, about 100 feet in height, commences, extending, with one break, about 18 miles, the extreme point being named Cliff end. To these cliffs succeed low land, with a shingle beach. Near and in-shore of the cliffs, to the westward of Bermejo head, are irregular ranges of hills, tolerably covered with rough verdure. Where there are no cliffs, the land is low and without hills. Rosas bay, and Tide creek, appear when first made from seaward, like openings in the land. The hills and cliffs on each side show like islands or headlands.

**VILLARINO POINT**, the east point of entrance to Port San Antonio, lying 21 miles from Cliff end, is low and sandy, with a few hummocks on it, nearly covered with verdure, and having, therefore, a dark appearance. The highest is 40 feet above the water.

From Cliff end to Villarino point the land is low, a few banks, partially covered with grass and stunted bushes, or low sand-hills, alone rise above the shingle beach. At high-water mark, the beach is either shingle, or coarse sand. At half tide it is composed of large shingle stone ; at low water tosca and rock are mixed with large stones.

**LOBOS BANK**, which extends south nearly 5 miles from Villarino point, is the only danger in the north-west bight of the gulf, either near the shore, or outlying. Deep soundings may be had anywhere, and when only a few miles from the land, soundings will be obtained with the hand lead.

**DIRECTION HILL and EL FUERTE**, on the north-west shore of the Gulf of San Matias, are seen when approaching Port San Antonio from the south-eastward, before any of the low land can be made out. If approaching from the southward, the hill called El Fuerte, or the Fort, is seen sooner than Direction hill. If from the eastward, the reverse. El Fuerte is 380 feet high, and it would be difficult to find a more singular resemblance to a regular fortification. Some call it the Castle ; but its appearance is that of one side of a regular polygon, the curtain and each bastion showing distinctly. Direction hill is 560 feet above the sea ; three small hummocks, close together, are visible on its summit. *See* View on Chart No. 1,288.

[S. A.]                                                    C

**ANCHORAGE.**—There is a good outer anchorage in from 10 to 30 fathoms between the end of Lobos bank and El Fuerte, as well as to the southward and eastward of that bank, the ground being quite clear, either fine sand or a soft, greenish, sandy mud. The shelter is good, excepting with south-east winds, which, it must be remembered, do not often blow, and still less often with violence.

**PORT SAN ANTONIO** is the best place on the coast for a large ship in want of serious repair. It may, for that reason, prove very valuable, notwithstanding its remote situation and the barren nature of the surrounding country. A ship may approach as near as may be convenient without getting into shoal water, or running any risk, until an opportunity offers favourable for running within the Lobos bank.*

On each side of the entrance is a bank, partly dry at low water, and steep at the edges, the eastern or Lobos bank shoaling suddenly from 7 to 3 fathoms. The western shoal is called the Reparo bank.

Port San Antonio, the best harbour on this coast, is a good port in a bad situation. Distant 30 miles only from the Rio Negro, perhaps the medium of future inland communication between Chile and the now barren coast of Patagonia, it is well situated for commercial intercourse with the interior. Should the Rio Negro become of importance to commerce, vessels of much draught ought to anchor here rather than in San Blas bay. The land carriage from each place to the town of San Carmen is about the same, supposing that from San Antonio the direct course is taken to the river.

Fuel is plentiful, but no fresh provisions, excepting game and fish, which are abundant, could be procured in 1833. Water is deficient, but may be obtained by digging wells. Dry docks might be built for vessels of any size. There are many places where vessels may be laid ashore for a tide, without the slightest risk, the harbour being perfectly sheltered from every wind, although the entrance is much exposed to south-easterly winds, and at such times it would be imprudent to approach without a probability of speedily entering, as the vessel would be embayed, and have to contend with a heavy rolling sea. In Escondido creek the keel of a line-of-battle ship might be examined, the rise of tide being from 24 to 32 feet.

**TIDES.**—It is high water, full and change, at 10h. 45m. The rise of tide depends much upon the wind, being between 3 and 5 fathoms. Between Villarino point and the Reparo bank the tide runs from 3 to 5 knots. The greatest rise during our stay was 25 feet, and the least 16; but from

* See Admiralty Plan of Port San Antonio, No. 1,327 ; scale, m = 1 inch.

marks on the shore it was evident that 30 feet would be too small a limit to name for a spring-tide, during south-east winds.

For a line-of-battle ship two-thirds flood, but for most vessels the last-quarter ebb, should be chosen as the time for rounding the tail of Lobos bank, so as, if possible, to arrive off Villarino point at low water. When there all is safe and easy. The water is always smooth, and the only inconvenience is the rapid tide. A sheltered berth, suited to the object in view, may be gained by warping when the tide suits.

When off the bank of shoal soundings, or in upwards of 20 fathoms water, the stream of tide is little felt, not more than, if as much as one knot; but as the vessel nears the Lobos bank its influence increases fast.

**DIRECTIONS.**—Vessels bound to Port San Antonio should keep Direction hill bearing W.N.W. until the Fort bears S. by W., when the hummock on Villarino point bearing N. by E., and Nipple hill showing over it on the same bearing will be seen. Nipple hill, the highest land hereabout, 600 feet above the sea, is on a range of down-like heights to the northward of the port, and carries a small hummock upon its summit resembling a nipple. With these bearings on, steer about north (allowing for tide), so as to keep the Nipple just open of the west extremity of Villarino point until the vessel is half a mile from that part of the point, from whence steer by the eye, close to the bank on the east side, round the point, proceeding to the berth which is most convenient. *See* View on Chart No. 1,288.

It must be remembered that the fair-way is not sufficiently examined, particularly along the west side of Lobos bank. Many more soundings are required on both sides, and in the bight near the west shore: although enough have been taken to ascertain and show the nature and practicability of the passage.

A large vessel intending to enter this port, must take the trouble of fixing two boats or two marks upon the shoal places of 2 and 3 fathoms marked in the plan. The fair-way should also be partially sounded by the person who is to pilot the ship into the harbour, to ensure his acquaintance with the marks, and to guard against any shifting of the banks, which may occur in so rapid a tideway.

**THE COAST.**—Continuing along the coast southward from El Fuerte, the Sierra de San Antonio next attracts notice. The highest part of this range of hills (they can hardly be called mountains) may be about 1,700 feet above the sea. It is peaked, and visible from a ship's deck when 20 or 30 miles from the land. This Sierra is not one overgrown mountain, as represented in some charts, but, as its name denotes, consists of several ranges of hills, in irregular succession. Amongst these hills are streams of fresh water; but the steep and exposed nature of the shore renders it impracticable to use them for ship purposes.

All this west coast of San Matias is bold and steep-to. From the point
of the Sierra, to the southward, the coast is chiefly cliffy, but with intervals
of low land. The cliffs are moderately high, from 100 to 200 feet nearly
perpendicular. They are composed of loose earth, or *diluvium*, but mixed
with shingle and vast quantities of fossil shells. At high tide the shore
is sandy ; at low water it is rocky.

PORT SAN JOSEF.—This great basin, though 26 miles in length and
11 wide, was called Port San Josef by the Spaniards; it is free from interior obstructions or danger, but the entrance has an unpleasant appearance
owing to a rocky ledge crossing it, over which the water ripples so much
that a stranger would hardly think it safe to enter. Eleven fathoms is the
least depth that has yet been found in that channel, but the tide sets so
strongly over the narrow ledge that sounding upon it is not easy.
Many vessels have entered at various times, and as no one has yet en-
countered danger it may be supposed that none exists, excepting within
half a mile of either point.

Both east and west heads are moderately high and show distinctly from
seaward. They are bold cliffs, rising abruptly to 100 or 150 feet above
the water. Their colour is the same as that of all this coast, where a
yellowish sun-burnt appearance continually meets the eye.* Under water,
from each head, the above-mentioned ledge of rocks extends across,
causing, when the wind opposes the tide, a heavy rippling. The best
anchorage is in the bight to the eastward of the eastern head. Northerly
winds send much sea into the southern bight.

Masters of vessels who have been some months at anchor in this gulf
speak very unfavourably of it. They say that a short heavy cross sea
gets up with any strong wind, although the gulf is nearly land-locked.
Fuel may be cut in the south-west part of the bay, from stunted shrubby
trees, but near the shore there is no appearance of fresh water. To the
eye, all is barrenness and desolation. How such a spot could have been
selected for a colony is surprising.†

From the east side of Port San Josef to Norte point, a distance of 27
miles, there is a continued cliff from 60 to 100 feet in height. No high
land appears in shore ; all looks low, bare, and sun-burnt. No danger lies
under water, but to small vessels the races of tide which are sometimes
met between San Josef and Norte point, are rather troublesome, if not
dangerous.

From Norte point to Valdes creek the land is low, mostly a shingle
beach, and off this piece of coast are the worst tide races, being occasioned
by the rush of water across shoals and rocky patches lying from 2 to 10

miles off shore to the south-eastward of Norte point, and to the north-north-eastward of Valdes creek.

There are no marks on the land by which the shoals can be avoided. The strength of tide is so great as to render the sails of doubtful effect. It is, at least, prudent to avoid going nearer to this shore than 15 or 20 miles. Sometimes the overfalls extend 15 miles from the land ; sometimes they are hardly 5 miles distant ; their locality depending upon wind and tide.

**Tides.**—It is high water full and change within Port San Josef at 10h. 0m. The tide rises from 20 to 30 feet, and the stream rushes between the heads 3, 4, or 5 knots.

**Tide Race.**—A vessel of 90 tons, from the Rio Negro, bound to San Josef, was beset by one of these races when about 4 leagues north-east of the entrance to that port. No bottom could be found with the lead ; the sails were almost useless, notwithstanding a strong wind, owing to the violent motion of the vessel ; some of the crew ran aloft to avoid being washed overboard ; others ran below. After being the sport of the waves during two hours, carried along with the tide, and unable to get clear, so much damage and inconvenience was caused, that getting out of the race was followed by a change of purpose, and the immediate return of the vessel to the Rio Negro without again attempting to enter Port San Josef.

The races to the eastward of Norte point and Valdes creek are yet worse. But the height and violence of these overfalls, races, or ripplings, vary as much as the winds and tides by which they are occasioned. Norte point (by some called Lobos point) is lower than the adjacent cliffs, and a reef extends under water about a mile. It is rocky and unfit to approach nearer than that distance. In the offing is deep water.

**Valdes Creek.**—The entrance, which is 27 miles to the southward of Norte point, may be known by the shingle beach ending ; all the land to the northward being low, all to the southward high and cliffy ; also by a line of cliffs commencing at Cantor point, on the south side of the creek, and continuing to Ercules point. There are no heights inland which can assist in showing a ship's place. *See* View on Chart No. 1,283.

This creek is a dangerously deceiving inlet of a singular character. The entrance is only a third of a cable's length in breadth at most times, but sometimes it is entirely blocked up by shingle. The tides run most rapidly through the opening, and up the long narrow passage, which extends to the northward at the rate of 4, 5, or 6 knots. A heavy surf breaks across the entrance, when there is any swell.

The depth of water on the bar is quite an uncertainty, depending upon the length of time that has elapsed since a south-east wind has heaped up shingle at the entrance, the strong tides scouring out a channel as repeatedly as the gales block it up. It is an unfit place for any vessel unless obliged to enter from some urgent cause. Being merely in want of a port should never be a reason for attempting it, as Nuevo gulf, with every advantage, is close at hand, and always easy of access. With the wind blowing from the south-east, no vessel under any circumstances should run for it ; since, if the entrance is found closed, she will inevitably be lost, being then too close to get out again.

**ERCULES POINT**, 8 miles to the westward of Valdes creek, is a white cliff, 220 feet high ; when first seen it appears to be perpendicular, or rather overhanging. When near, the upper and lower parts of the face of the cliff meet at an angle near the middle, resembling the back of the letter K. Close to the northward of it are two perpendicular cliffs of the same height, off which a shoal ledge of tosca extends 2 miles to seaward, and 3 miles along shore.

**DELGADA POINT**, the south-eastern point of the Valdes peninsula, is sloping and green, and 200 feet above the water. A tosca ledge runs to the eastward 1½ miles. Valdes peninsula is 54 miles long by 25 wide, and is joined to the main by an isthmus 20 miles long by 7 broad, on the north and south sides of which lie respectively Port San Josef and Golfo Nuevo.

**LOBO PEAK**, nearly 6 miles from Delgada point, rises only 15 or 20 feet above the table range. It appears from the northward like a small peaked sandhill, but from the southward it appears double ; a smooth round-topped sandhill to the northward, with a peaked one close to the southward of it. A rocky ledge off this point extends about a mile to the south-east.

Nearly all the cliffy points between Valdes creek and Nuevo gulf have rocky ledges extending nearly a mile from the high-water mark. In some places they extend still further, and should be allowed a berth of 2 miles. On the outer part of some of these rocky places there are not more than 2 fathoms water. The upper outline of the portion of coast between Cantor point and Nuevo head, the eastern point of entrance to Nuevo gulf, is nearly horizontal.

**NUEVO GULF** is easily known by Nuevo head and Ninfas point, the two well-defined headlands at the entrance. Nuevo head is steep-to, high, and bold, the highest part being 200 feet above the sea. Ninfas point, which bears S.W. ½ S. distant 7 miles from Nuevo head, is also 240 feet high, and makes as a double point, the two angles lying N.N.W. and S.S.E. of each other.

**Supplies.**—There is plenty of small wood for fuel in the south-west part of the bay, and there are some ponds of excellent water upon a clayey soil, which appear to last all the year round, good water may also be procured in abundance by digging wells. The soil is good and fit for cultivation on the south-west shore, and wild cattle frequent the neighbourhood for the sake of grass and the fresh water. From this gulf an easy communication might be kept up with a settlement on the River Chupat, which lies 40 miles to the south-west, by means of coasting vessels which could enter the river.

NINFAS POINT ought not to be approached nearer than 2½ or 3 miles, as there are rocky ledges extending from it, particularly from the N.N.W. extremity, where the reefs dry at low water, are 1½ miles from the land. Alongside of these rocks there are 7 fathoms water; over them the tide rushes into Nuevo gulf 5 or 6 knots, causing a heavy and dangerous rippling, the swell from which is felt entirely across the gulf, and with a westerly wind, causes a short hollow sea.

In mid-channel there are no ripplings, and the tide is sufficiently strong to carry a vessel to windward, while hove-to in a fresh gale. When once well inside the heads, the tide is much less strong, and the sea longer and more regular. On the south side of the gulf there are several good anchorages, off the shingle or sandy beaches, between the different heads.

There is no danger whatever in approaching these anchorages, only taking care to avoid being too near to the projecting points, as they have, like the rest of the projections on the coast, rocky, foul ground, extending to seaward some part of a mile. The cliffs here are like those further northward, consisting of sandy clay, or rather diluvial earth, containing prodigious quantities of fossil shells.

Anchorage.—The best anchorage in the gulf is in the south-west corner, lying between the first two white cliffy points. These points are between 40 and 50 feet in height. With the eastern cliff bearing E.N.E., and distant about a mile, there is a good anchorage, with 5 fathoms at low water over a muddy bottom. Westward of this anchorage there are many places where an anchor may be dropped with propriety, the bottom being sandy mud or clay.

The northern part of the gulf is lined with steep cliffs. There is a remarkable rock called the Pyramid at the north-east part. Pyramid road, near this rock, is sheltered, having good anchorage, except with southerly winds, the best berth being with the Pyramid bearing West, or W. by S. Ten miles within the heads of the gulf the tide is scarcely felt.

TIDES.—It is high water, full and change, in Pyramid road, at 7h. 0m., and the average rise is 10 feet.

**ENGAÑO BAY.**—From Ninfas' point the coast runs nearly straight to Engaño bay, a distance of 36 miles. This straight shore is cliffy, similar to the coast eastward of the gulf, and with no outlying dangers. The tide runs north-east and south-west, nearly in the direction of the shore, from 1 to 2 knots.

In Engaño bay the water is shoaler than that near the higher land. Around this bay, between North and South cliffs, the land is low and sandy, with many small sand-hills or hummocks near the beach.

**CHUPAT RIVER.**—The entrance to this river is at the south-west corner of Engaño bay, just where the beach ends and the rising ground begins. A reef of rocks, uncovered at half-tide, and dry at low water, extends across the mouth, and apparently shuts it up. Chupat bar is between the reef and the beach, and has scarcely a foot of water on it when the tide is out, but at high water there are from 7 to 12 feet, according to the state of the tide. The greatest rise of tide during the neaps was found to be 6½ feet.

About 18 miles up the river (measuring by the very serpentine course of the stream) is a place admirable adapted for a settlement. It is a rising ground, from 20 to 30 feet in height, close to the banks of the river, commanding a view of 5 leagues to the north and west, and an uninterrupted view to the eastward. Throughout this extended view the country is fertile in the extreme. The soil is of a dark colour, and very rich. Excellent grass covers it in every direction. Numerous herds of wild cattle graze in the plains. There are several lakes on the south side literally covered with wild fowl.

Sauce trees (a kind of willow) grow on the banks of the river in great abundance: some are large, 3 feet in circumference, and about 20 in height. They are chiefly the red sauce, which is much more durable than the white. It is fit for pumps, pump-boxes, boats, and many other purposes for which elm is frequently used. The very winding course of this river, and the rich soil through which it flows, affords facilities for isolating numerous peninsulas, and artificially watering a very large extent of country. The most certain defence against the Indians is a ditch filled with water, no matter how small, as they never willingly cross water in their attacks.

Had Sir John Narborough seen this part of the country, he would not have given so unpromising a report to his master. How the Spaniards could have overlooked it, having a settlement so near, upon the Valdes peninsula is surprising. Perhaps the Indians were numerous, and kept them within bounds. Southward to the mouth of this river is a range of table-land, from 50 to 60 feet in height, ending in very white chalky-looking cliffs. *See* View on Chart No. 1,288.

**TIDES.**—It is high water, full and change. in the Chupat river, at 5h. 30m., and the average rise is 9 feet. The ebb runs 2, 3, or 4 knots; the flood is not felt more than 6 miles from the entrance.

**DIRECTIONS.**—Vessels wishing to enter the river must examine the channel at low water, and place a buoy at the end of the reef, because the deepest water is close to it. Within the reef the water is at all times quite smooth. Near the mouth the river is not more than 60 yards wide, and is 5 feet deep at dead low water; higher up it gets broader, but not deeper. At low tide the water is quite fresh at the bar, and at high water 8 miles only above it. Boats or flat-bottomed barges might be tracked up by men or horses to a great distance. The river is free from obstacles, and the banks are firm and level. The best time to leave the river is before high water, with the last quarter flood, because, when the ebb makes, it sets directly upon the reef. At each side of the mouth are great quantities of drift wood. Many fragments of light volcanic scoriæ were picked up, possibly washed from the volcanic mountains on the western side of the continent.

**CASTRO POINT** is the north-east termination of the table-land which extends 5 miles from the Chupat river. The coast from Castro point to Delfin point, and thence to Lobos head and Union point, is high and bold, the water being deep, and free from danger. The cliffs hereabout have a chalky appearance, and consist of very light-coloured sand and clayey earth, with dark horizontal strata.

**DELFIN POINT,** 8 miles south-west of Castro point, has a small conical hummock seen above the table-land. This hummock appears double, saddle-shaped, and rugged, when seen from the northward. Eight miles south of Delfin point is Hidden islet, and so much like the cliffs near it that one might easily run past, even if looking out, and only a mile or two distant, without seeing it. Union point, 13 miles to the southward of Hidden islet, is rather low and rocky. Here the character of the coast changes, bold cliffs and extensive ranges of table-land no longer meeting the eye, and the shore becoming lower and more uneven.

**TOMBO POINT,** 13 miles from Union point, is low and rocky, with a rugged outline. For a quarter of a mile from the sea it is quite bare, rocky, and dark-coloured, and there are some rocks, and a rocky ledge extending half a mile to the north-eastward. Lieutenant Wickham sounded

7 fathoms, a cable from the shore; farther off the water appeared to be more shoal, but a strong tide rippling prevented him from ascertaining. On the north side of the point, inside the reef, there is an anchorage where small craft might find shelter from south-east winds. Sealers anchor there at times.

ATLAS POINT, 6 miles south of Tombo point, forms the northern head of the Bay of Vera, and shows a smooth slope, terminating in a low rocky point, when seen from the northward or southward; but in a rugged ridge of rocks, if seen from the eastward. There are rocks under water near, within 2 cables of the extreme point. Atlas point is 70 feet high, and to the westward between it and Tombo point is situated Mount Trista, 300 feet above the sea. *See* View on Chart No. 1,288.

VERA BAY, an anchorage fit for small craft, drawing less than 12 feet, lies to the south-west of Atlas point. It is formed by a reef of rocks, lying south-west and north-east nearly a mile in length, and half a cable broad. A vessel may enter from the north-east or from the south-west, remembering that in each entrance the bottom is rocky and uneven, and there are only 12 feet at low water in them. Inside, opposite to the shingle beach, is a good berth, in 15 feet at low water, over a bottom of coarse sand, shells, and sandy mud.

With a strong wind from south-eastward a sea may be thrown into this corner, over and around the natural breakwater, from the beginning of the last quarter flood to the end of the first quarter ebb, during which time the reef is covered; the beach, however, does not show the effects of much sea. The land is low close around Vera bay; but rather high ground is seen in-shore. Two islets, with rocks near them, lie nearly in the middle of the bay, and the tide sets through, between the reef and the shore, about a mile an hour.

CAPE RASO, the southern point of Vera bay, is level, and rather low, with a few rocks close to its extreme point. Two miles north-west of Cape Raso is Rasa cove, a good anchorage with all winds, excepting those between N. by W. and N.E. This cove is free from impediment of any kind; but on the east side the ground is hard and stony; in the middle, and near the west side, there is good holding ground, a stiff yellow clay. Sand is shown by the hand-lead, but underneath is clay. There is neither fresh water nor any firewood, excepting a few straggling bushes.

SALABERRIA REEF.—In approaching this part of the coast from the northward, there are several rocks near the shore, which are very little above

the water, and there is a considerable reef in the offing called the Salaberria. the extreme of which is a dry rock, situated 4½ miles E. by S.
from Cape Raso.  This ridge probably projects off from Cape San Josef
to the southward of Cape Raso, for there are two dry rocks in the same
line of bearing, one a mile and a half, and the other 3½ miles from that
cape, besides several patches which break.  The tide sets rather strongly
along the shore, which is fronted by reefs for 2 or 3 miles off.  Great
caution should therefore be used in approaching the coast, as the water
is deep, and, if becalmed, it may be necessary to anchor, which will be
in at least 30 fathoms water.

CRUZ BAY, about 6 miles across, lies to the southward of Cape Raso,
between it and Cape San Josef.  The Salaberria reef extends from the
latter Cape, and if as continuous as it appears, there ought to be good
riding in this bay.  New Cove, close to the northward of Cape San Josef,
is small, and exposed to east winds.  West Cove, in the south-west
corner of Cruz bay, has not been examined.

PORT SANTA ELENA is situated on the south side of Cape
San Josef.  The harbour may be easily known by some hummocky hills
120 feet high, on the north-east projecting point, on the easternmost of
which there is a remarkable stone resembling a monumental record, but it
is a natural production.  The best anchorage is at the north-west corner
of the bay, in 6 or 7 fathoms, but not too near to the shore, for when
the sea is heavy the ground-swell breaks for some distance off.  The
small low island of Florido lies half a mile, W. ¾ S., from the eastern point
of entrance, and a quarter of a mile from the island is a bank, with only
12 feet water on it, which must be avoided in working into the bay.  The
plan of this port published by the Admiralty is a copy of the excellent
survey by the officers of the Spanish ships *Atrevida* and *Descubierta*, and
is sufficient for the navigator ; there is also a plan in Weddel's Voyage
that is equally correct.*

Supplies.—The water that is contained in the wells on the western
shore of the bay is too brackish to be worth consideration ; nor is there
any fresh water to be obtained from any part of the harbour.  Of fuel,
a temporary supply may be procured from the small shrubby tree which
is described in the account of Port Desire (*see* page 54), and which
is tolerably abundant here.  Guanacos, ostriches, armadillos, and the
cavia or Patagonian hare, are to be procured, as are also wild ducks, par-
tridges, snipes, and rails, but fish seem to be scarce.  The guanaco affords
an excellent food, but it is a difficult animal to approach ; one shot

---

* *See* Plan of Port Santa Elena, No. 335; scale, m = 3·1 inches.

in 1832, when cleaned and skinned, weighed 168 pounds. The Indians sometimes visit this part of the coast, but principally for the purpose of burying their dead.

**TIDES.**—It is high water at full and change in Port Santa Elena at 4 o'clock, and the tide rises at springs 17 feet.

**CAMARONES BAY** extends from Port Santa Elena to Cape Dos Bahias, a distance of 22 miles ; the shore is rocky as far as Fabian point, where it changes to shingle, and so continues as far as the Cape. In the depth of Camarones bay, there is a high rocky islet, with two lower and smaller ones to the northward, all of which are quite white, and so named Blanca islets ; this whiteness is caused by numerous sea birds. Along this shore, and especially along the headland, the tide runs strongly from 1 to 3 knots, nearly N.N.E. and S.S.W., rising from 8 to 12, and sometimes even to 15 feet. A ledge of rocks, extending a mile to the N.E. of Cape Dos Bahias, should be carefully avoided. Moreno islet, 2 miles north-west of the same cape, and a third of a mile from the shore, is high, rocky, and of a dark colour.

**GREGORIO COVE**, situated about 7 miles to the southward of Cape Dos Bahias, is exposed to south-east winds, and somewhat difficult to enter, because of the strong tides setting past. There is, however, good shelter from the prevailing winds. The intervening coast is bold and steep-to, without an anchorage. In the offing are the islets of Sola, Arze, and Rasa, but, excepting the latter, they are not dangerous, being rather high, from 80 to 100 feet, with deep water near them.

**RASA ISLET**, situated 11 miles south-east from Cape Dos Bahias, and E. ¼ N. 8 miles from Leones isle, is dangerous, because it is low and far from the land ; the water round it is deep, excepting at the south side, where there are some rocks near the islet, extending rather more than a mile.

**LEONES ISLE** lies one mile to the southward of Gregorio cove. Between this island (sometimes called Ship island) and Harbour isle, on its western side, is an anchorage, good in point of safety, but somewhat disagreeable on account of the eddy tides. The eye is a sufficient guide in entering, as the shore is bold, and there is no hidden danger. There is generally a small stream of fresh water at the north-west part of Leones isle, but it cannot be depended upon during the summer months. A few bushes may serve for a temporary supply of fuel, but there are not enough for a store. There is no passage between the two islands except for boats.

The ground towards the narrow part of the passage is rocky and uneven. The tide sets directly through, causing eddies which will turn a ship round at her moorings against a strong breeze.[*]

In the best anchorage, abreast of the middle of Harbour islet, there are 4 fathoms at low water over a clean sandy bottom. In approaching this anchorage, the set of the tide should be considered, in order to determine whether to pass round the north or round the south side of the island. The flood-tide runs strongly to the eastward round the northern side, between it and the main land.

**SAN ROQUE POINT**, off the west end of Leones isle, is low and rocky, with a hummock upon it. While the flood-tide is running, a vessel ought to give it a good berth, as the tide sets rather towards it. There is no other danger in the passage between Leones and the main.

Off San Roque point the flood tide sets to the N.E., easterly, 3, 4, or 5 knots. From San Roque point to Cabo del Sur, or South Cape, which is 2 miles to the south-west, the tide sets to the eastward during the flood, passing the latter at the rate of 3 knots, and causing much rippling. In the bight to the northward of Cabo del Sur is the Oven, a concealed dock-like cove.

**GILL BAY** lies westward of San Roque point, formed by Cabo del Sur and the north shore. There is anchorage in the bight to the south-west, in 8 fathoms at low water. Southerly gales send a heavy swell into Gill bay, as well as into Port San Antonio. South Cape makes as an island, when seen from the eastward, the isthmus being very low.

**NEW HARBOUR**, on the western side of Cabo del Sur, between it and Valdes island, formerly called Port San Antonio, is one of the best ports on the coast. In entering, a vessel may pass on either side of Valdes isle, and steer by the eye, as there is no hidden danger. Southerly winds send in a good deal of swell, but cause no further inconvenience. The best anchorage is in the north-west corner, shutting the point of the main land in with Valdes isle, where there are 5 fathoms water over a stiff clayey bottom, the lead may show shingle or gravel, but there is stiff clay underneath.

Westward of Valdes island are the Cayetano islands, between which and the main is a bight which appears to offer anchorage and shelter ;

---

[*] *See* Sketch of Leones or Ship Isle, No. 552 ; scale, *m* = 1·6 inch.

but it is unfit for any vessel. About 1 mile south of the Cayetano islands are Los Frayles, which show as three distinct rocks at low water, and at high water there are always breakers upon them. They may be closely approached on either side. *See* View on Chart No. 1,288.

**PAN DE AZUCAR**, an island situated nearly 3 miles from Los Frayles, does not resemble a sugar-loaf; its summit is uneven and rocky, rising to 190 feet above the sea. There are rocks close around, and dangerous ones between it and the reef of San Pasqual, which lies nearly 5 miles to the south-west. To the northward of the westernmost of the out-lying rocks, off the eastern and western extremities of the Pan de Azucar, there is anchorage under the island in 7 or 8 fathoms over a muddy bottom. This anchorage might answer for a vessel wishing to enter Port Melo, and waiting for the tide.

**PORT MELO**, the entrance of which is nearly 8 miles north-west of the Pan de Azucar, is too rocky, and too much exposed to southerly winds to be valuable as a port. Off it are several rocks, lying in a direct line between San Pasqual or Molino reef, and Pan de Azucar.*

Between Point Castillos, which lies 4 miles to the westward of Port Melo, and Cape Aristazabal, a distance of 25 miles, are several bights and coves, though none of them are worth notice as fit places for anything larger than a decked boat.

**TIDES.**—It is high water, full and change, in Port Melo at 8h, 40m., the rise at springs being 15 feet. The tides here are strong, running along the land 2 or 3 knots. Off the projecting points, and in confined passages, their strength is of course increased, and causes heavy ripplings, when opposing the wind.

**TOVA ISLAND**, 5 miles to the south-west of Port Melo, is large and bold in appearance, with detached rocks and islands all around it: at the north-east side of it is Tova cove, a good anchorage for vessels drawing less than 15 feet. In entering, avoid the reef on the starboard hand, and close the opposite shore which is steep-to; there is anchorage in 5 fathoms muddy bottom, sheltered from all winds excepting northerly, which cannot raise much sea, as the main land to the northward is so near, and a strong tide is always setting through, between the island and the main. On Tova there is abundance of excellent firewood, and, in general, water may be procured.

* *See* Admiralty Plan of Port Melo, No. 553; scale, m = 2 inches.

The ROLANDO ROCKS and MEDRANO SHOAL, lie south-eastward of Tova, and must be carefully avoided. The Medrano, which is 4 miles from Tova, is a mere reef, but dangerous, from lying so far off; the sea breaks upon it with violence at most times.

The LOBOS and GALIANO ISLES, situated in shore, 14 miles to the westward of Tova, are beset with rocks. Inside these islands is Bustamante bay, open and exposed to the south-east winds; several rocks lie near the middle. The Viana isles, which lie 4 miles to the eastward of Cape Aristazabal, have a reef 1½ miles to the eastward of the largest on which the sea generally breaks; in other directions they may be approached with safety.

PORT MALASPINA, a mere rocky inlet, unfit for anything except a boat, is separated from Bustamante bay by Gravina peninsula. A glance at this port will show that it little deserves either the name of Malaspina, or to be described as a harbour. Upon the coast, between Leones isle and this port, which forms the north side of the Bay of St. George, a southerly gale drives a heavy sea, the land has everywhere a barren desert appearance, destitute of trees or any verdure. In height it is generally between 100 and 600 feet, and no mountains are visible.

CAPE ARISTAZABAL is situated about 3 miles to the southward of the entrance to Port Malaspina. There is deep water close to the cape, excepting at three-quarters of a mile to the south-east, where there is a rocky place which causes breakers at half-tide. Off the cape the flood tide sets to the north-eastward about 2 knots; with the ebb it runs to the southward.

From Cape Aristazabal to Cordova Cove, a distance of 46 miles, the land gradually rises. A range of table land is seen increasing in height, as it extends southward. There is no danger of any kind between these places, excepting a rock off the Quintano isles, which lie 7½ miles to the westward. The soundings are regular, the tides scarcely felt, the coast steep-to, and bold. After passing the Quintano isles, high light-coloured cliffs bound the sea; they continue nearly as far as Cordova cove.

SALAMANCA PEAK, situated 11 miles to the northward of Cordova cove, is remarkable. It is a regular-shaped cone, shows itself distinctly above the high ranges of table land, and is visible 40 or 50 miles distant.*

CORDOVA COVE is rocky and shallow, almost unfit for the reception of any vessel. From the northern point of the cove a reef extends half a mile, and 2 miles N.E. of the point lies the Novales shoal, a small

---

* See View on Chart No. 1,288.

rocky ledge under water.    There is also a ledge of rocks directly east of
the cove, and to the south-east of the north point.    Five miles southward
of Cordova cove, and one mile off shore, are the Ali rocks ; dry at low
water, and having breakers on them when covered.    There is no other
outlying danger between Cordova cove and Cape Tres Puntas, the
southern point of the Gulf of St. George.

TILLI ROAD, 17 miles to the south-west of Cordova cove, is a
tolerably good anchorage during westerly winds.    The beach is level
and sandy, but so much swell generally breaks upon it that landing is
difficult.    There is plenty of small firewood near the shore, and a
*salina* of fine white salt.    From north to south, by the east, Tilli road
is quite open.    The anchorage is in 5 or 6 fathoms water, over a clean
sandy bottom.    The heights near the shore are composed of a sandy light-
coloured diluvial earth, and great quantities of fossil shells, principally
oyster shells of a very large size.    The position of Tilli road may be recognized by its lying between the second and third prominent bluffs,
which are seen to the southward of the lower land about Cordova cove.

In the depth of the Gulf of St. George, from Tilli road to a cliffy bluff
named Cape Murphy, a distance of 35 miles, the shore is low ; generally
a shingle or sandy beach, without cliffs or rocks.    The soundings in
the offing are regular.    Thence for 28 miles to Casamayor point, the
shore is rugged and broken, with very little cliff.    Over Casamayor point
are Espinosa heights, forming a high ridge, fronted towards the sea by
precipitous cliffs.    Thence to Cape Tres Puntas the coast for 48 miles is
alternately cliff, rock, sand, or shingle, in disconnected portions.

CAPE TRES PUNTAS may be known from seaward, by observing
that it is the termination of a long level range of table land running
north and south.    A little to the south-east of the northern end of that
range is a remarkable conical hill, 250 feet high, like a sugar-loaf, attached
to the main range, though rather a straggler ; and there is a small sharp
peak rather northward of the fall of the range.    The cape shows three
distinct upright heads of a light-coloured earthy cliff ; off these heads are
ledges of rocks, extending half a mile to seaward, and the tide rushes and
ripples over them with violence.    One mile from the shore there is no
danger whatever.*

From Cape Tres Puntas to Cape Blanco, a distance of 8 miles, the
coast is low and rocky, with table land showing in shore.    A thick bed

---

* *See* Admiralty Chart :—East Coast of South America, Sheet XI., No. 1,284 ; scale
d = 4 inches.

of kelp lines the rocky coast. Salinas (salt lagoons or marshes) extend for many miles in shore of these points.

**CAPE BLANCO**, at a distance, appears to be an island. Three distinct masses of rugged rock 130 feet high are connected to the main land by a narrow low isthmus. On each side of the isthmus is a small cove. That on the south side is sheltered excepting from south to east. A very small vessel might obtain tolerable shelter from all winds, by anchoring close to the end of the kelp, in the north east corner of the cove.

A vessel intending to anchor here should make great allowance for the tide. The flood sets with force over the bed of rocks, which lie half a mile north-eastward of the Cape, and they would prove extremely dangerous, if drawn in among them. The beach around the cove is rather steep, and formed of shingle, here and there mixed with dark sand. The depth is from 4 to 6 fathoms. Plenty of excellent firewood, of a small sort, may be cut on the south-west side of the cove, a few yards only from the beach, but there is no appearance of fresh water.

**CAPE BLANCO SHOALS.**—From the summit of Cape Blanco, patches of shoal water are seen towards the north-east and south-east, some 10 or 12 miles off shore, on and about which are ripplings and overfalls more or less violent according to the time of tide, and the direction and strength of the wind. Over these shoals there is but little water in many places, and the soundings are very irregular. A vessel should entirely avoid them. H.M.S. *Adventure* passed over two of them, and had not less than 5 fathoms, but possibly at low water the depth may be considerably less; they are thrown up by the force of the tide, which sweeps round the Cape, into and out of St. George's Gulf, with great strength.

**Byron Shoal.**—The north and south ends of the Byron shoal bear respectively from Cape Tres Puntas and Cape Blanco, E. by N. ½ N., distant from the former 7 miles, and from the latter 5 miles; consequently it extends in a N.W. ½ N. and S.E. ½ S. direction for 6 miles; it is scarcely a quarter of a mile wide.

**Anne Shoal.**—The north end of this shoal bears E. ¾ N., 7 miles from Cape Blanco, and extends in nearly a southerly direction for 2 miles Between these shoals there is a passage 2 miles wide, and the depth gradually increases to more than 15 fathoms.

Within these shoals are two others; a small one with two fathoms on it, bearing E. ½ S. from the Cape, and distant 2 miles; and another 2 fathoms bank, which has been named after the Cutter *Susanna* which struck on it. It is 2½ miles in length, and nearly a mile broad; its direction is S.S.E. and N.N.W., and its northern end lies 3½ miles S.E. by S. from Cape Blanco.

There is probably more shoal ground to the N.E., for in the year 1829, having approached the land within 14 miles, with Cape Tres Puntas, bearing S.W. ½ S., the depth rather suddenly decreased from 40 to 14 fathoms, pebbly bottom, being then about 10 miles within the 50 fathoms' edge of the bank. On approaching the land, the quality of the bottom becomes irregular, and changing from ooze to sand, with pebbly shoal patches; so that by attention to the soundings and nature of the bottom, these shoals may be easily avoided.

A good mark to avoid them is, not to approach so near to the Cape as to see the rugged hillock of Cape Blanco, and to keep the high land of Cape Tres Puntas, which is visible from the deck about 20 miles, on the horizon.

**TIDES.**—The flood or northerly tide ceases in the offing at 4h. 15m., but in the neighbourhood of the Cape and among the shoals, the tides may be less regular; they produce strong ripplings, and set with considerable strength.

**The COAST,** from Cape Blanco southward, is low, and the beach regular, until near Port Desire, when it rises into a remarkable bluff. This coast line, however, was imperfectly seen; within the distance of 3 to 5 miles from the shore there are several small patches of rock, which uncover at half tide, but beyond that belt the coast is free from any known danger, and may be approached to not less than 15 fathoms; within that limit the ground is foul.

**PORT DESIRE,** at the mouth of a river of the same name, lies 33 miles to the southward of Cape Blanco; it has rather a difficult entrance, from the strength of the tide and its narrow breadth, and it is rendered still more confined from several rocky reefs that extend off the north shore, or that lie nearly in mid-channel. The north point of entrance of the port is a steep bluff, and is therefore remarkable as being the only point of that description along this part of the coast. At 4 miles N. by E. ½ E. from that bluff there are some rocks called Sorrell edge, a quarter of a mile without which the depth is 13 fathoms. The Tower rock, on the south side of the port, becomes visible after passing this ledge; it opens out when the North bluff bears S.W. ½ S.* The anchorage is off the ruins on the north shore, and the vessel should be moored; the tide sets in and out regularly.†

* *See* View on Chart, No. 1,309.
† *See* Admiralty Plan of Port Desire, No. 1,309; scale, m = 1·5 inches.

Some years since a Spanish colony was founded at Port Desire, but not answering the purpose, it was soon given up. The ruins of the edifices which are of stone, and the remains of a fruit garden, which in the year 1829, produced quinces and cherries, distinctly point out the spot.

The river was examined as far as a boat could go, and from a neighbouring hill it appeared at that time to be fed by a very small winding stream; but from the broad level ground, and muddy flats on either side of the stream, and the steep cliffs which bounded them, it appears probable that they form the bed of a large river at certain periods of the year, like that of the Samboronbon, which runs into La Plata.

**Supplies.**—Four miles above the ruins there is a small peninsula, connected by a narrow isthmus to the north shore; by sending a party up, and stationing men with guns on the isthmus, it is very likely that several guanacos may be shot as they are driven across it; for the peninsula is their favourite feeding-place. These animals are abundant, but unless stratagem be used, they are very difficult, from their shyness, to be approached. The easiest way of shooting them is by lying in wait, at break of day, near the places where there is fresh water. Guinea pigs are also numerous, and excellent eating.

There are some holes near the ruins, which generally contain water, but of so brackish a quality as scarcely to be worth notice. The wood is the same kind that was described in the account of Santa Elena, and it burned well; but on the islets farther up the inlet, and in many of the valleys, firewood of a superior quality may be freely obtained. The country appeared to be a parched barren desert, with some straggling tufts of brown grass and a few stunted bushes. Of edible vegetables there are few or none; good wildfowl are plentiful, and fish, especially shell fish, are abundant.

Once or twice in the year, a large body of Indians visit this place as if to reconnoitre, and therefore no straggling parties from the ship, much less individuals, should venture to any distance without having ascertained that the natives are not in the neighbourhood. "War to the knife" with all white men is now their maxim, in consequence of the treatment they have received from the Spaniards and their descendants.

**TIDES.**—At the entrance of Port Desire, it is high water at full and change, 10 minutes after 12 o'clock, and the springs rise 18 feet. It should be borne in mind, when approaching any part of the coast between Union bay and Port Desire, that there is a difference of half a tide, or 3 hours nearly, between the turn of the tide-stream in the

D 2

offing and the time of high water in the harbours. Three hours after high water in the harbours, or rather upon the shores, the tide ceases to run to the northward, and begins to run in a contrary direction. Also 3 hours after low water on the shore, the tide turns in the offing.

**DIRECTIONS.**—A vessel bound to Port Desire, or merely wishing to anchor in the bay which fronts it, may procure a good berth in 7 fathoms, at low water, well sheltered from N.N.W. round westerly to S.E. with the North bluff bearing N.W. ½ W., and Tower rock W. ¾ S. This situation being a little to the southward of the fair way of the port, and about 1½ miles from the nearest shore, is out of the strength of the tide. The bottom, however, being strewed with rounded stones, is rather foul for hemp cables, but the holding ground, although of such suspicious quality, seemed to be good.

Waiting for low water, all the dangers that exist will be seen, and the vessel easily dropped in with the tide, should the wind be, as it generally is, westerly. If it be fair, it will be advisable for the ship to be in the entrance at slack water; or, if the breeze be strong enough, a little before: as the water is deep on the south shore, there seems to be no real danger that may not be avoided by a careful look-out for kelp, which on that coast always grows upon, and therefore plainly indicates the existence of rocky ground. The course in is about W.S.W., and the distance from the entrance to the anchorage is 1½ miles.

Captain Fitz-Roy adds to the above account by Captain King, that vessels of 300 tons will not find easy access to Port Desire, the narrow and hooked entrance, the strong tides, the short interval of slack water, and the uncertain bottom, are sufficient obstacles to deter any large ship from making the experiment, unless urged by necessity. In the *Beagle's* last visit she knocked off a piece of her forefoot against the rock, to which she has given her name, and which lies N. ½ E. 3 cables' lengths from Chaffers point. Entering by a compass course is out of the question; two leads, a sharp eye, the braces in hand, and moderate sail are precautions absolutely necessary, as the tide hustles a vessel through in a few minutes, even against a strong wind.

**PENGUIN ISLAND,** 12 miles south-east from Port Desire, is bold on the outer side, and may be passed very close without danger, for the tide rather sets off than towards the shore. The tide is very rapid, and forms, even in a calm, strong ripplings, which in a breeze must be very dangerous for boats to pass through, and, indeed, not agreeable for vessels of any size.

**SEA BEAR BAY**, inside Penguin island, is a good anchorage, but from the strength of the tides troublesome to enter. The bottom, besides, is very- foul, and 30 fathoms deep ; and though an anchor might save a vessel from driving ashore, she would not be likely to regain it. In entering the bay, border pretty close to the Wells point, the low rocky point to the southward, in order to avoid May reef, which lies about a quarter of a mile N. ⅓ E. from it ; but as the sea always breaks upon it, the eye and a due consideration of the tides are the best guides.*

**May Reef** extends for some distance to the eastward of the breakers, and therefore the tide, when within it, sets in or out of the bay, but with little strength. Should a vessel not be able to enter the bay, there is anchorage off Wells point between it and the reef, on, it is said, tolerably clean ground. There is 12 or 13 fathoms off the reef ; then the depth shoals for one or two casts of the lead to 7 fathoms, after which it deepens again. From thence the vessel may haul across the bay, and anchor in 4 fathoms at low water, at about a quarter of a mile within Wells point, bringing it to bear E. by N., and avoiding the kelp which projects off from the sandy beaches. For further directions the Plan will be the best guide.

When once in, the anchorage is good, and protected at all points, except between N.E. and E. by N., but from the appearances of the beaches it is not probable that a heavy sea is often thrown in. There is no wood to be procured of any size. A few gallons of water may be collected in the wells situated immediately within Wells point : the passage to them lies over a small rocky bar, that a boat may cross at three-quarters flood ; there is also a small spring at the north end of the third sandy beach, which a herd of guanacos was observed to visit every morning, but as the water only trickles down in a very small quantity, it cannot afford more than a temporary supply. In short, besides a secure anchorage, this place affords no other advantage, though convenient for sealing vessels to anchor in while employed in their occupation upon Penguin island. Wells Point is in lat. 47° 57' 20" S., and long. 65° 45' 40" W.

**TIDES.**—It is high water, full and change, in Sea Bear bay at 12h. 45m., and the tide rises 20 feet. The flood sets to the northward, and during its strength at more than 3 knots; and the ebb has been known to set a ship 15 miles to the south in 5 hours. Off Penguin island the high water, or the termination of the northerly stream, takes place at about 4h. or 4h. 15m. after the moon's passage ; which is 3½ or 4 hours at least after it is high water at the shore.

---

* *See* Admiralty Plan, No. 1,309; m = 0·4 of an inch.

**SPIRING BAY** is contained between the south head of Sea Bear bay and Hilly point ; it forms a deep bight, 13 miles across, but is much exposed, being quite open to the south and east, and at the conclusion of a south-west gale, when the wind always veers to south and south by east, there is a considerable sea. The shore is skirted for some distance off with many rocks, and the bay altogether is quite unfit for anchorage. The land is of the same height as about Sea Bear bay, but has more lumps or nodules of rocky hills visible on the outline of its summit.

Off this bay, in the old chart, is laid down a rock called the Eddystone ; it would seem that this rock and the Bellaco rock, discovered by the brothers Nodal in 1619, are the same danger ; but the whole coast between Cape Blanco and Port San Julian is much strewed with shoals, which are the more dangerous from the strength of the tides which set between them. In navigating upon this part of the coast, the depth and quality of the soundings is a good guide, and, as a general rule, when the depth is more than 40 fathoms, there exists no *known* danger.*

In directing the ship's course by night near this coast, regard should be paid to the tide, which sets with considerably strength, the current parallel with the shore.

**The SHAG ROCK** is a whitish mass of rock, perfectly bare, lying about 1¼ miles off Hilly point ; 2 miles to the southward of it there are four small dark-coloured rocks ; and a mile farther there is rather a large rocky islet.

On the land, and at a short distance from the coast, there are three hills, which appear, when a little to the southward of Sea Bear bay, to be round-topped, but on reaching more to the southward, they extend in length and form into two hills, and when 9 miles to the southward of the Shag rock, they appear to form one mass of table land.

**CAPE WATCHMAN**, 22 miles from the Shag rock, is low, but may be distinguished by its bell-shaped mount ; at 6 miles from the point there is a shoal with kelp on it, and only 3 fathoms water. There are also many other shoal patches, but all are buoyed with seaweed ; the ship passed between several in 7 and 9 fathoms.

---

* On March 10, 1849, the *Sirius*, merchant ship, is said to have run on a reef of rocks about 10 miles E.N.E. from the southern point of Spiring bay. The captain, in his report to the owners, and by them transmitted to the Secretary of the Admiralty, states, that after striking about 20 minutes, by throwing all aback, the ship went off into 4 or 5 fathoms water, but before way could be gathered on her, she again struck on her port bilge; though there were at the same time 7 fathoms under her starboard fore channel. Soon after, however, the vessel got clear, without damaging her rudder, but making a great deal of water. The position of the reef, according to the same Report, is about 48° 7' S., 65° 37' W. *See* also Nautical Magazine for August 1849, page 433.

. The ground is very foul for more than 4 miles from Cape Watchman, and is so uneven that the tide ripples much. Though no positive danger has been discovered, it would always be prudent to give the cape a wide berth, and pass to the eastward of the Bellaco. When the cape has been passed, a hill will be seen to the N.W. of it, somewhat resembling Monte Video, in the river Plata, both in shape and colour, but not quite so high: it also is called Monte Video, and is in lat. 48° 13′ 40″ S., and long. 66° 25′ 50″ W.

**Desvelos Bay** lies to the northward of Cape Watchman, the shore falling back into a bight, which offers good shelter from westerly winds ; but in approaching or quitting it, due allowance must be made for the tides, which sweep along the shore from 2 to 3 knots.

**The BELLACO ROCK,** or the San Estevan shoal, was searched for in vain in the voyage of the *Descubierta* and *Atrevida*; but Captain Stokes, in the early part of 1828, on his passage down the coast, found it, and had an observation of the sun close to it for the latitude, which is 48° 29′ 20″ S., and the long. 60° 12′ 15″ W. It bears S. 40 E. *true,* or nearly S.E. by E. ¼ E. *magnetic,* distant 10½ miles from the extremity of Cape Watchman, and S.E. ¼ E. from Monte Video. The rock is a dark mass, about 6 feet above the water at high tide, and has the appearance of a boat turned bottom up : within half a mile of its south side the *Beagle* sounded in 12 and 15 fathoms, rocky bottom, and on its east side, at the same distance, the depth is from 20 to 24 fathoms. The ground around it being foul and uneven, the coast in its neighbourhood should be avoided.

**LOOKOUT POINT.**—From 4 miles to the southward of Cape Watchman to this point the land rises and the coast is safer ; but 5 miles to the eastward of Lookout point, there is a large patch of foul ground with much kelp. The land still rises, in advancing to the southward, till it attains the height of above 600 feet, and is then remarkable for its horizontal outline. Flat and Bird islets, 11 miles to the southward of Look-out point, though low, are too near the land to be dangerous to vessels that keep a fair offing.

**DAÑOSO REEF** lies 9 miles to the southward of Flat islet, and off the high table land of Cape Dañoso it is a dangerous reef, projecting 3 miles from the shore to the south-eastward, but it does not appear to be steep-to. From thence to Port San Julian, there is no known danger ; Mount Wood, near that port, and the above-mentioned long range of horizontal land, show so far out to sea, that they are unfailing guides for making that port.

PORT SAN JULIAN is 73 miles from Cape Watchman. Mount Wood, 951 feet high, and visible from the deck for at least 33 miles, is a good mark for this port, being flat-topped and much more elevated than the land about it ; the trend of the coast may also be useful as a mark ; and the land about the port being higher than that on either side of it, no mistake can be made.

The north head, Cape Curioso, 4½ miles to the northward of the entrance, is a low point jutting out to the northward, formed of cliffs horizontally stratified, of which the upper part is white brown, and the lower generally black, or with black streaks. The little monument erected by his shipmates to the memory of Lieutenant Sholl, close to the point of the same name, 1½ miles within Point Peña, the northern point of entrance, stands in lat. 49° 15′ 20″ S., and long. 67° 42′ W.*

An extensive bar crosses the entrance, and in the middle dries at low water, leaving a channel on either side, with something less than 2 fathoms in them and rather intricate. But the great range of tide (30 feet at the highest springs, and 16 in the dead neaps) makes the passing this bar comparatively easy. A reference to the plan will show the way across it better than any description, but great attention should be paid to the set of the tides, which run in and out sometimes at the rate of 4 knots. Half-flood is the best time for entering, as a vessel will then have depth enough on the bar and the parts that dry at low water will still be visible. The most convenient anchorage is off Sholl point in 4 fathoms.

No fresh water was found in any part of the inlet, its upper division being a chain of *salinas ;* but wood may be procured with ease on Shag island, and at other places. Abundance of seafowl may be killed, and fish of various kinds are plentiful.

TIDES.—It is high water at full and change, in Port San Julian, at 10h. 45m., and at high springs the tide rises 30 feet, at neaps to 23 feet, and neaps range 16 feet.

DIRECTIONS.—Having made out the two points which form the outer entrance, Curioso and Desengaño, which will be easily distinguished when at the distance of 6 or 7 miles, or more, according to the state of the weather, steer for the latter, keeping Mount Wood in a line with it, bearing W. by S. ½ S., and if the tide does not serve for going in, anchor in 8 or 9 fathoms about a mile N.E. of that point till a proper opportunity offers. But if the wind be S.E., or the weather threatening, stand off and on.

The COAST.—The land to the southward of Port San Julian is uniform, flat, and low : it is covered by scrubby bushes, and fronted by a shingle

----

* *See* Admiralty Plan of Port San Julian, No. 1,292 ; scale, m =1·5 inch.

beach. At 10 or 12 miles south of it, coming from the eastward, a small flat-topped hill is seen over the low coast hills.

In lat. 49° 29′ S., the character of the coast changes to a range of steep white clay cliffs, the average height of which was calculated to be about 315 feet. They rise like a wall from the sea, which, at high water, nearly washes their base; but at low water they are fronted by a considerable extent of beach, partly of shingle and partly of mud. Some short rocky ledges, which break at half tide, lie off several parts of this range, but none of the ledges extend for more than a mile from the shore. This cliffy range occasionally forms projections, but so slight as not to be perceived when passing abreast of them.

Anchorages along the coast may be taken up with the wind off shore, at from one to 2 miles from the beach, and in from 9 to 14 fathoms oozy bottom. In lat. 49° 58′ S., the range of steep white cliffs begins gradually to diminish in height, and terminates at 9 miles farther to the southward, in a comparatively low point, 180 feet high, forming the northern side of the entrance of Santa Cruz river. It is called in the chart North point, and is in lat. 50° 5′ 20″ S., and long. 68° 3′ W.

**RIVER SANTA CRUZ.**—The appearance of the coast about the entrance of this river of Santa Cruz is remarkable, and easy to be known, from the conspicuous manner in which it makes either to the northward or the south-eastward. From the latter direction a coast line of cliffs and downs of considerable height is seen extending from the southward, as far as the eye can reach, and terminating abruptly in the high, steep, flat topped cliff, Mount Entrance, of which the upper part descends vertically; the lower slopes off and appears to be united with some low land, which will be seen (according to the distance off) two or three points of the compass to the right of it. Mount Entrance stands on the south side of the river, 11 miles from North point, and is 356 feet high; the low land is on the northern side, and outside of the river. Twelve miles up the river, on the south bank, is Weddell bluff, 300 feet high, a conspicuous headland; and 6 miles farther, on the opposite shore, is another called Beagle bluff.[*]

If the object of entering this harbour be wood, water, or refit, a good berth will be found above Sea Lion island, and near the shore under Weddell bluff; but strangers should first anchor in the bight near Keel point, so that another ebb may expose to view the shoals that surround that island. If it be intended to sight the vessel's bottom, the sloping shingle beach at Keel point, 3 miles from the entrance, where the

* *See* Admiralty Plan of Santa Cruz, No. 1,308; scale, m = 0·5 inch.

*Beagle* was placed on the shore, in 1834, offers a most convenient spot; and the great rise of tide, and clean shore, renders Santa Cruz a most desirable place for that operation. The anchorage off Keel point is easily taken or quitted, but in moving all vessels should have their anchors ready, and a boat ahead, for the tides are sometimes strong, from 3 to 6 knots, and the banks are somewhat changeable. In bringing up, heavy anchors should be used, and plenty of cable veered, taking care not to anchor in the strength of the stream; and if near high water, the probable fall of the tide should be carefully considered.

Firewood may be cut near the anchorage, but water can only be obtained by sending the boats up the river ; during the last half of the ebb the river water is generally fresh above Weddell bluff.

**Northern Arm.**—At Weddell bluff the river divides into two arms ; the northern one, which passes under the east fall of Beagle bluff, was examined by Captain Stokes for 12 miles above its commencement, where it ceases to be navigable, even at high water. Its bed was divided by banks of sand into several little fordable streams, preserving, as far as the inequalities of the land would permit the eye to follow their course, a mean N.W. by N. direction. The stream at this part was quite fresh, but still subject to the regular ebb and flow. On the boat's return she was left dry for 6 hours, in the middle of the channel, above 2 miles about Beagle bluff. At half tide the boats took in their water at this place.

The shore on the south-west side is a range of clay cliffs, of the average height of 250 feet with grassy downs, and intersected with valleys and ravines. On the eastern side the land, for the most part, is low and level, with a shingle beach ; the aspect of the country is dreary, the soil gravelly, and the vegetation scanty, the largest production being bushes bearing berries, but none exceeding 7 or 8 feet in height. Many brant geese and ducks were seen, as well as the common sea-fowl of these parts, such as penguins, cormorants, gulls, ducks, and divers : several ostriches also made their appearance on the beach, and traces of guanacos were observed.

**The Western Arm,** which is far the more considerable of the two, was examined by Captain Stokes for 33 miles. It appeared to Weddell to be of such magnitude, as to be likely to communicate with some branch from the strait of Magellan. The first reach of the arm runs S.W. by W. 6 miles, with a mean breadth of 2½ miles. At 4½ miles up, the influence of the tides had altogether ceased, and the water was quite fresh. The stream ran beautifully clean and pure, with the velocity of at least 5 miles an hour, over a bed of pebbles mixed with dark sand ; its mean breadth being three-quarters of a mile, and depth in mid-channel 8 feet. It runs between two nearly parallel ranges of hills, about 4 miles asunder ; beyond this the reaches are short, seldom more than 2 miles long, and forming tortuous courses. The wind blew directly down, and the rapidity of the stream was so great that the boat was obliged to be tracked up the river.*

Captain Fitz-Roy undertook an expedition up the river, with three light whale boats ; they laboured against the stream, by rowing or tracking, for 16 days, when their provisions falling short, they were obliged to abandon this interesting exploration and to return to the ship, which occupied but four days. The utmost point they reached was 140 miles in a direct line to the westward of the entrance, or 245 by the course of the river ; and they were then within 30 miles of the foot of the snow-capped Andes. The temperature of the water was there much higher than that of the air, which proved that the sources of the river were to the northward ; and that they were still very distant, was equally shown by the continuous breadth of the stream, which had scarcely narrowed for the last 100 miles. The spot reached was 400 feet above the level of the sea, and, therefore, 1·6 feet per mile was the mean descent or fall of the river, though in many places it ran like a torrent.†

TIDES.—It is high water, full and change, in the river Santa Cruz, at 9 h. 30 m. ; spring tides range 40 feet, the neaps 18, and they run from 6 to 3 knots. In the offing they turn 2 hours later than in the harbour, and the flood runs to the northward. Well up the harbour the spring tides rose 42 feet, and ran sometimes 6 knots. The neaps rose only 18 feet, and with a much more moderate velocity. In the offing the tides were observed to flow regularly 6 hours each way, but to turn 2 hours later than the time of high water in-shore. The flood, as before, was observed to run to the northward.

DIRECTIONS.—The outer part of the bar, on which at low spring tides there are 14 feet, lies 4 miles outside of Mount Entrance. There are several shoal patches on it which dry, or at least break, at low water. Weddell bluff kept open to the north of Entrance point, and seen over Sea Lion island, on the bearing of N.W. by W. ¾ W., seems to offer the most direct route across the bar ; with this mark on, and at high water, the *Beagle* crossed the bar in 7½ fathoms ; and the Beagle bluff, a little open of the low points on the north side of the river, is also a leading mark to cross the bar. After passing the bar, which is about a mile broad, there is no impediment to a free course up the river, keeping midway between

———————————————————————————————

* The above description of Santa Cruz and the river, is taken from the late Commander Stokes' MS. Journal.

† See Captain Fitz-Roy's Narrative of the Surveying Voyages of H.M.S. *Adventure* and *Beagle*. (Colburn, 1839.)

the narrow points of entrance until reaching the shoals which project off the east point of Sea Lion island.

The above mark, however, does not agree with that given by Captain King and Weddell; and Captain Fitz-Roy remarks that where the tides are rapid, and a heavy sea frequent, it is is not probable that a bar of sand and shingle stretching across the mouth of a large river should long retain its position. Strangers are therefore advised to remain outside the bar, either at anchor or under sail, till low water, when its shallow parts will show themselves; and then, as the tide rises very high, to weigh at half or two-thirds flood, and steer directly in through the most convenient of the channels, of which there were three open in 1834.

The best anchorage seems to be that on the south side of Sea Lion island, where the water is shoaler, and the tide not so strong. At an anchorage outside the bar, Mount Entrance bearing W. ¾ N. 5 miles off, the *Beagle*, in 1827, rode out a gale from S.S.W. with a heavy sea, without driving. The soundings that are marked in the chart, outside the bar, were taken at low water, while the brig occupied the above anchorage.

COY INLET.—Between Santa Cruz and Coy inlet, a distance of 58 miles, the coast trends in, so as to form a considerable bight. It is a succession of cliffs and intervening low beaches, bounded by a ledge of rocks, which are either dry at half tide, or are then shown by a line of breakers; they extend as far off as 3 miles. On one occasion the *Beagle* anchored among them, and had some difficulty, and not a little risk, in escaping.

Coy inlet is conspicuous, as it is the only part of the coast that has the appearance of an inlet between Santa Cruz and Cape Fairweather. When within 7 miles of its lat. (50° 57'), as well to the northward as to the southward, a ship should keep at the distance of 4 or 5 miles off the coast. There can be no inducement to go nearer, as it affords neither fuel nor water, and if incautiously approached much trouble and danger may ensue, from the ledges of rocks, which project at least 3 miles, and perhaps more, from the coast.

There is no account of Coy inlet in Captain Stokes's Journal; what is here given has been collected from oral information. It is said to be a shoal basin of some miles in breadth at high water, terminating 19 miles from the entrance, and fronted by a bar of rocks, leaving a passage of only 6 feet water; inside there are little more than 3 feet water, and in most parts of the inlet the banks, which are of mud and sand, are dry at low water; it seems to be useless for any other purpose than to afford shelter to a small boat. The southern side of the inlet is cliffy, and at its termination receives the drains of an extensive flat country.

Between Coy inlet and Cape Fairweather the coast is similar to that to the northward of the inlet, but more free from rocky ledges, and good anchorage may be had from 2 to 6 miles off-shore, in from 7 to 12 and 14 fathoms muddy bottom, the water shoaling gradually to the shore. The beach is of shingle to high-water mark, and then of hard clay as far as 100 feet beyond the low water limit, where a green muddy bottom commences, and the water gradually deepens. The outer edge of the clay is bounded by a ledge of rocks, on which the sea breaks; it extends for some distance parallel with the coast.

TIDES.—It is high water at full and change in Coy inlet at 9h. 30m., and the tide rises 40 feet. The flood sets to the N.W. by N., and the ebb S.E. by S., 6 hours each way.

CAPE FAIRWEATHER, 38 miles from Coy inlet, is the southern extremity of the long range of clay cliffs that extends from Coy inlet, almost without a break. The cape resembles very much Cape St. Vincent on the coast of Portugal, and appears to be of the same character as Cape Virgins, at the entrance of Magellan strait, for which it has frequently been mistaken, notwithstanding that there is more than 45 miles difference in the latitude of the two headlands. This error was made by one of the ships belonging to the fleet under the command of Loyasa, in the year 1525 (see Burney's Collection of Voyages, vol. i. p. 131): and the brothers Nodàl, in their description of the coast, warn the navigators from mistaking the one for the other, "y venido de mar en fuera a buscar " la tierra, facilmente podian hacer de Rio de Gallegos el Cabo de las Virgenes" (and in making the land [the north point of entrance of] the river Gallegos may easily be mistaken for Cape Virgins.)—*Voyage of the Nodales*, p. 53.*

On the old charts of this part of the coast the shore is said to be formed of chalk hills, "like the coast of Kent;" the resemblance certainly is very great, but instead of chalk they are of clay. They are from 300 to 400 feet high, and are horizontally stratified, the strata running for many miles without interruption. The interior is formed by open plains of undulating country, covered with grass and plants, among which is abundance of wild thyme, but entirely destitute of trees: it abounds with guanacos, which may be procured by lying in wait at the water-holes.

WATER.—About 17 miles north of the cape there is a ravine containing abundance of fresh water, which may be obtained, when the wind is off shore, without any difficulty; it is standing water, and being much grown over with plants, may not keep, but for a temporary supply it seemed to be good. Besides this pond, there is no want of fresh water; it may be seen trickling down the face of the cliffs at short intervals.

---

* Viaje de Bartholomé y Gonzalo de Nodàl en 1618–19.

**PORT GALLEGOS.**—The entrance of this river and port is formed on the north side by the cliffy land of Cape Fairweather, and on the south by a low shore that is not visible at sea for more than 4 or 5 leagues, excepting the hills in the interior called the Friars, the Convents, and North hill. It is fronted by extensive sand-banks, most of which may be crossed at high water, but at half-ebb they are almost all dry. The entrance is round the south extremity of the shoals, which bears from Loyala point S.E., distant 10 miles. The passage in is parallel with the coast, taking care not to open the land to the northward, the most eastern trend of Cape Fairweather. The shore may then be gradually approached, but in the present state of the knowledge we possess, the ship should be anchored there in 10 fathoms, to wait low water, at 1½ miles from the shore, as soon as Loyala point begins to be observed to trend round to the westward; the anchorage there is good, and well sheltered from the prevailing winds.*

**TIDES.**—It is high water, full and change, in the entrance of Port Gallegos at 8h. 50m.; the rise of tide at springs is 46 feet, and the stream runs as much as 5 knots.

By anchoring, the passage in will be easily detected, and may be passed before the shoals are again covered, which will be a good guide. Anchorage may be taken up on the south side, for to the northward the banks are extensive. There is also a middle channel, and as it appears to be the widest, may be the best for crossing the bar. The outer part was not examined, but no doubt there is a sufficient depth of water at three-quarters flood for any vessel to pass it. The water of the River Gallegos is fresh at 25 miles from the mouth.

**The COAST** from Port Gallegos towards Cape Virgins extends in a south-easterly direction for nearly 50 miles, and, for the first half of the distance, is formed by a low shelving coast which, at a few leagues from the shore, is not visible, so that a stranger might readily suppose it to be the entrance of the Strait of Magellan. There are, however, some marks by which it may be known, even should the latitude not have been ascertained. In clear weather the Friars and the other hills, which are situated to the southward of Port Gallegos, will be visible, and in thick weather the soundings off the cape will be a sure guide; for at the distance of 4 miles off no more than 4 fathoms will be found, whereas at that distance from Cape Virgins the depth is considerable; the bottom also to the north of Cape Fairweather is of mud, whilst that to the north of Cape Virgins is of gravel or coarse sand; and the latter cape has a long low point of shingle running off for nearly 5 miles to the S.W. ; and, lastly, if the weather be clear, the distant land of Tierra del Fuego will be visible to the S.S.W.

At 18 miles to the southward of Cape Fairweather the cliffs again commence, and continue to Cape Virgins, with only one or two breaks, in one of which, 8 miles north of the latter cape, a boat might land, if necessary. There is good anchorage along the whole coast between the Gallegos and Cape Virgins, at from 2 to 5 miles from the shore ; but the bottom is rather stony, and might injure hempen cables. As the cape is approached, the ground becomes still more foul.

**WEATHER.**—Between the parallels of 40° and 50° on this coast much uniformity of weather prevails, those ten degrees of latitude causing less variation of temperature than could reasonably be supposed. The winds are also more regular than those about La Plata, and as the quantity of rain which falls during the year is beyond comparison less, the climate is at least as warm as that of Buenos Ayres, and so very dry that the land is generally parched and sterile, except near rivers. In some ports on this coast, San Blas, the Oven, San Antonio, and others, it is ruinous for a ship to lie moored during many summer months ; even weeks of delay are injurious, so powerful is the effect of the sun, rarely clouded, and acting throughout the whole day upon the wood-work, un-moistened even by dew.

**Frost and Hail.**—In winter there are sometimes sharp frosts at night, but they do not continue through the day, snow is rarely seen : hail with southerly winds is common and very large. During the summer months, while the air is in a settled state, the wind generally veers round the compass during the 24 hours. A moderately fresh sea breeze from the S.E. in the afternoon being succeeded by a land wind of similar strength from the N.W. during the night, light winds or calms prevail in the mornings and evenings.

In settled weather the wind always goes round with the sun from east to west by the north : when it takes the opposite direction bad weather usually follows. Gales from the south-east occur once or twice in a month, and generally it is said about the full or change of the moon. In summer these gales are heavy, and are very much felt on the coast, as they send a heavy sea into the harbours, and are sometimes accompanied by rain and thick weather. Other strong winds blow chiefly from the land, and bring clear pleasant weather ; N.E. winds sometimes bring rain, but they rarely, if ever, increase to the strength of a gale. During the winter season, or from May to November, southerly winds are more frequent, and last longer than in summer ; more rain is brought at that time by winds from N.E. and S.E., but the latter wind is not usually so strong as in summer.

On this part of the coast, as well as in La Plata and about Tierra del Fuego, bad weather with northerly winds will continue until the wind shifts to the southward, going round by the west. Squalls or gales of more or less strength from S.W. to S.S.E. soon clear the air, and the louder and longer the southerly wind blows the finer and the more lasting will the weather be afterwards. These southerly winds are dry, cold, and elastic ; they cause the mercury in a barometer to rise unusually high, and have very beneficial effects upon the human frame.

With northerly and westerly winds there is at times much lightning and thunder, particularly during the warm months. Winds from the northward begin and increase gradually; those from the southward are sudden, and at times they are instantaneously violent. Ships should be always ready for a sudden shift to the southward when the barometer is low, with a northerly wind blowing and the weather threatening.

**BAROMETER.**—Very thick gloomy weather, with northerly winds, and perhaps rain, with lightning and thunder, is sure to end in a sudden shift to the southward. If the mercury is low—that is to say, about 29˙60 in a barometer averaging 30 inches in settled weather—a gale may be expected. After falling, the mercury will rise shortly before the wind shifts, and therefore the time when the mercury ceases to fall and begins to rise should be carefully noticed. The mercury rises higher with S.E. than with S.W. winds of equal strength. Northerly winds cause the mercury to fall : it falls most with the wind at N.N.W. and rises most with winds from the S.E.

During settled and clear weather a S.E. wind will raise the mercury to near 30˙50 inches. With weather equally settled, and apparently equally clear, a N.W. wind will depress it to 29˙80. S.E. and S.W. winds affect the barometer as the N.E. and N.W. in England, while N.E. and N.W. have similar effects to those of our S.E. and S.W. winds respectively. A north-easterly wind causes a high glass in England; so a south-easterly one does the same in these latitudes ; and a N.E. or northerly wind depresses the mercury as the southerly winds do in England. It is necessary to attend to these points in judging of the wind and weather by the barometer, or by the sympiesometer, since the same height (29˙80) which would indicate only moderate breezes from the northward would accompany a heavy gale from the south-east. The sympiesometer is supposed to be constructed so as to correspond in its average height with the barometer.

The rising of the mercury always precedes, by an interval more or less short, this change from a northerly to a southerly wind. Northerly gales are preceded by gloomy overcast weather, by numerous small clouds apparently very high in the air, sometimes by a mistiness or a thick haze, and sometimes by much lightning. Southerly gales may be foretold by large masses of heavy clouds, with hard defined edges, rising in the southern horizon, those who distinguish the clouds by names will understand that the last mentioned are Cumuli ; the former, or those preceding northerly winds, as Cumulo-strati and Cirro-cumuli.

**WINDS.**—Westerly winds are the most prevalent throughout the year, and they generally bring clear fine weather. Gales of wind sometimes begin blowing from the N.E. while the mercury in the barometer is high; if moderate at first, the wind generally increases and draws to the northward as the mercury falls, until it reaches north and N.W., when it blows hardest. Having continued to blow for 12 or 24 hours, it moderates, perhaps falls entirely, particularly if there be rain, and in a few hours afterwards shifts to the southward, quickly increasing to a gale, which will be strong in proportion to that which preceded it ; or perhaps it may shift suddenly in a squall to the southward, and blow with violence.

**SQUALLS.**—Fogs occur during the winter months, but they are neither frequent nor are they of long duration. Squalls are less numerous, and give more warning than in most other parts of the world, but when they do rise they are not to be trifled with. Those from the southward sometimes require nearly all sail to be taken in ; and if the barometer has been very low, and the clouds look very heavy, and you cannot see underneath them, it will be prudent to furl almost every sail, and even to run before the first heavy blast, which seldom lasts many minutes. If attention is not paid to this advice, dearly-purchased experience will soon teach the propriety of this cautious prudence, and especially to those who navigate small vessels in this climate.

**CURRENTS.**—When more than 50 miles from the coast of Patagonia, very little current is found during settled weather and moderate winds : what there is sets sometimes north, and at other times south, about half a knot ; but before strong winds, and while they are blowing, the current runs a knot, or perhaps 2 knots, in the same direction as the wind. Generally speaking, the currents from the southward have more strength, and run longer than those from the opposite point ; they are, however, very irregular, and appear to be governed chiefly by the winds. Nearer than 50 miles from the land, the current sets more strongly from 2 to 3 knots, particularly near the projecting headlands. When nearer than 20 miles to the shore, the influence of the tides begins to be felt, especially if to the southward of Cape Corrientes.

**TIDES.**—Along the almost unbroken coast, extending from Cape Corrientes to Bahia Blanco, the stream of the tide is very weak, although the water rises and falls about 10 feet.

[S. A.]  E

The great southern tide-stream here appears to end, after sweeping along the southern half of South America. In the archipelago of Tierra del Fuego the flood-tide comes from the N.W., passes round Cape Horn, and through the strait of Le Maire, and then, from Cape St. John, sets strongly to the eastward and north-eastward. From thence the flood runs to the N.E., along the north side of Staten island and Tierra del Fuego, occasions very high tides at the entrance of the strait of Magellan, where it unites with the stream which has come directly through the Strait, and passing onward along the coast of Patagonia, produces high-water at each place in succession until it is lost near Cape Corrientes.

Near the coast between the dangerous banks of San Blas and Bahia Blanco, the flood and ebb streams set nearly north and south, from 1 to 4 knots, according to the wind and the age of the moon. Between the banks of San Blas and the Rio Negro, the tides are regular a little more than six hours each way, if not affected by the wind; but they are very strong, running from 2 to 5 knots, particularly along the coast between San Blas bay and the Rio Negro. But these strong and dangerous tides are not much felt at the distance of 15 miles from the land. Between San Blas and Cape Bermejo the tide stream sets N.E. and S.W. equally strong each way; if there is a difference, the flood is the stronger. In the depth of the Gulf of San Matias there is very little stream of tide, but a rise and fall of from 20 to 30 feet.

In the Bay of St. George there is not much stream of tide, nor more rise and fall than 12 feet. Off Cape Dos Bahias and off Cape Blanco, particularly the latter, the tides are again strong, and there are two or three races off Cape Blanco almost as dangerous as those off the Peninsula of San Josef.

Within the Gulf of San Matias, and near the entrance of the Bay of San Josef, there are races violent at times, but not equal in effect to those at the east side of the peninsula.

In moving along this coast the mariner should bear in mind that between Port Desire and Bahia Blanco there is a difference of half a tide between the turn of the tides in the offing and of high or low-water in the harbours and along the shore; the turn of the stream in the offing being three hours later than the corresponding turn of the tide in-shore. In other words, the northern or flood stream runs outside three hours after the tide has begun to ebb on the shore; and the converse.

TIDE RIPS.—Off the peninsula of San Josef there are dangerous tidal races; and so high and so violent are the waves at particular times of tide, that a small vessel might be most seriously injured if not totally destroyed by getting into them. Lieutenants Wickham and Stokes,

while surveying this part of the coast in two vessels, one of 9 and
the other of 13 tons burthen, were drawn during a calm within a mile
of one of these races while it was roaring and boiling furiously.   No anchorage could be had, for no bottom could be found with the deep-sea lead,
and they were fast approaching the fatal race when a breeze fortunately
sprung up, which enabled them to stem the stream, and after a struggle
with oars and sails, at last to overcome the tide and avoid the
danger.

**VARIATION.**—The variation at the beginning of the year 1860 off
the Rio Negro was 15° E. ; off Cala Chica, 16° ; off Cape dos Bahias, 17° ;
off Cape Blanco, 18° ; off Point Lookout, 19° ; at the river Santa Cruz,
20° ; at Cape Virgins, 21¼° E.   The curves of equal variation on this
part of the coast assume a S.E. by S. direction (true), and the degrees lie
about 90 miles apart.   The variation is decreasing at the rate of about
4′ annually.

# CHAPTER III.

## THE FALKLAND ISLANDS—EAST FALKLAND.

VARIATION 16° to 18° East in 1860. Annual decrease 6'.

---

**THE FALKLAND ISLANDS**, the *Malouines* of the French, and *Malvinas* of the Spaniards, form an island group in the South Atlantic, belonging to Great Britain, consisting altogether of above 200 islands, large and small. They lie off the coast of South America, about 350 miles due East, true, of the south-eastern entrance of Magellan Strait, between lat. 51° and 52½° S., and long. 57½° and 61½° W. Only two of the islands are of any considerable size; these are called respectively East and West Falkland, and are separated from each other by a Sound varying in breadth from 2¼ to 18 miles. East Falkland is about 90 miles in length, and about 40 miles in average breadth, and has an area of 3,000 square miles. West Falkland is 80 miles long, by a mean width of about 25 miles; area 2,000 square miles. The other islands vary from 16 miles long by 8 broad to mere islets of half a mile in diameter, and may comprise another 1,000 square miles, making a total of 6,000 square miles. The whole group is indented in a remarkable manner by sounds and bays, which form excellent harbours, and these, together with the varied outline of the mountains, constitute the principal features in the general aspect of the country.

These islands were seen by Dr. John Davis on the 14th August 1592, in Cavendish's second voyage. In 1690, Strong sailed through the channel which separates the islands, and called it Falkland Sound, which name afterwards was transferred to the whole group. In 1710 a French vessel from St. Malo touched at them, and named them Iles Malouines. Settlements were afterwards formed on them by the French, Spaniards, and English alternately, which will account for the foreign names of some of the ports and harbours; they have ultimately remained in possession of the English.

**SURVEYS.**—The Falklands have been surveyed by the Admiralty at different periods between the years 1834 and 1845; and the surveys are published in one general chart,* and 7 plans of sounds and harbours.

---

* *See* Admiralty Chart of the Falkland isles, No. 1,354; scale, m = 0·2 of an inch.

The East Falkland was examined by Lieutenant Wickham of H.M.S. *Beagle*, and Captain B. J. Sulivan, the West Falkland by Commander W. Robinson.   The following remarks are chiefly from Captain R. Fitz-Roy's "Voyages of the *Adventure* and *Beagle*."

**ASPECT.**—In the general appearance of the Falkland islands there is little remarkable.   Ridges of rocky hills above 1,000 feet high are seen traversing extensive tracks of moorland, without a tree, and bounded by a low rocky coast.   On the northern part of East Falkland the hills attain a considerable elevation, but the whole of the south portion is so low that it can barely be seen from the deck of a ship at 5 miles distant.   The principal range of hills are the Wickham heights, stretching east and west, and rising to a height of from 1,400 to 2,300 feet.   The average height of the western isle is greater than that of the eastern, the highest peak, Mount Adam, in the north-west part of the island, reaching 2,315 feet above the level of the sea.   On the western face of the island, and on some of the adjoining islets, there are some precipitous cliffs, exposed to the fury of the western seas.   The summits of the hills and mountains are rugged, terminating in points and ridges, are seldom rounded, and never tabular.

**HARBOURS.**—Excellent harbours, easy of access, affording good shelter, with the very best holding-ground, formed by the remarkable indentations of the coasts, abound among these islands, and with due care offer ample protection from the frequent gales.

**The TIDES** differ much as to strength and direction in different parts of this group, but the times of high water, at the full and change of the moon, only vary from 5 to 8 o'clock ; and the range is almost similar every-where,—about 4 feet at neap, and 8 feet at spring-tides.   The great tidal wave which pours its streams among these islands comes from the south-east, and therefore scarcely any stream is perceptible on the south-east coast of East Falkland ; while along the north, south, and west shores, it increases in strength, until among the Jason islands, it runs 6 miles an hour, and causes heavy and dangerous races.   Into Falkland sound the flood enters at both ends, and meets near the Swan islands, showing pro-bably that the principal wave impinges upon the coast considerably to the eastward of south.

Generally speaking, the sea is much deeper nearer the southern and western shores than it is nearer those of the north ; and to these local differences may be attributed the varying velocity of the minor tide streams.

**CURRENTS.**—Besides these movements of the surrounding waters there is a current setting past the islands from south-west to north-east ;

a current which continually brings drift-wood to their southern coasts. On all parts of their southern shores that are open to the south-west, the beaches or rocks are covered with trees which have drifted from Staten Land, or Tierra del Fuego. Great quantities of this drift-wood may be found between Cape Orford and Choiseul bay, an interval of coast in which a vessel may not otherwise find a good supply of fuel. On Breaker island and in the bays behind the southern Sea Lion islands, portions of Fuegian canoes have often been found; one consisted of an entire side (pieces of bark sewed together), which could not have been made many years. At sea, when north-eastward of the Falklands, great quantities of drift-kelp* are seen, besides water-worn trunks and branches of trees, near which there are generally fish, and numbers of birds. These sure indications of a current from the south-west have been met with upwards of 200 miles to the northward of Berkeley sound. There is not, however, reason to think that this current ever runs more than 2 knots under any circumstances, and in all probability its usual set is even less than a knot.

winds.—Wind is a great feature in the climate of the Falklands; a region more exposed to storms both in summer and winter it would be difficult to mention. The winds are variable, seldom at rest while the sun is above the horizon, and very violent at times. During the summer, a calm day is an extraordinary event. Generally speaking, the nights are less windy than the days; but neither by night nor by day, nor at any season of the year, are these islands exempt from sudden and very severe squalls, or from gales which blow heavily, though they do not usually last many hours.

It has been stated by Bougainville and others, that in summer the wind generally freshens as the sun rises and dies away about sunset; also, that the nights are clear and starlight. Such may be generally the case; yet it is also true, that there are many cloudy and many windy nights in the course of each year, or even month. The *Magellan* was driven from her anchors, though close to a weather shore in the narrowest part of Berkeley sound, and totally wrecked in Johnson harbour about midnight of the 12th of January 1833.

The prevalent direction of the wind is westerly. Gales in general commence in the north-west and draw or fly round to the south-west; and it may be remarked, that when rain accompanies a north-west wind, it soon shifts into the south-west quarter, and blows hard. Northerly winds bring cloudy weather, and when very light they are often accompanied by a thick fog; it is also worthy of notice, that they almost always occur about the full and change of the moon.

North-east and northerly winds bring gloomy overcast weather, with

---

* Sea-weed detached from the rocks, and drifting with the current.

much rain ; sometimes they blow hard and hang in the N.N.E., but it is more common for them to draw round to the westward. South-easterly winds also bring rain ; they are not frequent, but they blow hard, and as the gale increases, it hauls to the southward. During winter the winds are chiefly from the north-west, and in summer they are more frequently south-west. Though fogs occur with light easterly or northerly winds, they do not often last through the day. Gales of wind as well as squalls are more sudden, and blow more furiously from the southern quarter, between south-west and south-east, than from any other direction.

Wind from the east is rarely lasting, or strong ; it generally brings fine weather, and may be expected in April, May, June, and July, rather than at other times ; but intervals of fine weather (short indeed), with light breezes from E.S.E. to E.N.E., occur occasionally throughout the year. Neither lightning nor thunder are at all common ; but when the former occurs, easterly wind is expected to follow. If lightning should be seen in the south-east while the barometer is low, a hard gale from that quarter may be expected. South-east and southerly gales last longer than those from the westward, and they throw a very heavy sea upon the southern shores. In the winter there is not, generally, so much wind as in the summer, and in the former season, the weather, though colder, is more settled and considerably drier.

BAROMETER.—Every material change in the weather in the vicinity of these islands is foretold by the barometer, if its movements are tolerably understood by those who consult it, and if it be *frequently* observed.

The TEMPERATURE may be considered equable: it is never hot, neither is it very cold ; but the average is low, and in consequence of frequent rain and wind, a really moderate degree of cold is much more noticed than would probably be the case if the weather were dry and serene. Since 1825, the thermometer has only once been observed as low as 22° Fahr. at mid-day ; and but once above 80° in the shade. Its ordinary range is between 30° and 50° in the winter, and from 40° to 65° in the summer.

CLIMATE.—Captain Sulivan observes on the climate of the Falkland islands, that the dryness of the weather in summer is remarkable, on one occasion nearly two months having passed without any rain falling. Ice has not been known to exceed an inch in thickness ; snow seldom lies upon the low lands, or at any period exceeds two inches in depth. Although rain is so frequent, it does not continue falling for any considerable time ; and as evaporation is rapid, in consequence of so much wind, there are no unwholesome exhalations ; indeed, the climate is exceedingly healthy, and no disease whatever has been hitherto contracted, in consequence of its influence, excepting ordinary colds or coughs, or rheumatic affections brought on by unusual exposure to weather. It is said by those who have had the most experience there, that the climate of West Falkland is milder than that of the eastern island. Probably the west winds are chilled in passing over the heights, and upon reaching Stanley harbour become several degrees colder than when they first struck upon the western islands. In Tierra del Fuego, and other places, the case is similar, the western regions having a milder climate than is found about the eastern districts.

GEOLOGY.—The more elevated parts of East Falkland are composed of quartz rocks ; clay slate prevails in the intermediate districts. Sandstone, in which are beautifully perfect impression of shells, occurs in beds within the slate formation ; and upon the slate there is a layer of clay, fit for making bricks. A peculiar feature in the geology of these islands is presented in streams of stones, or fragments of quartz, which appear to flow down the sides of the hills. These streams are from 20 to 30 feet wide, and the stones vary in size from one to four cubic feet, and are spread out in the valleys to a great extent. The soil of the islands is chiefly peat, but near the surface, where the clay is of a lighter quality, and mixed with vegetable remains, it is good soil, fit for cultivation. Stone of two or three kinds, suitable for building, may be found in different parts of the islands. Lime may be obtained by burning the fossil shells brought from the coast of Patagonia, where the cliffs are full of them ; or by collecting shells scattered upon the Falkland shores.

VEGETABLES.—A remarkable feature in the botany of the Falklands is the entire want of trees, but there is a great variety of sweet-scented flowers, which in November and December nearly cover the ground. The tussac grass (*dactylis glomerata*), a gigantic sedgy grass, having blades 7 feet in length and ¾ of an inch in breadth, was formerly abundant on the mosses, but rarely extends more than half a mile from the sea beach. Anti-scorbutic plants are very plentiful in a wild state, such as celery (*apium graveolens*), scurvy-grass (*oxalis enuphylla*), sorrel, &c. ; there are also cranberries and what the settlers call strawberries, a small red fruit growing like the strawberry, but in appearance and taste more like a half-ripe blackberry. A little plant which grows like a heath in many parts of the Falklands, as well as in Tierra del Fuego, has long been known and used by the sealers as a tea plant (*myrtus nummuralia*) ; but it has a peculiar effect at first upon some people ; which is of no consequence, as it soon goes off.

CATTLE.—Animals increase here rapidly, and the quality of their hides or fur improves. It must be remembered that cattle are no longer wild, but are the property of some owner, and there is a penalty for shooting them. Cows give a large quantity of milk, from which good butter and cheese may be made. For much of the produce of these islands, such as salt meat, potatoes, oil, butter, cheese, tallow, &c., a ready sale would always be found on the coasts of South America. Should any accident happen to a vessel in doubling Cape Horn, obliging her to make for the nearest port at which she could obtain supplies, she would find all she required at the Falkland isles.

It may not be uninteresting or be unimportant to vessels returning from the Pacific to be informed, that, according to an official report which has been recently received from the Governor at Port Stanley, relative to the agricultural produce of the settlement, it appears that every useful kind of green crop or garden produce can be raised there, and of unexceptionable quality. At the first exhibition of the kind, held at Stanley by the "General Improvement Society" on the 17th April 1849, potatoes, turnips, cabbages, carrots, cauliflowers, beans, onions, &c. were shown, all of excellent quality. A few English flowers were also to be seen, such as carnations, stocks, mignionette, &c. Several milch-cows, butter, pigs, and poultry were exhibited of a very passable description. Hence the admirable position of the East Falkland, and the facilities it affords from its safe anchorages, together with the daily increasing facility of procuring the above mentioned refreshments, so essential to the health of the seamen, will most probably induce many vessels to touch there in their homeward voyage, in preference to Rio de Janeiro, or other ports in Brazil, especially when a great advantage will be derived from shortening the length of the voyage; for instance, Capt. J. B. Maxwell, in H.M.S. *Dido*, touched at Port Stanley instead of Rio de Janeiro, on his return from New Zealand, by rounding Cape Horn in December 1848, and thence direct to Spithead in 50 days. Had he pursued the usual mode, by touching at Rio de Janeiro, he considers it would have prolonged the voyage three weeks, the average time occupied in the passage from the Falkland islands to Rio de Janeiro.

SEALS (both hair and fur) and sea elephants were abundant along the shores of these islands in former years, and by management they might be encouraged to frequent them again, but now they are annually becoming scarcer, partly from indiscriminate slaughter, and possibly from the islands becoming more inhabited.

WHALES frequent the surrounding waters at particular seasons, and they are still to be found along the coasts of Patagonia and Tierra del Fuego (within easy reach from the Falklands), though their numbers are very much diminished by the annual attacks of so many whale

ships, which have made the Falklands their head-quarters during the last
20 years. There is also a description of blubber to be found in the waters
about the Falklands, from which large quantities of oil may be extracted
with much ease; but the art of fishing for it seems only known to the
Danes, who find the same ingredient in the northern seas, where it is in
great request, and affords occupation to a considerable number of vessels
that are engaged in securing it.

**FISH.**—A valuable source of daily supply, and, by salting, of foreign
export, is the inexhaustible quantity of fish which swarm in every harbour during the summer. The description which most abounds is a kind
of bass, from 2 to 3 feet long, and 6 inches in depth; it takes salt well,
and has been exported by cargoes to the River Plata, and to Rio de Janeiro; and there are small delicious fish in such shoals, that our boats'
crews were sometimes obliged to let a large portion escape from the net
before they could haul it ashore without tearing.

In the fresh-water ponds, so numerous on the large islands, there is a
very delicate fish, somewhat resembling a trout, which may be caught by
angling. The shell-fish are chiefly mussels and clams, both of which are
very abundant, and easily gathered at low water.

It may here be remarked, that the cod-fishery off Patagonia and
Tierra del Fuego might be turned to very good account by settlers at the
Falklands.

**KELP.**—In approaching any part of the Falkland Islands, and especially while entering a harbour, a careful look-out should be kept for
" fixed kelp," or the sea-weed which grows on every rock that is covered
by the sea, and not very far beneath its surface. Lying upon the water,
the upper leaves and stalks show, almost as well as a buoy, where there
is a possibility of hidden danger. Long stems, with leaves lying regularly
along the surface of the sea, are generally attached to rocky places, or
else to large stones. In passing to windward of patches or beds of kelp,
or rather in passing on that side from which the stems stream away with
the current, care should be taken to give the place a wide berth, because
the only part which shows when the tide is strong, lies on one side of,
not over the rocks. Where the stream of tide is very strong, this kelp
is quite " run under," or kept down out of sight, and can no longer be
depended on as a warning. When a clear spot is seen in the middle
of a thick patch of fixed kelp, one may expect to find there the least
water.

Drift kelp, or that which is floating on the surface of the sea, unattached
to any rock or stone, of course need not be avoided; and it may be known
at a glance, by its irregular huddled look.

**WEST FALKLAND.**—The two ports most easy of access to vessels from the East island are, White Rock harbour at the north entrance, and Fox bay at the south entrance of Falkland sound.   There is a little good grassy land near White Rock harbour, though quite cut off from the central valley; but Fox bay, though not a good port for large vessels, has a safe anchorage for small ones in the head of the cove in the north-west corner, and a break in the hills there would admit of an easy road to the centre of the island.   Vessels bound there from Port William would have sheltered anchorages all along the south side of East Falkland and through Eagle channel.

**MAKING the LAND.**—All vessels intending to touch at the Falkland islands, and coming from the northward, should endeavour to get soundings off Cape Corrientes, in about lat. 39° S.   If their longitude is incorrect, they would thus be able to correct their position; for the edge of the bank is so steep that in a distance of 10 miles the depth changes from 100 fathoms no bottom to 60 fathoms sand; and by sounding every 2 or 3 miles, until the edge of the bank is hit in about 80 or 90 fathoms, a vessel might obtain her longitude within a very few miles of the truth.   In the parallel of 39° S. the edge of the bank is in long. 55° 45′ W.; in lat. 41°, in about long. 56° 55′; in lat. 45°, in about long. 60°; and in lat. 46° S., in about long. 60° 15′.*

Should a vessel be unable to get in with the coast so as to strike the edge of the bank as far north as lat. 39° S., she should endeavour to do so as soon afterwards as possible; taking advantage of every northerly and N.W. wind to steer about S.W., in order to make up for what she is certain to be driven to the south-east when the wind draws to the S.W., which it does at least every second or third day.   By persevering in getting to the south-west, whenever the wind will allow it, until to the westward of long. 60° W., there will be no fear of being driven to the north-east of the islands; whereas, if a vessel make a straight course for the islands when the wind is fair, she will be certain of being driven to leeward by the frequent south-westerly winds, and find great difficulty in getting to windward again.   Having, if possible, kept as far to the westward as long. 60° W., until in lat. 49° 30′ S. soundings will be obtained on the bank to the northward of the Falklands, in about 80 to 85 fathoms, fine dark sand.   If the longitude can be depended on, a course may then be steered to make the land about 20 miles to the westward of Volunteer

* *See* Admiralty Chart of the South Atlantic Ocean, No. 2,203; scale, $d = 0.4$ of an inch.

point ; but if the position of the vessel is doubtful, or the wind drawing round from the N.W. towards the S.W., it would be better to keep to windward, so as to make the Eddystone rock.

Between the Rio de la Plata and the Falklands the bank of soundings will be found useful in approaching the islands from the northward ; soundings may be obtained in good time to prevent any danger except off Cape Carysfort, where a light would be of great service.

There is a current to the north-east, probably part of the Cape Horn drift, which has been found 500 miles from these islands, sea-weed, driftwood, and a commotion in the water, strongly marking its existence.

On the way to the Falklands, penguins may be seen and heard full 300 miles from the land ; they need not, therefore, cause any alarm ; one sign, however, is well worth noting, viz.: that of the diver bird called the " shag," which is rarely seen more than ten miles off the land, and often at a less distance.

EDDYSTONE ROCK, lying N.W. by W. $\frac{1}{4}$ W., distant 4 miles from Cape Dolphin, the northern point of the East Falkland, is seen well from a vessel's deck about 8 miles off, and exactly resembles a ship under all sail when seen at that distance.  It is visible in the darkest night, if the horizon is clear, before a ship would be in danger, as there is deep water close round it ; but if the longitude is uncertain, it would be better, in the night, not to run on, after shoaling to 50 fathoms, should the wind be towards the shore, as a vessel might pass the Eddystone, and become embayed in the deep bight to the westward of it.  The same rule should apply in thick weather, which is always the case with northerly and N.E. winds ; but if a vessel has had observations shortly before, and can depend on her position, she may run for the N.E. point of the island in any weather ; and if the land is not seen about Macbride head, or Cape Carysfort, when the water shoals to 40 fathoms, her head should be put off shore until daylight, or until a break in the thick weather enables the land to be seen; but the days of thick weather are very few, and it is not often that the land cannot be seen when 20 miles off.

The first appearance is very unfavourable ; rugged hills, the summits of which are stony and very light-coloured, have made many suppose that the high land is always covered with snow; but this is rarely the case from October to April or May, except patches in the hollows of the mountains, which sometimes remain till November.

CAPE DOLPHIN is a long, low, and narrow strip of land jutting out from the north-west part of the East Falkland ; there is a shoal about three quarters of a mile to the S.W. of it, marked by kelp.  Between this Cape and the Eddystone there runs a turbulent race, which would be often fatal to boats and very small vessels.

**CAPE BOUGAINVILLE** bears E. ¼ N., distant 28 miles from the Eddystone rock. Between this cape and Cape Dolphin the coast slightly indents, and an indraught was observed. The depth of water, at a distance of 4 miles, was found to be from 40 to 45 fathoms (a fine greenish-coloured sand, with small black specks), gradually decreasing to 12 fathoms close to the shore. Five miles to the westward of Cape Bougainville, off a headland, there is a cluster of rocks, with 10 fathoms a cable's length from them on the outside. With northerly winds, a heavy sea prevails all along the coast, from Cape Carysfort to Cape Dolphin.

**PORT SALVADOR** lies 7 miles eastward of Cape Bougainville, with reefs extending off each point of the entrance, which is difficult on account of its long and narrow channel, as well as from the rapidity of the tides, which sweeps the kelp under water, and causes in many parts of the channel a violent race; moreover, the water is deep and the bottom hard, consequently it is doubtful whether an anchor would hold, if necessary to let one go. The extent of the passage is 7 miles from Hut point to Plat point; it is more difficult to enter than to quit, as the wind generally prevails out, and it is absolutely necessary to have, on entering this port, a good commanding breeze.

**CAUTION** should be used when passing the entrance of Port Salvador, as the tide rushes in strongly, and the reefs on either side make it dangerous, if a vessel should get embayed in bad weather.

**TIDES.**—It is high water, full and change, in the lagoon at the entrance at 8h. 10m.; the rise and fall being 8 feet.

**DIRECTIONS.**—The usual passage is to the westward of Centre island, but after passing well to the southward of Mid rock, to cross over between it and the island, and work up on the eastern side, as the water is not so deep, nor the tide so strong as on the other side of the island; but, with a fair wind, perhaps the western channel is the best, though it requires a strong breeze, as the ebb runs out at the rate of 6 knots.

After passing Centre island, there is more working room, and anchorage for one vessel may be obtained at the mouth of the lagoon, on the west side, in 7 fathoms good bottom, but it shoals suddenly to 2 and 3 fathoms. Having cleared the entrance channel, good and secure anchorage abounds all over the port. The strength of the tide is trifling everywhere but in the channel and between some of the islands, and there it seldom exceeds 1¼ or 2 knots. The dangers are all visible, with the exception of two shoals, and they are marked in the chart, and carry not less than 2

fathoms. The coves and creeks abound with fish, and the shore with
cattle, rabbits, and wild fowl, heath fuel, and good water. The best time
to enter this spacious and magnificent port is at low water, or the early
flood, and to leave it at the last quarter ebb.

**MACBRIDE HEAD** bears E. ¼ N. distant 19 miles from Cape Bougainville. In coming from the northward, the most eastern hills seen
are those immediately over Berkeley sound; by steering for them,
when within 6 or 7 miles of the land, the cliffs of Macbride head and
Cape Carysfort will be plainly seen. Both these capes, and a projecting point between them, have small detached rocks off them, which
show plainly in coming from the westward.

**CAPE CARYSFORT** may be passed at a mile distance, and the low
land and rocky islets which form Volunteer point will then be distinctly
seen. Cow bay lies a couple of miles to the southward of Cape Carysfort
and affords clean sandy anchorage in 7 or 8 fathoms, but open to the eastward. It is easily known by its white sandy beach, and the bluff land
about the cape; and, at the close of the evening, vessels bound to Stanley
might find it convenient to drop an anchor here for the night.

**URANIE ROCK** lies E. by N. one mile from the rocky islets off Volunteer point. A berth of 2 miles, therefore, should be given to them in
order to clear the Uranie rock, on which a French frigate of that name
struck. It is the more dangerous, as with westerly winds the sea seldom
breaks on it, and it is the only rock of the whole group on which no kelp
grows. The best marks to clear it, particularly in the night, are the
bearings of Cape Carysfort and Mount Low; by keeping Cape Carysfort
to the westward of W.N.W., until Mount Low bears S.S.W. ¼ W. or
Cape Pembroke Light bears S. ¼ E., a vessel will pass nearly 2 miles
outside of it, and may then haul up for the light.

The most eastern high hill on the island, Mount Low, is easily seen, in
a clear night, when to the northward of Volunteer point; the summit,
which is 840 feet high, forms two peaks, and from the eastern one the
land slopes down to the point that divides Berkeley sound from Port
William.

**BERKELEY SOUND.**—The entrance to this capacious sound lies between Eagle point and Kidney island, opening out directly after passing
Volunteer point, and cannot be mistaken either by night or day. It is
4 miles wide at its entrance, and upwards of 5 leagues in length, terminating in the three excellent anchorages of Johnson harbour, Stag road,
and Port Louis. After passing Eagle point, from which a reef extends half
a mile to the eastward, the sound is clear of all danger up to Sea Lion

rocks. These rocks make at a distance like two or three small boats; and vessels bound to either of the above three anchorages should in the first instance steer for these rocks, the dangers round which are well marked by kelp.*

Berkeley sound may be entered by night, if the entrance has been made out before dark; and even worked into safely, by standing close to the shore on each tack till ·nearly abreast of Johnson harbour, where a ship can anchor in from 12 to 15 fathoms, outside the kelp patches off Long island, but rather to the southward of mid-channel, to avoid getting too close to the Sea Lion rocks, which cannot be seen in a dark night.

JOHNSON HARBOUR.—The entrance to this harbour is two-thirds of a mile in breadth, from kelp to kelp, which shoots up there in 5 or 6 fathoms, and well marks the limits of the channel. Off Lamarche point the kelp will be seen to run out a long way, with another large patch opposite to it, which together narrow the passage to a third of a mile. After clearing these, the vessel may boldly proceed up the harbour and anchor off Magellan cove, in from 5 to 6 fathoms, mud. The *Conway* found the ground there so tough that in weighing the anchor the head of the capstan was wrung. The watering-place is at the north-west corner of Magellan cove, but it is inconvenient, as the beach shelves out a long way, so that the people are obliged to roll the casks some distance, and are therefore constantly wet. If not in want of water, a better berth may be found farther to the westward, for the sake of shelter during heavy S.W. and southerly gales, which raise a heavy sea off Magellan cove. The landing, also, is more sheltered on the western shore; and that side of the harbour abounds with rabbits.

Stag Road offers a still better anchorage for large ships off Bougainville creek; the two large kelp patches may be passed as close as possible, and the largest ship may work in, and anchor in 4½ or 5 fathoms in any part of the road. The best berth and the best ground will be found in mid-channel between Hog island and the north shore; but it is 2 miles from the watering place, which is to the westward of the Carenage.

PORT LOUIS lies in the western extremity of the bay. The entrance is between Long island on the south and Peat and Hog island on the north; but a rocky patch nearly mid-channel contracts the passage to a little more than a cable's length. After passing this narrow gullet, keep to the northward of Round island. This anchorage is nearly land-locked; the most convenient berths for small vessels are off Carenage, in 3 fathoms, about a quarter of a mile from the shore, or farther over to the southward, in 3½ or 4 fathoms. The French made a settlement here in the year 1764.

---

* *See* Admiralty Plan of Berkeley sound, No. 1,326; scale, m = 1 inch.

The Carenage, at its entrance, is scarcely more than 100 yards across, but expands to a sheet of water of nearly a circular form, and nearly half a mile wide; from the shoalness of the water, however, it is only adapted for boats. The old settlement was on its western side. It is high water at Port Louis, full and change, at five o'clock, and the rise of tide at springs does not appear to be more than 7 feet.

**CAPE PEMBROKE LIGHT** is a *fixed white* light placed at an elevation of 114 feet above the mean level of the sea, and should be visible in all directions except from the westward, when bearing between N.E. ½ E. and S.E. ¼ S. In clear weather it should be seen from the deck of a ship at a distance of 15 miles. The illuminating apparatus is catoptric, or by metal reflectors. The light-tower is an iron circular building, 60 feet high, and painted white. It stands about 100 yards from the extreme pitch of the cape, in lat. 51° 40′ 42″ S., long. 57° 43′ 0″ W. The beacon which formerly stood on this cape is to be removed, it is said, to William point. Pilots are ready to board ships off the lighthouse.

**PORT WILLIAM**, the outer approach to Stanley harbour, is entered between William point on the north and Cape Pembroke on the south, and is well marked by the lighthouse on the latter, which is level land of about 30 feet high. William point is low and rocky, and was formerly distinguished from the rest of the land by tussac,† which has the appearance of low green bushes, but which is now gone. On it (it is said) stands a beacon, 84 feet high, and visible from sea at a distance of 10 miles.

The port is entered after passing William point, and there is good anchorage in the bay, between it and Cape Pembroke, in from 11 to 12 fathoms, sheltered from all the prevailing winds. The next point is Charles point, about 2 miles farther; it has two small detached rocks at its extremity, off which there is a kelp patch, extending about a third of a mile: there is deep water close to the edge of the kelp.

**The Billy Rock**, which always shows at half-tide, or when there is much swell, but at high water is covered, lies in a line with William islets and the Seal rocks, and with Cape Pembroke light bearing S. by W. ¼ W. distant 4¼ cables' lengths. This rock is more particularly to be guarded against in leaving Port William to pass round Cape Pembroke with an ebb tide. The passage between it and the east William islet should not be attempted.

---

* *See* Admiralty Plan of Stanley Harbour with Ports William and Harriett, No. 1,774; scale, m = 2 inches.

† Tussac is a very high coarse grass, which affords excellent food for horned cattle.

**The Seal Rocks** lie about three quarters of a mile from Cape Pembroke, and are clean on all sides.  The tide runs north and south 3 knots between the cape and the rocks, the flood stream setting to the northward and the ebb to the southward.

**The Wolf Rock** is of a triangular shape, each side being about 3 cables. Cape Pembroke bears from it N. ¼ E. distant nearly 3 miles.

**Yorke Point,** on the south side of the harbour, and the islets to the eastward of it, are all steep, and may be approached within 100 yards ; the entrance is wide enough for a line-of-battle ship to work in, and the edge of the kelp is a secure guide ; but the white sandy bay on the south side should not be entered, as it is shallow.  In standing towards it, a vessel should tack when in the line of the islets and of Yorke point. Immediately this point is passed, Sparrow cove will be seen open on the north side of the harbour, under Mount Low, and the entrance of Stanley harbour (where the settlement is now) on the south side.

**Sparrow Cove** forms a good anchorage for vessels that remain in Port William ; there is an unfailing supply of good water in its north-west corner. Nearly half a mile from the entrance of Sparrow cove is Doctor point, on which stands a sign post pointing to the Narrows : there is a similar one on Tussac point.

Sailing ships bound round Cape Horn should not enter Stanley harbour, as the wind which would be fair for them to sail, would be foul for getting through the Narrows ; they may anchor outside the entrance about a quarter of a mile north-west of them.  But those coming from Cape Horn may enter, as any wind which would be fair for them to sail, if bound to the northward, would also be fair to leave the harbour.

Large ships, when abreast of the entrance, in coming up Port William, not intending to enter Stanley harbour, should stand towards the Narrows till they shut in the entrance points of Port William, when they may anchor in from 6 to 7 fathoms, about a third of a mile from the shore, and can easily leave with any wind.

**STANLEY HARBOUR.**—The entrance to this harbour is little more than a cable broad, lying between Engineer and Navy points, both of which may be passed within 30 yards, and all dangers are buoyed by kelp. The harbour is excellent, being a large natural dock, 3 miles long by about one-third of a mile broad.  On the south shore stands the town of Stanley, situated on the slope of a hilly range.  The population of the islands was about 500 in 1858 ; and in the previous year 52 vessels of 20,672 tons entered Stanley harbour.

**Supplies.**—A reservoir has been constructed at Port Stanley, from which vessels can be supplied with water at a moderate charge.  Good and cheap beef and mutton may be obtained in any quantity, and wild fowl and fish are also very abundant. Coals may usually be procured from the Falkland Islands Company, as well as ships' requirements of all sorts.

Wood is scarce, but *peat*, which is a fair substitute for it, is plentiful, and when compressed is found to be a valuable fuel.

**Murrell River.**—Above Port William anchorage, a long creek winds through the hills to the westward, up to this river, its whole extent being about 3½ miles. It varies in width from a fifth of a mile to half a cable's length for the first mile, with a depth of 3 to 2 fathoms, beyond which it shallows rapidly, so that a boat cannot get up till the tide begins to flow. The landing on all the beaches is very bad, in consequence of the scattered fragments of quartz rock from one to three feet long, which may greatly injure a boat if she touches them ; the only safe landing is on the rocks where the shores are steep-to. There is a very good watering-place on the west side of the cove on the north shore, in a bight outside the entrance to Weir creek ; but care must be taken in landing on account of the stones. Nearly every hollow has a small stream running through it, and peat may be dug in several places.

**TIDES.**—It is high water, full and change, in Ports William and Stanley about 5h. 30m., the rise and fall being 4 feet at neap, and 7 feet at spring tides ; and about the same at all places on the south-east coast of the East island. It sometimes rises and falls 8 feet, but very seldom. On the south-east coast of the Falkland islands there is so little tide that it need not be considered, though a current will generally be found running with the wind, of from half a knot to one knot ; but after passing Port Harriet a strong tide begins to be felt. The flood runs to the north-east, past the Wolf rock, and becomes stronger as it approaches Cape Pembroke, round which it runs from 2 to 3 knots, according to the age of the moon. The flood runs directly to the northward of the Seal rocks to Volunteer point, while very little tide is felt within the heads of Berkeley sound or Port William. The ebb runs equally strong to the southward, and when there is a strong breeze, a heavy tide rip extends 2 miles off shore.

**DIRECTIONS.**—When bound into Port William with a fair wind, after rounding Volunteer point, steer for the lighthouse ; some white sand-hills will then be seen a-head, and close to Kidney island ; at the same time the beacon on William point will be seen just inside of Cape Pembroke. William point may be passed within a cable's length, and the entrance to the harbour will open.

With a flood tide it is necessary to guard against being swept too near Cape Pembroke or the Seal rocks. With a commanding breeze vessels have passed between the rocks and the Cape, but it is best to pass outside both the Wolf and Seal rocks.* Should a vessel find that she is setting. towards the Seal rocks, the only alternative is at once to try and pass. between them and the cape, as the tide sets through strongly; but there is the Billy rock to be avoided in this passage. There being no tide felt while running down the port till near the Billy rock, and then the outside tide being met running strongly to the southward, it is very likely to sweep a vessel towards the rock, unless allowance is made for it. In passing Cape Pembroke bound to the southward, the same rule applies as when bound into Port William; and in light winds, or much swell, it is better to keep well to the northward of the Seal rocks, in order to allow for the tide running to the southward.

In coming from the southward the lead will not be a guide, as Beauchêne island (*see* p. 98) is on the southern edge of the bank which surrounds the Falkland islands; and within 4 miles of the south point there is not any bottom with 100 fathoms. After passing outside Beauchêne, a N.N.E. course for 60 miles will clear all the islands off the south-east coast of the East Falkland, and the depth then will be from 50 to 56 fathoms, about 15 miles to the south-east of Lively island. This is supposing it to be night, or thick weather ; if daylight, and the weather clear, the high land in the central chain will be seen, and a course may be steered nearer the coast, passing about 6 or 7 miles outside the east Sea Lion island, and the same distance from the Shag rock (which shows high out of the water) and from Lively island. In either case, after passing Lively island, a course should be steered towards the easternmost of the hills ; if thick, or too dark to see the hills, a vessel will be in a very good position for waiting for daylight, and should endeavour to keep in about from 40 to 50 fathoms, not standing in shore into less than 30, nor off into more than 60 fathoms.

When Cape Pembroke light can be made out, bring it to bear N.N.E., and keeping it on that bearing it will lead into the shore, where it can be approached in perfect safety, just outside the entrance to Port Harriet. The Wolf rock will then be seen ; and, by passing a mile on either side of it, the Seal rocks, at the entrance of Port William, will show clear of the extreme eastern point of Cape Pembroke. If the wind is off the land, a vessel might pass inside the Wolf rock, but not nearer to it than three-quarters of a mile, as some rocks run out nearly half a mile to the westward of it ; nor should the kelp near the main shore be approached nearer than a quarter of a mile, as there is a sunken rock at its outer edge, which breaks when there is any swell.

---

* By neglecting this caution the American clipper ship *Russell* struck on a reef close to the Billy rock on the 4th of September 1859, and sank in deep water.

Coming from the northward, with westerly winds, make Cape Carysfort, or with easterly winds, Volunteer point ; when they are passed, steer for Cape Pembroke, on which the lighthouse will be seen, until Port William opens to starboard, when run in and anchor, or wait for a pilot. In case of darkness or fog, ships may anchor in the mouth of Berkeley Sound or of Port William, or stand off and on, as may be expedient, there being no danger that is not buoyed by the kelp.

If the wind is southerly, the passage into Stanley Harbour should not be attempted, except by a small quick-working vessel ; but, with the wind to the westward of south-west, it may be passed by large ships ; it is little more than a cable's length broad, but both points can be passed within 30 yards. If the wind is south-west, so as to make it necessary to pass very close to Navy point to fetch through, a vessel should work up well to windward of the entrance, and entering the passage under all sail and with good way on, directly the sails lift from the wind drawing out in passing the point, she should be kept a little higher, so as to shoot through with the sails shaking till she gets the steady wind inside the point.

When through the Narrows, the harbour may be traversed by any vessel under 20 feet draft ; there are $3\frac{1}{4}$ fathoms at a cable's length from the kelp on each side, and above 4 fathoms in mid-channel, close up to the site of the town, which is about 2 miles to the westward of the entrance, on the south shore. Large ships have plenty of room to round-to and anchor in mid-channel, in about 5 fathoms, as far up as they choose to fetch. After passing the Narrows, the bottom is excellent—a stiff mud, which often causes some trouble in getting the anchor up again.

In January 1859, H.M.S. *Cumberland*, drawing 24 feet water, sailed both in and out of Stanley Harbour, mooring just south of Navy point, with open hawse, to the north-west, having not less than 75 fathoms on the south-west anchor, and 50 on the north-east ; the winds being violent from S.S.W. to N.W.

PORT HARRIET is the first port to the south-west of Port William. The entrance being about $5\frac{1}{4}$ miles to the south-west of Cape Pembroke, it lies immediately on the south side of the ridge, which separates it from the settlement and Stanley harbour ; the distance across being about $2\frac{1}{4}$ miles. Ships working from the southward, intending to touch at Port William, and finding a strong northerly breeze, and a good deal of sea, for getting round Cape Pembroke, would find excellent anchorage in Port Harriet, about a quarter of a mile N.N.W. of the south entrance point, which is low; and has a round low mound off its extremity, to which it is joined by rocks ; off this mound a ledge runs out above half a mile, but the kelp marks its extent. There is a good stream of fresh water in a cove on the north shore, about a mile inside the entrance.

About a quarter of a mile N.W. by N. from the extremity of Seal point, there is good anchorage in from 6 to 7 fathoms, just outside a small kelp patch, which is nearly in mid-channel. From that anchorage supplies could be obtained from the settlement, while waiting for a fair wind. On the above-mentioned kelp patch, the least water found was 3 fathoms, with 5 fathoms on each side of it. At about half-a-mile within it, a bar extends across from Lake point on the north shore (off which there is a small detached rock) to the south shore, at about three-quarters of a mile from Seal point. The deepest water on the bar is $3\frac{1}{2}$ fathoms, fine sand, in mid-channel, and shoaling gradually to 2 fathoms, close to the kelp on each half side. It is about a mile wide, and its inner edge suddenly deepens from 4 to 7 fathoms, the bottom changing from fine sand to soft green or black mud.

The only wind that would raise any sea in the harbour would be from E.N.E. to S.E., which seldom blows ; and should it rise from that quarter, and be too strong to work out against, by running in over the bar a secure anchorage may be found in any depth, and quite sheltered. No marks are necessary for passing over the bar, the above description being sufficient.

TIDES.—In Port Harriet it is high water, full and change, at 5h. 0m. p.m. ; the rise and fall 6 feet. There is scarcely any tide felt.

DIRECTIONS.—After passing the bar Port Harriet affords excellent anchorage for an extent of 3 miles, the breadth being nearly three-quarters of a mile. All the dangers are marked by kelp, except in one place about half-way up the harbour, where a sand-bank, without kelp, extends about a cable's length off the south shore ; the best anchorage is in mid-channel, in order to have plenty of room for getting underway if the wind should be blowing strong off either shore; the depth varies from 5 to 8 fathoms, the bottom soft black mud. The head of the harbour is terminated by a creek which runs about 2 miles to the westward ; it is only a cable's length wide at the entrance, but it gradually widens, and near the head is about half-a-mile across ; for a mile inside the entrance there is anchorage for small vessels in from 2 to 3 fathoms ; but after that it becomes very shallow, being dry at low-water half-a-mile from the head, and the bottom strewed with fragments of rock. The land on the shores of Port Harriet is generally swampy, and it will probably never be a port of any importance, beyond affording a good stopping-place for a vessel that cannot reach her port before night.

CAUTION.—No vessel should attempt to enter any of these ports on the south-east coast, except with good daylight, the kelp being the only guide,

and that cannot be seen at night, except in bright moonlight, and then
it is difficult.   It should be an invariable rule never to pass through
kelp, unless perhaps through a few straggling stems outside a large bed,
where there is generally sufficient water for all ships.   In general, by
keeping clear of kelp you keep clear of danger, but this must not prevent
attention to the lead, as the rule sometimes fails; for the *Arrow*, after
passing the kelp patch at the entrance of Port Harriet, ran aground on
the north side of the harbour at the distance of about the vessel's length
outside the kelp on a bank of white sand extending more than a cable's
length from it.   Parallel ridges of kelp have generally deep water between
them, nevertheless a good look out should always be kept aloft; for the
shallow sandy bottom can be plainly seen, on which no kelp grows.

**The COAST** from Port Harriet trends to the westward to Beach point,
and there is no danger outside the kelp that fringes the shore; but between this point and East island, which forms the south entrance point of
Port Fitz Roy, there are numerous kelp patches.   From Beach point the
coast trends W.S.W. for 4 miles, terminating in a bay open to the
eastward, in the middle of which there is a rock dry at low water and
surrounded by kelp.   The south horn of this bay is Bold point, the north
entrance to Port Fitz Roy.   At the north side of the bay, there is a
narrow inlet, leading to North basin, and passing through a gorge in the
ridge of low hills: the depth in the narrow part is from 1 to 1½ fathoms.
The basin is a mile long and very shallow.

**MOUNT KENT.**—It is difficult to make out the entrance to Ports Fitz-
Roy and Pleasant, when standing in direct from the sea, as the land is low,
and the points cannot be made out distinctly till close to.   The best guide
is the high range of hills north of the port, with three peaks near each
other, the middle one showing a broad flat summit.   The westernmost and
highest of these, Mount Kent, bearing W.N.W., will lead direct to East
island, and when near it the kelp will lead into either channel of Port
Fitz Roy.   In approaching all these shores a good look-out should be
kept for detached patches of kelp lying some distance off, which are
numerous.

**PORT FITZ ROY** lies 8 miles S.W. by W. of Port Harriet; the coast
between them is bold and rocky, and nearly straight, except about half-
way, where there is a small bight with a white beach open to the eastward.   There is a bar extending across the harbour, from the west

* *See* Admiralty chart, Ports Fitz Roy and Pleasant, No. 1,956; scale, *m* = 1·7
inches.

end of East island ; the deepest water is close to the south edge of the kelp patch, where there are 3 fathoms gradually shoaling to 2 fathoms close to the kelp off East island : to the northward of the kelp patch there is a narrow passage with 2¼ fathoms, but it can only be passed through with a fair wind.

**East Road.**—Between the west end of East island and the point of the mainland forming the south shore of Port Fitz Roy, there are several small islets, between which there is a deep, but narrow and winding channel into the port ; it is well marked by kelp : the depth of water in it varies from 4¼ to 7 fathoms ; and some rocks nearly awash lie very close inside the kelp edge, on both sides of the channel.

**White Point.**—For 3 miles above the bar, the port is fully a mile wide, and quite clear of danger to within 2 cables of the shore, and to near that distance the kelp extends.  There is, however, a sandy point on the north shore, with a small knob on its extremity, from whence a sand spit, without kelp, and with less than a fathom water on it, projects about 2 cables' lengths : it is easily seen, as the water looks quite white.

**Fitz Basin.**—This wide part of the port affords excellent anchorage for the largest vessels ; the depth varies from 6 to 4 fathoms, and the bottom is mud.  On the north shore there is a remarkable gorge in the ridge of low hills, through which a very narrow inlet runs for about half a mile, with a depth of from 1½ to 2 fathoms ; it then opens out into Fitz basin, large and shallow, like that of North basin.

**Tussac Island.**—About 3 miles inside the bar the port is divided into two arms by a point with a low island formerly covered with tussac. The southern arm is very shallow, and nearly all covered with kelp, except close to Tussac island, where there is a small patch of clear ground with 2 fathoms water.

To the northward of Tussac island, and directly in the middle of the entrance to the northern or main arm, there is a large patch of kelp, with only one fathom on it.  On the north side of this patch there is a channel a quarter of a mile wide, with 3½ fathoms water for about a mile ; and further on will be found the best anchorage for small vessels in 2½ fathoms soft mud, abreast of Garden point, on the north shore, where the harbour turns to the northward.  Between this point and the point on the opposite shore there are some rocks dry at low water, but the kelp extends well outside of them, leaving a passage in mid-channel nearly 2 cables' lengths wide ; the harbour then opens into a wide space, but all shallow except in the channel, which is about a cable's length wide, with a depth of 2 to 3 fathoms.

On the north side of this open space there is a very narrow opening, through which the channel runs close to its eastern shore, and so steep-to that large vessels may lie alongside the rocks, which would only require levelling to make excellent wharfs; but the channel is hardly wide enough for swinging a large vessel, the opposite side being low, with rocks lying some yards from the shore. The depth of water close to the steep shore is 4 fathoms. Inside of this passage it opens out again to a wide creek which runs above 2 miles to the westward, and ends in a small fresh-water river. Much of this upper space is dry at low water, so that a boat cannot get within a mile of the river, which, though rather large for this island, is too shallow at the mouth for the smallest boat to enter. There is good anchorage for small vessels for half a mile inside the narrows, in 2 and 3 fathoms muddy bottom.

**TIDES.**—There is scarcely any tide to be felt in Port Fitz Roy, except in the small entrance through the islands, where it runs about 1½ knots, and in the narrows near the head, where it is rather stronger. It is high water on full and change days at 4h. 45m.; the rise and fall 6 feet.

**DIRECTIONS.**—The best course into Port Fitz Roy, coming from the eastward, is to the northward of all the kelp patches between Beach and Bold points, keeping close to the kelp on the main shore, where the passage is, in the narrowest place, above a quarter of a mile wide; but if the wind should be blowing hard from the southward, it would be advisable to keep to the southward of all the kelp patches, and run for the large kelp patch that extends above a mile off the east end of East island. Following the inner edge of this kelp patch will lead right into Port Fitz Roy; but when abreast of the island you must cross over to the north shore, looking out for another kelp patch which lies one mile inside of Bold point (or the north entrance point) and nearly in mid-channel.

A large ship should not attempt the channel through East road, unless the wind be between South and East, or to come out unless between North and West, as nothing but a small craft could work through it, and in some parts it is too narrow to bring up a large vessel; the middle part of the channel opens out to about a-third of a mile in breadth, and forms a nice anchorage for small vessels.

**PORT PLEASANT** lies immediately south of Port Fitz Roy. It has two entrances, formed by a long narrow island with some small islets: these islets are surrounded by a large and thick kelp patch of 2 miles in length; its edges are the best guides into both entrances. Mount Kent bearing N.W. by N. will lead directly south of this patch, and is a good mark for the port.

**Bars.**—This port also has a bar in both channels; the north one is abreast of the second of the small islets: the south one is a little inside the east end of Pleasant island ; the deepest water on both is 2½ fathoms, but ships drawing 17 feet might enter with a leading wind, during the last quarter flood. The north bar carries its 2½ fathoms right across to the kelp on each side; but the south bar has only 2¼ fathoms close to the island kelp; and from these shoals to 1¼ fathoms on the south shore. The north channel is therefore the best for large sailing vessels; but they must have a fair wind, as a little above the bar the channel is contracted to about 100 yards by one of the small ialets: but there are 8 fathoms water there, and it is quite steep-to on both shores. When inside of this, the harbour expands to about three-quarters of a mile, and continues that breadth till it is joined by the south channel, round the west end of the island ; the bottom in all this space is soft mud, the depth varying from 10 to 6 fathoms. After passing the bar at the south entrance, there is a depth of 4 fathoms to where it joins the north channel; but about a mile inside the bar, off a remarkable white sand patch on the south shore, there is a low projecting patch of rocks, nearly covered at high water, from which a sand bank extends, with less than a fathom water on it. The water over it is quite white, and it may easily be seen; but the best way is to keep close to the island kelp when entering by the south channel, and directly you pass the west end of the island stand over to the northward, as along the south shore there is a shallow bank.

**Island Harbour.**—Round Turn point there is a narrow opening to a large inlet, which winds through the hills for about 3 miles to the S.W., and then leads into Island harbour. For the first mile, the inlet is about a third of a mile wide ; it then opens out to a space about three-quarters of a mile each way : the deep channel runs close to the south shore of this space, and then turns to the southward, into the narrow part of the inlet, which for 2 miles is scarcely 100 yards across. There are 4 and 5 fathoms water in the channel, and though it occasionally crosses from side to side, no small vessel would find any difficulty in sailing through it with a fair wind. Island harbour is almost all shallow, but there is space for small vessel to moor across it with one anchor close to the kelp on the south shore, and the other on the edge of the northern mud bank. A mile from the head it is dry at low water, as well as in all its small creeks.

Off the south entrance point of the port, Pleasant Point, a kelp patch, extends above a mile, and on it there are some rocks which break very heavily after southerly gales.

**Tides.**—The tide in both entrances of Port Pleasant runs nearly a knot at the springs ; and in the narrow pass of the north entrance it runs nearly 2 knots : it is high water on full and change days at 5h. 0m.; and the rise and fall 6½ feet.

**PLEASANT ROAD**, situated to the southward of Port Pleasant, is well sheltered from the southward and south-eastward by a bed of kelp, which extends 3 miles off the shore, and is 3 miles wide; in it are three small islands about a mile from the shore. The road is of course exposed to N.E. gales, but they are not frequent; and the holding ground is good, being sand with a stiff clay under it. The *Arrow* rode out a N.E. gale here, without bringing home the stream anchor by which she was riding.

**KELP LAGOON**, is a shallow piece of water, about 3½ miles long and from 1 to 2 wide; it has several islands in it, and two entrances, but both of them are blocked up by kelp, which extends nearly 2 miles off shore, and is a continuation of the large kelp patch south of Pleasant road.

**CHOISEUL SOUND** extends 26 miles from east to west; it is 5 miles across from Fox point to Lively island, gradually becoming narrower towards its head, where it is about 2 miles wide. Fox point, about 9 miles S.W. of Pleasant road, forms its north entrance, which is best known coming from the eastward by the long white sandy beach to the north-east, with a small dark islet that shows plainly on the white ground behind. Fox point is the southern extremity of this long beach, with a small islet off it; and the kelp extends about 2 miles farther to the eastward.*

The entrance to the sound, from the main land to Phillimore island, is 3 miles wide; but divided into two channels by Middle island, which lies within 1½ miles of Phillimore island. The best channel into the sound is to the northward of Middle island, between it and the two Black rocks, which are several feet above water, and lie about half a mile off the north shore, with a long tail of kelp to the eastward. There is a passage with 4 fathoms water between the rocks and the shore, but only small vessels can work through it. The channel between Middle island and the Black rocks is about 2 miles wide, and clear of all danger, except two or three small patches of kelp inside which are easily seen. A long kelp patch runs 2 miles to the eastward of Middle island: there are two small islets in it, with a large black rock near its outer extremity. Between this rock and Phillimore island there is a wide and deep passage, but full of large kelp patches.

**Mare Harbour.**—After passing the Black rocks, in the north channel, the first opening in the north shore leads into Mare harbour, one of the finest on the coast, and easy of approach for the largest ships. The entrance is nearly 2 miles wide, but the kelp extends a long way off the western side, from Seal island, contracting the channel to less than half a mile; but inside it opens into a clear piece of water, about 2 miles long

---

* *See* Plan of Choiseul sound and Bodie inlet, No. 2,671; scale, m = 1·4 inches.

and 1 mile wide, with excellent anchorage for the largest ships, in from 6 to 10 fathoms, muddy bottom.

The whole of the country round is covered with herds of cattle. The soil appears good, and there are no swamps, except in the bottoms of the valleys : in some parts of Swan inlet, the shores are so steep that a vessel may lie alongside the rocks. Water can be found in most of the hollows, but the smaller streams are dry in the middle of summer.

**East Cove.**—On the east side of this harbour, an opening a quarter of a mile wide leads to East cove, another very fine harbour, and the best for a vessel that intends to remain a long time, but with westerly winds a large ship could not work out. It is nearly a mile wide, and its east end terminates in two coves, with sufficient water for small vessels. On the western side of Mare harbour there are also two coves ; one of them, West cove, is nearly 3 miles long, and for nearly 2 miles affords good anchorage for small vessels, in 3 fathoms water.

**Swan Inlet** stretches 7 miles to the N.W. from the northern side of Mare harbour, and terminates in a small river. For 5½ miles this narrow inlet has a depth from 2 to 1½ fathoms ; it then opens into a wider space, and becomes so shallow that a boat cannot go higher at low water.

**DARWIN HARBOUR.**—In the north-west corner of the sound there is a good harbour, called after Captain Fitz Roy's scientific companion, Mr. Charles Darwin, but the entrance is narrow, between two large clusters of islands and the east shore. Outside the entrance there is also excellent anchorage between the two clusters of islands and the shore. To the southward of these islands there is also another good anchorage in Arrow harbour, inside the Arrow islands ; the best entrance to which is between the islands and the south shore.

**Bodie Creek.**—Choiseul sound terminates, at its south-west extremity, in a long and narrow inlet, called Bodie creek, which runs about 4 miles to the westward of the head of the sound, with 3 fathoms water at a mile from its western end, and varying in breadth from a quarter of a mile to 100 yards.

**VICTORIA HARBOUR.**—The whole of the sound is studded with islands which form well-sheltered anchorage for small vessels ; there are also numerous creeks and coves, in many of which they may lie securely ; large ships can bring up in any part of the sound in from 12 to 18 fathoms water : the bottom is all mud, but generally covered with shells and weed, which give it, on the arming of the lead, the appearance of a rocky bottom, but all the rocky dangers are marked by kelp. On the south side of the sound about half way up, a large roomy inlet has been honoured with the name of Victoria harbour ; it is formed by a long peninsula running parallel to the shore, and extends about 7 miles; the first 4 miles of which has good anchorage for large ships, in from 10 to 5 fathoms water.

The TIDES run nearly a knot in the entrance of Mare harbour; it is high water full and change at 6h. 0m.; the rise and fall 6 feet. There is but little tide in the sound, except in the entrances to creeks and between the islands. In the south entrance to the sound also, both ebb and flood are strong; at springs about 1½ knots, and when blowing hard they cause a ripple off Pyramid point, which is dangerous for a boat. The flood sets to the northward in this entrance, but in the north entrance there is scarcely any tide, and the little flood there is, runs to the eastward out of the sound.

LIVELY ISLAND is a large island 6 miles long by 5 broad, lying in the entrance of Choiseul sound; reefs and kelp passages extend for 4 miles off its north-east point. The south-east and south-west points have also reefs extending nearly 3 miles from them.

LIVELY SOUND.—To the westward of Lively island there is good anchorage for a vessel for the night, on the west side of Lively sound, to the northward of Motley island; and any vessel bound to the westward, and not able to reach Bull road before dark, would find this the best anchorage; but if it should be blowing hard from the southward or S.E., a heavy swell would set into Lively sound; in which case it is quite safe to run up the sound, taking care of the shoal off Sal island, and either haul into Seal cove, or pass further on, and anchor on the north side of Pyramid point, where there is very good anchorage, in from 5 to 8 fathoms, the bottom as usual sandy, with a stiff clay beneath.

Seal Cove.—For vessels going to remain any time in this part of the island there is no spot superior to Seal cove; it is sheltered from every wind, and it is well watered. There are large rivulets running into the head of the cove, and several small streams on the north shore close to where vessels would anchor. Seal island is the best guide for the cove. There is deep water close round the kelp, on the north side of the island; and though long kelp reefs project from it, there is plenty of room for a square-rigged vessel.

LOW BAY.—The entrance to this bay lies between the north point of Breaker island and the rocky Triste islands on the opposite shore, 3 miles apart, after which the bay expands to a breadth of 8 or 9 miles, with a depth of water from 15 to 22 fathoms.

It is indented with several bights and bays, and although inviting in appearance they are not advisable anchorages, as the ground is rocky and foul in many parts, and a heavy swell rolls in with southerly gales. There is no danger in it except a patch of rocks, dry at low water and fringed with kelp, lying midway between Bluff head and Turn island: its outer edge bears from Bluff head south 1½ miles. That head may be easily recognised, being a dark bluff cliff of 60 feet in height, and the most conspicuous object in the bay, with a small islet close to it. The character of the land in this part of East Falkland and to the southward is low, few places being of greater elevation than 150 feet.

**TIDES.**—It is high water, full, and change in Low bay at 5h., but accelerated and retarded by southerly or northerly winds; its velocity at the entrance of the bay is about a knot.

**BLEAKER ISLAND** is a long, low, and narrow island, lying at the entrance of Adventure sound. Between the north end of this island and North Point island, there is a passage for small vessels of about 5 or 6 fathoms depth.

A bay on the north-east side of Bleaker island, between Lively sound and Bull road, forms the most convenient stopping-place; the water is deep, a vessel has to work close to the head to get as little as 12 fathoms, but she is then well sheltered from almost every wind. The anchorages are very good inside of Bleaker island, but as there is no passage out between the south-west end of the island and Driftwood point, they are not convenient for vessels bound to the westward.

**Salt Island** lies three miles S. by E. from North Point island, with a long kelp spit running off it, but affording for small vessels a fair anchorage in 6 or 7 fathoms, over a bottom of stiff mud. There are no other anchorages round Bleaker island. To the westward long spits run off from the points, and there are several kelp patches, but the water is deep between and around them. There are ponds of fresh water on the island.

**The SHAG ROCK,** 5 miles S. by E. from Motley island, is an excellent guide in running from the southward for either Lively or Adventure sound. It is a high peaked mass, situated in the north part of a kelp patch 3 miles long by one broad, which can be seen 5 or 6 miles from a vessel's deck, and there is no other island that resembles it; while the low land is all so much alike, that it is almost impossible for a stranger to recognise his landfall, particularly as the high range of hills is seldom seen so far south: but the Sea Lion islands, or the Shag rock being made out, the chart will show the bearings of the points. A vessel coming from the S.W. and wishing to get into a port before dark, should endeavour to make the west end of the Sea Lion islands, in order to haul up for Bull road; or, if daylight allows, by making the Shag rock she may run for Lively sound. If bound to Port William, but not caring to run during the night, she may run through the south entrance of Choiseul sound, so as to anchor inside Pyramid point; and in the morning go out through the north entrance of the sound by Middle island.

**SEA LION ISLANDS** consist of one large and three small islands lying 11 miles to the southward of Bleaker island; they extend W.S.W. and E.N.E. 12 miles, and have an outlying reef distant 3 miles N.E. by E. from the eastern islet. There is a safe passage between them to the eastward of the large island; but a long reef which breaks heavily extends 3 miles to the southward of that island, for which a good look-out must be kept, in running for the opening.

**BEAUCHENE ISLAND** is in lat. 52° 54' 45" S., long. 59° 12' W., distant 30 miles S. ¼ E. from the Sea Lion island; the south and east sides form high cliffs, but the west side slopes gradually to the sea. There is no danger within a quarter of a mile of the shore, so that it may be run for in safety. Vessels from the southward, wishing to call at Stanley harbour, should make this island, which is about 2 miles long and half a mile wide. Its northern point rises to a green mound about 200 feet high; the southern end is less than half that height, and is all rocky.

**ADVENTURE SOUND** is 20 miles in length from Bleaker island to its north-west extremity, and its general breadth between 3 and 4; it contains several good harbours, and various creeks and coves. Those on its south-west side are to be preferred, as they are sheltered from the prevailing winds. It has several islands formerly clothed with tussac, and its shores are fringed with kelp. The two best harbours are in the southern part of the sound; the principal one, Adventure harbour, the other Moffat bay, 1½ miles to the southward of it.

**Adventure Harbour.**—In proceeding to this harbour, after having rounded Bleaker and North Point islands, a W.S.W. course 4 miles will carry the vessel abreast of Little island, which is a small dark-looking mound of tussac; from thence a S.S.W. course, nearly 3½ miles more, will lead into the entrance of the harbour, which is clear of all dangers, and in which a good berth may be taken in from 12 to 5 fathoms, stiff mud. A look-out must be kept in going up the sound for the kelp patch, which lies S.E. by S., distant 1¼ miles from Little island; there is, however, plenty of water close to the kelp. This fine harbour is fit for vessels of any class; and excellent fresh water may be obtained from the ponds, which are frequented by quantities of wild fowl.

Fuel (a great desideratum to vessels that have not time or opportunity to dry the peat) may be had here for daily use in the dry season, by gathering the heath, the resinous qualities of which cause it to emit a very powerful heat, and it cooks excellently. It was constantly used in the *Arrow.* Drift-wood also may be procured from the sea-coast, but it lies at a great distance, and requires much time and labour to collect and embark it.

**Barrow Harbour.**—In steering for Barrow harbour, after having passed Turn and Large islands, which have both deep water close to them, little direction is necessary, as no danger exists in the passage to it. The course from Turn island is S.W. by W. ½ W. 6 miles, passing close to a little kelp island, which is steep-to on the northern side, and about three-quarters of a mile to the northward of Large island. On gaining the harbour, a berth may be taken in any depth of water from 10 to 4 fathoms, good bottom. For small vessels either of the arms will be found equally convenient. Good water and fish may be procured here as plentifully as in Adventure harbour.

**Fox Harbour** is the next anchorage to the northward of Barrow harbour, but not so desirable, as it has less water, and there are some shoals in it ; it is, however, a good place for small craft, and fish abounds in its creeks in the proper season, that is, from November to February.

**Great Island** is half a mile to the northward of Fox harbour. Between it and Law point there is a passage for boats and small craft, with 3 fathoms water; the shore from thence to the promontory is only fit for small vessels to navigate, several kelp patches lying off it. Nearly half a mile E.S.E. from Great island there is a patch of kelp, with 5 and 6 fathoms close to its edge, and also a small round island, named the Button, lies E.N.E. of Great island ; there is, however, deep water between them, but the passage is narrow. From the North point of the island runs a rocky spit half a mile to the northward, but partly dry at low water, carrying 4 fathoms close to the kelp. Vessels proceeding to the upper part of the Sound, after passing Turn island and Shell point, must keep a look-out for a small kelp patch nearly midway between Shell and Button islands, and bearing from the former W. by S. rather more than a mile. Then, passing a quarter of a mile to the north-eastward of Great island spit, a W.N.W. course of 3 miles will bring the vessel between Promontory and Saturday points.

**Sullivan Harbour** has a good anchorage, with 5 to 3 fathoms, muddy bottom. Up the creeks and arms there is abundance of fish, wild fowl, and good water. Small vessels may proceed up the West arm, by keeping close to the north shore when abreast the little spit that runs off from the opposite side ; the passage there is very narrow and the tide runs out strong, but there is plenty of water. Small craft may also anchor inside North island. The creek and coves on the north-east side of the Sound are only fit for boats and small vessels.

**TIDES.**—It is high water, full and change, in Sulivan harbour, at 5h. 30m., and nearly the same in all parts of the sound. With the exception of the narrow passage in the west arm, the tides are very weak. The rise and fall is generally from 5 to 6 feet.

**The BAY of HARBOURS**, from Cow point to Bull point, is 7 miles wide : nearly midway is Middle shoal, with a rock awash at low water : there is, however, plenty of water between it and Small island, if necessary to go between them : it bears S.E. from Small island 1½ miles. The bay extends 20 miles in a north-west direction, contracting to 1¾ miles between West and Cattle points.

**Bull Road** is by far the most convenient anchorage in the southern part of East Falkland ; the largest ships can work into it, and by anchoring close to the shore, on the south side of the bay, off the entrance of Bull cove, the large kelp reefs on the north side of the point will completely shelter them from easterly winds. The bottom is excellent, and the depth from 7 to 10 fathoms. Except during a very dry summer, there is water in all the bights on the south side of the roads, but the best watering-place is in a cove on the western side, off which there is good anchorage.*

The guide for Bull road is to make Bull point, which is low, and off which there is a rocky islet, surrounded by kelp. Follow the edge of the kelp to the north-west, and it will lead into the roads. The chart and the kelp will be the best guides in this, and in every part of the Falklands.

**Fanny Road** is the only other advisable anchorage to the southward of West point. It is secure, with a good depth of water over a bottom of sand and mud ; it is formed by the Fanny islands and West point, and shows three beaches of white sand. No direction is necessary for it beyond a look-out to be kept for a small kelp patch, which lies half a miles outside of the islands.

Large vessels bound up the bay, with a leading wind, should keep a mid-channel course between Cow point and Small island, to clear the kelp spit, which runs off Kelp island. A course of 5 miles W. by N. ¼ N. will carry them past West and Cattle points, and clear of 3 kelp patches, which lie W.N.W. 1¼ miles from West point.

Northward of West point there is good anchorage for large vessels between the 3 kelp patches and the cliff off Snug cove, as there is an

---

* *See* Admiralty Plan of Bull road, No 1,935; scale, m = 1·7 inches.

easier depth of water there than there is in the other parts of the bay, and it is the weather shore; the bottom is mud and sand, 15 or 16 fathoms, and it will be necessary to moor.

The arms and creeks present for vessels of moderate tonnage and small craft secure and snug berths; the only direction necessary for them is to look out for the kelp running off the points and islands, and which is generally bold-to. Here, as in Adventure sound, the hills seldom exceed in height 150 or 180 feet, and are of an uniform appearance. Fish and water are procurable.

TIDES.—It is high water, full and change, in the Bay of Harbours at 6h. The rise and fall of the tide is 5 feet, with little or no velocity.

EAGLE PASSAGE lying between the East Falkland and a chain formed by Speedwell, George, and Barren islands, is not recommended for large vessels, as there is a tide race of 3 knots. Vessels passing through may go on either side of the kelp patch which lies 1½ miles to the north-westward of Blind island. The only other obstruction in the passage is Mid island, with its two reefs marked with kelp, and extending three-quarters of a mile N.W. ½ N. and S.E. by E. The sea breaks heavily on them in bad weather, as it does on all the coast from Porpoise point to Mid island. After passing that island the channel becomes narrow but clear of dangers.

TIDES.—It is high water, full and change, at 8 hours, in the Eagle passage, rising 4 feet, the flood setting through to the northward, and the ebb to the southward, at a velocity of from 2 to 3 knots, but greatly influenced by the winds.

OWEN ROAD.—Of the group of islands that form this passage, Barren and George islands are the southernmost, and between them there is good anchorage for any kind of vessel in Owen road. But in coming from the southward care must be taken not to pass nearer to Barren island than 2 miles, as there are two reefs off it full that distance in a north-easterly direction. From the western point of George island a reef runs off about 3 miles to the south-west, with kelp extending 2 miles farther in the same direction.

DIRECTIONS.—Vessels should not haul up for the anchorage until the passage between Barren and George island is opened, or until the north point of Barren island bears S.S.W. ½ W., and the north point of George island W. ½ N.; then a S.W. by W. course will lead to a good berth. There is also a kelp-covered reef, half-way between the eastern points of George island and Barren reefs; the passage lies to the southward of it, more than

a mile wide and quite clear.  The best anchorage in Owen road is off
George island in 8 or 10 fathoms; straggling stems of kelp will be seen,
but they are of no consequence.

With a S.W. gale, a vessel may work up to this anchorage, as the
water is perfectly smooth; and with a S.E. gale, which causes the
heaviest sea on the coast, she will ride in smooth water, if the north-east
part of Barren island bears east.  The bottom is chiefly stiff clay covered
with a crust of broken shells.  Large vessels should anchor before shutting
in the points of the passage between the islands.  This passage is only fit
for boats.  These islands are well supplied with water from many lakes :
the largest is on George island, and is nearly a mile in extent.

**SPEEDWELL ISLAND** is to the northward of George island, and is
the largest of the group, being 9 miles in length and about 3 miles broad ;
it is low and flat, about 70 feet high, and scarcely visible more than 3
leagues from a vessel's deck.  It has two good anchorages for small
vessels on its north-eastern side, where fresh water is procurable.

**Halfway Cove** is the first, and lies immediately to the southward of a
small rocky islet ; but it is small and requires quickness in coming-to.
There is a great deal of kelp on both sides, but not less than 5 or 6
fathoms water on its edge ; the holding-ground is excellent, and the depth
from 5 to 8 fathoms.  The second or northernmost anchorage is about a
mile within the north point of the island, in the first sandy bay seen in
standing along the island to the southward.  Between George and Speedwell islands there is a passage for small vessels, but it requires a good
pilot, as there are some rocks in it.  It is sometimes used by small craft,
in running for the southern anchorage of Speedwell island.  In making
for either of these anchorages from the northward, keep about 2 miles to
the northward of the Elephant cays, and of the islets between them and
Speedwell island, as there are patches of kelp off them both, but outside
that distance all appeared clear.

**The ELEPHANT CAYS** are low sandy islets surrounded by reefs and
kelp : there is no passage between them and Speedwell island.

# CHAPTER IV.

## FALKLAND SOUND AND WEST FALKLAND.

### VARIATION 18° East in 1860.

---

**FALKLAND SOUND** is a narrow strait, which separates the two main islands of East and West Falkland from each other. It extends 45 miles in a northerly and southerly direction, and varies in breadth from 18 to 2½ miles. At its southern end there are many flat islands, and some shoals; the dangers are, however, generally visible. It ought not to be navigated by night; and as good anchorages may be obtained in almost any part of it, and good harbours abound, a safe position may always be selected before dark; the water being always smooth, and thick weather seldom occurring, its navigation is rendered easy.

The eastern side of the sound, after passing the N.W. islets and Grantham bay, changes its aspect to a low country, with gently undulating hills, which seldom exceed 150 feet in height; and it maintains this character to the southern extremity of the sound. Its shore is indented with excellent harbours and creeks, affording good shelter in all weathers, and fronted with flat islands, particularly at the S.E. part.

The western side, on the contrary, is high and bold, forming a singular ridge, varying from 300 to 500 feet, nearly the whole length of the Sound, but reft asunder in three places, and thus forming Port Howard, Shag harbour, and Hill gap. From Bold point again it is a continuous ridge, lowering gradually to White rock, South point, separated midway by the entrance to Manybranch harbour. These gaps, or fissures, in the southern portion of the sound, form excellent guides to the opposite harbours and islands on the flat side of the sound, by determining their position by bearings from them. The Hornby Hills extend in a parallel line immediately behind this ridge as far as Hill gap, from whence they take a westerly direction, and range from 1,800 to 2,300 feet in altitude.

The ports in Falkland sound need but few written directions, as the shoals are all buoyed by kelp. The chart is the best guide. Tho harbours on the East Falkland will be first described.

**FOUL and MIDDLE BAYS** lie between Cape Dolphin and Race point; they present no inducement to enter, being both lee-shore bays,

G 2

and the first is encumbered with shoals; it has, however, a shallow harbour in its bight, but with a shifting sand-bar, very difficult, and only fit for very small craft.

**PORT SAN CARLOS** is one of the finest harbours in the Falklands, being capacious, secure, and clear of all dangers. Anchorage may be taken either in the bay formed by the long tussac islands, or in the south or west arm; and the river is navigable for small vessels 3 or 4 miles up, or for boats about 6 miles. Just inside the Narrows, on the north side, there is an excellent cove for placing a vessel on the ground, and examining her bottom.

**TIDES.**—It is high water, full and change, in the harbour at 7 hours, and one hour later in the river, rise and fall 8 feet.

**PORT SUSSEX**, in Grantham bay, is a snug and good harbour for small vessels; the best anchorage is round the flat stony point, about a mile up on the southern side, in 4 and 5 fathoms, stiff mud. It has plenty of good water, and is an excellent port for careening or refitting, but there are no inhabitants. There is a small kelp patch at the entrance of the harbour, with no less than 4 fathoms on it; but the kelp extending from the East point must be avoided. Keep therefore near the West point in entering.

**Brenton Loch** has a narrow entrance and a strong tide, and is therefore only fit for small vessels or boats, and its navigation is impeded by several rocks and shoals; but it is admirably sheltered, affords a water communication of 9 miles in length from the centre of the island to the sea, and nearly joins Choiseul sound, from which it is separated by a narrow neck of land little more than a mile in breadth. From the number of islets within this great inlet it would appear to be a good resort for seals.

**TIDES.**—It is high water, full and change, at the entrance of the loch, at 8h. 15m.; rise and fall, 8 feet; in Port Sussex there is but 6 feet.

**NEWHAVEN** is a little port well suited to small vessels on the south side of Grantham bay, in which there is good anchorage in 6 to 4 fathoms stiff clay, under a surface of sand. It is open to the N.W., but little sea sets in. Anchor rather on the western shore, near a small cove, about a mile from the outer point.

**CYGNET, KING, WHARTON,** and **FINDLAY HARBOURS** are situated inside the Tyssen islands; all excellent and secure, particularly the three latter, having no shoals in them of any consequence. The first has several small kelp patches, which would render it inconvenient for a square-rigged vessel to work in ; but it is a capital harbour for small vessels. There is good anchorage in the bay on the eastern side of Great island in 12 to 15 fathoms; and it is a good starting point for vessels bound to the southward. It is high water, full and change, in these four harbours about 8h. 30m.; rise, 5 and 6 feet.

RUGGLES BAY lies abreast of the Calista islands, and the passage to it is between Ruggles and Wolf islands. Very good anchorage may be had in Danson and Moffat harbours ; they are clear of dangers, and the bay is a good starting port when bound to the southward ; there is, however, a scarcity of fresh water in the summer season. It is high water, full and change, at 7h. 30m. ; rise and fall, 5 feet.

Returning now to the northern end of the sound, the several ports which lie along its western shore will be briefly described.

WHITE ROCK BAY is an excellent port, and can be entered or left with any wind ; but care must be taken to avoid the rocks off the entrance. If strangers are entering from the northward, with a foul wind, they should work up on the eastern shore, near Race point, till they can weather the dry rock in the centre of the passage, and then they may stand across into the harbour. If leaving the harbour, and bound to the northward with a northerly wind, they should stand across, passing either side of the dry rock, and work out on the east shore, in order to keep them clear of the Sunk rock, and the rock a-wash off White Rock point. This rock is surrounded with kelp, which only shows at slack water, being run under by the tide. With a fair wind a vessel may run in or out close past the large white rock at the extremity of White Rock point. Whether bound through the sound, or into White Rock harbour, should a vessel have to wait outside for the change of tide to enable her to work to the southward, she should keep a little to the westward of the entrance, under easy sail, where there is scarcely any tide.

MANY-BRANCH HARBOUR can only be sailed into by fore-and-aft craft, in consequence of its narrow and crooked entrance between its high heads, which flaws the wind in all directions. For large square-rigged vessels, warping or towing must be resorted to in the mornings or evenings, when the wind is generally moderate. It is a good harbour inside. It is high water, full and change, at 7h. 40m. ; rise and fall, 7 ft. 6 in.

PORT HOWARD has also a narrow entrance, but the harbour opens out immediately inside the heads, and a vessel would shoot well in by keeping close round the south head, with the wind to the southward of west. With a northerly wind a square-rigged vessel would scarcely succeed ;

and as there is good anchorage outside in 11 and 12 fathoms, she should wait there for the morning and evening calms. It is a very narrow harbour, but very secure, and there is plenty of good fresh water, fish, and geese. The best anchorage for large vessels is just inside the heads, in 4 and 5 fathoms. High water, full and change, at 7h. 15m.; rise and fall, 8 feet.

**SHAG HARBOUR** is fit for small vessels only; the gusts of wind down the ravine are very violent, but the holding ground is excellent. There is good anchorage ground between the Swan islands and the West Falkland.

**FOX BAY** is wild, and exposed, with south winds, to a heavy sea, but should a vessel be caught there, there is a good retreat in the north arm, which is quite secure.*

The passage to it is between the kelp, which extends from either side, and which always shows itself; and a berth in 3 or 4 fathoms stiff mud, with the first flat island inside the Narrows, bearing south about 5 cables' lengths, is secure and free from danger, with perfectly smooth water. This inner harbour is small, and can be entered by square-rigged vessels with a fair wind only; but it can be easily left with all northerly or westerly winds.

We shall reserve the description of Ports Edgar and Albemarle till we come round to the southward of the West Falkland, as vessels are most likely to have recourse to them when making this great group of islands from the south-westward; and we shall now, therefore, proceed to the north coast of the West Falkland.

**TIDES.**—It is high-water, full and change, at Race point on the shore of the northern entrance of Falkland sound at 6¾h., but the flood stream continues to run in till 10 o'clock; its velocity here is about 4 knots, but to the southward of Port San Carlos its rate diminishes to less than 2 knots. At the southern entrance it is high water, full and change, at 7h.

The tide sets to the westward during the flood along the whole south shore of East Falkland; its strength is from 1 to 2 knots, but near Porpoise point nearly 3 knots, and, with westerly gales, forms a strong race. The stream turns when it is high water by the shore, which at full and change is about 6h. 10m.

The tides in both entrances are strong, and between the islands, but in the main stream they are moderate. The stream of tide at the north entrance does not run into the sound until 2½ hours before high water on

* *See* Admiralty Plan of Fox bay and ports in West Falkland, No. 1,874; scale, m = 1·0 inch.

the shore, which at full and change is 6¾ hours.  At the south entrance the stream sets in at 7 hours ; it will, however, sometimes vary more than an hour.  Among the islands in the south-eastern part of the sound the tides are very irregular in their set.

There appears to be tide and half-tide all through Falkland sound. The flood-stream commences by running to the northward when it is half-ebb by the shore, and runs till half-flood ; it then turns and runs to the southward until it is half-ebb again.  In the northern entrance to the sound it is very strong, running from 4 to 5 knots at the springs.  But the tides among these islands require further investigation ; Captain Fitz-Roy states that the tide flows into both ends of Falkland sound, and that the two streams meet near the Swan island.*

DIRECTIONS.—In coming from the northward or eastward, and intending to enter the sound, steer for Fanning head, the high double peak on the eastern side of the entrance ; or, if it be obscured, make a S. by E. course, from the Eddystone rock, until within a couple of miles of Race point, when the eastern shore must be kept on board to avoid Tide rock, Awash rock, and Sunk rock ; the two latter are lurking dangers, but as the Tide rock shows itself, and is steep-to, it forms a good guide to avoid the others, by never bringing it to bear to the eastward of south.

The soundings from the Eddystone to the entrance of the Falkland sound are pretty regular, from 22 to 26 and 18 fathoms, over a bottom of fine sand and black specks.

Off Race point there is a ledge, which does not, however, extend beyond a cable's length from the point ; and from thence to Fanning head is all clear, with a depth of 20 fathoms.  If the wind should be blowing strong from the southward, and the tide running out, there will be a great difficulty in entering the sound until it changes inwards. Small smart-working vessels may, however, in that case, accomplish the passage by getting close under the north shore of the West Falkland, where the water will be smooth, then stand for the White Rock, luffing close round the margin of kelp which surrounds it, and which is not in less than 10 and 11 fathoms water ; make short tacks, not standing to the eastward more than 4 cables' lengths or half a mile, and work into White Rock bay, where the influence of the tide is soon lost.  The current sets very strongly past the White rock, but when once to the southward of it, a secure anchorage may easily be reached, in the upper part of the harbour, where a vessel may wait till the tide turns.

* Narrative of the Surveying voyages of H.M. Ships *Adventure* and *Beagle*, vol. ii. p. 242.  This book is recommended to the attentive perusal of all persons sailing along either shore of South America.

A South course for 9 miles from White Rock bay will bring the vessel to the North-west islets, which lie off Grantham bay, where the high land of the East Falkland terminates. Starting therefore from these a S.S.W. course of 17 miles will take her abreast of High Cliff island, which, although small, is the highest in the Sound, being nearly 100 feet above the sea. It is remarkable for its white cliff, somewhat discoloured by birds, and is first seen from the deck when abreast of Grantham bay; afterwards the Swan islands will be seen; they are low and flat.

MAIN PASSAGE is between High Cliff and Swan island. Vessels should round the south-east point of the latter island, and then borrow on the West Falkland to clear the large kelp patch off the Tyssen islands, which has from 4 to 5 fathoms close to its edge. A good leading mark from this point, to clear the patch, is a horizontal line of white sand, deposited in the ridge of the West Falkland, and bearing S.W. about 9 miles. The course from between the Tyssen patch and the West Falkland is about S. by W., so as to pass to the eastward of West island, and to the westward of the Calista islands.

SWAN PASSAGE.—In many cases, however, it may be more advantageous to pass west of Swan island, and the Swan passage is straight and clear, the water is smoother, and there is better anchorage near it than in the main passage; where, with a strong gale against the tide, there is a heavy race. In both passages with a foul wind the tide is too strong for a vessel to get through; and in this case, or if waiting for the night, if bound to the southward, the best anchorage is near the north end of West Swan island, and if bound to the northward, in the corner between West Swan island and Hill gap. There is a passage only for small vessels between West Falkland and the West Swan island.

TYSSEN ISLAND PASSAGE.—The passage between the Tyssen islands and East Falkland to the several harbours of East Falkland is clear and good, the narrowest part being to the eastward of Sandbar island; but even there it is rather more than a mile wide, with a good outlet to the main passage, round the north point of Great island. The only shoals in it are, a reef about 2 cables from the north-eastern point of Great island, which always shows the kelp; and a spit extending about 4 cables from the south end of Sandbar island, which must be attended to, as the tide sweeping down the kelp, frequently conceals it. Tickle Pass, a passage for small vessels, between the south-east part of Great island and the next island, is very narrow, and the tide runs rapidly through; but the least water is 7 fathoms.

**FROM the SOUTHWARD.**—In entering the sound from the southward, after passing Wood shoal, which is marked by kelp, keep on the West Falkland side, at a moderate distance from the shore; and then the converse of the above directions will be a sufficient guide to any vessel. With northerly and N.W. winds a passage to the northward through the sound, as the water is always smooth, is sooner effected than by proceeding round by the east coast.

Vessels, at either extremity of Falkland sound, intending to proceed to sea, should the day be at all advanced, would do well to anchor for the night, and start at daylight; as in that case they would have the whole day to get clear of the entrances, and thus save some anxiety and risk, as the wind generally becomes light after sunset. Good starting posts from the northern entrance are Port San Carlos and White Rock bay; and from the southern Anchorage bay, on the east side of Great island, and Ruggles bay, but well up towards Danson harbour, as the ground is hard near the entrance.

**TAMAR HARBOUR** is the first port on the coast of West Falkland, but it is not advisable for any strange vessel to use it. The entrance or pass is very narrow, and a reef extends nearly half-way across from the west entrance point. The kelp on this reef is run under by the extreme strength of the tide, and the eddies are very dangerous. There is also a kelp patch, which shows at slack water, in the centre of the channel, between the outer and inner entrance; but the least water we found there was 3 fathoms, and the sealing vessels never avoid it. The tide sweeps so rapidly through, that a vessel wishing to bring up in Tamar harbour must haul out of the stream very quickly, to prevent her being carried through into Pebble sound.

Small vessels, when well acquainted with the passage, may find it very useful to pass through it into Pebble sound, if bound to the westward, as it will give them smooth water and good anchorage all the way to Port Egmont; but the North-west Pass out of Pebble sound is almost equally dangerous, being only a cable's length wide, with a furious tide; and nothing but a small, quick-working vessel should attempt to work through it.

**TIDES.**—It is high water, full and change days, at 9h.; the rise and fall is 8 feet. Running along the north coast of the islands to the westward, the flood rushes through Tamar and Whaler passages and sweeps round the West Pebble islet into Keppel sound, filling that sound, and Port Egmont, two hours before it has ceased running to the westward. Rushing through the North-west pass at 5 or 6 knots an hour, it sweeps along part of Pebble sound, meeting the flood-tide that comes in with equal velocity through the Tamar pass, and thus causes whirls and eddies in several quarters.

The water having attained its height remains quiet only a little while and then ebbs with similar fury ; the result is, that if bound through the Passes to seaward, a vessel should run through the North-west pass with the flood and out of Tamar pass with the ebb.

**PEBBLE SOUND** is formed by Pebble island and the north shore of West Falkland. It is about 15 miles long and 9 wide, and is full of islands, with good anchorage in every part of it. The islands are low, and except Golding and Middle islands, are not well supplied with water through the summer. One of the largest streams in the Falklands, the Warrah, runs into the south side of the Sound, and at high tide boats can go up into the fresh water.

**Anxious and Creek Passes.**—Besides the Tamar pass and the North-west pass, there are two other passages into and out of Pebble sound among the islands, Anxious passage, between Golding and Passage islands, and another very narrow one to the southward of all the islands, close to Creek point ; but the approaches to them are so intricate that they are not likely to be used, except by coasting vessels. The chart is the only guide to them, for the kelp, as in the other channels, is always run down by the tides. There is no good anchorage on the north side of Pebble island ; but with the wind off shore, temporary anchorage may be found on the west side of Elephant bay, though very much exposed if the wind should veer to the northward. The peaks on Pebble island are the best marks for making out this part of the coast, and are seen very clearly when coming in from the northward.

**PORT EGMONT CAYS.** — Two islands surrounded with rocks and kelp extend 5 miles W.N.W. from the west point of Pebble island, 3 miles farther to the west are Port Egmont cays of a similar nature. Wreck island is long and low, with a reef to the northward of it, lying North distant 5½ miles from Elephant point. Sedge island is larger, lying W. by S. 5 miles from Wreck island ; it has a reef and kelp patch extending one mile from its north-east side.

**KEPPEL SOUND** lies under the west end of Pebble island, between it and Keppel island. From a peninsula which forms the east point of the latter island a reef extends about 1½ miles to the north-east ; south of this peninsula lies Committee bay, in which is the Patagonian Mission station. The passage into this sound is clear, and there is another from it into Port Egmont, round the south side of Keppel island, which is quite clear of danger, and well marked by kelp, but it would only be used by vessels that had passed through Pebble sound.*

**PORT EGMONT,** in which was the old settlement, is situated to the southward of Keppel island, between it and Saunders island. There is good anchorage in every part, and the tide is not strong. The north, and usual entrance, is between Saunders and Keppel islands, nearly a mile wide, and clear of danger; but it would take a very fast-sailing vessel to work into it against the tide. The best anchorage is off Old Settlement cove, close to the southward of a kelp patch, or in Sealers cove, a little further to the south-west, which is better sheltered from the southerly winds. Fresh water is abundant in both these coves.

**Burnt Harbour.**—There is a narrow and winding channel out of the south side of Port Egmont into this harbour, and through it to the head of Byron sound; but it is not fit for large vessels. Burnt harbour is the only good anchorage in the head of Byron sound, and on the north side there is a good watering-place; but a long reef in the western entrance forms two narrow channels, through which a small vessel only can work. There is a very narrow channel round the east end of Burnt island, but it is only fit for coasting vessels.

**TIDES.**—It is high water, full and change, at 7h. 30m., in the Old Settlement cove, and the rise and fall is 11 feet, which makes it the best place in the islands for beaching a vessel.

**DIRECTIONS.**—Vessels bound to Keppel sound and Port Egmont from the northward should endeavour to make the high land of Mount Harston, on the western peninsula of Saunders island, which will be seen in clear weather long before Sedge and Wreck islands, which are low islands, and lie about 6 miles to the northward of it. Passing between Wreck island and Egmont cays, the entrance to Port Egmont will be plainly seen. There is a clear passage on either side of Wreck island, as well as of Egmont cays; and between them and the entrance of Port Egmont, there is no danger, except a small kelp patch, which always shows about 1¼ miles to the N.W. of Gull point.

The **RACE ROCKS** lie W.N.W., distant 1½ miles from Elephant point, the north-west point of Saunders island; 1 mile to the northward of them is a patch of 4 fathoms, generally marked by a tide rip.

**BRETT HARBOUR,** on the west side of Saunders island, is of little

---

* *See* Admiralty Plan of Port Egmont, Keppel sound, &c. No. 2,438; scale, *m* =1·7 inches.

service; it has a long and narrow entrance, and is shoal with reefs and kelp patches.

CARCASS ISLAND is the largest of a chain of islands running W. by N., 16 miles from the west point of Saunders island; it is easily distinguished by the high double peak in its centre. Off the north-west end lie two islands called the Twins; to the south-east are the Needle rocks, and off the south-west side is Carcass reef, a narrow ridge about 1½ miles in length.

BYRON SOUND.—This sound is much exposed to westerly gales, which send a heavy sea up to its head; but there is good anchorage for small vessels in the south-east corner, in the entrance to Hill cove, or the cove itself.

HOPE HARBOUR is situated at the north-west point of the West Falkland; there is a bank to the northward of the entrance, but not less than 5 fathoms could be found on it. At the head of Hope harbour there is a stream, in which, at spring tides, quantities of fish may be taken. Vessels going into that harbour should anchor off Grave cove, on the south side, where there is a good watering place.

West Point Island lies at the mouth of Hope harbour; there is a cove on the north-east side with a large kelp patch off the entrance, but with not less than 8 fathoms on any part of it. Should a vessel intend remaining for a few days, this cove is the best place to adopt, as it is not exposed to the heavy squalls off the island which rush up Hope harbour. Rabbits are abundant on the island, while there are not any on the main land.

Gibraltar Reef is an extensive ledge covered with kelp, running W. by N. 7½ miles from West Point island; there is a white rock about 2 miles from the extreme. The tides set directly across this reef, and it is to avoid the risk of being set towards it that vessels are enjoined to pass to the eastward of Carcass island.

TIDES on the NORTH COAST.—In Hope harbour it is high water, full and change, at 8h. 10m., though it is not slack water in the passage till 10 o'clock. The stream of tide sets strongly round the west end of Carcass island towards Gibraltar rock, and also through all the passages into Byron sound; but there is very little tide in that sound, nor is a vessel sensibly affected by it until as far west as Carcass reef. There the flood or western stream will be found setting very strongly towards Hope harbour, and through it to the southward. The ebb, or eastern stream, sets from Hope harbour towards both ends of Carcass island, and through the channels east of the island to the northward; but it, likewise, is scarcely felt in Byron sound. The flood tide sets to the westward along the north side of the West Falkland for about 2 hours after it is high water in Port Egmont ; the springs run nearly 3 knots off the points, and round the islet, causing strong tide-rips in heavy weather. After passing the north-west point of Saunders island, it runs to the S.W., between Carcass island and the islets near it, and through all the channels between West point and the several Jason islands. But here again the tide becomes tide and half-tide, as in Falkland sound, for, meeting the southerly tide-wave round the West island, the last 4 hours' flood and the first 2 hours' ebb run to the north-east.

**DIRECTIONS.**—Vessels bound to Hope harbour, or to any of the ports to the southward of it, should avoid getting to the westward of Carcass island, as between it and the Jasons there are several reefs, the kelp on which is drowned by the tide. But from Sedge or Wreck islands, to the east end of Carcass island, there is no danger, except the foul patch, with 4 fathoms water, outside of the Race rocks. With a strong breeze there is a heavy race for nearly a mile from those rocks ; but by giving them a berth of a full mile, which all vessels should do, there will be no danger, and a direct course may then be steered for the Needle rocks, or for the eastern end of Carcass island.

The passage to the eastward of these rocks is the best, as they may be almost touched. From thence to the entrance of Hope harbour there is no danger but what shows well above water ; and though vessels may have to make two or three tacks to reach Hope harbour, it is much better to do this than to pass through the reefs to the north-west of Carcass island, with the chance, if the wind fails, of having to anchor, in order to avoid being swept through between the Jasons and West point. This is not a well-sheltered anchorage, if it should be necessary to wait for the tide ; whereas, by steering for the eastern end of Carcass island, good anchorage may be taken in the bay at the eastern end of that island, except with northerly winds, and then vessels may anchor off the bay, on the south side, to the north-eastward of Carcass reef. H.M.S. *Philomel* never anchored in either of these places ; but according to a sealer, who knew them well, the anchorage to the east end of the island is very good, and much used as a halting-place by the sealers. The depth from 7 to 12 fathoms.

**CAUTION.**—The kelp off Hope point being overrun by the tide, the point should not be passed nearer than a cable's length. No sailing vessel can enter against the tide, except with a good leading wind ; and if the wind is between N.W. and S.W., it would be useless for her to attempt going out through the southern channel against the tide, as the wind would be so baffling ; but she may bring up in the entrance of the cove in West point island, and wait there for slack water.

In passing to the northward through West Point passage, which lies between West Point island and Hope harbour, it is necessary to wait for the tide, unless the wind should be to the southward of S.W., and even then it may be difficult to get through the northern part of the channel ; should this be the case, a vessel ought to bring up in the entrance to Hope harbour, out of the stream of tide, and wait till slack water.

A vessel bound from Hope harbour to the northward, should again pass to the eastward of Carcass island and the Needles, and to the east-ward of Sedge island, thus avoiding all the reefs and tide-rips to the west-ward and north-westward of the Twins and Carcass island. Should it be necessary to pass to the westward of Carcass island, the best passage is near the north-west point of the island and round the Twins ; but to the southward of a rock which always breaks, and which lies nearly 2 miles N.N.W. of Carcass island. To the westward of this rock there are some reefs, the kelp on which only shows at slack water, and vessels should cautiously avoid them, as we have no clearing marks to offer.

JASON ISLES or the Sebaldines form a chain extending W. by N. 34 miles from the north-west point of the Falklands. The extreme, Jason West cay, is low and less than a mile long ; the East cay is of a similar nature. Grand and Steeple Jasons are lofty, the former being 1,210 feet in height ; inshore of these lie the Flat, Elephant, and South Jasons. The latter has a ledge termed the Hope reef, extending 4½ miles from its eastern point ; the space between this reef and Carcass island is full of ledges and tide rips. The passages through these islands are not fit to be generally used by any vessels ; those between the Flat Jason and West point are full of dangers, the tide setting violently across them.

To the westward of the Flat Jason the passages are clearer, but there can be no necessity for vessels passing through them. Should the tide, in light winds, set a vessel towards them, by keeping near the middle of the passage she will be taken clear through ; but having passed in this manner, between the Grand and Flat Jasons to the southward, she should be very careful that, on the turn of tide, she is not swept towards the South Jason and the reefs off West point.

The TIDES between West point and New island, and in the bays within that line, are not very strong, and, except near the points, would not prevent a vessel working either way. They run to the N.E. from half-ebb to half-flood by the shore, and to the S.W. from half-flood to half-ebb.

**PORT NORTH,** the first harbour after passing West Point passage, is a deep bight on the northern shore of King George bay, with a moderate depth of water, but rather exposed to westerly and south-westerly gales. There is a good stream, and a valley, with pretty good land, at the head of the port. At Pickthorn point, separating Port North from King George bay, there is a good harbour for small vessels, between the two Bense islands and the mainland, with a clear passage through it ; but the land there is very steep and mountainous, and not likely ever to be of much value.

**KING GEORGE BAY.**—The passage into this bay is quite clear on either side of Split island, and the different openings through the Passage islands are also plainly seen.

**Hummock Island** is the best guide in running up King George bay; its peak forms a cliff on the north side, and slopes off to the southward. Half-way between Bense harbour and Hummock island, stands Rabbit island, the west side of which is a high and conspicuous cliff. Between it and Hummock island there is a chain of small islets, through which there are three good passages : one, close to the south point of Rabbit island, and the other two on each side of the islet nearest to Hummock island. They are all clear of danger ; but the tide sets rapidly through them, and a vessel should endeavour to keep nearly in mid-channel.

**Whaler Bay,** in the corner inside of Rabbit island, offers a safe and good anchorage with coves for small craft, and good watering-places. There is a narrow channel, between Rabbit island and the main, into this anchorage, but it is scarcely fit for the smallest vessels, being very narrow, and the tide strong. On the north shore, abreast of Hummock island, a very secure and deep creek, called Roy cove, was much used by the American whalers, some years since, for trying-out their oil. The rise and fall in it is 10 feet at spring-tides, so that it is a good place for beaching small vessels; and, being deep close to the rocks, it answers also for heaving down a vessel. There are several good streams of water in Whaler bay, and it is a good place for fish, a weir having been built at its head by the sealers and whalers.

**CHRISTMAS HARBOUR.**—King George bay narrows into this harbour, which, one day, is likely to be of much importance, as it leads to the heart of West Falkland ; at its head there are creeks, and large fresh-water streams running through some of the best watered land in the whole group. The entrance to it is well marked by Town point, which, after a vessel has passed Hummock island, appears as if covered by scattered houses ; this appearance being caused by numerous patches of white sand spread over the side of the hill above the point. The three islands between Hummock island and this point may be passed on either side; and there is good anchorage inside them, off the mouth of the harbour, where the *Philomel* rode out a heavy westerly gale in smooth water.

In the entrance of Christmas harbour there is a long narrow reef, covered with kelp, and forming two channels. The southern one is only fit for small craft, there being only 2¼ fathoms at low water on a bank which extends from the west end of the reef to Town point; but to the northward of the reef the channel is quite clear, with 4½ fathoms, as far as Tide islet, which is small and green. Abreast of this islet is the best anchorage for large vessels. About half-a-mile above it, where the harbour widens, there is a bar, or flat extending across the harbour, the deepest water over which is 2½ fathoms, near the south shore; but in mid-channel, and from thence to the north shore, there is only one fathom; and this shallow water, having a sandy bottom, is not marked by kelp. The best guide is the kelp on the south shore, which should be left about 2 cables' lengths distant, until the water deepens to 4 and 6 fathoms, where there is good anchorage for a space of 2 miles with that depth. At half-tide, vessels drawing 18 feet can pass the bar; and above it there is anchorage for a great number of large vessels. About 7 miles from the entrance the harbour branches off into several creeks, the southern one extending 5 miles up the country; but it is only fit for boats, which, at high water, can go into the fresh water of Chartres river. This is the largest stream in the Falkland islands, and drains the fine central valley of the West islands.

**Water.**—There are several good streams of water in different parts of this harbour; but the easiest for vessels is that marked in the chart, on the south shore, nearly abreast of Tide islet.

**PASSAGE ISLANDS.**—King George and Queen Charlotte bays are only separated by a long narrow slip of land, the extremity being named Dunnose head, off which lies the four Passage islands; the two eastern passages are very good and clear of all danger. The streams of tide set directly through them, and turn when at half-flood and ebb by the shore. No very exact observations having been made in these channels, it may be found that the time of tide-turn may be half an hour before or after half-tide by the shore; but, as in the other channels, and in Falkland sound, it is at or about half-tide. The third or West passage, is good with a leading wind; but so narrow, that it was with difficulty the *Philomel* worked through. The fourth or False passage is completely blocked up by kelp, with water only for a boat.

Three-quarters of a mile N.W. of the fourth island there is a kelp patch, which may be passed on either side. There are 14 fathoms close to its inner edge ; and, as it is not thick, there is probably deep water through it ; but it was not sounded, and should, therefore, be avoided. There is a tide-rip off this point with strong breezes.

All the shores of the Passage islands are steep ; and there is no outlying danger but what shows plainly.   Round island and the Sail rock near it, are good marks for Whale passage, the easiest for vessels to pass through ; but with a foul wind it would be always necessary to wait for the tide.

There is no good anchorage near the islands ; but vessels detained by a southerly wind, or caught in a southerly gale, might anchor under the lee of Dunnose head near the entrance of Rous creek, a small cove, in which there is water for vessels drawing 14 feet ; and vessels of any size, if damaged, might safely make for this cove, and run on the sand at its head, unless at the top of high water ; the cove is well sheltered, and the water always smooth.

QUEEN CHARLOTTE BAY.—This large and extensive bay, 10 miles wide at the entrance between Dunnose head and Swan point, is generally clear of shoals and possesses several good harbours.   On the south side of Dunnose there is no safe anchorage short of Shallow harbour, which is about 9 miles to the S.E. of that headland.

Philomel Road and Shallow Harbour may be easily known in coming from the westward, by the steep bluff about 2 miles to the westward of them.   When abreast of this bluff, a small island, which lies off Shallow harbour, will be plainly seen.   This island may be passed close to its south side ; and a vessel can then haul up directly into the road, on the north side of which lies Shallow harbour.   The chart, and a look-out for kelp, of which there are several patches, will be better guides than any directions.   A small islet stands in the entrance of the harbour, and may be passed on either side ; but if to the westward, care must be taken not to pass too close to Shallow point, on the main land, off which the shoal water extends a cable's length outside the kelp.   All those dangers are marked by kelp, and can easily be avoided.

There is a narrow channel into Philomel road to the northward of Green island.   A kelp reef runs a long way off the island, and nearly joins the kelp off Dick point ; but there is a narrow clear spot near the point at the end of the reef, through which there are 6 fathoms, and, with a leading wind, this is a perfectly safe channel ; but it requires a good look-out from aloft to find the clear spot.   There was a small spring of very good water at Spring point, on the south side of Philomel road, but it was nearly dry ; and none was found in any other part of the south shore.

PORT PHILOMEL.—At the north-east end of Philomel road a long narrow channel leads to the large land-locked harbour of Port Philomel.

The tide is so strong in this channel, which is nearly 6 miles long, that it is hardly safe for any but small quick-working vessels to venture up it, particularly through the Narrows between the two small islands at its inner end; but there is a small bay on its north side below these islands, called Halfway cove, in which there is excellent anchorage, out of the tide, for vessels of any size. All vessels going through should anchor here, even if going up with the tide, and wait till slack water to shoot through the Narrows.

So great is the area of this splendid harbour, that the entrance is not sufficiently large to admit the water as fast as it rises outside; and consequently it is high water inside 2¾ hours later than in Philomel roads and Shallow harbour, and rises 2 feet less.

When inside the Narrows there is no danger in any part of Port Philomel, except what is marked by kelp; there is excellent anchorage in every part of it, in from 7 to 10 fathoms; and in all the creeks there is good anchorage for small craft, in from 2 to 4 fathoms. All the surrounding shores of Port Philomel are well watered; except on the peninsula between Port Philomel and Queen Charlotte bay, where there was scarcely a drop of water to be seen during a dry summer. On the inner shore of the isthmus, in Edye creek, there is a watering-place marked in the chart. In the south-east corner of Symonds harbour there is a deep creek, where abundance of fish may be caught by shooting a seine across it at high water, about the distance of half-a-mile from its head, and then waiting for the tide to fall.

**Port Richards** lies in the south-east corner of Queen Charlotte bay: a deep inlet, the head of which nearly joins the inner basin of Port Edgar. But it offers no very good anchorage, except in a cove on the north shore. The water is deep in every part of the port, from 15 to 20 fathoms, and it is completely out of the track of vessels.

**Antony Creek and Carew Harbour** are two small well-sheltered harbours at the mouth of Port Richards; the water is rather deep, from 12 to 17 fathoms, except in their southern coves, where there is excellent anchorage in from 4 to 7 fathoms, and very good watering-places.

Vessels requiring anchorage in the south-east part of Queen Charlotte bay, should run into one of these ports in preference to Port Richards. The western one is the best, as the water is not so deep as it is in the other, and 10 or 12 fathoms may be found off a cove on the west shore, a little inside the entrance. The land is high on all sides of both creeks.

**Double Creek.**—A little to the eastward of Carew Harbour there is very good anchorage for small vessels, inside the islets at the mouth of Double creek.

**TIDES.**—It is high water full and change days in Shallow harbour at

9h. 30m., and the rise at springs is 6 feet. At Halfway cove in the narrows leading to Port Philomel it flows till 11¾h., and the rise is 8 feet.

**NEW ISLAND.**—Vessels bound into any part of King George or Queen Charlotte bay, from the westward, should make New island, which is very easily known, and, in fact, cannot be mistaken, as it is the northernmost of the high cliffy islands which form the south-west portion of the Falkland group; and the lofty cliffs at its north-west points are very remarkable. If coming from the south-west, these high cliffs will be the extreme land seen; but from the westward, two small but high islands, North and Saddle islands, will be seen to the northward of New island. Between these islands and New island there is a clear passage; but as it is narrow, and the winds often baffling under the high cliffs, vessels should pass outside of North island, which is bold close to.

**Ship Harbour.**—If a vessel is bound into New island, she should haul round the north-east point, and passing close outside Cliff island, haul up for Small islet at the head of Ship harbour. Inside this islet the best anchorage will be found, if a vessel is going to remain any time, or requires water and peat. There is working room on either side of the islet, and deep water close to the kelp in every part of the harbour. The best watering place is on the beach at the head of the bay : the water is very good ; and it is impossible to find a place better calculated for ships to touch at for the purpose of procuring a supply, if passing to the westward of the islands. There is, moreover, abundance of excellent peat on the island, and any quantity can be obtained, fit for immediate use, from having been charred by fire. Rabbits are also very plentiful.

**TIDES.**—It is high water full and change days in Ship harbour at 10h. 30m., the stream making to the westward of Grey channel at 7h. 30m., when it is half-flood by the shore.

**GREY CHANNEL,** to the southward of New island, is clear of all danger on either side of the Seal rocks; but the tides are very strong; and with westerly winds there is a heavy race, so that it would be always better for ships to pass round the North island. But vessels going to the southward from Ship harbour with the wind northerly, may always run out through Grey channel by waiting the turn of tide.

**DIRECTIONS.**—A stranger who has the chart will have no difficulty in making out any part of this western coast. Though the bays appear so much exposed to westerly winds, there is much less swell in them than might be expected. The tide in the offing, running across from point to point of the bays, makes their navigation much easier ; while they are so

full of well-sheltered harbours and anchorages, that a stranger, if caught
in a westerly gale and unable to weather the headlands, need have no
hesitation in running up either King George or Queen Charlotte bays :
in the former he will find good anchorage inside of Hummock island ; and
in the latter, hauling round Swan point, and anchoring under the lee of
any part of the east side of Weddell island ; or, if the weather is quite
clear, running for Shallow harbour on the north side of Queen Charlotte
bay.

New island might be found most useful to vessels caught in heavy gales
to the south-west of the islands, instead of knocking about for days and
nights, with the chance of getting on a lee-shore, or being drifted towards
the Jasons.   Directly the high cliffs of New island, or Beaver island, are
seen, a vessel should bear up, and if the tide suits, and the wind fair,
might run through Grey channel, keeping in mid-channel to the northward of the Seal rocks, and, hauling in close round the shore of New
island, along the kelp, might anchor in either South harbour or Ship
harbour without having to tack.   There she might lie quietly, saving all
the wear and tear of the vessel and crew, and filling up with water and
fuel, besides giving the men a fresh meal of rabbits ; and as soon as the gale
is over, and the wind draws round, she can run out either to the northward or southward, and proceed on her voyage.   Those who have neither
dogs nor guns for procuring rabbits, can always get a few by watching the
holes they run into, and digging them out, either with a spade or a handsaw,
the latter being generally the most useful.

**WEDDELL ISLAND** is a large island, forming the west side of Queen
Charlotte bay ; to the westward of this island lies Beaver island, and in
the channels between them are numerous anchorages.

**Beaver Harbour,** on the east side of Beaver island is the best ; and
the chart and the kelp will guide any vessel into it.  The northern entrance
to these channels is well marked by two singular rocks, called the Colliers ;
they look exactly, in some views, like vessels under sail.   They lie in the
western entrance to these channels, and may be passed on either side close
to.   But as all these channels are too intricate to be generally used, little
more need be said about them.   The tide sets powerfully through them,
and at the southern entrance forms a heavy race.   Vessels caught off the
southern shore of the group in a S.W. gale, and unable to weather
Beaver island, might run for either Governor or Tea channels in safety.

**Staats Island,** between Beaver and Weddell islands, is easily known by
the remarkable detached cliff at its south entrance.   Immediately to the
westward of this bluff is the entrance to Governor channel, which is
clear of all danger, and may be entered during the heaviest gales in

safety ; and by hauling close round the north end of Staats island, good anchorage, sheltered from every wind, will be found off its N.E. point ; but the water being rather too deep (15 to 18 fathoms) for merchant ships to get up their anchors quickly, it would be better for them to run through Governor passage, past Middle island, and then French harbour will be seen a little on the starboard bow.

FRENCH HARBOUR.—Into this any vessel may run in safety, furling her sails if possible before she gets in, should it be blowing strong from the westward ; and anchoring in the south corner of the harbour, in 5 to 7 fathoms. The only disadvantage of French harbour is, that the channel is rather narrow to work out again, but it is quite clear from shore to shore, and a vessel may tack in the edge of the kelp on either side.

Tea Channel.—The passage on the east side of Tea island is also very good, being clear right across ; but there is kelp all over the entrance, though it shows only at slack water. The least water in the kelp is 6 fathoms ; but there is a heavy race across the entrance, with a southerly gale. This passage may be easily found, coming from the southward, by the singular high islet or rock, called the Horse Block, which is about 2 miles to the S.E. of it.

In the narrow channel between Staats and Tea islands there is hardly water for a boat at low water ; and that to the westward of Governor island is also very narrow, and only fit for very small vessels. There is kelp across the entrance, with only 2 fathoms water through it.

BALD ROAD lies inside Bald island, on the north side of Weddell island. Bald island is small and round-topped, with a high cliff on its west side, and lies close off Beacon point, on which there is a hill, about 300 feet high, with a small stone beacon, which can be seen several miles. A vessel can run between this point and Bald island, and haul up into excellent anchorage under Beacon point, in Bald road. It might happen, that a vessel endeavouring to enter New island or Ship harbour, either from the strength of the gale or from previous damage, might be unable to fetch into safety ; in which case she should not hesitate to bear up, and, running along the group of islands off the N.W. side of Weddell island, she will soon make out Bald island at the entrance of Chatham harbour.

Chatham Harbour. — If she passes outside Bald island, she must give it a good berth, to clear a detached kelp-patch which lies three-quarters of a mile to the eastward of it, and she may then either anchor in Bald road, or run up to Chatham harbour. All the dangers there are marked by kelp ; and the best anchorage is in Elephant cove, where there is a good watering place.

**TIDES.**—It is high water full and change days, in Chatham harbour, at 9 o'clock ; rise and fall 8 feet.

**SMYLIE CHANNEL.**—There is a safe entrance through this channel into Queen Charlotte bay, between Weddell island and the main. Cape Orford, which forms the south entrance point to this channel, is cliffy, and about 100 feet high ; with a small island called the Sea-Dog, in the offing. There is a passage between them, but it is better to pass to the westward of all. There is a heavy race in Smylie channel with westerly winds, when the tide is running out ; and the kelp, which extends nearly across, shows only at slack water ; but under the kelp there are 4 fathoms ; and the deepest water is on the Weddell island side of mid-channel. When inside this entrance, two openings will be seen on either side of Dyke island ; the northern one leads directly into Queen Charlotte bay ; and the other into South harbour, in every part of which there is good anchorage for the largest ships.

**Stop Cove** is also a very good anchorage on the north side of Smylie channel, formed by a low green tussac island, which is joined to the shore by a spit, dry at half-tide. Vessels bound to the westward through Smylie channel should anchor in this cove, to wait either for the tide or daylight ; and it is the best anchorage in any part of the channel.

**Water.**—There is a good watering place in Penguin cove, about a mile north of Stop cove ; and the geese are very abundant. The reefs at the east entrance of Smylie channel always show, and can be passed on either side in safety ; but the best passage is that close to the shore of Weddell island.

**New Year Cove and Gull Harbour,** on the eastern side of Weddell island, are both excellent ports, but, until the islands become settled, are not likely to be frequented by any vessels.

**House Cove.**—There are some good coves on the eastern side of Dyke island, with good watering places, but they are out of the way of vessels going through Smylie channel. One of these, House cove, is an excellent place to lay a vessel ashore, the rise and fall of tide being 10 feet.

**RODNEY COVE** is between Cape Orford and Port Stephens, the only place of shelter on the south side of the West Falkland ; it is a secure and safe anchorage, but can only be entered with a leading wind. The kelp extends across the entrance ; but there is water through it in mid-channel for the largest ships. A small lane of clear water, near the middle of the kelp, shows the channel ; and when inside the kelp anchorage may be found in any part of the cove in from 7 to 3 fathoms.

**PORT STEPHENS** is the first port to the westward of Cape Meredith, and may be easily known by reference to the views. The land is very remarkable : Bird island and the Castle rock lie to the westward of the port ; and a notable hill called the Three Crowns, with three distinct masses of bare rock, lies to the south-eastward. Stephens bluff is a remarkable cliffy head, appearing like an island.*

**TIDES.**—It is high water full and change in Port Stephens at 7h. 45m.; the tide rises between 7 and 8 feet.

**DIRECTIONS.**—In making Port Stephens from the north-westward, the Castle rock and the outer part of Stephens bluff have a very similar appearance—two masses of rock leading to the southward. When running for the port, vessels may pass close to them, as they are steep-to ; but if there is much swell from the southward, it would be better to open the entrance well out before standing in. It is easily seen between two cliffy points, 60 or 70 feet above the sea. The entrance is 400 yards wide, and there is no danger if the kelp is avoided ; the tide seldoms runs more than 2 knots.

A small rocky islet lies N.E. of Stephens bluff, off which there is kelp, and a rocky bottom, for nearly half-a-mile in an easterly direction ; but by passing Stephens bluff to the eastward, so as to bring the entrance of the port to bear N.N.E., a ship may steer directly in, clear of all danger. Some stalks of kelp may be seen, but they can be avoided by keeping a little more to the eastward.

**ANCHOR INLET**, opposite the entrance, is the best anchorage. By standing across, leaving the two small rocky islets on the starboard-hand, and keeping to the middle of the inlet, which is about three-quarters of a mile wide, until opposite a sandy beach on its eastern side, a good berth perfectly sheltered, may be taken in 12 or 14 fathoms, sticky mud. Some kelp lies off the eastern point of the inlet, but the water is deep to the edge of it. There is also a fair anchorage to the south-east of the rocky islets.

Vessels not wishing to enter the port may find a convenient stopping-place between Stephens bluff and Pea-point. Steer as if intending to enter Port Stephens, and when about a quarter of a mile outside Pea-point haul up to the westward, keeping towards the starboard shore until past the small rocky islet on the port-hand, when you may near Stephens Bluff island, and anchor where most convenient; the bottom is sandy, mixed with mud, and the depth from 6 to 10 fathoms. This is a convenient place for vessels anchoring for a night, as it is easier left than Port-Stephens. In entering it will be necessary to pass through some

---

* *See* Admiralty Plan of Port Stephens, and ports in West Falkland, No. 1,874; scale, *m* = 1·0 inch ; and views on Chart of the Falkland islands, No. 1,354.

straggling stems of kelp, but the water is deep. There is a passage to this anchorage between the north-west part of Stephens Bluff island and the main land, but very narrow and dangerous.

Port Stephens is not easily quitted, excepting with a fair wind, particularly after a breeze from the southward, as then a heavy swell sets in.

A good anchorage may also be found by passing between Cross island and the small islands south-east of it; but it is difficult to leave, as westerly winds blow directly into the entrance, and it is too narrow for a vessel to beat through. The small islands to the south-eastward of it are joined together by rocks and thick tree-kelp through which it is difficult to force a boat.

THE ARCH ISLANDS are remarkably rugged, with upright light-coloured cliffs. They take their name from a natural archway at the western end of the largest island, through which a boat can pass.

Arch Road, the next port to the eastward, is a good anchorage, and much frequented by whalers. It is entered by passing to the westward of the two large islands, between them and a low dark-coloured island. The anchorage is formed entirely by the Arch islands, and is preferred to that of Port Albemarle, from being nearer the open sea, and with a more convenient depth of water. Vessels lie here well sheltered from the swell, but they feel the full force of the wind.

Albemarle Rock, which is a good guide to Port Albemarle, is bold, upright, and about 150 feet high; it is saddle-shaped at the top, and whitened at its sides by innumerable birds which frequent it. It bears from the N.E. end of the large Arch island E. by S. $\frac{3}{4}$ S 1$\frac{1}{2}$ miles, and may be passed on either side at the distance of a quarter of a mile.

PORT ALBEMARLE.—To enter this port vessels may pass on either side of the Arch islands. The water in the road is inconveniently deep, but free from danger; and leads to a good anchorage in Lucas bay in 10 fathoms, to which the largest ship may work up with ease after rounding Lucas reef.*

Chaffers Gullet lies close to the eastward of Lucas point, with deep water branches, which extend several miles inland; but it is too narrow to call for a more detailed description at present.

From the Arch islands to Port Edgar the coast is bold and clear, and the tide does not exceed 2 knots.

PORT EDGAR is easily known after making Cape Meredith or the Arch islands; and it is the nearest opening westward of Fox bay, about 5 miles

---

* *See* Admiralty Plan of Port Albemarle, No. 1,874; scale, m = 1·0 inch.

distant. The entrance is between two bluff heads, very narrow, and, with the wind northerly, very difficult; indeed, it would be almost impossible for a square-rigged vessel to enter at such a time without warping.

**Wood shoal.**—In going to Port Edgar, it would at all times be advisable to make the land well to the westward, as Wood shoal is only 9 miles S.E. by E. from the heads; it extends 3 miles in an east and west direction, and in some places is a mile wide. The soundings to the northward of it are regular, 15 and 16 fathoms, with straggling kelp about a quarter of a mile off. The reef is well marked by thick kelp, and the least water found was 11 fathoms; but having been seen to break during southerly gales, it may be much shoaler in some places; and as thick patches of kelp always indicate danger, it had better be carefully avoided.

**TIDES.**—There is but little tide at the entrance to Port Edgar. It is high water, full and change, at 7h. 15m.; the rise is 6 feet.

**DIRECTIONS.**—When the wind is between west and south-west, it is very baffling and squally, therefore good way should be kept on the vessel, and the western shore well closed, even to the edge of the kelp, where there are 4 fathoms only a few fathoms from the rocks. By these means a vessel may shoot in past the heads, so as to get the steady breeze; and with fresh way, this may be easily done, as the entrance between the two heads is not much more than half a cable in length, and about a cable broad. When once within the heads, the harbour opens out suddenly. The rocks on both sides the entrance are bold-to; there are from 15 to 17 fathoms in mid-channel.

As a secure harbour, Port Edgar is second to none. It has a great advantage over all the harbours to the northward and westward of it, having a moderate depth of water for some distance outside. When blowing hard from the northward or westward, so as to prevent a vessel from entering the port, an anchor may be dropped in from 15 to 20 fathoms sandy bottom, anywhere under the ridge that forms the western point of entrance, and which extends N.N.W. and S.S.E. about 4 miles. With these winds the water outside is quite smooth; and if they should veer to the southward, so as to make it necessary to weigh, the port will be open to leeward.

Port Edgar should not be quitted after a southerly gale without a commanding breeze, as the swell is heavy without the heads, and continues some time. But northerly winds generally following a breeze from the southward, there is little fear of a vessel being detained.

# CHAPTER V.

## STATEN ISLAND, AND THE OUTER OR SEA COAST OF TIERRA DEL FUEGO.

VARIATION 20° to 23° 30′ EAST in 1860. ANNUAL DECREASE ABOUT 2′.

STATEN ISLAND lies off the south-eastern extremity of the American continent, and, is so deeply indented by bays as to form nearly four different islets. It it 38 miles long in an E.N.E. and W.S.W. direction, and is separated from the mainland by Le Maire strait which is 18 miles wide. It is composed of lofty hills, the peaked summits of which rising to the height of 3,000 feet, are usually covered with snow, and it offers a good departure to ships bound into the Pacific, as well as a landfall when returning. The harbours are the continuation of the valleys, preserving nearly the same direction, and are surrounded by high land, the water in them deepening rapidly towards the centre. The coast consists every where of rocky cliffs elevated from 200 to 500 feet, which have generally from 15 to 20 fathoms water close to their bases.*

TIDES.—It is high water, full and change, in Port Vancouver on the south-eastern side at 4h. 30m., and is very nearly the same in all parts of the island; the rise and fall is from 7 to 9 feet. The velocity of the tides is very great, and as they meet with obstacles from the manner in which the headlands jut out at right angles to the direction of the streams, there is produced, when the wind is strong and contrary, a rough cross-breaking sea which is impassable by a boat, and even dangerous to a ship of considerable size.

There is also reason to believe that the meeting of the two streams of flood tide coming round Cape Horn, and through the various channels of Tierra del Fuego, contributes to the unusual agitation of the sea in the vicinity of Staten island. The flood comes from the eastward along the northern shores of the island, and continues its course southward through the Strait of Le Maire, varying in velocity from 5 to 7 knots. To the southward of the island but little amount of tide is perceptible;

---

* The description of Staten island is by Lieutenant Kendall, by whom it was surveyed, in the year 1828 under the orders of the late Captain Henry Foster, in H.M.S. *Chanticleer*. —*See* the Admiralty Chart of the island, No. 1,332; scale, *m* = 0·6 inch.

there is, however, a remarkable undertow, which renders it dangerous for boats to stretch across the mouths of the deep bays, as it is difficult to close again with the land ; and for this reason the sealers invariably follow the circuitous route of the shores.

The times of high water, full and change, at Cape St. Bartholomew at the south-western end of the island and at Cape Horn are 4h. 30m. and 3h. 50m. respectively, thus showing that the great tide wave of flood comes from the south-west. The flood stream after having passed Cape St. John, follows the direction of the coast of Tierra del Fuego to the westward, where it meets the flood coming down the eastern side of the South American continent, and also that proceeding from the Strait of Magellan, together with those streams that find their passage between the islands of Tierra del Fuego.

**The Harbours** of Staten island are with one exception confined to its northern side : they are named St. John, Port Cook, New Year, Port Basil Hall, Port Parry, Port Hoppner, and Port Vancouver. There are also two or three small bays in the Strait of Le Maire, but they are rendered unsafe by their exposure to the prevalent westerly winds. All these anchorages, though well protected when once gained, are more or less difficult of access from the force with which the tides set across the mouths of the inlets, the depth of water, and variableness of the wind, which in every instance, except that of blowing immediately lengthways of the port, finds its way down the ravines of the mountains in various directions, according to the peculiarity of their form.

**The NEW YEAR ISLANDS** afford a sort of protection from westerly winds, and there is anchorage under the north-easternmost in 17 fathoms ; but it cannot be recommended, being open to the influence of all winds between north and E.S.E., the bottom rocky, and the tides rapid, while the eddies are so uncertain in their distance from the island, that it is by no means easy to keep a ship clear of her anchors.

**ST. JOHN HARBOUR** is the easternmost in the island, and may be easily recognized at a distance by Mount Richardson, at the base of which it is situated. On nearing it, a remarkable cliff, like a painter's muller, appears on the eastern shore, which is high and steep, the ridge over it being 850 feet above the sea. Allowance must be made in steering towards the harbour for the set of the tide, which at all times runs rapidly across its mouth ; it is, however, less sensible when within the headlands that form the north-west bay, in which, in case of necessity, or to await the turn of the tide, an anchor may be dropped in from 20 to 30 fathoms. The mouth of the harbour is wide, having 25 fathoms in the centre, with a rock standing off at some distance from the western point, to which a berth must be given. The shores, with this exception, are bold, and immediately within the west point there is a small bay where anchorage may be had in 10 fathoms.

**Supplies.**—Wood and water are plentiful, and easily procured ; celery and wild fowl (race-horse or steamer duck?, kelp and upland geese) may also be obtained, and in the proper season (October) a good supply of penguins' eggs may be insured by having men in attendance at a *rookery* about a mile to the eastward of the harbour's mouth, whither they could walk along the eastern hills from the vicinity of the Painter's Muller, and remain to collect daily the eggs as they are deposited, and secure them until a favourable opportunity offers of embarking them from the foot of the cliff on which the rookery is established.

**Anchorage.**—The most sheltered situation is at the head of the harbour, distant from the entrance 3 miles S.S.W., where a vessel may choose any depth between 20 and 5 fathoms, with sandy bottom, and moor with an open hawse to the S.W., from whence the gusts that come from the mountains are violent. The wind, anything to the westward of N.N.W., or even N.W. outside, will be found to draw out of the harbour on nearing its head, and if at all strong, it will be impossible to beat farther, as it follows the direction imparted to it by every ravine in the hill as it passes ; and therefore warping will be found the only means of advancing, taking care to have hands by a bower anchor ready to let go, and the cable stopped at a short scope, in the event of the hawsers being carried awry. A ship might readily heave down on a beach of sand at the head of this harbour.

The shores of St. John harbour are lined with kelp, which is an excellent indication of its un-navigable part, its edge being almost invariably in 8 fathoms, and generally near to the shore ; the depth rapidly increases towards the middle of the harbour, until near its head, where the soundings gradually decrease to the beach.

**PORT COOK** is the most eligible harbour in Staten island for a ship in want of shelter, from the good anchorage at its entrance not being in too deep water, the greater regularity of the prevailing winds, and the facility of communication with the south side of the island, by means of the low isthmus separating it from Port Vancouver. The mouth, however, is narrow, with a small island in it ; there is a rock awash lying nearly a mile north from Wales point at the east side of the entrance.

**PORT BASIL HALL** is separated by a narrow neck of elevated land from the head of New Year harbour, though their mouths are 2¼ miles apart. The former is a most convenient anchorage when once attained, and well sheltered from all winds, though the depth of water, and contraction of its mouth by two detached rocks, render it difficult of entrance without a commanding breeze and favourable tide. There are three rocky patches nearly in the middle of the space within the entrance, but all the danger is pointed out by the kelp, and by the tops of the rocks which show at half-tide : they must be passed on their eastern side, where there is a space sufficiently wide to admit of beating in moderate weather.

**Supplies.**—Wood and water are abundant, and a fine sandy beach abreast of the anchorage affords facility for a small vessel in want of repair. Fish of a large size were seen among the kelp, and might be caught by a hook and line, for the patches of seaweed near the beach would prevent the use of the seine.

**Anchorage.**—The best anchorage is between a small green island on the western shore, and a fine sandy beach to the northward of the island ; it gives from 7 to 10 fathoms, and the island may be passed on either side, taking care to avoid a rocky ledge that runs off its south-east extremity. Although the whirlwinds common to all the harbours of Staten island are found to exist here, yet the comparative lowness of the south-western shore renders them less violent than in many of the others ; and the New Year islands afford some protection to the harbour from the sea, by stretching, though at a distance, across its mouth.

There is anchorage on the eastern side also of the bay, just within the entrance, but it cannot be compared with the other, being exposed to the swell that rolls in from seaward, and open to the influence of westerly, the prevailing winds. The best directions for entering the port are, after having passed the rocks in its mouth, to steer for a remarkable peaked hill, the easternmost of two hills near the head of the harbour, until another peak on the western shore, which stands by itself, comes into view, after which the course may be shaped so as to pass the island on its northern side, where the best anchorage will be found.

**PORT PARRY.**—The entrance of this port may easily be distinguished by its being the first opening to the westward of New Year islands, and by Mount Buckland on its eastern side, the quoin-shaped appearance of which renders it a very remarkable object. There are detached rocky islets off both the points of entrance, but they are bold and steep, and there is no danger in approaching them.

The harbour itself is divided into two parts by the near approach of its opposite shores forming a gorge or throat about 2¼ miles from the entrance, after which it again expands, affording a well-sheltered and secure anchorage. The depth of water in the gorge of the inner harbour is 8 fathoms, and the breadth about 150 feet ; and after having passed it, the eastern point of the outer entrance should be brought in one with the eastern point of the gorge, in order to lead clear of two rocky patches that lie on either side within it, and are pointed out by the kelp growing on them.

To this harbour the sealing vessels are accustomed to resort when in want of repairs, or as a place of shelter while their boats are absent in the pursuit of furs ; in this respect it is, however, very inferior in point of convenience to Port Cook, for though the distance across the island from the head of the harbour hardly exceeds the breadth of the isthmus at Port Cook, and there is a boat harbour on the southern shore nearly opposite, yet the elevation of the land is too great to admit of a communication being kept up without two separate establishments. From the height of the land also the squalls are frequent and violent, though not sufficient to cause apprehension to a ship well moored.

**DIRECTIONS.**—After entering the outer harbour, there is no bottom with 30 fathoms, until near the gorge ; and the western shore should be kept on board if the intention be to anchor in the western arm of the outer harbour. If bound into the inner harbour, the eastern shore should be hugged tolerably close until abreast of the white ravine. Unless with a leading wind, it would be impossible to sail through the gorge. If the wind is found to baffle on nearing it, the sail must be taken off the ship and preparations made for towing or warping ; indeed it would be perhaps better in all cases to adopt those precautions, as the height of the surrounding land frequently causes flaws that might, in so narrow a channel, drive the ship on shore before the sails could be trimmed.

**Anchorage.**—There is good anchorage in 9 or 10 fathoms to the southward of a small grassy island on the eastern shore of the inner harbour, with sandy bottom, after which the water again deepens to the head of the harbour, where a vessel may be moored with an open hawse to the S.S.W., opposite to a sandy beach—the place in which observations for position were made.

The deepest water is on the western shore, more particularly abreast of a perpendicular cliff opposite to the small island before mentioned.

There is a rocky patch having 4 fathoms on it nearly in the centre of the outer port, and as the water gradually shoals up it, a vessel might find there a temporary anchorage. The depth of water in the western arm is an objection to it as a place of long continuance for a ship ; but the bottom is good, and both wood and water may be procured, though the swell that rolls in on the rocks would be very destructive to the boats on landing ; there is, however, a convenient place for the employment of a hose for watering. The hawse should be open to the westward, in the event of mooring.

**PORT HOPPNER** is the next opening to the eastward of Cape St. Anthony, and is separated from Port Parry by a peninsula of 2 miles. This harbour is also divided into two parts, like Port Parry; it has a high rocky island in its mouth, by which it is protected from N.W. winds, and within which anchorage may be taken for the space of half a mile in from 20 to 8 fathoms. To the south-east of this island, however, there is a continuation of rocky patches, two pinnacled rocks of which peep above the water's edge, and are surrounded by abundance of kelp.

Near the head of the outer harbour there is also anchorage close to the shore, which on the western side is high and precipitous. The gullet is not more than from 25 to 30 yards broad, with from 2 to 4 fathoms in it; yet the sealers sometimes haul their small schooners through, into the inner basin, where they are completely secure from all winds in from 20 to 6 fathoms, and moor by hawsers to the rocks. The tide rushes through the narrow opening with great velocity; it would be prudent therefore not to commence warping in until the last quarter flood.

**Water.**—A tolerably full stream falls into the head of the inner basin from the mountains by which the whole of the harbour is surrounded. In the outer harbour there are no soundings in the middle, with 40 fathoms of line. Wood and water are here tolerably plentiful; but the harbour is by no means so convenient or so secure as either of those previously mentioned.

**The COAST** on the west side of Staten island forms the eastern side of Strait Le Maire; it is high, rugged, and deeply indented by Flinders, Crossley, and Franklin bays. Anchorage may be found in the small coves on the eastern side of the two former; they are, however, both open to the prevalent winds, unprotected from the sea, and can by no means be recommended.

**Tide Rips.**—Heavy tide rips are found off Capes St. Anthony, Middle, South, and St. Bartholomew, which form the high projecting horns of the above named bays; they extend 5 or 6 miles to seaward, the flood setting to the southward with a velocity of 5 to 7 knots. A reef is said to exist in the rip just north of Cape South, distant 3 miles from the coast.

**PORT VANCOUVER.**—On the south side of Staten island, the only well-sheltered harbour is Port Vancouver, immediately opposite to Port Cook, from which it is separated by a low isthmus. A vessel may ride in security there in 16 or 17 fathoms, sandy bottom, close to a rivulet, and near a convenient wooding-place in the western arm. There is a rocky island on the eastern side of the entrance, which is always visible, and a reef extends some distance from the south point of the western arm, which may be avoided by not hauling in for the anchorage until a remarkable white ravine on the south shore becomes visible; these are the only dangers in this harbour, while it affords a free communication with Port Cook and the other side of the island by means of the isthmus, and possesses the important advantage of affording a place of refuge during a S.W. gale. It may easily be recognized from seaward by its being the first opening to the eastward of Dampier islands, from the southernmost of which it is distant 4½ miles. A ship must moor with an open hawse to the westward.

**BACK HARBOUR** cannot be recommended, on account of the heavy rolling sea that sets into it with S.W. winds, the scarcity of wood, and difficulty of procuring water, in consequence of the surf that breaks on the beach. The holding-ground, however, is good, and small vessels go in there occasionally in N.W. winds.

**NATURAL HISTORY**—The otter, the rat, and the mouse are the only quadrupeds on Staten island. Birds are more numerous, and comprise three species of penguins, gulls, albatrosses, and shags, of which the crested shag form extensive flocks, building their nests in the loftiest trees on the hills. The black oyster-catcher, the Johnny-rook, the mountain hawk, a small owl, a few thrushes, two species of linnet, the humming-bird, occasionally ducks, and three species of geese, form the ornithological catalogue of the island. The loggerheaded duck, the *racer* of old voyagers but now called the *steamer*, is the finest bird for the table in these regions.

**GEOLOGY.**—The geological structure of Staten island is chiefly of quartz-rock, greywacke, clay-slate, and micaceous schist. The greywacke and quartz-rock are intermixed and alternate, forming the principal mountain hills of the island, and unstratified. The quartz in many parts is disposed in vertical veins.

**VEGETATION.**—With the most boisterous and humid climate on the globe, and with a low but very uniform temperature, vegetation flourishes with such surprising beauty and luxuriance, that the rugged aspect which the island appears to wear at a distance is changed into perpetual and unbroken verdure. Every spot is clothed with plants, the hills are crowned with evergreens; and every season finds them much the same. The antarctic beech is the common and prevailing tree. The name is neither appropriate nor distinctive, as there is another beech of these regions. As the antarctic beech is an elegant evergreen, might it not be called the *fagus sempervirens?* The tree in its young state is handsome and elegant. It grows to the height of thirty or forty feet, with a girth of from three to five feet; and sometimes, doubling these dimensions, it forms majestic trees.

The next tree to the beeches, both in frequency and in size, is the *winterana aromatica*, an evergreen with a complete laurel aspect, attaining sometimes to a very considerable magnitude, even to that of twenty feet in circumference. Its general height is eight or ten feet, and girth small. The *arbutus aculeata*, or the arbutus with sharp pointed leaf, is the pride of these regions. It is an elegant and most pleasing evergreen, with so much of the appearance of a fine myrtle, as generally to obtain from the seamen the name of the myrtle bush. It stands from three to four feet in height, leaves small and terminated by a prickle, from whence its name. The *dactylis glomerata*, or tussac grass, grows in very large mounds or tufts; the stems of it were as big round as the middle finger, and the lower parts being blanched and having a sweet flavour, were boiled in imitation of asparagus, but it was too stringy. The ground was thickly covered with a creeping plant like a strawberry with small bright red coral berries, each containing one small seed. The *rubus geoides,* or strawberry plant, produces a very pleasant and agreeable berry. A small cranberry plant, with an insipid berry, was abundant in some districts.

**SEA-WEEDS** entangle the harbours and shores. The sea teems with them, especially in the rough and open bays, while they are comparatively rare in the still sequestered creeks. They are gigantic in form, some being 300 feet in length—some of singular strength and sturdy stems, and becoming as it were oaks in the sea; others again spread their tough and leathery substance, like hides, so that buckets, bowls, and cups, may readily be made of it. Some form ropes and cables to moor boats with; some yield a jelly pure and tasteless, like isinglass, far exceeding that of the *fucus crispus*, or carrageen moss of our shores; and one pretty tinted green fucus was very acid, being the only species, I am aware of, that has the property.

---

**TIERRA DEL FUEGO**, the archipelago which constitutes the southern extremity of America, consists of one large island, four others of a moderate extent, and a great number of smaller islands and rocks. The shores are intersected by deep but narrow arms of the sea, on the sides of which rise the mountains, the summits of which for the greatest part of the year are covered with snow, while their steep and rocky declivities are partially overgrown with evergreens. The natives of this archipelago are low in stature, and live in a barbarous condition; having frequently no other

covering than a scrap of hide, which is tied to their waists. They have
no government, and the neighbouring tribes, who speak different dialects,
are almost always hostile. Cannibalism, it is said, is practised. They
never cultivate the soil; but, occupying only the sea-shore, live chiefly
on shell-fish. Almost the only vegetable production which they eat is a
peculiar fungus (*Cytharia Darwinii*) which grows on the beaches.*

**The COAST** from Cape Horn to Cape Pillar is very irregular and
much broken; being, in fact, composed of a large number of islands.
It is generally high, bold, and free from shoals or banks; but there
are many rocks nearly level with the surface of the water, distant 2 and
even 3 miles from the nearest shore, which make it very unsafe for a
vessel to approach nearer than 5 miles, except in daylight and clear
weather. The coast varies in height from 800 to 1,500 feet above the
sea. Farther inshore there are ranges of mountains always covered with
snow, their heights being from 2,000 to 4,000 feet, and in one instance,
Mount Sarmiento, 6,800 feet.

**Rocks buoyed by Kelp.**—With daylight and clear weather a vessel
may close the shore without risk, because the water is invariably deep,
and no rock is found which is not so marked by sea-weed (or kelp, as it
is generally called), that by a good look-out at the masthead, its situation
is as clearly seen as if it were buoyed. By avoiding kelp you are sure of
having sufficient water for the largest ships on any part of this coast. At
the same time it must be remembered that kelp grows in some places
from a depth of 30 fathoms, and that on many parts of this coast you may
pass through thick beds of sea-weed without having less than 6 fathoms
water; still it is always a sign of danger, and until the spot where it
grows has been carefully sounded, it is not safe to pass over it with a ship.
As an instance :—After sounding a large bed of this weed in one of the
*Beagle's* boats, and thinking it might be safely crossed, a rock was found,
not more than four feet in diameter, and having only one fathom of
water over it.

**ASPECT.**—Viewing the coast from a distance, it appears high,
rugged, covered with snow, and continuous, as if there were no islands;
but, when near, one sees many inlets which intersect the land in every
direction, and open into large gulfs or sounds behind the seaward islands.
The high land, covered with permanent snow, now disappears, and the
hills close to the sea will be seen thickly wooded towards the east, though
barren on their western sides, owing to the prevailing wind. These
hills are seldom covered with snow, because the sea winds and the rain
melt it soon after it falls. A range of high mountains continues uninterruptedly from the Strait of Le Maire to the Barbara channel. Mount

* Tierra del Fuego was discovered and named by Magalhaens in 1520.

Sarmiento, 6,800 feet above the sea, is in this range. Southward of these mountains is a long extent of broken land, intersected by passages or large sounds. A boat can go from the eastern entrance of the Beagle channel to the Week islands without being once exposed to the outside coast, or to the sea which is there found.

ANCHORAGES.—Opposite to the eastern valleys, where the land is covered with wood, and water is seen falling down the ravines, good anchorage is generally found. But these valleys are exposed to tremendous squalls, which come from the heights. The best of all anchorages on this coast is where you find good ground on the western side of high land, and are protected from the sea by low islands. It never blows nearly so hard against high land as from it, but the sea on the weather side is of course too formidable, unless stopped by some barrier, cape, or islet.

Where the land is chiefly composed of sandstone or slate, anchorages abound; where of granite, it is difficult to strike soundings.

The difference between granite, and slate and sandstone hills, can be distinguished by the former being very barren and rugged, and of a grey or white appearance; whereas the latter are generally covered with vegetation, are dark-coloured, and have smoother outlines. These slates or sandstone hills show few peaks, and the only rugged places are those exposed to wind or sea.

SOUNDINGS extend to 30 miles from the coast. Between 10 and 20 miles from the land the depth of water varies from 60 to 200 fathoms, the bottom almost everywhere a fine white or speckled sand. From 10 to 5 miles distance the average depth is 50 fathoms; though varying from 30 to 100, and in some places no ground with 200 fathoms of line. At less than 5 miles from the shore the soundings are very irregular, generally less than 40 fathoms, but in some places deepening suddenly to 100 or more: while in others a solitary rock rises nearly to the surface of the water. If desirous of entering an inlet, after carrying 50, 40, 30, or 20 fathoms, towards it, possibly the water will suddenly deepen to 60 or 100 fathoms at the entrance, and in the large sounds, behind the seaward islands, the water is considerably deeper than on the outside.

Outer Bank.—A bank of soundings extends along the whole coast, 20 or 30 miles in breadth, and which appears to have been formed by the continued action of the sea upon the shore, wearing it away and forming a bank with its sand. There is in fact, much less risk in approaching this coast than is generally supposed. Being high and bold, without sandbanks or shoals, its position accurately determined, and a bank of soundings extending 20 or 30 miles from the shore, it cannot be much feared. Rocks, it is true, abound near the land, but they are very near to the shore, and out of a ship's way.

Between the islands where there is no swell or surf worth notice, the
water is deep, and the bottom very irregular.  East of Cape Horn the
water is not so deep as to the westward, neither is the land so high.
A small vessel may run among the islands in many places, and find good
anchorage ; but she runs into a labyrinth, from which her escape may
be difficult, and, in thick weather, dangerous.

CATHERINE POINT is the north-eastern extremity of Tierra del
Fuego ; it is formed of shingle, very low, and precisely similar to Dungeness at 18 miles distance upon the opposite coast of Patagonia. Between
Catherine point and Cape Orange, 26 miles to the westward, there is a
large bay, called by Sarmiento, Lomas bay.  The land around it is very
low, and the space which, in the chart, appears to be water, is chiefly
occupied by extensive shoals ; some of them are visible at low water.

Before describing Magellan strait it is proposed to follow the outer coasts
of Tierra del Fuego and lead the mariner round Cape Horn to Cape
Pillar returning to the description of Magellan strait in chapter VI.*

CAPE ESPIRITU SANTO, 8 miles to the S.E. of Catherine point is a
steep white cliff, 190 feet high, somewhat resembling the gable end of a
large but low barn ; it forms the seaward termination of a range of rather
high land, varying from 200 to 600 feet, lying nearly east and west, and
corresponding somewhat in height and position to the opposite range
in Patagonia which ends at Cape Virgins, but it is not so flat in outline.
A reef of rocks lies 2 miles off the cape.

NOMBRE HEAD.—From Cape Espiritu Santo, cliffs from 100 to 300
feet in height, extend, but with few breaks, about 23 miles south-east to
Nombre head : the land is 300 or 400 feet high, irregularly rounded in
outline, and quite destitute of wood.  From thence a low shingle beach,
or rather a long ridge of shingle, forming a spit, extends 9 miles, ending
in a narrow steep point, called Arenas point.  Westward of this point,
and between it and Cape San Sebastian, 10 miles farther south, there is
a spacious harbour, secure from all but easterly winds, which seldom blow
home ; or with any strength.  Coasting along this shingle spit, the depth
is not more than 10 fathoms, but it deepens suddenly near the south-east
extremity.  Within the shingle point, which is steep-to, or nearly so, the
bottom is uniform, but the depth gradually decreases.

SAN SEBASTIAN BAY is an excellent anchorage, as respects shelter,
good bottom, and ease of access, but without wood, or a good watering

* *See* Charts of Magellan strait, Nos. 554 and 1,316 ; scales, m = 0·13 and 0·2 of an
inch.

place, though water may be procured.    There is no hidden danger on the north side of the bay, the shingle is steep-to, the shores of the bay shoal gradually, the bottom is clean, and the soundings are regular.    On the south side, off Cape San Sebastian, it is otherwise ; a shoal rocky ledge extends under water to the north-eastward, and requires a berth of 3 miles: there is no kelp upon it.    On its edge, the water shoals suddenly from 12 to 4 fathoms ; the ebb tide sets rather strongly over it, about 2 knots; the bottom is hard, and affords bad holding.

**CAPE SAN SEBASTIAN** is a bold cliffy headland, of a dark colour ; inshore of it the land rises to near 1,000 feet above the sea, and becomes more irregularly hilly.    From Cape San Sebastian a short range of cliff extends ; then low land ; and then another small cliff, off which there is a rock above water, about a mile off shore.

**CAPE SUNDAY** is a prominent headland, of a reddish colour, rising 250 feet above the sea ; the shores near it are free from danger until near Cape Peñas, near which are some dangerous rocks.    Between Cape San Sebastian and this cape, a distance of 22 miles, the shore is rather low, irregularly hilly, and fronted by a shingle beach.

**CAPE PEÑAS** is not more than 100 feet above the sea : around it, to a distance of 2 miles, there are dangerous rocks ; the sea generally, if not always, breaks upon them; but they should be carefully avoided, especially at night.    The bay lying to the southward of Cape Peñas appears to afford anchorage ; but the appearance is deceitful, as it is shallow and strewed with rocks.    The hills hereabout are higher and partially wooded, and the view of the country is pleasing.*

**CAPES SANTA INEZ, MEDIO, and SAN PABLO** are high and bold; they are fronted by steep cliffs, 200 or 300 feet in height.    Hence to Cape San Diego there is no outlying danger ; the water is rather deep near the shore, but not so deep as to prevent a ship anchoring during westerly or southerly winds.

**The TABLE of OROZCO,** lying 4 miles inland, is a remarkable table-topped hill, about 1,000 feet above the sea.    Between it and Cape San Diego there are three remarkable hills, called the Three Brothers, and the westernmost of these hills is very like the Table of Orozco: they are from 1,000 to 1,400 feet in height.    Vessels bound through Strait Le Maire should make these peaks.

---

* *See* Chart of the south-eastern part of Tierra del Fuego, No. 1,373; scale, m = 0·13 of an inch.

**POLICARPO COVE** is a deceiving place, it looks like a harbour, but is fit only for a boat. False cove will hardly give a boat shelter. Cape San Vicente is a dark-looking low bluff point, backed by woody hills, 200 or 300 feet in height. Some rocks and foul ground extend half-a-mile from the point of the cape.

**THETIS BAY,** lying between Cape San Vicente and Cape San Diego is an anchorage which might be taken by a vessel intending to go through the Strait of Le Maire, but detained by wind or tide; the bottom is mixed, rocky in some places, in others sandy, with mud and stones; and the tide sets strongly from 1 to 3 knots across the bay, in a line between the heads. There is much kelp in the bay, but no dangers could be discovered amongst it. When the wind opposes or crosses the tide, a great sea rolls into this bay, which can only be recommended as an anchorage fit for remaining at during a few hours. In leaving it to pass Cape San Diego, a considerable offing must be obtained in order to pass round the very heavy tide-race which extends from the cape to a distance of 3 miles into Le Maire strait, and on no account should a small vessel go through it.

**CAPE SAN DIEGO** is low, with a smooth outline ending in a small bluff. The rocky ledge, extending to the eastward from this cape, is more dangerous than has been supposed. So violent was the race of tide upon it, as to prevent soundings being taken so frequently and accurately as was desirable; in the middle of the race, they were found to vary suddenly from 60 and 70 to 9 and 5 fathoms; it will be, therefore, but prudent to give this cape a berth of at least 3 miles. This race is at times very dangerous; a vessel was once seen to founder in it, but whether from striking on a rock, or from being swamped, was not known. Along this north-east coast of Tierra del Fuego the soundings are regular, and the bottom in most places clean and fit for anchorage. During westerly and southerly winds, a vessel may lie at anchor under the lee of the land very conveniently; but a northerly or easterly wind sends a heavy swell upon the shore.

**TIDES.**—Near the Strait of Magellan, as well as at Cape San Diego, the stream of tide is much felt, but there is very little along the intermediate coast. From Cape San Diego to the northward the flood tide sets north and west along the shore from 1 to 3 knots. The ebb sets in a contrary direction, but not so strongly.

**GOOD SUCCESS BAY** is 7 miles south of Cape San Diego. It is a good anchorage, perfectly safe, provided that a vessel does not anchor too far in towards the sandy beach at its head; for, during south-east gales, a heavy swell with dangerous rollers sets right into the bay. The best berth is shown in the plan. Heights, of about 1,200 feet above the sea, surround the bay; therefore with strong winds it is subject to squalls, which during westerly gales are very violent. It is an excellent anchorage for vessels of any size to stop in to get wood or water, but it would not answer if a vessel required to lie steady for repairs, as a swell frequently sets in. It is quite safe; but in the winter season, when easterly winds are common, no vessel should anchor so near the head of the bay as she might in summer. The "Broad Road," mentioned by Cook, is a good mark for the bay, if the inbend of the land does not sufficiently point out its situation. It is a barren strip of land on the height outside the harbour. Cape Good Success is high and bluff; some rocks lie close to it above water.* †

TIDES.—It is high water, full and change, on the shore in Good Success bay, at 4 hours, and slack water in the strait. The tide slacks in the offing at 10 in the morning. The rise of tide is from 6 to 8 feet, according to the wind. In Le Maire strait the flood tide runs from 2 to 4 knots near Cape San Diego, and from 1 to 3 in mid-channel more or less, according to the strength and direction of the wind. The tides in the strait are regular, and will assist a vessel materially in her passage, if taken at the right time.

STRAIT of LE MAIRE.—The soundings in this strait are regular near their southern entrance, 70 to 30 fathoms over a sandy bottom; towards the north the depths decrease; and 2 miles from Cape San Diego there are not more than 30 fathoms water over a rocky bottom. The eastern side of the strait, already noticed (see page 127), is formed by the irregular bays and rugged capes of Staten island : surrounding the latter are heavy tide rips, which extend outward to a considerable distance; in the one off Cape South a reef has been reported. As the strait is wide, free from obstacles of any kind, the tide rips excepted, and the soundings regular, vessels may pass through without difficulty or risk, Good Success bay being close at hand, in case the wind or tide should change. Neither Valentyn bay, Aguirre bay, nor Spaniard harbour is fit for more than temporary anchorage during northerly or westerly winds, all being much exposed to the south. The chart is a sufficient guide for that purpose.

The Bell Mountain situated between Valentyn and Aguirre bays, is remarkable : it is seen far at sea, from the northward as well as from the southward; it is high, and in shape resembles a large bell.

---

* See Plan of Good Success bay and Lennox harbour, No. 1,376 ; scale, m = 3 inches.
† See View on No. 1,373.

**LENNOX and NEW ISLANDS** lie S.W. ¼ W. distant 41 miles from Cape Good Success. These islands, and indeed any part of the coast in their vicinity, may be approached with confidence, using the lead and looking out for kelp. Richmond road, the first good anchorage after leaving the Strait Le Maire, lies between Lennox and New islands, and good temporary anchorage, during westerly winds, may be obtained under the latter, or near the shore to the northward. At the east side of Lennox island there is excellent anchorage; small vessels may go into a cove, where the *Beagle* lay moored, but large ships must anchor in the road, which is quite secure, and sheltered from all but south-east winds.[*]

**BEAGLE CHANNEL** is a narrow passage running about W.S.W. for 120 miles, in nearly a direct line between ranges of mountains, always covered with snow; the highest being between 3,000 and 4,000 feet above the sea. Its eastern entrance lies to the north-west of Lennox and New islands, on either side of Picton island.

This channel averages 1¼ miles in breadth, and in general has deep water ; but in it there are many islets with rocks off them. Although easy of access it is useless to a ship, but boats may profit by its straight course and smooth water. At 45 miles from Picton island is the first opening to the southward, leading into Ponsonby sound ; 27 miles farther the channel divides, the south-west arm falling into Cook bay, the north-west into Darwin sound, and thence through Whale-boat sound, and Desolate bay to the Pacific. The tide sets about a knot an hour through the Beagle channel, the flood to the eastward and ebb to the westward.

**GOREE ROAD,** on the west side of Lennox island, is an excellent place for ships, very easily entered or quitted, and able to furnish wood and water with little trouble. It should be particularly remarked here that the kelp in Goree road, as well as that which extends out from Guanaco point partly across the entrance to the Road, does not, as far as we have been able to discover, grow upon rock but upon loose detached stones, and need not, therefore, be so much avoided.[*]

**NASSAU BAY.**—Nassau bay extends to the north and north-west into the Beagle channel, through Ponsonby sound. It is very accessible and free from dangers. Anchorage may be found on each coast, and the only dangers are some rocks (or islets) above water, shown in the chart, and visible at a distance by daylight. The northern shore is low, particularly towards Guanaco point, where the coast changes its level land, and low earthy cliffs, for rocky heights.

---

[*] *See* Plans of Lennox cove and Goree road, Nos. 1,376 and 1,321 ; scale, m = 2·0 inches.

**TERHALTEN and SESAMBRE** are two small but high islands at the entrance of Nassau bay ; they lie 8 miles S.E. by E. from Guanaco point, off Sesambre, the southern one, there is a reef marked by kelp.  If bound to the westward of Cape Horn, it might be preferable to work through this bay, and stand out from False Cape Horn, instead of making westing in the open sea, as is usually done.  There are no dangers but those which are shown in the chart ; the water is comparatively smooth, and an anchorage may be taken at night ; for this purpose Goree road, or North road, or Orange bay may be chosen.  When it blows too hard to make any way to windward, it is at least some satisfaction, by lying quiet, to save wear and tear, and to maintain one's position, instead of being drifted to leeward, and perhaps damaged by the sea in the offing.  There is less current through the bay than in the offing near Cape Horn.

**CAUTION.**—In Nassau bay the compasses are much affected ; they become very sluggish, and might cause serious errors if not carefully attended to constantly.  During the survey of the *Beagle* the magnetic needle was remarkably affected on many of the neighbouring islands, although no great difference was observed when on board the ship.  On one occasion, on ascending the summit of Maxwell island, Franklin sound (between the Hermite islands), the compass was placed for convenience upon the rock, where the needle was found to be so much influenced by its ferruginous nature, being composed of quartz with large and numerous crystals of hornblende, that its poles became exactly reversed.  And afterwards, on taking a set of bearings of a distant object, from several stations, at fifty yards from the above magnetic rock, the extreme difference amounted to 127°.  The block upon which the compass was placed in the first instance is now in the museum of the Geological Society.  No sensible difference, however, was found in the valley at the bottom of St. Martin cove, where the variation of the compass was observed by several different instruments, and compared with astronomical bearings, and the needle was found not to be locally affected.

**WOLLASTON ISLANDS.**—The northern shores of these islands afford several anchorages, which are mostly free from outlying or hidden dangers.  The most eastern are Scourfield and Hately bays, but the water is deep for anchorage ; 4 miles north from Hately bay is Middle cove, which, though small, is secure, but when it blows, the squalls from the high land around are furious.*

**Gretton Bay** is an extensive bay formed to the west of Cape de Ros, the north head of Middle cove.  It is open to the north-east, but it has a

* *See* Plan of anchorages in Wollaston island, No. 1,385 ; scale, m = 2·0 inches.

convenient depth for anchorage. Mr. W. P. Snow, Commander of the
Mission yacht *Alan Gardiner*, explored the opening in the bight of this
bay, and found that it led into a harbour about 5 miles long, by 2 broad,
with a narrow but deep passage on its southern shore into Franklin sound,
thus dividing Wollaston into two islands. The flood tide sets to the
northward through this channel. North road, in the northern part of
Gutton bay, is easy of access, quite sufficiently sheltered, and a very good
stopping-place for a vessel working through Nassau bay.

**DÆDALUS ROCK.**—This dangerous rock, discovered in 1850 by
H.M.S. *Dædalus*, lies N.N.E. three-quarters of a mile from Dædalus
island, in the north part of North road. It is not larger than a ship's cutter,
having only 3 feet water on it, with 7 fathoms within 20 yards all
around, and 15 to 11 between the rock and Dædalus island. It is buoyed
by kelp. Franklin sound, lying between the Wollaston and Hermite
islands, is clear of obstruction, and has no other dangers than those
which are shown in the chart.

**The HERMITE, or CAPE HORN ISLANDS** are composed of greenstone, in which the hornblende and felspar are more or less conspicuous,
and the presence of iron very apparent, as already stated. Their shores
are bold ; and the mountains are peaked, rising with a steep ascent to an
elevation of from 1,000 to 1,700 feet above the level of the sea ; and,
being thickly clothed to within 200 or 300 feet of their summits with
different sorts of shrubs and evergreen trees, render them difficult of
access.

This group consists of four large, with several small islands, and some
outlying rocks ; Hermite island, the largest, is high and rugged at the
east end, but sloping down towards West cape which is low. Kater peak
at the back of St. Martin's cove is 1,740 feet above the sea ; there is
also a ridge on Herschel island. The passages between these islands are
deep, and free from dangers ; what few rocks there are show themselves
above water, or are thickly covered with kelp. Some rocks lie off Chanticleer island, at the entrance of St. Martin's cove, but they are too close
to be of much consideration. No dangers exist to the southward ; in
approaching these islands they may be closed without hesitation.*

**SUPPLIES.**—Wood and water are in abundance in every part of St.
Martin cove, but cannot always be procured, from the steepness of the
shores, and the heavy swell that sometimes sets in. The water is highly
coloured by the vegetable matter through which it percolates, but we found
no other inconvenience from its use than that of giving to our tea a deeper

* The remarks on the Hermite islands are by Captain Foster, of H.M.S. *Chanticleer*.

colour, and a somewhat unpleasant taste. The wood was very much twisted and stunted in growth, and did not appear fit for any other purpose than fuel.

The shores of St. Martin cove are skirted with kelp, which serves to protect the boats in landing, and amongst which fish are to be caught with a hook and a line abreast of the rills of fresh water that discharge themselves into the sea. It was from the Indians that we obtained a knowledge of this most valuable supply, by observing them in the act of fishing, which they manage ingeniously : to the end of the line they fasten a limpet, which the fish eagerly swallows, and not being able whilst in the water to disgorge it, is thereby drawn to the surface, and taken by the hand. In this manner they have been known to catch several dozen in the course of a few hours. Our people were immediately furnished with hooks and lines, and on favourable occasions (for it was found that with a swell setting into the cove the fish did not so readily take the bait) they would bring on board a sufficiency for the supply of three or four messes.

At the head of the cove, and a few feet beyond the reach of high-water spring-tides, abundance of celery is to be found, as also in many other places in the cove. During the whole time of our stay here (viz., two months), which was at the latter end of the autumnal season, a sufficiency was daily procured for the use of the ship's company, and although not of so luxuriant a growth as we found it in December, it was nevertheless considered wholesome.

PORT MAXWELL is a secure anchorage, and untroubled by mountain squalls (or williwaws), but it is rather out of the way. Though it has four openings, only two are fit for vessels—those to the north and east. The best berth in it has 16 fathoms water, over a clear, sandy bottom. This harbour is decidedly good, though it requires a little more time and trouble in the approach.

St. MARTIN COVE on the east side of Hermite island bears from Cape Horn W. by N. ½ N., distant about 10 miles, and is further distinguishable by Chanticleer island, that lies about a mile N.E. from South head. In this direction there are no dangers but what show themselves, and the cove is of easy access with N.E., E., or S.E., winds ; but with the westerly winds that prevail here it is quite the reverse, and ships then should anchor off the entrance in about 22 fathoms, and warp into the cove, where there is a convenient berth in 18 fathoms, sandy bottom, midway from either side, and about half a mile from the head of the cove. This anchorage is safe, although the gusts of wind in westerly gales (which are of frequent occurrence at all seasons of the year) rush down the sides of the mountains in various directions with impetuous violence, and may be very properly called hurricane squalls. They strike the ship aloft, and have more the effect of heeling the vessel than of bringing a strain upon the anchors, which, when once imbedded in the sandy bottom, hold remarkably well, and will cost a heavy heave in weighing.*

**DECEIT ISLAND** is the eastern one of the group, and from the eastward its appearance is not unlike Diego Ramirez. Off Cape Deceit, its south-east point, are several rocks, and 2 miles south-east of the cape there is a cluster of pinnacle rocks rising 30 or 40 feet above the sea. St. Francis bay divides the Cape Horn island from the rest of the group. A strong current sets, at times, along the outer coast of the Hermite islands and through the Bay of St. Francis, varying from half a knot to two knots, according to the wind and tide; and in the bay changes its direction with the change of tide.

**CAPE HORN** is the southernmost point of the Hermite islands. There is nothing very striking in the appearance of this promontory, as seen from a distance; but in passing near it is more remarkable, showing high, black cliffs towards the south; it is about 500 feet above the sea. One mile to the westward of the cape there are three rocks, generally above water; the sea always breaks on them. Also, off the east point of Horn island, there are some small rocks and breakers, but all above water.

**TIDES.**—It is high water, full and change, in St. Martin cove at 3h. 50m. and off Cape Horn at 3h. 30m; rise about 8 feet; but the swell that sets into the cove rendered the observations both on the times and on the rise and fall very uncertain. There was no very decided direction observed in the stream of either tide, which was very slack; but it appeared that the flood came in from the southward amongst these islands.

**The CURRENTS** off Cape Horn are as strong as on any part of the coast; but between it and Cape Pillar they are by no means regular; sometimes with a strong wind and flowing tide they run 2 knots or more, at others it is hardly worth notice. During the survey the current was never found to set to the westward at any time of tide, or with any wind.

It appeared that while the water was rising upon the shore, the tide stream sets along shore from the north-west towards the south-east at the rate of a knot, or more, according to the wind. During six

* *See* Plan, St. Martin cove, No. 1,841; scale, m = 3·0 inches.

hours of falling water, or ebb tide, there was little or no current setting along shore.

In the channel between False Cape Horn and the Hermite islands, a current is found setting into Nassau bay, and rather towards the Hermite islands, at the rate of 2 knots, with the flood tide, and about half a knot with the ebb. As this current sets rather towards West cape, a good berth must be given to it in passing.

THE BARNEVELT ISLANDS lie 11 miles N.E. by E. from Cape Deceit. The chart and sketch are a sufficient description. For the Evouts isles, and for the appearance of this part of the coast from Cape Horn to Cape Good Success, the mariner is referred also to the chart and the accompanying views. Navarin and Hoste are two large islands forming the south shore of the Beagle channel; they are separated by Ponsonby sound, which opens into Nassau bay.

HARDY PENINSULA, a projection of Hoste island, affords several anchorages on its eastern side. Its southern extremity, False Cape Horn, is a very remarkable headland, and from the east or west resembles a large horn. Packsaddle bay on the north-east side of the peninsula, sheltered from N.W. winds by a curious island resembling a packsaddle, is safe and roomy, but not so convenient as Orange bay, which is considered the best anchorage on the coast; it is somewhat open to east winds, but they seldom blow strong. No sea can be thrown in because of the Hermite islands. The best watering-place is in a small cove at the north side, called Water cove. This harbour is fit for a fleet of line-of-battle ships, and could supply them with any quantity of wood and water. High water at full and change 3h. 30m., rise of tide 6 feet.*

ANCHORAGE.—Off Orange bay anchor-soundings extend to 2 miles from the land. The opening of the bay is 3 miles wide, and in that part are 18 or 20 fathoms, over fine speckled sand. Two islands, the largest having a smooth, down-like appearance, lie in the middle; behind them is the harbour, a square mile of excellent anchorage, without a singe rock or shoal. In the two creeks at the south side is good anchorage for small vessels, on a bottom of fine speckled sand, and the depth of the water varying gradually from 5 to 20 fathoms. The land hereabouts is low, comparatively speaking, and you are not annoyed by the violent squalls which come from the heights in other places. A vessel may go close to the shore in every part, therefore no directions are necessary to point

---

* *See* Plans of Packsaddle and Orange bays, Nos. 1,322 and 559 ; scales, m = 2 and 1·5 inches.

out the way to the best berth, which is marked in the plan. Off the north point there are several small islets, which must not be approached too closely ; they are, however, out of the way.

**SCHAPENHAM BAY** is a mile and a half wide ; there is a small black rock, above water, rather to the northward of its middle. A great deal of kelp, lying over a rocky bottom, is seen at the head of the bay, and a large waterfall marks the place distinctly. There is anchorage in from 10 to 15 fathoms, near its south point ; but I should not recommend a vessel to use it, when by going further she may get into an unexceptionable harbour (Orange bay), or anchor off its entrance in perfect security. The land behind is high and rugged ; two singular peaks show themselves which resemble sentry-boxes. Near the shore the land is low, compared with other parts of the coast, and has not the iron-bound, forbidding appearance of the more westerly shores. From the heights sudden and very strong squalls blow during westerly winds. Being generally a weather shore, and regular soundings extending along it, there is no difficulty in choosing or approaching an anchorage. Lort Bay lies 5 miles S.S.E. of Schapenham bay, and is about 2 miles wide. A vessel may anchor there, if necessary, in 8 or 10 fathoms, sand ; but some rocks above water, lie off its northern shore.

**THE DIEGO RAMIREZ ISLANDS** lie nearly N.N.W. and S.S.E., and extend over a space of 5 miles. From the North rock, which is in lat. 56° 25′ S., and long. 68° 44′ W., Cape Horn bears N.E. ½ N., distant 56 miles. The highest part of these islands is about 150 feet above the sea. There is no hidden danger near them. A ship may pass between the northern cluster and that to the southward. Detached rocks lie off the southern island : all the outer ones are above water. The southern, or Boat island, has a cove at its N.E. corner, in which boats may land, and where water may be procured on the point close to the eastward of the landing-place. Between the Hermite and Diego Ramirez islands, there is no danger of any kind.

**The ILDEFONSOS,** a large group of rocks and islets, bear N.W. by W. distant 37 miles from Diego Ramirez. They extend 5 miles in a N.W. and S.E. direction, are very narrow, and about 100 feet above the sea. They appear to be the remains of a mountain ridge, broken through in many places by the sea. Vessels may pass close by them, for there is no danger. Sealers have much frequented them for seals.

**NEW YEAR SOUND,** situated on the west side of Hardy peninsula, next presents itself in returning to the coast. It is a large tract of water studded with islands, and extending to the north-west. There may be good anchorage between Morton and Henderson islands at the entrance of the sound. There was no time to examine some coves on the eastern side of Morton islands, but their appearance seemed to promise shelter and holding-ground. On Henderson island there is a high sharp-pointed hill, visible at a great distance. From its summit the Diego Ramirez islands were seen though 50 miles distant. Between False Cape Horn and Cape Weddell, at the eastern side of New Year sound, there is a tract of broken land, which has not been properly examined. It is, however, a lee shore during S.W. and southerly winds, and therefore unfit for anchorage. Indian cove, about 12 miles from the entrance of the sound, and on the western shore, is not a place to be recommended : vessels must go far among the islands to reach it, and when there, they will find a bad rocky bottom, with deep water.

CLEARBOTTOM BAY is at the north end of Morton island, and a good anchorage. It is thus described in Weddell's journal, in 1837. This anchorage, by being close to the coast, is convenient for a vessel to touch at for wood and water : to sail into it from sea, bring the eastern Ildefonso island S. $\frac{1}{2}$ E., and steer N. $\frac{1}{4}$ W. for Turn point. About 1$\frac{1}{2}$ miles E.N.E. of this point is the anchorage, and at the distance of 3 cables' lengths from the shore, in 22 fathoms on a bottom of sand and clay, is the most eligible berth.

Leading Hill, on Hind island, is a very remarkable double-peaked height, and may be seen from a distance of 6 or 7 leagues. It points out the entrance of Duff bay. Neither Rous sound nor Trefusis bay afford anchorage. The Wood islands afford no good anchorage. Passages and broken land lie behind them to the northward.

CHRISTMAS SOUND is situated on the west shore of Hoste island, between it and Waterman island. Captain Cook's description of Christmas sound is as accurate as his accounts of other places. His Great Black rock and Little Black rock show themselves as you enter. There is no hidden danger ; the chart and plan are sufficient. Adventure cove (in which he anchored) is the easiest of access, but it will only hold one vessel.*

WATERMAN ISLAND is soon known by its remarkable heights, the southernmost of which was named by Captain Cook, York Minster, from its fancied resemblance to that building. He well describes it as a " wild-

---

* *See* Plan of Christmas sound and Chart of Magellan strait, Nos. 559 and 554; scales, m = 1·5 and 0·13 inches.

looking rock." Close to the eastward of York Minster there are several
rocks and islets ; one, on which the sea breaks violently, lies 2 miles
E.$\frac{1}{2}$ N. from the extremity of the Minster, a vessel may pass it quite close.
Off the Great Black rock there are two or three breakers, caused by rocks
under water. But little current sets among these islands. Eight miles
west of York Minster, and 4 miles south of the western point of Waterman island are the Capstan rocks, about 20 feet above water. There are
no other dangers to seaward of a line from York Minster to the Phillips
rocks.

MARCH HARBOUR on the east side of Waterman island is large, with
good holding-ground, but there are many rocky places ; and on a sunken
rock marked by very thick kelp, there is only one fathom. The *Beagle*
worked through the narrow passage, round Shag island, from Adventure
cove, and into the innermost corner of the harbour without using a warp ;
larger vessels would of course find themselves more confined. No vessel,
however, of more than five hundred tons should attempt to enter Christmas
sound. The *Beagle* lay moored in this harbour all the month of March
in safety ; but her chain cables became entangled with the rocks, and
were not hove in without much difficulty and delay. Port Clerke, about
one mile to the northward of March harbour, is a bad place for any
vessel, though quite secure when inside ; access is difficult, and, from
its situation, it is exposed to very violent squalls.

COOK BAY is a large space between Alikhoolip and Waterman islands
Broken land, islets, and breakers surround and make it unfit for the
approach of vessels. Its shores were explored by the *Beagle's* boats.
To the eastward of Cook's bay there is an entrance to the Beagle
channel, and to the north-west a passage to Whaleboat sound, but both
are unfit for sailing vessels, excepting with a fair wind.

The LONDONDERRY ISLANDS are a large group which nearly fills
up the space between the two last-mentioned passages. Treble island, off
the western shore, is remarkable, having three peaks, and is visible from
a considerable distance ; near it are some straggling rocks, shown in the
chart. Phillips Rocks are about 4 miles S.S.W. $\frac{3}{4}$ W. from Alikhoolip, the
south-west cape of these islands. They are dangerous, though above
water, because so far from the shore, and so low.

GILBERT and STEWART ISLANDS are 4 miles farther to the westward. Lying between these two islands is an open space, called Adventure passage, with deep water, and clear of danger. Cape Castlereagh

their western promontory, is high and remarkable ; north of it is an excellent anchorage called Stewart harbour. Doris cove, at the north-east side of the eastern Gilbert isle, is a safe anchorage for a small vessel, where the *Beagle* lay moored for a week. There are no hidden dangers hereabouts ; the eye and the chart will guide a vessel safely.*

**STEWART HARBOUR** is not large, but for small vessels is an exceedingly good place, being easy of access with any wind, having three openings. A sailing vessel may anchor in the entrance and warp in ; there is nowhere more than 16 fathoms, generally from 6 to 12. Two rocks just awash at high water lie nearly in the middle of the harbour. The plan shows their place exactly. A rock on which the sea breaks, lies one mile nearly W.S.W. of the middle opening to the harbour. There is no other danger. Wood and water, as in every Fuegian harbour, are plentiful and easily obtained.†

**DESOLATE BAY** is a large space of water between the Stewart islands and Tierra de Fuego ; it leads into Courtenay, Thieves, and Whale-boat sounds. Cape Desolation, the south point of Basket island, and north entrance to Desolate bay, is a very remarkable headland ; it is rugged, with many peaks. Rocks and breakers abound, and make these sounds quite unfit for shipping ; no doubt small vessels might, in clear weather, traverse any of these passages ; but it would always be with much risk, and should not be attempted without an adequate object. Such an object does not now, nor is it likely to exist.

**The CAMDEN ISLANDS** are a large group lying off Brecknock peninsula, the western point of the main island of Tierra del Fuego. Between, and to the northward of these islands, there are several passages with deep water, and anchorages opposite to most of the valleys, or between the islands, in which small vessels could lie securely, if necessary. Brecknock passage, between the Camden islands and Tierra del Fuego, is wide and clear of all danger. Vessels entering or leaving the Barbara channel should use this passage in preference to passing the Fury rocks.

**TOWNSHEND HARBOUR** is in London isle, the largest of the group. It is a safe anchorage, situated at its east end. The Horace peaks point out its position. Some rocks, on which the sea breaks violently, lie off the islands, and near the entrance of Pratt passage. They are exactly laid down in the chart. As there are no soundings in less than 50 fathoms after passing these rocks, and getting into the

---

* *See* Plan, No. 599.   † *See* Plan, No. 549.

passage, you must depend upon the wind lasting to carry you into or out of the harbour. The holding ground in it is excellent, and though you have tremendous squalls off the high land to the westward, there is no fear of an anchor starting. The *Beagle* lay here, moored, during the worst weather she had on the coast. A very high sea was raised outside by a violent southerly gale, but she remained in security without moving an anchor.

The lee side of high land is not the best for anchorage in this country. When good holding can be found to windward of a height, and low land lies to windward of you, sufficient to break the sea, the anchorage is much preferable, because the wind is steady, and does not blow home to the heights. Being to leeward of them is like being on the west side of Gibraltar rock when it blows a strong Levanter.

**EAST** and **WEST FURIES.** — The entrance of the Barbara and Cockburn channels lies between the Camden and Magill islands. This entrance is strewed with clusters of rocks, of which the Furies are the most remarkable and important, as the passage into the Cockburn channel lies between them. They are situated on a west bearing from Cape Schomberg, the west point of London island; the East Furies being 4, and the West 9 miles from that promontory. In a line between them, 3 miles from the latter, is an insulated rock. N.E. $\frac{1}{4}$ E., 4$\frac{1}{2}$ miles from the West Furies, lie the Tussac rocks, two in number, with no outlying dangers. Vessels entering with a westerly wind should pass near the West Furies and steer for these rocks. After passing them there are no known dangers in the entrance of the Cockburn channel. A reference to the chart will show everything else that need be noted. They have been much frequented by sealing vessels' boats, fur seals being numerous upon them at times.*

**MAGILL ISLES** are a large group, lying in Melville sound, at the entrance of the Barbara channel, having several coves and anchorages among them. Mount Skyring, on the island of the same name, is a very conspicuous object. It rises to a peak to the height of 3,000 feet; and was very useful in connecting the triangulation of the strait with that of the outer coast. It was seen from Field bay, at the north end of the Barbara channel; and, from its summit, Captain Fitz-Roy obtained a bearing of Mount Sarmiento.

**Port Tom,** on the south-east side of Mount Skyring island, is good

---

* *See* Charts of the Barbara channel, Nos. 2,113 and 1,306; scales, m = 0.28 and 1.0 inch.

and well-sheltered, excepting from the violent squalls off the high land, but they are frequent everywhere among the coves of Tierra del Fuego. For sealing vessels, it is more safe and secure than Fury harbour, the place they usually frequent ; and everything that a Fuegian harbour can afford is to be obtained in it. The *Adelaide* anchored here when exploring these parts.

**Fury Harbour,** on the south-east side of Fury island the central island of Magill group, is a wild anchorage, with little shelter and bad ground. From its contiguity to the East and West Furies, and the Tussac rocks, on which seals are found, it is, however, much frequented by sealing vessels.   In the winter of 1826–27 the *Prince of Saxe Coburg* sealer was wrecked in Fury harbour, but the crew were saved by the *Beagle's* boats.[*]

**North Cove,** on the north-east side of Fury island, is a snug temporary anchorage for small vessels.   When there, they are in security ; but it must be remembered that there is no anchorage in the channel, nor inside the cove, unless they close the weather shore.   H.M.S. *Beagle* anchored in this cove during the survey.   Bynoe island affords an anchorage on its north-east side ; and Hewett bay, on the opposite shore, is a good stopping-place either for entering or quitting the channel.

**BARBARA CHANNEL,** leading into the strait of Magellan at English reach, has its southern entrance so much occupied by the above islands and rocks, that no one direct channel can be specially recommended. The chart must be referred to as the best guide for its navigation. For small vessels there is neither danger nor difficulty; there are numerous anchorages that they may reach without trouble.   The situation of the rocks off the entrance of this channel, as laid down in the chart, is accurate ; but no vessel should attempt to pass them without daylight and clear weather, so that she may steer more by a good eye at the masthead than by any chart.   Four remarkable mountains point out the entrance to this channel very distinctly.   The Kempe island peaks are high, and show three points.   The Fury island peaks are high and divided.   Mount Skyring is high, and has a single peak.   Mount St. Paul, from near Fury island, appears very like the dome of the cathedral, the name of which it bears, St. Paul's, London.

**The AGNES ISLANDS,** a group situated to the north-west of the Magill islands, and those in their neighbourhood, do not require any description.   They are so fortified by outlying rocks as not to be fit places for the approach of any vessel.   The south-west entrance to the Barbara

---

[*] *See* Charts of the Barbara channel, Nos. 2,113 and 1,306 ; scales, m = 0·25 and 1·0 inch.

channel lies between the Agnes and Magill islands.   No vessel ought to entangle herself in these labyrinths ; if she does she must sail by the eye. Neither chart, directions, nor soundings would be of much assistance, and in thick weather her situation would be most precarious.

**NOIR ISLAND,** the outlying islet of this labyrinth bears W. by S. distant 34 miles from Cape Schomberg.   It is about 600 feet above the sea, and with a remarkable neck of land to the south-west, ended by a rock like a steeple or tower.   One mile south of this point there is a sunken rock, over which the sea occasionally breaks ; two other breakers are in the bight close to the point.

Between Cape Schomberg, on London isle, and Noir island, lie many reefs, and a great number of detached outlying rocks, which render this part of the coast extremely dangerous and unfit for vessels.   No chart could guide them ; they must trust to daylight and clear weather, with a good look-out, if necessary to enter or leave the Barbara channel, which opens into this bay.

**Noir Road** under the east side of Noir island, affords good anchorage Several ships may lie there secure from all westerly winds, over a clear sandy bottom.   Wood and water plentiful, and easily obtained. There is a cove at the south part of the island, where boats would be safe in any weather, but the entrance is too narrow for larger vessels.*

**Jupiter and Neptune Rocks.**—The large space between Noir island and the Agnes islands is extremely dangerous for shipping, being scattered with rocks, some just awash, some showing themselves several feet above, and others under water. Still there is abundant room to go round the island in security, therefore no ship need fear being hampered by an easterly wind, in the event of anchoring in Noir road.   A rock lies in the road, and another, which is very dangerous, 4 miles to the eastward; they are exactly laid down in the chart.

**MILKY WAY.**—This name has been given to the space between Kempe and Noir islands, as in every part of it rocks are seen just awash with, or a few feet above, the water.   On them the sea continually breaks. The *Beagle* passed in-shore of them all, close to the Fury, Kempe, and Agnes islands ; but no vessel should follow her track, nor is there any probability of its ever being attempted.   This part of the coast only requires to be known to be the more avoided.   The Tower rocks are 7 miles S.S.E. of Noir island ; they are high, and steep-to.   A ship may pass close to either side of them.

---

* *See* Plan of ports in Desolation Island, No. 558; scale, m = 1·5 inches.

**TIDES.**—In the Barbara channel the stream of flood was found to set to seaward, or to the southward, as was also the case in Cockburn channel; but the whole system of tides in this great archipelago requires a careful and patient investigation.

**THE GRAFTON ISLANDS.**—Isabella, the south-east island of this group, lies 13 miles N.N.W. from Noir road.   They are high, extend about 20 miles in a north-west direction, and the remarks on the general character of the coast are applicable to them.   Between them are several anchorages, but the best and easiest of access is Euston bay.   Behind them lies the Wakefield passage, through which a sealing vessel has passed.   To the north-east of it is a mass of land, broken into islets and rocks.   Hope harbour on the east shore of James island, is one of those formerly used by sealing vessels.

**Isabella Island** contains an anchorage, fit for a sealing vessel, but for no other.   Rocks lie in the way to it, as the chart shows; but the *Beagle* passed a night there, though not by choice.   There are several outlying rocks off the south side of Isabella island, terminating in the Kennel rocks, which lie 3 miles S.S.E. from the south point of the island.   A rock awash lies 3 miles W. by N. from the Kennel rocks.

**CAPE GLOUCESTER,** the western extremity of Charles island, the largest and westernmost of them, is a very remarkable promontory, and cannot be mistaken.   At a distance it appears to be a high, detached island; but, on a nearer approach, a low connecting isthmus is seen.   A rock, on which the sea breaks, lies nearly a mile to the north-west; there is no other danger.   The cape may be passed quite close, being steep-to.

**EUSTON BAY.**—Cape Gloucester is a guide to this bay, one of the best anchorages on this coast, as it can be approached and left with any wind, without risk, and in which a fleet may lie in perfect security from all but the S.E. winds, the least prevalent of any on this coast.   If coming from the westward, on passing Cape Gloucester, a high island will be seen to the south-east, distant 7 miles; this is Ipswich island, between which and Cape Gloucester lies Maria bay, in which are many rocks and breakers.

Rounding Ipswich island, a good berth must be given to the rocks under water, which lie a mile from its south-east extremity.   The sea does not always break upon them, but it does generally, and there is no other hidden danger.

**LAURA HARBOUR.**—After clearing these rocks, pass close to Leading island, and steer for the opening of Laura harbour, which will be seen under a high peaked mountain.   Choose a berth by the eye, if intending to anchor in Euston bay, or work as far up the passage to the basin as convenient, then anchor, and warp to the berth marked in the plan.*

The *Beagle* worked up all the way against a fresh wind blowing directly out.   There is water for a frigate in the basin, but it is better suited to a small vessel.   Large ships should anchor in the bay ; and as the bottom is even and good, and the bay capacious, exposed only to south-east winds, which come on gradually, and seldom blow hard, it may be considered a fit place for ships of any size, or for a squadron.   Wood and water are plentiful, and easy to be obtained.   The depth of water in the bay varies from 5 to 20 fathoms, the bottom generally fine speckled sand.   A large patch of kelp lies across the entrance of the harbour, but there is no danger beneath it, except for a line-of-battle ship, as in one spot there are 4 fathoms only.   This kelp was very closely examined, and its safety satisfactorily proved.

**THE FINCHAM ISLANDS** may be next noticed in passing to the north-westward.   Between those islands and Cape Gloucester, lies the Breaker coast ; a large wild extent of 20 miles, full of rocks and breakers, and exposed to all the strength of westerly winds ; utterly unfit for the approach of a vessel.   The shore is broken into islands, islets, and rocks almost innumerable.

**CAPE TATE** is rather high, and rounded at the summit ; there are several clusters of rocks off it, the two southern parcels are called the College rocks : they are only seen when near the land.   As a reference to the chart will show, there is no good anchorage hereabout : the coast is very dangerous and unfit to be approached.   The *Beagle* tried to anchor in Deepwater sound, but failing to find a proper depth of water, was obliged to drop her anchor upon the shelving end of a small island, being too far up the sound to get out again before dark.

**LANDFALL ISLANDS** 21 miles to the north-west of Cape Tate, were so named by Captain Cook, from seeing them first when he visited this coast ; their south extremity is Cape Schetky, a remarkable double-peaked height ; some rocks just awash lie a mile off it.   Cape Inman is a very remarkable headland at their western extremity.   Off this cape there are several detached rocks, on which the sea breaks violently, and gives them a formidable appearance.   The outermost one is not 2 miles from the shore, and shows itself plainly.

---

* *See* Plan of Ports, Desolation Island, No. 558 ; scale, *m* = 1·5 inches.

**LATITUDE BAY** is behind Cape Inman, an anchorage decidedly good, though somewhat exposed to a swell thrown in by heavy north-west winds. The *Beagle* rode out there a heavy gale from that quarter, though having anchored too far in, she was exposed to rollers. The plan shows the best anchorage. Between the Landfall islands there is a snug berth for a vessel not drawing more than 12 feet in security, and with smooth water; she should not moor in less than 10 fathoms, but keep as close to the western shore as possible, with an anchor to the eastward, in the event of the wind coming from that quarter. Water and wood are plentiful, as is the case in every Fuegian harbour. High water on full and change 1 hour.*

**OTWAY BAY** is inside and to the eastward of the Landfall islands, an extensive sheet of water, surrounded by broken land, islets, and rocks. Many of the latter are scattered about, and render it unfit for navigation. It is probable that passages lead from thence to the Straits of Magellan, as Dynevor and other deep inlets run in that direction as far as the eye can reach from the Landfall islands; but they could not be explored for want of time. Captain King remarks: "It seems probable that a communication may exist between Dynevor inlet and Sarmiento's opening of Abra, in the strait opposite to Playa Parda."

**WEEK ISLANDS** lie north-west of the Landfall islands, separated from Desolation island by Murray passage. At their south side is a roadstead, with good holding in 18 or 20 fathoms, coarse gravel and sand with patches of rock. It is exposed to southerly winds and to those from the west, therefore an improper place for a vessel to anchor. Between the islands a small vessel may find a snug berth in Saturday harbour, quite secure, but difficult of access. The *Beagle* lay at anchor there a week, in 24 fathoms, good holding ground. As before said, the eye must be the chief guide in entering most of these places; they are of one description,—inlets between high ridges of land,—having, generally, deep water, and kelp buoying the rocky places. Flaws of wind and violent gusts off the high land, render the approach difficult to all, and to a large ship impracticable. Graves Island is the largest of this cluster. Cape Sunday, its western point, is high and prominent. Two islets and two dangerous rocks lie off it, and which are shown in the chart. Barrister Bay opens out after passing Cape Sunday, but it is an exposed place, full of islets, rocks, and breakers, and unfit for any vessel.†

---

\* *See* Plan of Ports, Desolation Island, No. 558; scale, *m* = 1·5 inches.

† *See* Plan of the Week Islands, No. 1,330; scale, *m* = 1·5 inches.

**CAPE DESEADO,** lying 5 leagues to the north-westward of the Week isles, is the highest land hereabout, and remarkable. A rocky islet lies a mile off the cape ; and 2 miles to the south-eastward there is an opening which has not yet been examined, from whence to the cape the coast runs high and unbroken. The 50 fathoms edge of soundings appears to extend about 20 miles off shore along this south-western coast of Tierra del Fuego, the bottom being coarse sand.

**DISLOCATION HARBOUR** is situated 4 miles N.W. ½ W. from Cape Deseado. It is a place of refuge for an embayed or distressed ship but unfit for any other purpose ; its entrance is rendered difficult to the eye by rocks, on which the sea breaks violently ; and by two rocks under water, on which the sea does not always break, but their place is accurately shown in the plan of the harbour. The position of Dislocation harbour is pointed out by the heights called Law and Shoulder peaks; they are the most remarkable on that part of the coast, and immediately over the harbour. Water may be obtained very easily ; the boats can lie in a stream which runs from the mountains, and fill alongside. Wood is plentiful.*

**Weather and Lee Rocks** are 2 outlying dangers off this port, lying respectively W. ½ N. and S. ¼ E., 2¾ and 3½ miles from the south entrance, which is narrow, exposed to the prevailing wind and swell, which might, for days together, prevent a vessel from getting out to sea.

**DIRECTIONS.**—To find the entrance, steer for the peaks, look out for the Weather and Lee rocks, both several feet above water, the sea breaking violently on them, and when within 4 miles of the shore you will distinctly see the opening from the mast-head. In going in, avoid the two rocks at the entrance, and anchor in the innermost part. Only a small ship can get out again without a fair wind. The prevailing winds send in a swell, but the place is quite secure. Four small vessels may lie in security, the bottom is very even, from 15 to 25 fathoms, fine white sand.

**The JUDGE ROCKS and the APOSTOLES** show themselves in proceeding from Dislocation harbour to the north-westward; they are from 5 to 50 feet above the sea, but many breakers near them indicate an extensive reef. The outer rock is 4 miles from the land.

**CAPE PILLAR** the north-western extremity of Tierra del Fuego, forms the western point of entrance to Magellan strait. It is a bold cliff and as seen from the westward, the land rises in four peaks, the most southern

---

* *See* Plan of Dislocation Harbour, No. 558.

being the highest. The shore is steep to, there being a depth of 60 fathoms within one mile of the pitch of the cape, which is in lat. 52° 43′ S., and long. 74° 43′ W. *See* View on Chart, No. 554.

**TIDES.**—It is high water, full and change, at Cape Pillar, at 1h., and at York Minster, Christmas sound, about 200 miles to the south-east at 3h. At the intermediate places the time gradually changes between those limits ; and the rise varies from 4 to 8 feet, as noted in each plan.

**GENERAL OBSERVATIONS.**—Fogs are extremely rare on this coast, but thick, rainy weather prevails with strong winds. The sun shows itself but little, the sky even in fine weather being generally overcast and cloudy ; a clear day is a rare occurrence.

**WINDS.**—Gales of wind succeed each other at short intervals, and last several days. At times the weather is fine and settled for perhaps a fortnight, but those times are few.. Westerly winds prevail during the greater part of the year. The easterly winds blow chiefly in the winter months, and very hard, but they seldom blow hard in summer.

Winds from the eastward invariably begin gently, and with fine weather ;—they increase gradually,—the weather changes,—and often end in a determined heavy gale. More frequently they rise to the strength of a treble-reefed topsail breeze, then die away gradually, or shift to another quarter. Gales of wind from the southward, and squalls from the south-west, are preceded and may be foretold by heavy banks of large white clouds rising in those quarters, having hard edges, and appearing very rounded and solid.

North and north-west winds are preceded and accompanied by low flying clouds, with a thickly overcast sky, in which the clouds appear to be at a great height. The sun shows dimly through them, and has a reddish appearance. For some hours, or for a whole day, before a gale from the north or west, it is not possible to take an altitude of the sun although it is visible, the haziness of the atmosphere in the upper regions causing its edges to be quite indistinct. Sometimes, but very rarely, with the wind light between N.N.W. and N.N.E., a few days of beautiful weather may occur ; but they are generally succeeded by gales from the southward, with much rain.

This wind always begins to blow moderately, but with thicker weather and more clouds than from the eastward, and it is generally accompanied by small rain. Increasing in strength it draws to the westward gradually, and blows hardest between north and north-west, with heavy clouds, thick weather, and much rain. When the fury of the north-wester is expended, which varies from 12 to 50 hours, or even while it is blowing hard, the wind sometimes shifts suddenly into the south-west quarter, blowing harder than before. This wind soon drives away the clouds, and in a few hours you have clear weather, but with heavy occasional squalls. In the south-west quarter the wind hangs several days (generally speaking), blowing strong, but moderating towards its end, and granting two or three days of fine weather.

Northerly winds begin generally during the summer months; but change continually from north to south, by the west, during that season, which would hardly deserve the name of summer, were not the days so much longer, and the weather a little warmer. Rain and wind prevail much more during the long, than the short days. It should be remembered that bad weather never comes on suddenly from the eastward, neither does a south-west or southerly gale shift suddenly to the northward. South-west and southerly winds rise suddenly and violently, and must be well considered in choosing anchorages, and preparing for shifts of wind at sea. The most usual weather in these latitudes is a fresh wind between north-west and south-west, with a cloudy, overcast sky.

SEASONS.—It may be as well to say a few words respecting the seasons in the neighbourhood of Cape Horn, as much question has arisen respecting the best time for making the passage round the cape—in winter or in summer. The equinoctial months, and especially March, are the stormiest in the year, generally speaking. Heavy gales prevail at those times, though not, perhaps, exactly at the equinoxes.

August, September, and October, are the coldest months. Westerly winds, rain, snow, hail, and cold weather, then prevail. December, January, and February, are the warmest months: the days are then long, and there is some fine weather; but westerly winds, which often increase to very strong gales, with much rain, are frequent even throughout that season, which carries with it less of summer there, than in almost any part of the globe.

In April, May, and June, the finest weather is experienced, and though the days shorten, it is more like summer than at any other period. Bad weather often occurs even during these months, but not so much as at other times. Easterly winds are frequent, with fine clear settled weather. During this period there is some chance of obtaining a few successive and corresponding astronomical observations. To endeavour to rate chronometers by equal altitudes would be a fruitless waste of time at other seasons. June and July are much alike, but easterly gales blow more during July.

PASSAGE round CAPE HORN.—The days being so short, and the weather cold, make June and July unpleasant, though they are, perhaps, the best for a ship making a passage to the westward, as the wind is often in the eastern quarter. The summer months, December and January, are the best for making a passage from the Pacific to the Atlantic, though that passage is so short and easy that it hardly requires a choice of time. For going to the westward, April, May, and June are preferable. Further directions for making this passage are given in chapter VIII.

**LIGHTNING and THUNDER** are very rare, indeed scarcely known, except in very bad weather, when violent squalls come from the south and south-west, giving warning of their approach by masses of clouds. These storms are rendered more formidable by snow, and hail of a large size.

**BAROMETER.**—Much difference of opinion has prevailed as to the utility of a barometer in these latitudes; but Captain FitzRoy, during twelve months' constant trial, found its indications of the utmost value. Its variations correspond with those of high northern latitudes in a remarkable manner, changing south for north.

**CURRENTS.**—There is a continual current setting along the south-west coast of Tierra del Fuego, from the north-west towards the south-east, as far as the Diego Ramirez islands. From thence the current takes a more easterly direction, setting round Cape Horn towards Staten island, and off to seaward about E.S.E.

Much has been said of the strength of this current, some persons supposing that it is a serious obstacle in passing Cape Horn to the westward, while others almost deny its existence. From the experience of the *Beagle* it appears to run at the average rate of a knot. Its strength is greatest during westerly, and least, or insensible, during easterly winds. It is strongest near the land, particularly the projecting capes or detached islands. This current sets rather from the land, which diminishes the danger of approaching this part of the coast.

**CAPE HORN CURRENT.**—At the distance of 3 or 4 leagues to the southward of Cape Horn, however, there is a current running to the E.N.E., at the rate of about one mile per hour; but in what manner this current may influence the tides near the shore, or what changes may be produced in the direction and the strength of the current itself by the flood and ebb-tides, will require a very extensive series of observations to ascertain.

The circumstance of there being no well authenticated account of the existence of a current to the southward of Cape Horn, induced Captain Fitz-Roy to collect the observations upon that subject carefully, during the passages of H.M.S. *Chanticleer* in the years 1828-9, in the summer and autumnal months of those regions.

Whence it appears that in the voyage from Staten island to Cape Horn the average of five days in December 1828 gave a set in the direction of N. 80° E. true, or N.E. by E. mag. 11½ miles in the 24 hours. Between Cape Horn and South Shetland in January 1829 an average of six days gave a current of S. 65° E. true or E. mag. of 11 miles a day. On the return voyage from South Shetland to Cape Horn in the month of March an average of 16 days gave a current of N. 49 E. true, or N.N.E. ½ E. mag. 21 miles a day. From Cape Horn to Staten Land in the month of May the current was found to set N. 51 E. true, or N.E. by N. mag. 54 miles in 23 hours. The result of these observations seems to point out distinctly that the usual easterly set is produced by the greater prevalence of the S.W. West and N.W. winds.

The last voyage run from Cape Horn to Staten Land in the month of May is exceptional. Possibly, if no error has crept in, it would point to the conclusion that the set of the flood-tide round Cape Horn comes from the S.W. ; for, at the time of the *Chanticleer* taking her departure from Cape Horn, it was ascertained to be nearly low water ; and on her arrival off Cape St. John the flood-tide had just made its mark. The passage from Cape Horn to Staten island was performed in 23 hours, in which interval the ship had felt the whole influence of two flood-tides, while that of one ebb only had been experienced ; and on comparing the ship's place, ascertained by bearings at the time of departure from Cape Horn, with the dead reckoning on arriving off Cape St. John, and kept in the most unexceptionable manner under very favourable circumstances, viz., fine weather, a free though side wind, and the ship's way through the water measured by a self-registering log, a set N. 51° E. (true), at the rate of 54 miles in 23 hours, was experienced ; from which, if 24 miles be deducted for the effect of the previously established current at this season, we have 30 miles for the set of the flood-tide at *neaps*, or about 3 miles per hour ; which seems improbable.

How far the strength of these tides may have operated in producing some of the irregularities in the north easterly set of the sea, deduced from the previous observations, when near in shore, by having been influenced by either tide for a longer period in the interval between the observations, has not been ascertained ; but from some notes which were made at the time, there is every reason to believe that the tides caused part of the irregularities in question.

CAUTION.—Under any circumstances the mariner should be on his guard against the possible existence of a much stronger north-easterly sea than he might be disposed to suspect.

# CHAPTER VI.

MAGELLAN STRAIT—CAPE VIRGINS TO THE BARBARA CHANNEL.

VARIATION 21° to 22° East in 1860.   Annual decrease about 2'.

MAGELLAN STRAIT.—The eastern entrance of the Strait of Magellan lies between the two headlands, Cape Virgins on the north, and Cape Espiritu Santo on the south, about 20 miles apart.   Both capes present white cliffs, forming the seaward termination of ranges of hills of moderate height, extending into the interior.   Both capes, too, have low shingle points connected with them, Dungeness on the north, and Catherine point on the south, and these points limit the real width of the entrance to 14 miles from beach to beach.*

CAPE VIRGINS is about 160 feet high, and in approaching the Strait from the eastward is the first land usually seen, and the best to make.† It bears some resemblance to Cape Fairweather, above 20 miles to the north, and they have frequently been mistaken for each other. There are, however, some marks by which they may be distinguished, even if the latitude should not be ascertained.   In clear weather, some hills in the interior, to the south-west of the Gallegos river will be visible, whereas in the same direction from Cape Virgins is the strait, and beyond the low shore of Tierra del Fuego.   In thick weather the soundings off the respective capes will be some guide, the water off Cape Virgins being much deeper ; the bottom, also, to the north of Cape Fair-weather is of mud, whilst that north of Cape Virgins is of sand.

DUNGENESS, 5 miles to the southward, is a low flat, formed entirely of shingle, extending nearly 4 miles to the southward, from the foot of the low range of hills between Cape Virgins and Mount Dinero, varying from 160 to 240 feet in height.   Like its namesake on the south coast of England, Dungeness is steep-to, having 7 fathoms close to the Ness.

* For Directions for making the passage through Magellan Strait see Chap. VIII.
† See Charts of the Eastern Entrance and of Magellan Strait, Nos. 1,316 and 554 ; scales respectively, m = 0·2 and 0·13 of an inch.

**CAPE ESPIRITU SANTO,** on the opposite coast of Tierra del Fuego, is the north-eastern extreme of a range of white hills, which extend across the northern part of the main island of this group, varying from 200 to 600 feet in height. From its foot the low land of Catherine point extends 5 miles to the northward, and, as before mentioned, meets the opposite point of Dungeness within 14 miles, the greatest depth between being about 40 fathoms. A dangerous reef extends 2 miles off the points of Cape Espiritu Santo, and must be carefully avoided.

**The SARMIENTO BANK** lies off Cape Virgins, and probably has been too little considered by those who have hitherto passed that cape. After half-flood, or before half-ebb, a ship may pass Cape Virgins at any distance not less than a mile and may cross the Sarmiento bank without hesitation ; but when the tide is low, 10 miles is not too far for a ship to keep from the cape, until it bears N.W. by W., when she should steer W.N.W., to close Dungeness. A ship might often pass over this bank without touching, even at low water, because the bottom is uneven ; but there are places which no vessel drawing more than 12 feet could pass at a low spring-tide without injury. The 10-fathoms edge of this bank extends about 14 miles E.S.E. from Cape Virgins, and appears to average 4 miles in breadth. The soundings on it are shoal, and irregular at a less distance from the cape than 10 miles ; but beyond that distance, as they increase, they became more regular, although there is another patch of 14 fathoms, at 5 miles farther to the south-east.

**Virgins Reef,** which at half-tide is scarcely observable, projects more than a mile from Cape Virgins, and must be carefully avoided ; this and Sarmiento bank are the only dangers on the north side of the approach.

**SOUNDINGS.**—In crossing the Sarmiento bank when standing to the southward, the fine dark greyish-brown sand changes to coarse slaty sand, with small stones and shingle ; the stones chiefly slaty. Some casts, after crossing the ridge, were found to be entirely of coarse sand, while others were all shingle. As the water shoals, the bottom is coarser and more mixed. When to the northward or eastward of the Sarmiento bank, the lead brings up fine brown-grey sand while near the latitude of Cape Virgins ; but when N.N.E. of the cape, the sand is like steel filings. In standing eastward from Cape Virgins, the bottom is very brown sand, without shells or stones. When by standing more southerly, the water shoals upon the Sarmiento bank, the sand becomes much coarser, and is mixed with slate pebbles, or broken stones of all sizes : the sand is slaty.

This rule continues till the water deepens to 30 fathoms, or more, to the southward, when shingle only is found ; and when it begins to shoal in approaching Tierra del Fuego there is coarse dark sand, mixed with stones of various sorts, chiefly slaty.  Between the shoal parts of the bank and the deep water, or from 16 to 30 fathoms, the sand is coarse, particularly near the deep water.  In standing to the southward after bringing the cape to bear West, the bottom is a very fine grey sand, until near the ridge or bank, with Cape Virgins bearing W.N.W.; with the cape in this bearing, the sand is coarser, and mixed with large and small shingle.

**TIDES.**—It is high water, full and change, between 8h. and 9h., in the vicinity of Capes Virgins and Espiritu Santo; while the stream of flood is still running to the westward into the strait and to the northward past Cape Virgins.  Until near noon, the principal stream continues running to the westward, though the water is falling by the shore, making what is called tide and half tide.  About noon, the direction of the main stream changes; runs to the eastward until past 6 o'clock, or until 3 hours after low water by the shore.  Spring tides rise from 36 to 42 feet vertically; neap tides to about 30 feet.  In the First Narrows it is high water about 9h.; the water rises as above mentioned, and runs through the most confined parts of the Narrows at the rate of from 5 to 6 knots or more.

Before approaching the land, the state of the tide stream should be well considered.  Upon the knowledge of its movements may depend the safety of the vessel, and the quickness of the passage (as far as Elizabeth island), more than upon the wind or weather.  As already mentioned, at the full and change of the moon the main stream of tide begins to run to the eastward about noon; but in the bays on each side of the Strait, such as Lomas bay on the south, and Possession bay on the north, the times and direction of the stream vary much.  It should be borne in mind that it is high water at nearly the same time over all the eastern entrance.

**MOUNT DINERO** will appear when to the southward of Cape Virgins; it is a sloping pointed hill, 240 feet in height.  Thence to Cape Possession 15 miles the land continues between 200 and 400 feet in height, rather level topped, and generally covered with grass.  A few broken cliffy places show themselves near the water.  Three miles east of Cape Possession there is a remarkable bare patch, serving as a fixed mark for bearings.

**WALLIS SHOAL,** with only 2 fathoms on it at low water, lies W. by S. ¼ S. nearly 10 miles from Dungeness; it is of small extent, and correctly placed on the chart.  Dungeness may be passed closely, as may all the northern points as far as Possession bay.  The soundings in shore are irregular, varying from 13 fathoms to 6¼.

**CAPE POSSESSION** is a bold cliffy headland, 360 feet above the sea. The coast between it and Cape Virgin is fronted by a shingle beach, which dries to about half a mile from high-water mark. On the opposite, or Fuegian coast, the nearest land is low : the hills, the tops of which are seen in the horizon, are 10 miles inland, and not, as they appear, near the water. This range is from 200 to 600 feet above the sea.

**Possession Bay**, extending 15 miles from Cape Possession to the entrance of the First Narrows, curves in to the northward round the cape ; on its west side there is good anchorage for sailing vessels bound to the westward to avoid the ebb stream.

**Water.**—There is a convenient watering-place near the centre of Possession bay, on the shore under Mount Aymond ; a ship may go very near it, and lie in a safe berth out of the strength of tide, if she crosses the Narrow bank after half-flood, or before half-ebb.

**Mount Aymond,** on the north shore of Possession bay, 1,000 feet in height has near it, to the westward, 4 rocky summits, called Asses ears. Direction hills on the west shore of Possession bay show and become serviceable as marks when to the westward of Cape Possession and northward of the Orange bank ; they are, like the hills about Cape Orange, low, rather peaked, and of a yellowish-brown colour. Sometimes the largest Direction hill looks as if it were covered with verdure.

**Narrow Bank** lying off Direction hills is most dangerous when the tide is falling, and the stream setting to the eastward ; near the Narrows, the stream, at that time, sets strongly over the banks, and, therefore, a vessel going to the eastward should give them a wide berth.

**LOMAS BANK.**—On the southern shore of the strait is an extensive bank, projecting from Catherine point 4 miles to the westward, and skirting the shore round Lomas bay ; it should be carefully avoided, as it shoals very quickly, and the tide runs near it with much strength. To keep clear of this and other shoals the ship's position must be ascertained by cross bearings or angles of the several capes and hills laid down on the charts.

**CAPE ORANGE,** and some low hills near it, will show themselves on advancing westward. The great Orange bank, dry at low water, extends from 10 to 12 miles to the eastward of the Cape, joining the Lomas bank, and affording shelter to Lomas bay. Cross bearings must now be used carefully, particularly in approaching the Orange bank or reaching across to Possession bay. To recommend any leading course for a sailing vessel could be of little service, because the wind is generally from the westward.

**ANCHORAGE.**—There is anchorage along the north shore of the strait, and also to the eastward of the Orange bank; the latter is preferable, being much more sheltered from south-west winds, with a less depth of water, and less strength of tide. Lomas bay between the Lomas and Orange banks also offers anchorage, but the ship's position must be well attended to, as the banks shoal suddenly.

A berth for anchoring to the eastward of the Orange bank is with the following bearings :—Cape Possession, N. ¾ W.; Catherine Point, E. ⅓ S.; the summit of Cape Orange, S.W. by W. All these banks are composed of shingle mixed with coarse sand and mud ; where they are exposed to a strong tide, shingle only is found.

The **FIRST NARROWS** are about 9 miles long by 2 broad with cliffy shores, and a depth of 40 fathoms. There is no anchorage. After emerging from the Narrows a ship should be allowed to drift to the south-west with the flood stream for at least 3 miles before hauling up for Cape Gregory in order to avoid the Barranca ledge, which is dangerous, consisting of rocks covered with kelp, extending 6 miles S.W. by W. from Barranca point. The northern cliff on the southern shore kept open of Barranca point, a flat-topped sand hill, will carry a vessel to the southward of this ledge, but neither mark is easily seen in rainy weather.

**TRITON BANK** is a narrow shoal 3 miles long in a S.W. ¼ W. and N.E. ¼ E. direction; its east end bears S. by W. distant 7 miles from Barranca point; when the water is low, there are not more than from 2 to 3 fathoms on this shoal, it must therefore be carefully avoided. Ships should pass to the northward of it by steering so as to shut in the First Narrows as soon as Direction hills come open westward of Barranca point. (See page 214.) The ship's position should be repeatedly ascertained, and she should not go into less than 6 fathoms.

**PHILIP BAY.**—After passing the First Narrows the strait opens out to a width of 15 miles, forming two bays, Philip bay on the south, and St. Jago (in which there is a watering place) on the north. The land on the south shore of Philip bay is low, but gradually rises towards a range of high ground, which extends from Cape Espiritu Santo to Cape Boqueron. Anchorage may be taken, by the aid of the chart and lead, in almost any part of the bay, in some few places only the bottom is rocky. On this south side the tide is felt less than on the Patagonian shore.

**VALLE POINT** in St. Jago bay has anchorage off it, out of the tide, in from 6 to 10 fathoms. This point is known by some remarkable peaked hillocks, and when abreast of it the land and bay north of Cape Gregory will easily be distinguished; the cape will be first seen making like an island, and then a conspicuous hummock half way between it and the flat table land. This hummock is marked on the chart.

**GREGORY BAY ANCHORAGE** at the north entrance of the Second Narrows is from 2 to 2½ miles N.N.E. of Cape Gregory, abreast of the north end of the sand-hills that form the headland, and at about a mile from the shore, in from 13 to 15 fathoms. The bottom is excellent, a soft, but tenacious mud, which nearer the shore, is of a stiffer quality. At low water a sand-spit extends off for a third or nearly half a mile from the shore, close to which there are 7 fathoms water. Care should be taken not to approach too near.*

**TIDES.**—In Gregory bay the tide stream turns to the south-westward, towards the Cápe, for 2½ or 3 hours before it makes through the Second Narrows; which should be attended to, for a ship will lose much ground by quitting her anchorage much before the tide has turned in the main channel.

**ISIDRO POINT,** low and sandy, with a reef off it, forms the south point of entrance to the Second Narrows; under the point, and in the bight to the south-eastward of it, there is good anchorage; but the bank near that point shoals very suddenly, and the western tide sets strongly towards the shoal point, which should be well guarded against.

**The SECOND NARROWS,** lying 20 miles S.W. from the First Narrows, are about 9 miles long by 3 miles broad; in passing through them, a small shoal of 3 fathoms, which lies one mile from the north shore, and half way between Cape Gregory and Gracia point, should be avoided.

**Sweepstakes Foreland** † is a cliff on the south side of the Narrows. When to the westward of this head, a vessel should keep well to the westward of a line joining Cape St. Vincent and Elizabeth island, in order to insure safe anchorage, and to avoid being swept by the tide amongst the ripples and shoals between the islands of Santa Marta and Quarter-Master, which lie to the south-east.

Not less than 6 fathoms has yet been found between Silvester point and Cape St. Vincent, but the tide ripples much in some places: at all events, the anchoring ground is bad; and as the flood stream sets directly and strongly to the southward and eastward over this uneven ground, it should be avoided, particularly if the wind is light, or likely to fail.

---

* See Admiralty Plan, No. 1, of Ports in the Strait of Magellan, No. 555; scale m = 0·5 of an inch.

† Sweepstakes was the name of one of Narborough's ships.

**ELIZABETH ISLAND,** so named by Sir Francis Drake, is a narrow island, 8 miles long, lying nearly parallel to the shore of the strait, which here takes a S.S.W. direction ; its north-east end is called Silvester point. It is composed of ranges of heights extending in ridges in the direction of its length, the north-eastern hill being 170 feet high.   There is good anchorage out of the strength of the tide at a mile north of Silvester point, and it is convenient for a ship to leave with the intention of round-ing Elizabeth island.   A vessel drawing less than 16 feet may pass round to the westward of this island, betweeen it and Cape Negro, but the eastern passage is the shorter; and if in a steamer or there is wind enough to ensure maintaining your position and keeping close to Elizabeth island until past Walker shoal, it is also far the easier.   No sailing vessel ought to pass Silvester point without a commanding breeze, because the water there is very deep; and as the tide sets directly towards the islands of Santa Marta and Magdalena, much inconvenience, if not danger, might be caused by the failure of the wind.   The land hereabouts is low, not exceeding 200 feet in height, and without wood.

**ROYAL ROAD,** into which the Second Narrows open, lies west of Elizabeth island, and is 12 miles long by 3 broad ; in it there is no diffi-culty, and only one danger, the middle ground of $2\frac{1}{2}$ fathoms, half way between Silvester point and Peckett harbour; in every other part of this road a vessel may anchor in security and out of the tide, as the stream sets strongly from the Narrows to the south-east towards Santa Marta, and hence to the southward.

**OAZY HARBOUR,** so called by Narborough, is 3 miles to the west of Gracia point, the western end of the Second Narrows on the north side ; it is a secure place for small vessels.   The entrance is nearly 2 miles long, and too narrow for large ships, unless the weather be moderate, when they might drop in or out with the tide : the depth inside is from 3 to 10 fathoms.   There is neither wood nor water to be had, and therefore no inducement to enter it.*

**PECKETT HARBOUR** is 8 miles to the west of Cape Gracia, and although very shoal, offers a good shelter, if required, for small vessels, but the space is very confined ; the anchorage without is almost as safe, and much more convenient.   The entrance lies between the south-west point and Plaid island, and is rather more than a fifth of a mile wide.   Half a mile outside the anchorage is good in 7 fathoms : shoal-ground extends for a quarter of a mile off the point.

---

* *See* Plans of Oazy and Peckett harbours, No. 555 ; scales, $m = 0 \cdot 75$ and $1 \cdot 5$ inches.

**SANTA MARTA** is a small group of islets or rocks lying E.S.E. about 2 miles from Silvester point; a reef extends half a mile off its north end, and is dangerous, particularly during the flood tide.

**SANTA MAGDALENA,** 4 miles south of Santa Marta, is shoal on both sides, Walker shoal extending 3 miles to the south-west, and Adventure bridge reaching nearly to Quarter-Master island on the eastern shore of the strait. On this bridge there may be water enough for a vessel in some places, though not in others; the tide ripples violently over it, and the water looks very light coloured.

**LEE BAY,** lying southward of Cape St. Vincent, the south point of entrance to the Second Narrows, is a bad place for a vessel; it is exposed to wind and sea, besides having very shoal water in some places, and should be avoided.

**GENTE GRANDE (or TALL PEOPLE) BAY** is shoal and unfit to enter; and the shore around it is very low and dangerous. Quarter-Master island lies in the mouth of it. Southward of Gente point, along shore to Cape Monmouth and Cape Boqueron, there is no danger; the water is deep, and the coast safe to approach. Cape Monmouth is a cliffy point: Cape Boqueron, the abrupt termination of a range of high land extending across the country from Cape Espiritu Santo, is a precipitous headland.

**At CAPE NEGRO,** on the coast of Patagonia, the land becomes high and wooded. From this cape the Strait runs S.S.E. through Broad and Famine reaches into Froward reach, a distance of 60 miles. Its average width may be 12 miles; the water is deep, and there are no dangers.

**LAREDO BAY** under Cape Negro offers a secure anchorage when there is any westing in the wind; and with easterly winds, which are not common, and which seldom blow with violence, no danger is to be apprehended if the ground tackling be good. The holding ground is strong and the depth of water easy all over the bay. Do not borrow too much on the south side, which is shoal and foul. The north-west corner is a very snug berth. This is the first port in the strait, in coming from the eastward, where wood is to be procured. The distance over land from Lands bay to the large expanse which has been named Otway water, appears to be less than 8 miles.*

**TIDES.**—When to the southward of a line drawn from Laredo bay to Gente point, the tide streams are scarcely felt; but to the northward of

* Remarks of Commander A. S. Hamond, R.N., 1843.

that line they are strong, and must be carefully guarded against during the night, or in light winds. A vessel in mid-channel, between Gente Grande and Laredo bays, would be set by the ebb-tide, if the wind failed her, directly amongst the dangers surrounding Santa Magdalena island.

There is some discrepancy in the times of high water and in the rise of tides in the Strait, but we believe the following to be nearly correct :—

It is high water, on full and change, in the First Narrows at 9h., and the rise is 30 feet ; in Philip bay, on the east side, at 9h. 30m., and the rise about 24 feet ; in Gregory bay at 9h. 45m., and the rise is 23 feet ; in the Second Narrows at 10h., and the rise 20 feet, when it rapidly decreases ; and at Laredo bay the rise is only 9 feet, at 11h. 30m ; at Port Famine, rise 6 feet at 12h.

**CATALINA BAY** of Sarmiento, lies between Cape Negro and Sandy point, in which there is good anchorage from 1 to $2\frac{1}{2}$ miles from the shore. Here the country begins to assume a very picturesque appearance, particularly in the vicinity of Sandy point. At 5 miles inland a range of mountains is seen, reaching from 1,000 to 1,200 feet in height, while to the south-west are several rugged peaks covered with snow.

**SANDY POINT** projects for more than a mile from the line of coast, and should not be passed within a mile. A shoal projects off it in an easterly direction : the mark for its southern edge is a single tree, on a remarkable clear part of the country (a park-like meadow) near the shore on the south side of the point, in one with a deep ravine in the mountain behind. One mile and a half from the point there is no bottom with 18 fathoms. The Chilian Government has removed the penal settlement first established at Port Famine to Sandy point, as being more suitable for agriculture and rearing cattle, and the troops are employed in clearing land for the plough.

**Supplies.**—An ample supply of fish may be caught on the beach near the mouth of a rivulet a little to the northward of the houses, the best time to haul the seine being in the evening, when it generally falls calm, and a fire should be lighted on the beach. There is excellent wood for steaming, the authorities affording every assistance in their power. The settlement is often in great distress, and therefore few supplies will be found beyond wood and water. It is reported that coal exists in the mountains west of the settlement.

**ANCHORAGE.**—There is good anchorage off the settlement at Sandy point in 13 fathoms about $1\frac{1}{2}$ miles off shore; with the flag staff W. by N. $\frac{1}{2}$ N. Extreme north point N. $\frac{1}{2}$ W. Extreme south point S. $\frac{3}{4}$ E. A vessel intending to wood and water, may safely stand into 10 fathoms.*   To the southward of Sandy point, as far as St. Mary point, good anchorage may be had at three-quarters of a mile from the shore, in 11 and 12 fathoms, sand and shells over clay.   At the edge of the kelp, which fronts the shore, there are 5 and 6 fathoms; so that, with the wind off-shore, a ship may anchor or sail along it very close to the coast, by keeping outside the kelp.   The squalls off the land are very strong, sometimes so much so as to lay a ship on her broadside ; it is not prudent therefore to carry much sail in coasting here; and it is necessary to have the quarter boats secured with gripes, because the wind, for a moment, blows with the force of a hurricane.   These land squalls are denominated by the sealers *williwaws*.

**ST. MARY POINT** is 12¼ miles to the southward of Sandy point, and may be known by the land suddenly trending in to Freshwater bay.   It has also a high bank close to the beach, with two patches bare of trees.   All the points to the northward are low and thickly wooded.   As the bay opens, the bluff points at its southern extremity become visible.   There is also a remarkable round hill a short distance to the westward of the bay, and a valley to the southward of it, through which a small river falls into the sea.

**FRESHWATER BAY** is a convenient place for wooding, but from the river being blocked up by much drift timber, watering is difficult ; the proximity, however, of Port Famine renders this of no material consequence.   When the wind is from the northward, a swell is thrown into the bay ; but no danger need be apprehended from its being open to the eastward, for the wind seldom blows from that quarter, excepting in the winter, and then rarely with great strength.   If it does, the holding ground is good, and with good anchors and cables there is no danger.   Should the day be advanced, it is better to anchor in Freshwater bay than run the risk of being under way all night ; unless it be in the summer, with moonlight and the weather likely to be fine.   In this climate, however, the latter is very doubtful, for the weather changes so suddenly that little dependence can be placed upon appearances.

**DIRECTIONS.**—In standing into the bay from the northward, keep within three quarters to half a mile from the coast, in 10 or 11 fathoms ; and, after passing St. Mary point, a course towards the bluff southern points of the bay, until the south pitch of Centre Mount bears W.S.W., will clear the kelp that extends from the north side of the bay ; among which, though there may be a sufficiency of water, the ground is foul:

---

* Remarks of Commander F. L. Barnard, R.N., 1852.

round the edge of the kelp there are 6 and 7 fathoms. Having the mount bearing as above, steer for it, or a little to the southward of it, and anchor in 9 fathoms, sandy mud over clay, which will be with the following bearings; extremity of St. Mary Point N. ¾ W.; Centre Mount (south pitch) W. by S. ¼ S.; and the entrance of a river S.W. b. S.

A good berth may be had much nearer the shore in 6 fathoms, towards which the depth gradually decreases. If the anchorage is used merely as a stopping-place, the 9 fathoms berth is the best; for the wind near the shore is apt to flaw and veer about.

**FAMINE REACH.**—Rocky point on the mainland, and Cape Valentyn, the northern point of Dawson island, form the northern entrance of Famine reach, width 14 miles across, and it rapidly narrows and between Cape San Isidro and St. Joachim is only 5 miles wide, the land on either side being mountainous. Between Freshwater bay and Santa Anna point the coast is very bold, and so steep-to as to offer no anchorage, excepting in the nook that is formed by the reef off Rocky point; but it is small and inconvenient to weigh from, should the wind be southerly.

**SANTA ANNA POINT** extends far into the strait, with a clump of trees at its extremity. It bears from Cape Valentyn S.W. ¼ W. On approaching it, the distant Cape of San Isidro will be seen beyond it; but there can be no doubt or mistake in recognizing it. Three miles to the north-west of Santa Anna point is Mount San Felipe, 1,308 feet above the level of the sea.

**TIDES.**—It is high water, full and change, at Santa Anna point, at 12h. The strength of the tide is not great, but frequently after a southerly wind there is, in the offing, a current to the northward independent of the tide. In winter the tides occasionally rise very high, and on one occasion, in the month of June, they nearly overflowed the whole of the low land on the west side. Along the whole extent of the coast, between Santa Anna point and Elizabeth island, the flood sets to the southward and the ebb to the northward.

**PORT FAMINE,** in which was San Felipe, the old Spanish colony of Sarmiento, and since then the penal settlement of the Chilian Government, now removed to Sandy point, is situated south of Santa Anna point. Landing may be almost always effected, excepting in easterly gales, on one side or the other of this point. The position of Captain King's tent and observatory is indicated by the stem of a tree 16 inches in diameter, placed upright about 3 feet above the ground and banked up by a

mound, as shown in the plan ; it stands in lat. 53° 38′ 12″ S., and long.
70° 57′ 44″ W.* High water, on full and change, at 12h. ; rise 5¼ feet.

**Supplies.**—There is firewood in abundance on the beaches, and wells
containing excellent fresh water were dug by the *Adventure* at the north-
west extremity of the clear part of Santa Anna point, on the bank above
the third, or westernmost, small shingle bay. The water of the river, as
well as of the ponds, of which there are many upon the flat shore of the
western side of the port, was very good for present use, but did not keep,
in consequence of its flowing through an extensive mass of decomposed
vegetable matter ; but the water of the wells drains through the ground,
and not only kept, but was remarkably clear and well tasted. The situation of the Watering-Hole is marked on the plan.

**The River Sedger** in the south part of Port Famine is fronted by a
bar that dries at low water, but can be entered by boats at half tide,
and is navigable for 3 or 4 miles ; beyond which its bed is so filled up by
stumps of trees that it is difficult to penetrate farther. The water is fresh
at half a mile from the entrance, but to ensure its being perfectly good it
would be better to fill the casks when the tide is quite out. The low
land near the mouth of the river, as well as the beach of the port, is
covered with drift timber of large size, which was found very useful, and
serviceable for repairing the boats.

This river was called by Sarmiento, Rio de San Juan. Nathaniel
Peckett, Sir John Narborough's "ingenious lieutenant," calls it Segars
river, and his boat is described to have gone up it for 9 miles ; but was
there stopped from going farther by "reason of the trunk-timber and
shoaliness in the water." Byron describes the river, which he calls the
Sedger, in glowing terms, and gives rather a more flattering account of
the timber growing on its banks than it deserves ; but "the fallen trees
rendered it impossible" for him to go farther up than 4 miles.

**DIRECTIONS.**—To enter Port Famine with a leading wind, round
Santa Anna point at the distance of 2 cables' lengths in 17 fathoms ; but if
the wind is scant, do not get too near, on account of the eddy tide, which
sometimes sets on the point. Steer in towards the head of the bay, for
the summit of Mount San Felipe, keeping it half way between the rivulet
(which will be easily distinguished by a small break in the trees) and the
north-west end of the clear bank on the western side of the bay. This
bank being clear of trees, and covered with grass, is very conspicuous.
Keep on this course until the mouth of Sedger river is open ; and upon
shutting in the points of its entrance, shorten sail and anchor in 9, 8,
or 7 fathoms, as convenient.

* *See* Admiralty Plan of Port Famine, No. 555 ; scale, *m* = 1·5 inches, and note to
Table of Positions, p. 376.

The best berth, in the summer, is to anchor over towards the west side in 9 fathoms, with Cape Valentyn in a line with Santa Anna point ; but in the winter season, with north-east winds, it is better to anchor more in the centre of the bay. The strongest winds are from the south-west. It blows also hard sometimes from south, and, occasionally, a fresh gale out of the valley, to the southward of Mount San Felipe. Unless a long stay be meditated, it would be sufficient to moor with a kedge to the north-east ; the ground is excellent all over the port, being a stiff tenacious clay.

**VOCES BAY** is immediately to the southward of Port Famine, and of the River Sedger. A ship may anchor in from 7 to 10 fathoms, off Second river, but the shelter is not so good as in Port Famine. Second river has a shoal entrance, but comes from some distance up the valley ; between this bay and Cape San Isidro the water is too deep for anchorage, even close to the beach.

**CAPE SAN ISIDRO,** with a low but conspicuous rounded hillock covered with trees at its extremity, forms the termination of a ridge, the summit of which is Mount Tarn, the most conspicuous mountain near this part of the strait. A rocky patch, covered with kelp, stretches off the cape for 2 cables' lengths, with a rock at its end which is awash at high-water.*

**MOUNT TARN** is readily distinguished from abreast of Elizabeth island, at 50 miles to the northward, whence it appears to be the most projecting part of the continental shore. When viewed from the northward its shape is peaked, and during the summer it has generally some patches of snow a little below its summit ; but in the winter months its sides are covered with snow for two-thirds down. From abreast, and to the southward, of Port Famine, it has rather a saddle-shaped appearance ; its summit is a sharp ridge, extending nearly a mile, north-west and south-east, with a precipitous descent on the north-east, and a steep slope on the south-west sides. The highest peak near its north-east end is 2,602 feet above the sea by barometric measurement. The coast from Cape San Isidro trends S.W. by S., about 7 miles to Glascott point. Between these two points are several small bays, in which a vessel may find temporary anchorage.

**EAGLE BAY,** under Cape San Isidro, is a recess of three-quarters of a mile, with anchorage at the head of the bay, in from 20 to 12 fathoms. A small reef reaches out a cable from the south-west point of the bay, and on

* See Admiralty Plan, No. 2 of Ports in Magellan strait, No. 556 ; scale, m = 0·75 inch, m = 1·5 inch.

it there is an islet.  Eagle bay is useful only for a small vessel that can be towed in, and it will be necessary to steady even her by warps to the shore.  The squalls, or williwaws, at times, are very violent.  Two streams fall into it ; but the water being very much impregnated with decomposed vegetable matter, cannot be preserved long.  The woods here abound with Winter-bark, of which there are many large trees.

**GUN BAY,** the next to the south-westward, although small, affords anchorage for a single vessel near the shore, at its south-west part, in from 8 to 9 fathoms.  Two rivulets discharge themselves into it, from which water is easily procured.  The bottom is a stiff clay, and good holding ground.  A round hill of moderate elevation, and thickly wooded, separates it from Indian bay.  From the east point the shore runs due west, and then curves round towards an inlet covered with trees ; between it and the shore there is only sufficient depth for a boat to pass.  A rock about 12 feet high lies to the south-east, on either side of which there is an anchorage, sufficiently sheltered from the prevailing winds, on a good bottom, and in from 7 to 9 fathoms.  The north side of the bay is shoal, caused probably by the alluvial deposit from a stream nearly in the centre.  A patch of kelp projects from the south-east point 2 cables' lengths, but carries 9 fathoms over its middle.  Neither Gun nor Indian bays are noticed in Cordova's description of the strait, although they are quite as good as any others in the neighbourhood for stopping-places.

**BOUCHAGE BAY** is small, and the water very deep, except near the head of the bay, where anchorage may be obtained in 8 fathoms, clay.  It is separated from Bournand bay by Cape Remarquable, which is a precipitous, round-topped, bluff projection, wooded to the summit.  At 2 cables' lengths from the cape no bottom was found with 30 fathoms of line ; but at the distance of 50 yards the depth was 20 fathoms.

Bournand bay is more convenient than its northern neighbour, being somewhat sheltered from the southerly winds by Nassau island; and there being a rivulet of good water at the south-west end of a stony beach, off which there is good anchorage in 8 fathoms, stiff mud.

**BOUGAINVILLE COVE,** known to sealing vessels by the name of Jack harbour, forms a basin almost as snug as a wet dock, in which a vessel might careen with perfect security.  From its small size, great depth of water, and equal height of the land, it is rather difficult of access, and therefore vessels will find it almost always necessary to tow in.  On entering, the anchor should be dropped in 12 fathoms, and the vessel steadied by warps to the trees, along the sides of the cove.  It is completely sheltered from all winds, and an excellent place for a vessel to

remain at, particularly if the object be to procure timber, which grows here
to a great size, and may be both readily cut down and easily embarked;
it was here that M. de Bougainville cut timber for the French colony at
the Falkland islands. A rivulet at the head of the cove affords a moderate
supply of water; and if more be required, the neighbouring bays will
afford an abundance.

Nassau island lies less than a mile S.S.E. from Bougainville cove; it
terminates in a point of needle-shaped rocks, and is the only island in
the neighbourhood not covered with trees. In the Nassau channel,
between Nassau island and the main, the least water is 7 fathoms, over
a stiff clay bottom, gradually deepening on each side. But the winds
being baffling, and the tides irregular and rippling in many places, a
vessel should not attempt it but from necessity. Unless close off the
shore, the passage between the island of Nassau and the main cannot be
seen.[*]

SAN NICOLAS BAY is of larger size than any of the bays to
the southward of Cape San Isidro, and offers the best anchorage between that point and Cape Froward; as well as from its being more
easily entered and quitted, as from its moderate depth of water and
extent of anchoring ground. Nearly in the centre stands a small islet
covered with trees, between which and the shore there is a passage with
9 fathoms water and stiff clay bottom. The shore is, however, fronted for
its whole length by a shoal bank, which very much reduces the apparent
extent of the bay. This bank stretches off to the distance of a quarter of
a mile from the shore, its edge is steep-to, and is generally distinguished
by a ripple, which, with a moderate breeze, begins to break at half
tide. In passing through the strait, this bay is very useful to stop at, as
well from the facility of entering and leaving it, as from its proximity
to Cape Froward.

Into the middle of the bay issues the De Gennes river, 100 yards across,
and apparently flowing in a winding direction from a long way up the
valley. From its entrance being fronted by a shoal or bank, the form of
which must be constantly shifting, and from the bank being strewed with
trees which drift out of the river during the winter freshes, it is far
from being an eligible place for procuring water.

The *Beagle* anchored in the bay at 3 cables' lengths to the north-
east of the small central islet in 12 fathoms, pebbly bottom; but the
best berth is a quarter to a third of a mile to the south-west of the
islet, in 10 or 11 fathoms, muddy bottom. Captain Stokes recommends,

* Remarks of Commander F. L. Barnard, R.N., 1852.

in his journal, in coming in, to keep sail upon the ship, in order to shoot into a good berth, on account of the high land of Nodales peak, which becalms the sails ; and to avoid the drift of the stream out of the river, which would set the vessel over to the eastern side of the bay.  It is not, however, probable that the stream of the river would much affect a vessel between the islet and the peak.  In taking up an anchorage, much care is necessary to avoid touching the bank.  Less than 10 fathoms is not safe.

GLASCOTT POINT, the southern boundary of San Nicolas bay, and also of these small anchorages, forms the extreme of a high range of hills which run back for some distance ; the most conspicuous being the peak of Nodales.  From Glascott point the coast pursues nearly a straight line to Cape Froward, a distance of 7 miles, the interior land continuing mountainous and woody.  A point, formed by a beach of shingle, covered with trees to within 20 yards of the water's edge, and distant nearly 3 miles from Cape Froward, is the only projection.  Between this point and the entrance of a rivulet which waters the only valley in this space, there is an anchorage at a quarter of a mile from the shore, in 11 fathoms, which might be occupied during a westerly wind ; but with the wind more southerly, it would be too much exposed to be safe.  The *Beagle* anchored there at 2 cables' lengths off the sandy beach, in the above depth.

CAPE FROWARD, the southern extremity of the continent of South America, rises abruptly from the sea.  At its base stands a small rock, on which Bougainville landed, as did Lieutenant Graves also, for the purpose of obtaining angles and bearings.  The hill that rises immediately above the Cape was called by Sarmiento, the Morro of Santa Agueda.  Cape Froward is in lat. 53° 53' 43" S., and long. 71° 18' 15" W.  It is high water, at full and change, at 1h. ; the ebb tide setting to the northward, and the flood to the southward ; but with very little strength.

---

Having now reached the middle point of Magellan strait, round which it suddenly turns to the north-west, we will cross over and describe the inner shores of the main island of Tierra del Fuego, with Dawson and Clarence islands, including the Magdalen, Cockburn, and Barbara channels.

USELESS BAY, to the eastward of Cape Boqueron, was examined, in the hope of its communicating with the supposed San Sebastian channel of the old charts ; but it proved to be terminated by low land, reaching, perhaps, across the country towards San Sebastian bay on the east coast of the island.  It is more than 30 miles deep, and from 12 to 20

wide, and entirely exposed to the south-west.  The northern shore affords
no shelter, but on the southern side there is an indentation of the coast
line under the hill, called Nose peak, which may possibly afford a sheltered
anchorage.

This country abounds with guanacos, and the Indians are probably
more dependent on hunting than fishing for their subsistence, for we
observed their fires upon the hills, at a distance from the coast.

**DAWSON ISLAND**, which fronts both Useless bay and the deep inlet
called Admiralty sound, is 46 miles long and about 20 broad.  Its northern
extremity, Cape Valentyn, is low, but becomes visible in passing down
the opposite shore, between Sandy point and Freshwater bay.    Mount
Graves, however, which is 1,498 feet in height, is seen from a much greater
distance.   On the western side of the island there are but two places in
which vessels can anchor, viz., Lomas bay and Port San Antonio, but both
being on a lee shore, they are not to be recommended.  Lomas bay is a
deep bight, sufficiently sheltered from south-west, but quite exposed to
the north-west and westerly winds, which, during the winter, are the most
prevalent.

**Cape Valentyn** is the northern extremity of Dawson island.   It is low,
and has a small hummock near the point.   Between the two points which
form the cape, there is a slight incurvation of the shore which would
afford shelter to small vessels from any wind to the southward of east or
west ; but the water is shoal, and the beach, below high-water mark,
consists of large stones.   The coast to the south-west is open, and
unsheltered ; it is backed by cliffs : the beach is of shingle.

Lieutenant Graves remarks that Lomas bay, although only tolerably
sheltered from the prevailing winds, would, from its extent (6 miles) and
from the nature of the bottom, a stiff blue clay, afford good shelter for
vessels of any draught or burthen.  The appearance of the shores also seem
to favour such an opinion, for scarcely any drift wood was found thrown
up, even in those parts which were most exposed to the surf.  Wood is
sufficiently plentiful and, water very abundant.   This bay appears at
certain seasons to be much resorted to by the Indians, for upwards of
twenty wigwams were seen near the beach.

Between Lomas bay and Cape Valentyn, there is no landing even for a
boat, excepting at Preservation Cove, which affords only just room enough
to beach one of small size.

**PORT SAN ANTONIO**, on its western coast, opposite to San Nicolas
bay, has the appearance of being well sheltered, but during a fortnight that
the *Adventure* was there so much inconvenience was experienced, and

even risk, from the violence of the squalls, that they were obliged to secure the ship with three anchors. Some difficulty was also found in leaving it, on account of the baffling winds, as well as the narrowness of the passage, as she went out by the north entrance. This place was called Port San Antonio by Cordova; it is scarcely a third of a mile across, and deserves the name only of a cove. It is a very unfit place for a ship or indeed, for any vessel to enter, especially as there are so many much better places on the opposite or continental shore.*

The harbour is formed by the channel between Dawson island and the two adjacent islands of North and San Juan, in the latter of which, particularly at the north end, are several islets. The anchor may be dropped in from 10 to 15 fathoms, off a small beach in Humming-Bird cove, which lies on the inner side of, and about half a mile from, the south end of San Juan island. From the west end of North island a reef extends for a quarter of a mile, and to the southward there are two small islets, which may be passed on either side. North island is separated from San Juan island by a narrow and impassable strait.

The southern entrance is, perhaps, the best, although with a northerly wind the northern should be preferred. There is no danger but what is apparent; the ground, however, is not very clean until you reach Humming-Bird cove; in entering, haul round the south point of San Juan island, for near the shore of the eastern side there is a rock under water. Opposite to Humming-bird cove, in a small bight, there is a stream of fresh water.

**Port Valdes** is a deep inlet fronting W.N.W., and not at all inviting to enter; from the appearance of the hills, which on both sides of this port rise to an elevation of from 2,500 to 3,000 feet, squalls must be very frequent, and blow with the greatest violence, for trees are seen torn up by the roots, in long lines, evidently caused by the destructive force of the wind.

**EAST COAST.**—The eastern side of Dawson island is much intersected by deep inlets, particularly Brenton sound, and its termination, Port Owen, which reaches within 4 miles of Lomas bay on the west; the dividing land being low and marshy. Wickham island, in Brenton sound, is high, and there is a remarkable sharp-peaked hill on it, which is seen in clear weather from Port Famine. Non-entry Bay was not examined, though it appeared to offer snug anchorage; the depth between the points of entrance was from 9 to 19 fathoms.

**Fox Bay.**—The south side of this bay is shoal, but the banks are indicated by kelp. A rapid stream of water empties itself into the bay.

---

* See Admiralty Plan of Ports in the Strait of Magellan, No. 556; scale, m = 1·5.

The anchorage is in from 3 to 5 fathoms.  The north head, Tree Bluff, is of bold approach : within 20 yards of the shore the depth is 9 fathoms. Harris bay is an indenture of the coast, 2 miles deep.  Willes bay, west of Offing island, by which it may be known, although of small extent, affords excellent anchorage, upon a mud bottom, in 9 or 10 fathoms.  At the bottom of Willes bay is Gidley cove, where a small vessel may lie in perfect security.  There are not less than 3 fathoms in the entrance, and inside, in most parts, there is the same depth.  It is high water, full and change, about 12h. ; rise of tide 6 feet.

GABRIEL CHANNEL separates Dawson island from Tierra del Fuego. It is merely a ravine of slate formation, into which the water has found its way and insulated that island.  It is 25 miles long, and from half a mile to 1½ miles wide ; extending precisely in the direction of the strata, with almost parallel shores, the narrowest part being in the centre.  The northern shore is a ridge of slate, rising abruptly to a sharp edge, and then as abruptly descending on the opposite side, where it forms a valley ; which, had it been a little deeper, would have been filled by water and have become another channel like the Gabriel.  At its south east end it divides, one part leading in Admiralty sound, the other opening into Fitton harbour.  The ebb tide sets to the northward through the channel.

MOUNT SARMIENTO.—The south side of the Gabriel channel is formed by a high mass of mountains, probably the most elevated land in Tierra del Fuego. Among its many high peaks are two more conspicuous than the rest, Mount Sarmiento and Mount Buckland.  The first, situated at the south-east angle of Magdalen sound, is 6,800 feet high, and rising from a broad base, terminates in two peaked summits, bearing from each other N.E. and S.W., and about a quarter of a mile asunder.  From the northward they appear very much like the crater of a volcano ; but when viewed from the westward, the two peaks are in a line, and the volcanic resemblance ceases.  It is noticed by Sarmiento, as well as by Cordova, in the journals of their respective voyages.

Mount Sarmiento is the most remarkable mountain in Magellan strait; but, from the climate and its being clothed with perpetual snows, it is almost always enveloped in condensed vapour.  During a low temperature, however, particularly with a N.E. or S.E. wind, when the sky is often cloudless, it is exposed to view, and presents a magnificent appearance.  From its great height and situation it served our purpose admirably to connect points of the survey.  It was seen, and bearings of it were taken, from the following distant stations, viz., Elizabeth island, 96 miles to the north, Port Famine, Cape Holland, Port Gallant, and Mount Skyring, at the southern entrance of the Barbara Channel.

Mount Buckland, on the west shore of Fitton harbour, is, by estimation, about 4,000 feet high. It is a pyramidal block of slate, with a sharp-pointed apex, and covered with perpetual snow. Between these mountains the summit of the range is occupied by an extensive glacier ; the constant dissolution of which feeds innumerable cascades, which pour large bodies of water down the rocky precipices overhanging the southern shore of the Gabriel channel.

ADMIRALTY SOUND extends for 43 miles to the eastward into the island of Tierra del Fuego. It is 9 miles wide at the entrance between Useless bay and Dawson island, and gradually diminishes to 3 miles. On its north side the shore is straight, but the south side has 3 deep inlets, named Brookes, Ainsworth, and Parry harbours. The sound terminates in a bay; affording anchorage in from 10 to 15 fathoms, but much exposed to north-west winds, which, from the funnel shape of the sound (probably), blow with furious strength. On the north side of the bay is Mount Hope, a lofty insulated mass of rock, but to the southward of it lies a considerable tract of low land ; over which the view is unobstructed for a considerable distance, being bounded only by a distant mountain, in the direction of Captain Basil Hall's volcano, in lat. 54° 48' S., long. 68° W. If that volcano does exist, it is most probably in the above mountain, but nothing was seen during Captain King's visit to indicate the appearance of its being in an eruptive state. It is placed on the chart from Captain Hall's authority, as seen in rounding Cape Horn in the year 1820.

Port Cook is a very convenient and useful port, on the south shore of the sound. It is sheltered by a high wooded island. The anchorage is off the rivulet on the west side, in 9 fathoms. Brookes harbour is spacious, but not good as a port, for the water is deep, and the anchorages being in coves, are not easy of access without the labour of towing. In Ainsworth harbour there is anchorage at the bottom, on the west side. The mountains at the back of the harbour are capped by an enormous glacier that descends into the sea. Parry harbour is about 5 miles deep and 3 miles wide ; at the entrance, on the west side, there are two covse, either of which offers a convenient stopping-place for a small vessel.

MAGDALEN SOUND.—The opening of Magdalen sound was first noticed by Sarmiento. Coming from the northward, it appears to be a continuation of the strait, and it is not until after passing Cape San Isidro that the true channel becomes evident. It extends in a southerly direction for 20 miles, and is bounded on either side by high and precipitous hills, particularly on the western shore. The eastern entrance of the sound, Anxious point, is a low narrow tongue of the land, with an island off it. Opposite to it, on the Clarence island side, there is a steep mountain, called by Sarmiento the Vernal (or summer-house), from a remarkable lump of rock upon its summit.

**Hope Harbour** lies under this mountain ; a convenient stopping-place for small vessels passing through Magdalen sound. The entrance is narrow, with kelp across it, indicating a rocky bed, but on it not less than 7 fathoms were found. Inside it opens into a spacious basin, with good anchorage in 4 fathoms, sheltered from all winds, excepting the squalls off the high land, which must blow with furious violence during a south-westerly gale. This little port is much frequented by the Indians, for many wigwams on the south side were occupied by the women and children of a tribe, the men of which were absent on a fishing excursion.

**Stokes Inlet** lies to the southward of Hope harbour, between the Vernal and Mount Boqueron. It is 3 miles long, with deep water all over ; there is a cove on its north side, but neither so good nor so accessible as Hope harbour. In the entrance of the inlet lie the three Rees islets.

**Mount Boqueron**, the extremity of which is Squally point, is a very precipitous and lofty mountain, about 3,000 feet high, and carrying on its summit three small but remarkably conspicuous peaks. It is the eastern ridge of Stokes inlet, and forms a part of the western shore of Magdalen sound. The squalls that blow down the sides of this mountain during a south-west gale are most furious, and dangerous unless little sail be carried. On one occasion the *Adventure's* decked sailing-boat was seven hours in passing it. The sound here is not more than 2¼ miles wide. On the opposite shore, within Anxious point, an inlet extends to the south-east for 2 or 3 miles, but it is narrow and unimportant.

**Shell Bay** is a small bight of the coast line, 5 miles to the southward of Squally point. There is a reef off it, which is pointed out by the kelp. Keats sound, on the opposite shore, extends to the eastward for 6 or 8 miles, and is between 4 and 5 miles wide. In the middle of Magdalen sound, off Ariadne point, there is a rocky islet ; and a little farther to the southward, on the western coast, a bay containing the Labyrinth islands, among which small vessels may find good anchorage.*
Transition sound is deep, but of little import. Four miles farther at Cape Turn the channel narrows to 2 miles, and the shore turns suddenly to the westward. Here Magdalen sound terminates, and Cockburn channel commences.

On the eastern shore, to the southward of Keats sound, there are no objects worth noticing, excepting Mount Sarmiento, which has been

---

* *See* Admiralty Plan, No. 556 ; scale, *m* = 0·75 of an inch.

already described, and Pyramid hill, which was found to be 2,500 feet high. The southern shore of Tierra del Fuego being much broken by many sounds penetrating deeply into the land, the isthmus between Thieves sound and the bay under Pyramid hill is only 7 miles across; and 11 miles more to the westward, at Courtenay sound, its breadth is not more than 3 miles.

COCKBURN CHANNEL, between Clarence island and Tierra del Fuego, with its many spacious inlets and clusters of islands, begins at Cape Turn, and runs in a westerly direction for 33 miles, where it enters the Pacific between the Magill and Camden islands (see page 150). In working down the channel, the south side should be preferred, as it is usually a weather shore, and seems to be better provided with coves and harbours in which vessels might find it convenient to anchor.

Warp Bay lies at its commencement, which, though small, and exposed to southerly winds, is a convenient stopping-place. Stormy bay, on the north shore, is a very wild, unsheltered place, unfit for any vessel to enter. At the anchorage the water is deep, 17 to 20 fathoms, and the bottom rocky. The bay is strewed over with shoals, but though they are all marked by kelp, they narrow the channel so much as to render the entrance and exit both intricate and difficult for any but a small and handy vessel.

Park Bay, at the entrance of Mercury sound on Clarence island, is both very snug and secure, with good anchorage in 12 fathoms, sand and mud. It has the same disadvantage as Stormy bay, it being on the lee side of the channel, and is therefore difficult to leave. There is, however, more room to beat out, and no dangers to encounter but what are visible. At the N.E. angle of the bay a narrow isthmus, not more than 500 yards across, separates it from Mercury sound, which was not examined, it being represented on the chart from an eye sketch. King and Fitz-Roy islands, in mid-channel, are of bold approach, as are also Kirke rocks, more to the westward.

Prowse Islands, on the south shore, are very numerous, and skirt the coast for several miles. There are several anchorages among these islands, behind them the land trends in, and forms a deep sound. The *Adelaide* schooner anchored in a bay on the north side of one of these islands, opposite to Barrow head, in 6 fathoms; but there are many places of a similar nature equally convenient and secure. A vessel in want of anchorage should hoist a boat out, and wait in the offing until one answering the purpose be found; and when about to enter one of these deep-water bays, a boat should always be ready to carry a hawser to the shore. It will frequently be necessary to tow up to the head of the harbours; for, from the height of the land, the wind generally fails or becomes baffling. The distance across the channel, between Prowse islands and Barrow head, is scarcely 1¼ miles.

**Dyneley Sound** extends for more than 9 miles in a north-west direction into the interior of Clarence island. On the west side of its entrance there is a group of islands, affording several anchorages. On the western shore, Eliza bay offers shelter and security from all winds. The bottom of Dyneley sound was not examined.

**TIDES.**—The flood tide sets to the southward, or to seaward, but was not found to run with sufficient strength to benefit or impede a vessel beating through. The rise and fall is also inconsiderable, not being more than 6, or at most 8 feet, at spring-tides.

**ADELAIDE PASSAGE,** lying between the broken land of Clarence island and the Magill group, is about 13 miles long, connecting the Cockburn and Barbara channels. The Magill islands, as well as the entrances to the above channel, have been described in page 150.

**BARBARA CHANNEL** is about 38 miles long from the Magill islands in the Pacific to Charles island in the Strait of Magellan. It separates Clarence from St. Ines island, and is strewed with many rocks and shoals, some of which, although covered with kelp, only show at half-tide. Much caution is therefore necessary, and all patches of kelp should be carefully avoided.*

**Hewitt Bay** is the first anchorage on the western shore of this channel; there is anchorage in 9 fathoms in the north part of this bay. Browns bay, 2 miles to the northward, is extensive, and also affords good shelter, in a small cove at the north entrance, in 8 fathoms sand, among some kelp. Nort bay, on the same side of the channel, for a small vessel, is tolerably secure, but not to be recommended. The tide, to the northward of Nort bay, which, to the southward, was not of sufficient strength to interfere with the navigation of the channels, is so much felt here as to impede vessels turning to windward against it.

The country here has a more agreeable appearance, being better wooded with beech and cypress trees; but the latter are stunted, and do not attain a greater height than 15 or 18 feet. They are, however, very serviceable for boat-hook spars, boats' masts, &c. The wood, when seasoned, works up well.

**Bedford Bay,** on the western side of the south narrows caused by Browell island, is a good anchorage. Its depth is from 20 to 8 fathoms,

---

* See Chart of the Barbara channel and ports, Nos. 2,113 and 1,306; scales, m = 0·25 and 1 inch.

good holding-ground, and sheltered from the prevailing winds. At its entrance there are several patches of kelp, the easternmost of which has 4 fathoms on it. As it is a place likely to be frequented by vessels navigating the strait, the plan may be of service.*

**Nutland Bay,** 5½ miles to the northward, has 8 and 15 fathoms over a sand and mud bottom. It may be known by two rocks, named the Hill islands, which lie a mile N.N.E. from the anchorage. Between Bedford and Nutland bays, and, indeed, as far as the Shag narrows, the channel is open, and may be navigated without impediment. There are many bays and inlets which, though not here described, might be conveniently occupied ; but all would require to be previously examined, for though they all trend far enough into the land to afford good shelter, yet in many the bottom is foul and rocky, and the water too deep for anchorage. The western, being the windward shore, should of course be preferred.

**Broderip Bay** lies to the northward of Nutland bay. At its northern part there are some good coves, and a very convenient one at its eastern extremity. This cove extends to the northward for about an eighth of a mile, affording good anchorage in 10 fathoms, and sufficiently well sheltered and distant from high land to be free from the mountain squalls, or williwaws. Icy sound lies round the cape that bounds this cove. It is a deep inlet with a glacier of considerable extent, from which large masses of ice are constantly falling, and drifting out of the inlet. The water is deep, and the anchorage should not be chosen when there are so many better places.

**Dean Harbour.**—Dean harbour is another considerable inlet trending in under the same glacier, and extending to the head of Smyth harbour as well as to a great distance into the interior. If of a favourable depth it might afford good anchorage, but the *Adventure* did not enter it. Field bay is an open bay at the head of the channel, too much exposed to southerly winds to be recommended as a stopping-place, unless the wind be northerly. Nutland bay is a more convenient place to start from with a view of passing the Narrows.

**SHAG NARROWS** is the only navigable communication that exists between the Barbara channel and Magellan strait, along the western side of Cayetano island. The breadth of the opening is at least 1¾ miles, but the eastern portion is so filled with rocky islets and shoals, that the actual navigable passage at the northern end is only 100 yards across; and the widest part at the southern end, scarcely half a mile. The whole length

---

* *See* Admiralty Plan of Ports in Barbara channel, No. 1,306 ; scale, *d* = 1·0 inch.

of the passage is rather less than 2 miles. It is formed on the western side by a projecting point of high land, which gradually trends round to the westward; and on the opposite side by three islands, the northernmost of which is Wet island, and the southernmost Mount Woodcock, one of the points of the triangulation; all the space to the eastward of Mount Woodcock is so thickly strewed with islands and rocks, that the passage is as difficult as it is dangerous. Between Wet island, where the Northern Narrows commence, and the western shore, the breadth is not more than 100 to 150 yards, and perhaps 300 yards long.

TIDES.—In these Narrows, at full and change, the stream commences to set to the southward at 12 o'clock. Through them the tide sets as much as 7 knots, but the sides of the rocks being steep-to, an accident can scarcely happen to a ship in passing them, notwithstanding the want of room for manœuvring. At Wet island, the stream of the ebb divides —one part sets to the eastward, round Wet island, while the principal volume runs through the Shag Narrows. And in the same manner a part of the southern tide, or flood, after passing Wet island, runs to the south-east, round the eastern side of Mount Woodcock.

DIRECTIONS.—To avoid the danger of being thrown out of the Narrows, it is only necessary to keep the western shore on board : where there are no indentations, the tide will carry a vessel along with safety. At the northern end of the Narrows, on the western side, there is a shelving ledge with only 5 fathoms; and there is an eddy, but as soon as the vessel is once within the Narrows (that is, inside of Wet island), the mid-channel may be kept. In shooting this passage, it would be better to furl the sails and tow through, for if the wind be strong, the varying and violent squalls would be very inconvenient, sometimes baffling, and sometimes almost laying the vessel up on her beam ends, though every sail be furled. It will be necessary to have a couple of boats out, ready to tow the ship's head round, and also to prevent her being thrown by the tide into the channel to the southward of Wet island.

If anchorage be desirable, on leaving the Narrows, there is none to be recommended, until the coves between Smyth harbour and Cape Edgeworth be reached. There is another passage into the strait through St. Michael's channel, on the south side of Cayetano island, into Toms narrows and Simon bay. (See page 188.)

SMYTH HARBOUR, the first opening after passing the Narrows, is about 4 miles long in a westerly direction, and from half a mile to a mile wide and is surrounded by high land. The water is deep, excepting in Earl cove, on the north side, where vessels might lie, if necessary; but it

would probably be a very wild place in bad weather.   The hills at
the head of the bay are capped by glaciers that communicate with those
at the head of Icy sound ; and all the mountains between this place and
Whale sound in the strait appear to be entirely covered with a coating
of ice.*

**Dighton Bay and Warrington Cove,** to the northward of Smyth harbour,
offer also good shelter and anchorage, but both are exposed to easterly
winds.   Dighton bay is the best.   The anchorage is off the sandy beach
in 20 fathoms.   Edgeworth shoal lies half a mile south-east of Cape
Edgeworth, and is so thickly covered with kelp as to be easily seen
approaching it; there are not more than 2 feet water over its shoalest part.
To pass through the Barbara channel, from the northward, it would be
advisable to stay at Fortescue bay, on the north shore of the Strait of
Magellan (see page 191), until a favourable opportunity should offer; for
with a south-west wind, it would not be safe, even if practicable, to pass
the Shag Narrows.

* *See* Plan of Ports in the Barbara channel, No. 1,306.

# CHAPTER VII.

STRAIT OF MAGELLAN—BARBARA CHANNEL TO CAPE PILLAR.

VARIATION 22° to 23° East in 1860.   Annual decrease about 2.'

---

HAVING described the southern channels leading into the Pacific, we will now return to Froward reach, and proceed through the Strait to the north-westward.   North-west winds prevail more than any others in the western portion of the strait, in consequence of the reaches trending in that bearing.   It seems to be a general rule hereabouts that the wind either blows up or down them; and seamen should be aware that between Cape Froward and the western entrance of the strait, the wind is generally from the north-west, although at sea, and in the Cockburn channel, it may be in the south or south-western quarters.

CLARENCE ISLAND, on the south side of Froward reach, extends from Magdalen sound to the Barbara channel; and the whole length of its northern coast is indented by sounds stretching deeply into the island. Port Beaubasin, at the western entrance of Magdalen sound, is sufficiently pointed out by the small rocky islet called Periagua, and by Mount Vernal. The outer part of the port decreases in breadth gradually to the entrance of the harbour, which is formed by two projecting points, a very short distance apart, and is very shoal, the deepest water being only 2¾ fathoms. Inside in the basin, there are 5 fathoms.   It is a very snug place when once in, but possesses no advantage, since it is on the wrong side of the strait for vessels bound to the westward, as the northerly wind, which would be favourable · to proceed, would prevent a vessel sailing out. Inman bay, Hawkins bay, Staples inlet, and Port Sholl, are all deep inlets, surrounded by high precipitous land.

LYELL SOUND penetrates 9 miles into Clarence island to the westward of Greenough peninsula, and is separated from Sholl harbour by a ridge of hills only 1½ miles wide.   In the entrance of the sound, there are two conspicuous islands, though one of them is very small.   They are called the Dos Hermanas (Two Sisters), and bear from Cape Froward S.S.W. 5½ miles. Kempe Harbour, 1¼ miles within the entrance of Lyell sound, on the western side, is rather difficult of access, but perfectly secure, and would hold six ships. Stokes creek, on the same side, and more to the south-ward, also offers good anchorage; but from its being out of the way, can be of no utility.

**Mazaredo Bay and Cascade Harbour** are of less size, and therefore more attainable, but of the same character with Lyell sound: viz., deep water surrounded by high land. The former is known by the cascade which M. de Bougainville describes, from which it derives its name. On the headland that separates these harbours from Lyell sound there is a sugar-loaf hill, the position of which was well determined to be in 53° 57′ 32″ S., and 71° 27′ 58″ W. Hidden harbour has a narrow entrance; but, if required, offers a good shelter.

**San Pedro Sound** is the most extensive inlet with which we are ac-quainted in Clarence island. It extends in a southerly direction for nearly 13 miles, and has three other inlets branching off into the land, two to the westward and one to the eastward. There is a good, although a small anchorage on its western side, 1½ miles within the entrance, called Murray cove, and another close to it; which is even more sheltered. Freshwater cove, on the shore of the strait, is a confined and indifferent place.

**Bell Bay** has one very convenient anchorage, Bradley cove, on its western side, bearing S.W. ¾ W. from Taylor point, the eastern head of the bay. It will be readily distinguished by a small, green, round hillock that forms its north head. The anchorage is in 17 fathoms, and the vessel may be hauled in by stern-fasts or a kedge into 9 fathoms in perfect security. Pond bay to the northward has good shelter, but it is not of such easy access; for it would be necessary to tow both into and out of it. Mount Pond, a double-peaked hill over the harbour, 2,500 feet in height, is conspicuous, visible from the eastward when open of Cape Froward, on which bearing it makes as one peak.

**Simon Bay**, situated between Mount Pond and Cayetano island, is studded with islands and rocks. To the southward it communicates with the Barbara channel on both sides Burgess island; the easternmost of these channels is called Toms narrows, and is tolerably large, but from the irregularity and force of the tides it is not to be preferred to the more direct passage through the Shag narrows on the western side of Cayetano island; for there is no good anchorage in St. Michael channel (which leads to it, between St. Michael point and Cayetano), and it is bounded by a steep and precipitous coast. Gonsalez narrows, on the western side of Burgess island, is not more than 30 yards across; and from the force of the tide, and the fall of the rapids, would be dangerous even for a boat to pass.

**Millar Cove**, on Cayetano island, is the only good anchorage in Simon bay, lying about 3 miles within Elvira point, with three rocky islets off its entrance, and a conspicuous mount on its eastern point. The anchorage is in five fathoms, a good bottom, and entirely sheltered. Wood and water are plentiful.

**Port Lang'ra** is immediately round the above eastern mount of Millar cove. It is rather more than a mile long and two-thirds of a mile wide, and trends in a W.N.W. direction. The water is deep, excepting at the head of the port and in a cove on its northern shore, in either of which there is good anchorage. At the former the depth is 8 fathoms, and in the cove 5 fathoms. On the eastern side of the bay are Shipton and Mellersh coves. Both are surrounded by high land, and the water being very deep, neither of them affords anchorage. Off the head that divides them are the Castro isles; on the north side of the largest of which there is a very convenient cove with a moderate depth of water. In the Castellano group there five good-sized islands; they lie in the centre of the bay but have no anchorage among them.

To return to the north side of the channel.

**The COAST** from Cape Froward to Jerome channel, a distance of 40 miles, is very slightly indented. The anchorages, therefore, are few in number, but they are of easier access, and altogether more convenient than those of the southern shore.

**SNUG BAY.**—Taking them in succession, Snug bay, 5 miles W.N.W. of Cape Froward, is a slight hollow of the coast at the outfall of a small rivulet, the deposits from which have thrown up a bank near the shore, on which anchorage may be had in 8 and 9 fathoms. The best anchorage is half a mile to the E.S.E. of the island, in 9 fathoms, black sand; the rivulet mouth bearing N.N.W. three-quarters of a mile. It is much exposed, being open from W.S.W. by South to S.E.

Commodore Byron anchored in Snug bay in 1765, in *H.M.S. Dolphin*, with the following bearings: Cape Froward, E. ½ S., 5 miles; the islet in the bay, W. by S. half a mile; and the river's mouth, N.W. by W. three-quarters of a mile; having shoaled suddenly from 17 to 9 fathoms. He described it as being fit for his purpose, and it is certainly a convenient stopping-place in fine weather.

**CAPE HOLLAND**, 14 miles to the westward of Cape Froward, is bold and high, and although projecting but slightly, yet a very conspicuous headland. It is precipitous and descends to the sea in steps, plentifully covered with shrubs.

**WOODS BAY,** situated under the lee of Cape Holland, is a convenient stopping-place for ships, but only small vessels should anchor inside the cove. The anchorage is very good to the eastward of the river's mouth, at half a mile from the shore in 17 and 13 fathoms water. Small vessels may enter the cove by luffing round the kelp patches that extend off the south point of the bay, on which there are 2½ fathoms.*

**DIRECTIONS.**—Entering Woods bay, steer for the gap or lowland behind the cape ; and as you near the south point keep midway between it and the river's mouth ; or, for a leading mark, keep a hillock or conspicuous clump of trees at the inner end of the bay, in one with a remarkable peak 1 or 2 miles behind, bearing N.W. ¾ W., and anchor in 17 fathoms immediately that you are in a line between the two points. Small vessels may go farther in to 12 fathoms. The western side of the cove may be closely approached, and the depth will not be less than 5 fathoms, except upon the two-fathoms patch which stretches off the east point, and the extent of which is sufficiently shown by the kelp; but on the eastern side the bank shoals suddenly, and must be avoided, for there are 13 fathoms close to its edge, upon which there is not more than 2 feet water. Near Cape Coventry and in Andrews bay anchorage may be had near the shore if the weather be fine. To the westward of the former, at half a mile from the shore, there are 13 fathoms.

**CORDES BAY,** 12 miles to the westward of Woods bay, may be known by Mussel island, small, bright, and green, which lies in the entrance, and also by a hill to the northward with three hummocks, about 1,750 feet high, called Mount Three Peaks, and standing detached from the surrounding hills at the head of the bay. The western entrance which lies between the west point and the reef of Mussel island, is two-thirds of a mile wide ; within it, the bay continues a mile, but is much contracted by shoals covered with kelp ; between them, however, the anchorage is very good and well sheltered. The bottom is of sand, and the depth 5 to 7 fathoms.

**PORT SAN MIGUEL** is a large lagoon at the inner extremity of the bay, trending in a north-east direction for 2 miles, and two-thirds of a mile across ; the entrance to the lagoon is both narrow and shoal, and not safe for a vessel drawing more than 6 feet ; but inside the depth is from 3 to 13 fathoms. With Fortescue bay and Port Gallant so near, the probability is, that it will never be much used ; but in beating to the westward it would be better to anchor there than to lose ground by returning to

---

* *See* Sketch in No. 556.

Woods bay. By entering the western channel and steering clear of the kelp, a safe and commodious anchorage may be easily reached.

FORTESCUE BAY is the first good anchorage to the westward of San Nicolas bay ; it is spacious, well sheltered, easy of access, and of moderate depth. The best berth is south-east of the small islet outside of Wigwam point, in 7 or 8 fathoms.

PORT GALLANT forms the inner harbour to Fortescue bay; having the entrance open, small vessels may sail into the port, but the channel is rather narrow. The banks on the western side off Wigwam point may be distinguished by their kelp. When within, the shelter is complete ; but Fortescue bay is quite sufficiently sheltered, and much more convenient to quit. In this part of the strait, as the channel becomes narrowed by the islands, the tides are much felt, and run 3 knots an hour. One mile to the westward of Port Gallant is Mount Cross, rising to an elevation of 2,290 feet.*

PASSAGE POINT ROCK.—Passage point lies 8 miles from Cape Gallant, the West entrance to Fortescue bay. At a short distance E.S.E. of the point there is a shoal with 2 fathoms on it. There are two other anchorages before reaching the entrance of the Jerome channel, namely, Elizabeth bay and York Road river ; they are, however, only fit or stopping places.

ELIZABETH BAY, just north of Passage point, has a sandy beach, and a rivulet emptying itself into it. Cordova recommends, as the best anchorage, Passage point, bearing E.S.E., distant half a mile, in 15 fathoms, and about 3 cables' lengths from the river ; this being to the north-west of a bank on which there is much kelp. Commodore Byron, in the *Dolphin*, anchored here in 10 fathoms, with the following marks: Rupert island, S. by E. 2 or 3 miles ; the western part of the bay, W. by N. 2 miles ; and a reef of rocks about a cable's length from the shore, N.W. by W. a quarter of a mile. The reef is quite covered at high water.

YORK ROAD, or Bachelor bay, at the entrance of the Jerome channel, is a good and convenient anchorage. The best berth is half a mile off a wooded point (just to the westward of the river), bearing N. by W. ¾ W., and the mouth of the river N. by E. ¾ E. three-quarters of a mile, because there is plenty of room to weigh and space to drive should the anchor drag ; the bottom is good in 10 or 12 fathoms, but not in less depth. The

---

\* *See* Admiralty Plan, No. 556 ; scale, m = 1·5 inches.

shore is a flat shingle beach for 2 miles, the only one in this part of the strait. Cordova recommends the following as the best anchorage at half a mile from the beach, the river bearing N. by W. ¾ W., and the west point of the bay N.W. ¾ W.

**BACHELOR RIVER** is accessible to boats only ; and in going into or out of its entrance they must be very careful to follow exactly the course of the stream, for a bar lies outside ; large boats cannot enter at half-tide. Three-quarters of a mile to the eastward of Bachelor river lies a shoal which has not more than 6 feet upon it at low water, and 14 feet at high water ; it is about half a mile from the shore, and shows itself by the sea-weed. To the islands in the middle of the strait, which form the eastern limit of English reach, Captain King, in his survey, has restored the names that were originally given to them by Sir John Narborough.

**TIDES.**—The set and change of the tide here are very uncertain on account of the meeting of the Jerome channel tides with those of the strait through English and Crooked reaches, which occasion many ripplings, and would require a long experience to explain them correctly. Captain Fitz-Roy says, that "the tide along shore, near Bachelor river, changed an hour later than in the offing. In Bachelor bay, by the beach, during the first half or third of the tide that ran to the south-east, the water fell ; and during the later half or two-thirds it rose. In the offing it ran very strong." By the same officer's observations, the time of high water at the entrance of the river would be at full and change, at 1h. 46m.; but according to Captain Stokes, two years previously, it was 2h. 13m. ; and the stream at the anchorage ran 3 knots.

**The JEROME CHANNEL,** connecting those singular inlets, the Otway and Skyring waters, with the strait of Magellan, was only slightly examined by Cordova's officers; for their object being merely to confirm or disprove Sarmiento's statement that the land between it and the Gulf of Xaultegua was an island, the Lago de Botella was alone explored by them. The continuation of the Jerome, named in the old charts Indian sound, having never been traced ; and being, therefore, an object of great interest, it was investigated by Captain Fitz-Roy as carefully as could be done in the middle of winter in an open boat. The period of his absence from the ship, however, thirty-two days, not being sufficient to complete the service, the western shores of the Skyring Water were not visited.

The Jerome channel is narrow, but throughout free from danger. The western shore is high and steep, and covered with trees : the eastern shore is lower and less wooded. In mid-channel, near its western end here are two islets, called in the Spanish chart the Terran isles.

Commodore Byron, in the *Dolphin*, anchored on a bank in 15 fathoms, "which lies half a mile from the north shore," and 2 miles S.S.W. from the south point of the Jerome channel, with Cape Quod bearing W.S.W., 8 miles; but after veering two-thirds of a cable the ship was in 45 fathoms.

**Wood and Seal Coves** lying on the western side of the Jerome channel may be used with advantage by small vessels. On the eastern shore, also, Three Island bay and Coronilla cove, appeared to be commodious. Arauz bay is open and exposed to the N.W.

The *Dolphin* also anchored at 5 or 6 miles from Cape Quod, the cape bearing W.S.W., and the south point of Despair island (the largest of the Ortiz isles off Borja bay) just in one with the pitch of the cape, at half a mile from the shore; the depth was 45 fathoms, and close in shore there were 75 fathoms. Here the tide was found to run 8 hours to the eastward and 4 to the westward, at the rate of 1½ to 2 knots.

**Cutter Cove.**—Where the Lago de la Botella joins the Jerome channel, the latter winds round to the north-east. On its eastern side, behind the False Corona isles, is Cutter cove, affording anchorage for a small vessel. Opposite to it is Nuñez creek, with deep water.

**Sullivan Sound** lies abreast of the Corona isles, one of which, the Sugar Loaf, is about 200 feet high. It penetrates 5 miles into the land on the western side of the channel; and, at a league to the northward of the Sugar Loaf, there is another opening to the westward, on the northern shore of which is Bending cove; which, with Cutter cove, are the only stopping-places between Cape Forty-five and Child bluff.

**OTWAY WATER.**—Between Child bluff and Stokes point, 25 miles from the entrance of the Jerome channel, the Otway water commences, extending 46 miles in a N.E. by N. direction. The western shore affords several commodious anchorages; the eastern, apparently low land and lagoons, was not fully examined. An isthmus, 6 to 10 miles across, separates the Water from the strait near Elizabeth island. From an elevated station on the north-eastern side of Fitz-Roy channel, this narrow neck appeared to be low and much occupied by lagoons. The southern shore of Otway water is formed by high land, with three deep openings that were not examined; and behind them Brunswick peninsula, a mass of high mountainous land, forms the most southern termination of the continent.

Off Villiers point, in lat. 53° 09′ S., on the west side of Otway water, at a quarter of a mile from the shore, there are from 10 to 30 fathoms; and this depth decreases in advancing more northerly. There is anchorage all across the north-east part of the Water, in from 5 to 20 fathoms, the bottom of sandy mud. Inglefield and Vivian islands, at the western end of Otway water, are low but thickly wooded.

FITZROY CHANNEL forms the communication between the Otway and the Skyring waters, by a winding course to the N.W. for 11 miles. Its south-eastern entrance is in lat. 52° 47′ S., and long. 71° 22′ W.

TIDES.—Fitz Roy channel is easily navigated, but a strong tide runs, even during the neaps, at the rate of 5 or 6 knots in the entrance, and of 2 or 3 farther in, and sets through it 6 hours each way. The rise and fall, however, were scarcely distinguishable. The Spanish account says, "The current is always in the direction of the channel, but rarely sets to the N.W., particularly in mid-channel and on the western shore; on the opposite side, the tide sets 6 hours each way to the N.W. and S.E."

The following observations were made by Captain Fitz-Roy for the time of high water at full and change, in the Jerome channel and in the interior waters. At the entrance of the Jerome, near Arauz bay, at 1h.; near Bending cove, at 3h.; at Cutter cove, at 4h.; on the south shore of Fanny bay, at Gidley island, as also at Martin point at 5 h.; at Inglefield island, at 4h.; and at the same hour at the south-eastern entrance of Fitz-Roy channel; but at the north-western end of it at 1h. 15m.

The VARIATION of the compass in 1833 was found to be at the

|  | | | | | ° ′ |
|---|---|---|---|---|---|
| Point of Islets | - | - | - | - | 23 58 E. |
| Donkin cove | - | - | - | - | 23 40 |
| Wigwam cove | - | - | - | - | 23 34 |
| Inglefield island | - | - | - | - | 23 56 |
| Martin point | - | - | - | - | 23 58 |

The mean of which is 23° 49′ easterly, and with an annual decrease of 3′ would make the variation in 1860 to be 20° 28′.

SKYRING WATER is 10 leagues long from east to west. Its shores are low. At the western extremity two openings were observed to wind under a high castellated mountain (Dynevor castle), which were supposed by Captain Fitz-Roy to communicate with some of the sounds of the western coast, as through Euston opening, the southern one, no land was visible in the distance; but on a subsequent examination of the termination of Obstruction sound, by Captain Skyring, no communication was detected. The chart will give a sufficient idea of these two large inlets, which, as they are of no importance to the general navigator, will not be further described.

CHARLES ISLANDS, in the Strait of Magellan, lying between Fortescue bay and the Barbara channel, consist of three principal, and some smaller islets, and the centre forms a good roadstead, having anchorage within the islets in 13 fathoms. It has an outlet to the north-west, and one to the south-west; and also a Narrow communicating with the strait to the south-east.

**Secretary Wren Island,** the south-eastern of this group, is a small rocky islet, rising abruptly on all sides, and forming two summits. Near it, to the south-eastward, are two groups of small rocks; and at a mile E.S.E. there are two single rocks above water, called the Canoas. Opposite to Cape Gallant, on the eastern island, and near its north-west end, there is a conspicuous white rock, called the Wallis Mark. Next to the westward in succession are Monmouth and James islands, then Cordova islet, and lastly Rupert island. When the wind blows fresh there is a hollow sea between these islands and the northern shore, which very much impedes ships beating to the westward.

**CARLOS III. ISLAND** lies to the westward of these, and was so named by Cordova; it is separated from Ulloa peninsula on St. Ines island by the navigable channel of David sound, and from the northern shore of the Strait by English reach. To the northward of Whale point, the eastern extremity of Carlos III. island, there is a cove with an anchorage in 15 fathoms close to the shore, on a steep bank, but bad ground; the *Beagle* and *Adelaide* both dragged off the bank, from the violence of the squalls off the high land. From the north point of the cove a reef extends to Rupert island, called Lucky Ledge, over which the tide sets with considerable strength. The *Beagle* having dragged her anchor in the cove was brought up by its hooking a rock on the ledge, but it was found broken on being hove up. While there, the tides set past her in a north and south direction, at the rate of 3 knots.

**Mussel Bay,** to the westward of Cape Middleton, or Narborough, in English reach, has deep water, and an uninviting appearance. Cordova describes it to be a mile wide, with unequal soundings, from 12 to 40 fathoms, stones. The bay is not to be recommended, although it appears to be well sheltered. There is an anchorage on the northern side of this island, in from 15 to 30 fathoms, in Bonet bay, of Cordova, under the south-east side of some islands just opposite to Bachelor river. Tilly bay, at a short mile to the eastward of Cape Crosstide, the north-west end of Carlos III. island, has nothing to recommend it, particularly when the much better anchorage off Bachelor river, on the opposite side of the strait, is so close at hand.

**CHOISEUL BAY,** on St. Ines island, just to the southward of Monmouth island, is not in the least inviting; Captain Fitz-Roy describes it to be a large, deceiving bay, full of islets, and patches of kelp under which there are rocks, while between the islets the water is deep and unfit

for anchorage.   Nash bay, 4 miles to the westward, is equally un-
serviceable.

WHALE SOUND, also on St. Ines island, south of Ulloa peninsula,
is a large inlet trending 8 miles into the land, and terminating in a valley
bounded on each side by high mountains.   There is anchorage only in one
part of it, on the western side of Last harbour; and although this harbour
appears large, the anchorage is small, and close to the shore.

DAVID SOUND separates Carlos III. island from Ulloa peninsula.   At
its northern end the water is deep, but where it begins to narrow there
are soundings; and an anchorage might be found there if necessary, but
such a case is not probable.

CROOKED REACH.—English reach and David sound lead into Crooked
reach, which commences after passing the entrance of the Jerome channel.
In the navigation of this part Wallis and Carteret suffered extreme
anxiety; and no one that has read their journals would willingly run the
risk of anchoring in any port or bay on its southern shore.   The chart
will show several inlets which might tempt any navigator to trust to
them; and probably for small vessels, some sheltered nooks might be
found, but they have all very deep water, and when the wind blows strong
down Long reach, they are exposed to a heavy sea, and a furious wind.
This reach is the narrowest part of the strait, being little more than a
mile across from Cape Quod.

Cape Quod, a high projecting point on the northern shore of the reach,
has a very bleak rugged appearance; the almost perennial westerly wind
seems to forbid all vegetation on the heights exposed to its action; and
hence the desolate look of these shores.   Long Reach may be said to
begin at Cape Quod.   Narborough thus describes this cape:—"It is a
steep-up cape, of a rocky greyish face, and of a good height before one
comes to it: it shows like a great building of a castle; it points off with
a race from the other mountains, so much into the channel of the strait,
that it makes shutting in against the south land, and maketh an elbow in
the strait."   Abreast of Cape Quod, Captain Stokes tried and found the
current setting to the eastward at 1½ knots.

BORJA BAY (the island bay of Byron) is on the northern shore of Crooked
reach, 2 miles to the eastward of Cape Quod.   Its position is pointed out
as well by the Ortiz islets (vulgarly called Big and Little Borja), which
lie off its west point, as by its situation with respect to El Morion, the
helmet-shaped point formerly called by the English St. David Head. The
entrance to the bay is to the eastward of the larger of the Ortiz islets,

and presents no dangers ; all the islets and shores of the bay may be approached to half a cable's length, even to the edge of the kelp. Ortiz islets should be avoided in entering or leaving the bay.    The only impediments in the entrance are the baffling winds and violent gusts that occasionally come off the mountains and down the deep ravines which form the surrounding coast, and the utmost vigilance must be exercised in beating in under much sail to guard against their effects. Peaked mountain, on the north side of Borja bay, is 1,809 feet above the sea.*

The ANCHORAGE is sheltered from the westerly and south-westerly gales, which usually prevail there, and is open only to south-easterly winds, which very rarely blow here, and still more rarely with violence ; and as the holding-ground is good (small stones and sand), and the depth of water moderate (14 to 16 fathoms), and any fetch of sea prevented by the narrowness of the adjacent strait, the greatest breadth being only 3 miles, it may be pronounced a very good and secure port. The best plan is to anchor with the bower, and steady the vessel to the shore by a hawser or a kedge.    No surf or swell obstructs the landing anywhere ; good water and plenty of wood are easy to be embarked, and the trees, a species of beech, are of a considerable size.    The shores are rocky, and the beach plentifully stocked, as indeed are all parts of the strait to the eastward, with barberries and wild celery.    Carteret anchored in Borja bay, in the *Swallow.*    Byron attempted to anchor there, but was prevented by the strength of the tide ; he gives, however, a plan of it.    Between Borja bay and Cape Quod are two coves, too small to be of any use when Borja bay is so much superior.

CROOKED ROCK, lying about three-fourths of a mile from Ortiz island, has only 5 feet water on it ; is marked by kelp, and has, close to its sides, 3 fathoms, then 10, 20, and 40 fathoms.    Vessels running during the night, or by day in fogs or snow storms, should keep well over on the southern shore, where no danger seems to exist, and keep Little Ortiz to the southward of Big Ortiz until El Morion bears to the eastward of south.    In going to the eastward, when Borja bay opens out, they may be certain of having passed the rock.†

Crooked rock is thus described by Lieutenant Simpson, one of Byron's officers in H.M.S. *Dolphin,* 1765 :—" At not a league to the eastward of Cape Quod is a rock which has not more than 9 feet upon it ; but shows itself by the weeds growing upon it : it is a good distance from the north shore, and in the fair way of working to the westward round the cape."

---

* See Admiralty Plan No. 3, of Ports in the Strait of Magellan, No. 557 ; scale, m = 1·5
† Remarks of Commander J. A. Paynter, R.N., 1848.

**EL MORION,** or St. David Head, on the southern shore of Crooked reach is a lofty granitic rock, of which the outer face is perpendicular and bare, and of a light grey colour, distinguishable from a considerable distance both from the eastward and the north-westward, and forming an excellent leading mark to assure the navigator of his position.

**LONG REACH** extends from Cape Quod to the entrance of the Gulf of Xaultegua, a distance of 39 miles in the direction of W. by N. ¼ N. The weather here is generally so thick, that although the distance across the reach is only from 2 to 3 miles, yet one shore is frequently concealed from the other by the mist ; on which account Captain Stokes found it impossible to form any plan of this reach on his passage through it. When leaving Stewart bay, he says in his report, " We continued our progress to the westward, having westerly and S.W. winds, with thick weather and drizzling rain. The coasts on both sides were very rarely visible to us by reason of the thick mist by which they were capped." It is, however, a bold coast on each side, otherwise the strait would be utterly unnavigable in such weather.

The shores on either side are certainly much less verdant than to the eastward of Cape Quod; but scarcely so dismal as Cordova's account would make them appear; for, he says, "As soon as Cape Quod is passed, the strait assumes the most horrible appearance, having high mountains on both sides, separated by ravines entirely destitute of trees, from the mid-height, upwards." To Captain King it appeared only that the hills were indeed bare of all vegetation above, but that below they were not so deficient ; though the trees and shrubs, he confesses, were very small in size. For the purposes of fuel abundance of wood is to be obtained. In the winter months, the hills are covered with snow, from the summit to the base; but in the month of April, when the *Adventure* passed through, no snow was visible.

**CORDOVA PENINSULA.**—The following descriptions of the bays between Cape Quod and Cape Notch, on the northern shore of Long reach, are taken principally from the Appendix to Cordova's Voyage to the strait. Barcelo bay, the first to the westward of Cape Quod, seems to be large but incommodious, and strewed with small islets. Orsono bay follows, and, according to Cordova, has very deep water all over ; there being 40 fathoms within a cable's length of the beach, excepting the western side, where there is a rocky ledge with 10 fathoms. Langara bay is next, to the westward, and trends in for about a mile to the north-east, and has 10 to 12 fathoms, stony bottom. It is more sheltered than the two former bays.

**POSADAS BAY** is, most probably, Lion Cove of Wallis. Its western point is formed by a high, rounded, and precipitous headland, resembling, in Captain Wallis' idea, a lion's head ; and though Cordova could not discover the likeness, yet it is sufficiently descriptive to point out the bay, were the anchorage worth occupying, which it is not. Wallis describes it to have deep water close to the shore ; his ship was anchored in 40 fathoms. Arce bay, which Cordova describes as having anchorage in 6 to 17 fathoms, stones. It divides at the bottom into two arms, each being half a mile long. The outer points bear from each other W.N.W. and E.S.E., half a mile apart. Flores bay is, probably, the Good Luck bay of Wallis. Cordova describes it to be very small and much exposed, with 6 to 20 fathoms, stone and gravel. At the bottom there is a rivulet of very good water. Villena cove has from 15 to 20 fathoms, and is very open and quite exposed.

**GUIRIOR BAY** is large, and open to the south, but probably affords good anchorage in its coves. Cordova says it extends for more than a league to the northward, the mouth being 2 miles wide. Its western point is Cape Nótch, which will serve to recognize it. Near the entrance there are several rocks and an island ; and within them, on the western side, two coves, with from 15 to 30 fathoms, stones. Farther in is the port, which has a narrow entrance, and into which a river falls from a considerable height. The rapidity of the stream has formed a channel through the ooze in the direction of the entrance, and in this channel there is good anchorage in 20 to 26 fathoms ; on either side of the channel the bottom is stony. The port is too difficult to reach to make this anchorage an object of any value. Should, however, a strong gale from south or south-west oblige a ship to run in, she should avoid passing too near the west side of the narrow ; for a reef extends from it nearly a cable's length. There is also a bank outside the narrow, but it is pointed out by kelp.

**CAPE NOTCH** is a projecting point of grey-coloured rock, about 650 feet high, having a deep cleft in its summit. It is a conspicuous headland, and cannot be mistaken. Captain Stokes remarks, that the mountains in the neighbourhood of Cape Notch spire up into peaks of great height, and are connected by singularly sharp saw-like ridges, as bare of vegetation as if they had been rendered so by the hand of art. About their bases there are generally some green patches of brushwood, but upon the whole nothing can be more sterile and repulsive than this portion of the strait. This account agrees with Cordova's ; yet on examining the low grounds about the coves, they were found so thickly clothed with shrubs, brushwood, and small trees, that it was difficult to penetrate beyond a few yards from the beach.

From the above brief description of the bays between Capes Quod and Notch, occupying a space of 12½ miles, none seem to be either convenient, or safe; and the best port for shelter, for a ship in this part of the strait, is Swallow bay, on the opposite shore. On the other hand, small vessels may find many places that a ship dare not approach, where every convenience may be had; for if the water be too deep for anchorage, they may be secured to the shore at the bottom of the coves, where neither the swell nor the wind can reach them.

**SNOW SOUND**, in St. Ines island, is a deep inlet, formerly supposed to communicate with Whale sound, insulating Ulloa peninsula; but this was disproved by Captain Fitz-Roy's careful examination. It is unimportant to the navigator, and not worth the trouble of entering, though there are two anchoring-coves, one about a mile, and the other 2 miles, within its western head. It extends in for 10 miles, and terminates in two inlets, surrounded by high, perpendicular, black rocks.

**SWALLOW BAY** is 1¼ miles to the westward of Snow sound. It is a better anchorage for ships than any in the neighbourhood, and all the dangers are well buoyed by the kelp. It was first used by Captain Carteret in the *Swallow*; and Cordova gives a short description of it:—

" To the westward of Snow sound there are two bays, Swallow and Condesa, formed in a bight by an island. The eastern, Swallow bay, has in its mouth three islands and a rock; besides being strewed with kelp, which serves to point out the dangers in entering. Within, it is very well sheltered from all winds. The depth is from 40 to 16 fathoms, stones, and in some parts ooze. This bay lies to the southward of Cape Notch; and to recognize it, there is a cascade falling down the centre of a mountain at the bottom of the port, to the westward of which are two higher mountains; the summit of the eastern being peaked, and the western one rounded." At about a cable's length off the west point of the entrance of Swallow bay, Captain Fitz-Roy saw a rock just awash. This danger should be carefully avoided.

The anchorage is under the east side of the island which separates the harbour from Cordova's Condesa bay. Wallis describes the harbour to be "sheltered from all winds, and excellent in every respect. There are two channels into it, which are both narrow, but not dangerous, as the rocks are easily discovered by the weeds that grow upon them." The bay to the westward of the island is Condesa bay. It is full of islets and rocks, and the channel behind the island, communicating with Swallow bay, is very narrow.

**STEWART BAY** is less than a league from Swallow bay. On this place Captain Stokes makes the following remarks :— "Stewart bay afforded us a quiet resting-place for the night, but it is by no means to be recommended as an anchorage; for though it is sufficiently sheltered from wind and sea, yet the rocks, in different parts of it, render the passage in or out very hazardous : every danger in it is pointed out by rock-weed, but it is so much narrowed thereby as to require the utmost vigilance. A plan of it was made and connected with the coast by bearings and angles to Cape Notch, and to other fixed points ; but the description of the place by Cordova cannot be improved."

The account in Cordova is as follows :—" Stewart bay (la Bahia de Stuardo) follows Condesa bay. It has an islet, besides several patches of kelp, which are indications of the many rocks that exist. Even the best channel is narrow and tortuous ; the depth is from 12 to 16 fathoms, stones. At the bottom there is an islet, forming two narrow channels leading into a port or basin, 2 cables' lengths wide : the eastern channel is the deepest, having 15 to 20 fathoms. In the basin, on the eastern side, the depth is 6 and 9 fathoms, mud. A reef extends half a cable's length to the westward of the south end of the islet. It would be difficult and dangerous to enter this small basin."

**SNOWY CHANNEL** of Sarmiento next follows. It is an extensive channel, of which we know only that it extends to the southward for 5 or 6 miles, and probably terminates like Snow sound. To the westward of Snowy channel there are several inlets affording, apparently, good shelter, but those that were examined were found to have very deep water.

**GLACIER BAY**, on the east side of Long reach, is only re-markable for its glacier ; it was not examined, nor the next bay to the eastward, which is full of islands. Between Glacier bay and Playa Parda (grey beach), the shore is bold and uninterrupted, except by a small cove, about 2 miles to the eastward of Playa Parda, which seemed likely to afford shelter for small vessels. Off the western inner point there is a reef, but within that, there seemed to be a basin of at least half a mile in extent.

**PLAYA PARDA COVE** is the next place to the westward of Cape Notch, that can be recommended for an anchorage. It is well sheltered, and for chain cables has a good bottom, being of sandy mud, strewed with stones; it is half a mile wide at the entrance, and about a quarter of a

mile in length.  Round the western side of Middle point, there is a
channel, a quarter of a mile long and 150 yards wide, with 6 and 7
fathoms water, communicating with a good little harbour for a small
vessel, of about a quarter of a mile in diameter.*

The cove is easily known by Shelter island, that fronts the inlet of
Playa Parda.  The inlet is 1½ miles long, and half a mile broad, but with
very deep water all over.  By luffing round the island, a ship will fetch
the anchorage in the cove ; and, although sail should not be reduced too
soon, yet the squalls, if the weather be bad, blow down the inlet of Playa
Parda with great violence.  Anchor a little within, and half-way between
the points of entrance, at about 1½ cables' lengths from Middle point, in
5¼ and 6 fathoms.

Playa Parda cove is snug enough when inside, but a large vessel
should moor for fear of S.W. gales.  The *Gorgon* anchored with the
extremes of Shelter island S.W. ½ W. and W. by S., and the eastern
extreme of the cove E. by S.  The wood hereabouts is too stunted to
make it worth the while of a steam-vessel to stop ; and the anchorage,
in my opinion, is bad for a large vessel, as we constantly swung into
4 fathoms in the kelp.†

**ABRA** is a deep opening on the southern shore of the Reach, opposite to
Playa Parda, which has much the appearance of a channel leading through
Tierra del Fuego.  It is evidently the inlet noticed by Sarmiento, and thus
described by him :—"A great bay (*gran ensenada*) which enters the
land in a W.S.W. direction for more than 2 leagues, and has an island
at its mouth ; we called it the opening (*Abra*), because we could not see its
termination.  Within the Abra the land is low and hummocky; half a
league to the eastward of the Abra there is a cove (*Ancon*) on the right
hand shore ; and on the left hand shore, a league across, there is another
cove or creek (*Caleta*) which forms a harbour, called by the Indians
Pelepelgua, and the Ancon they call Exeaquil."  Pelepelgua may probably
be Glacier bay, and Exeaquil must of course be one of the unexamined
coves to the eastward of the Abra.

To Captain King, the opening of the Abra seemed to be about 1½ miles
wide, and had an island in the entrance.  Within, it appeared to take,
first a S. then a S.W. course, and afterwards to bend round a projecting
low point on its eastern shore ; when winding under the base of a high
precipitous ridge on the opposite shore, towards the south-east, its course
could not longer be observed.  On the seaward coast there is a deep open-

---

* *See* Admiralty Plan, No. 557; scale, *m* = 3 inches.
† Remarks of Commander J. A. Paynter, R.N., 1848.

ing in Otway bay, called Dynevor sound, with which, Captain King thinks, there is a probability that this Abra may communicate.

**MARIAN COVE**, 1½ miles to the westward of Playa Parda, is a convenient anchorage; at the entrance it is about a third of a mile, and more than half a mile deep; a plan was made of it, which will be a sufficient guide. Captain Stokes observes, that it affords shelter from the prevailing winds; the anchorage is in 22 fathoms, good holding-ground; but vessels may run into less water, there being 8 fathoms within 60 yards of the beach, at the bottom of the bay. In entering, the western shore should be kept aboard. This cove is about midway between Cape l'Etoile and Playa Parda, and is a very advantageous place to stop at. Opposite to Cape l'Etoile there is a bay, with anchorage in 17 fathoms in a well-sheltered situation. From Cape l'Etoile to the entrance of the Gulf of Xaultegua, the shore is straight and precipitous, and the hills are barren and rocky.*

**HALF PORT BAY** is the most useful on the opposite shore of the few inlets. It lies immediately round the south side of a deep inlet, rather more than a league to the eastward of Cape Monday, and is merely a slight indentation of the coast. The *Beagle* anchored here on two or three occasions, and found it to be an excellent stopping-place; the anchorage is within two-thirds of a cable's length of the western point, in 16 fathoms, muddy bottom. The position of this cove was ascertained by observation to be in lat. 53° 12' S., and long. 73° 18' 45" W., or 2° 21' W. of Port Famine.

The land on the south-west side of the anchorage is high, and thickly wooded from its summit to the water's edge. On the eastern side it is lower, the vegetation more scanty, and the trees crooked and stunted, and pressed down to the north-east by the prevailing winds. S.W. by W. from the anchorage there is a remarkable cleft in the summit of the high land, from which a narrow stripe cleared of brushwood descends to the water's edge, apparently formed by the course of a torrent, or possibly of masses of rock forced down by the prodigious gusts of wind which sweep everything before them in various parts of these straits. The anchorage is well sheltered from the prevailing winds and the holding-ground is good; water and fuel are abundant.

**TIDES.**—Little has been said of the tides in this part of the strait, for they rise and fall only 4 feet. It is high water at full and change in all parts, within a few minutes of noon. The ebb tide has little or no strength. The current sets constantly to the eastward; between Capes Notch

---

* *See* Admiralty Plan, No. 557; scale, *m* = 1·5 inches.

and Quod, it set the *Adventure* 2 miles to the eastward in 3½ hours ; and from Cape Quod to Port Gallant it carried her 6 miles in 3½ hours.

**CAPE MONDAY** lies about 3 miles to the north-west of Half-Port bay. There is an anchorage under this cape for small vessels, in which Byron anchored, and rode out a heavy gale of wind. With the exception of a shoal midway in the entrance, on which there are 4 fathoms, it seems to offer good shelter from the prevailing winds. On the western side of the cape lies Medal bay, of which a very full but rather florid description is given in the Appendix of Cordova's Voyage.

It has, according to that description, an island in the entrance, which forms two channels, the easternmost of which is only deep enough for boats, but the western is 25 fathoms wide ; it is strewed half way across with kelp ; but between the kelp and the island there is a good and clear passage with 6 fathoms, sandy bottom. In the kelp there are not less than 4 fathoms, and inside of it the depth is 9 to 7 fathoms, sandy bottom. To enter this port there are no dangers that are not visible, and those are easily avoided ; they consist only of the islet in its entrance and some patches of kelp, over which, however, there is plenty of water.

**The GULF of XAULTEGUA,** at the northern end of Long reach, is a deep opening, trending into the land in an easterly direction for 28 miles, approaching within 2 miles of Manning bay, one of the inlets on the north-west side of Jerome channel. The entrance is not 4 miles across, but it afterwards expands to a breadth of nearly 15 miles. At the entrance is St. Anne island, between which and the south point there is a navigable channel half a mile wide. St. Anne island is about two 2 miles long, W.N.W. and E.S.E. ; off its N.W. end there is an islet, and another close to its S.W. extremity.*

The plan that was made of the gulf is little more than an eye sketch. Captain Fitz-Roy, who passed through it in a boat, in order to examine its eastern termination, says :—" This gulf is utterly useless, as from the appearance of its shores there seems to be no anchorage : should a ship be so unfortunate as to make a mistake and get into it, she must keep under way till she gets out again, there being no thoroughfare." The triangle of land intervening between the strait of Magellan, the Gulf of Xaultegua, and the Jerome channel, is named Cordova peninsula, as that navigator explored and described most of the small bays along its south-western face.

---

* The name of Xaultegua is from Sarmiento, who very correctly describes it.—*Sarmiento, page* 208.

**SEA REACH,** forming the western entrance to the Strait of Magellan, runs W. by N. and E. by S. for 52 miles. On the northern shore, between Capes Tamar and Phillip, it opens into the channels which lead into the Gulf of Peñas, inside Queen Adelaide archipelago. On entering Sea reach a heavy swell at once comes in from the westward. Hitherto, in passing through Magellan straits, little sea and no swell had been experienced, but on quitting the western end of Long reach, the proximity of the Southern ocean soon makes itself felt. It was strongly felt by the *Beagle* when beating to the westward, immediately on reaching Cape Providence ; but there seems to be no other danger for vessels beating through the strait thereabouts, the shore being bold-to. Byron passed a night, and a very tempestuous one, here ; as did also the *Beagle*, as she was not able to choose anchorage before night. Captain Stokes, upon this occasion, writes :—" We continued beating to windward, the wind squally and weather rainy. The coast on both sides is bold. Our boards were directed during the night, which was very dark, by the sight of Cape Upright when near one shore, and of Cape Providence when nearing the other. We commonly tacked at the distance of a mile from either side."

**PORT ANGOSTO.**—A league to the westward of Cape Monday there is an inlet, which was supposed to be Sarmiento's Puerto Angosto. Upon its western end there is a conspicuous round mount, and to the northward, between the mount and a projecting point, there is a confined but very snug and commodious cove for a small vessel, in 17 fathoms, at a quarter of a mile within the head.

**Round Head.**—In consulting the Appendix to Cordova's voyage, it would seem that this Round head is an island. The description runs thus : —" The figure of the island, in the eastern part of its mouth, is triangular, and its north-east point lies in the line of bearings of Capes Lunes (Monday), and San Ildefonso (Upright). To the eastward of the island an inlet runs to the south-west, 1⅓ miles wide and a league long, to the bottom of the bay; the S.E. side of the island being 1½ miles long. To the westward, the distance between the shore and the island is much more, and the direction of the second channel is N. ¼ W. The bay, the greatest breadth of which is 2 leagues, has at its bottom, and towards the S.E., an inlet, the course of which disappears behind the mountains, in a S. ¼ E. direction. There appeared to be a good anchorage between the island and the eastern shore, but we had no bottom with 30 fathoms."

There seems to be no doubt that the island above described is the projecting point 4 miles to the westward of Cape Monday (named in the chart Round Head), and that Sarmiento's Puerto Angosto insulates it ; but his

Spanish chart is so vague, and our own so imperfect in this part, that Captain King thought it better to leave it for future examination, rather than insert an island without more direct evidence ; although, from the Spanish account, there seems no reason to doubt its existence.

Of UPRIGHT BAY little is known.  The *Adelaide* rode out a gale from the eastward with her stern in the surf on the beach, and the *Beagle* anchored under the east side of the cape, at about half a mile S.W. of the rocky islet, and for shelter from westerly winds, found it to be very good.  Of this Captain Stokes says :—"We anchored at a cable's length of a small patch of light-coloured shingle beach, on the western side of the bay, in 22 fathoms, sandy bottom.  This anchorage, though affording excellent shelter from the prevailing wind, is bad with a southerly one ; for the steepness of the bottom requiring a vessel to anchor close to the shore, sufficient scope is not left for veering cable.  There is a plan of the bay in Hawkesworth from Byron's account, who anchored in the southern part of the bay, perhaps under the lee of the islands to the south-east of the cape.

CAPE UPRIGHT (San Ildefonso of Cordova) bears S. by E. 5 miles from Cape Providence.  It has a rocky islet a quarter of a mile off its eastern extremity, surrounded by kelp, which extends also for some distance from the cape towards the islet, to 7 fathoms.

CAPE PROVIDENCE is a rugged mountain, on the northern shore, distant 13 miles from the entrance of the Gulf of Xaultegua ; it is higher than the adjacent coast, deeply cleft at the top ; and when bearing about north, the western portion of its summit appears arched, and the eastern peaked and lower.  When the cape bears E. by S., distant about $4\frac{1}{2}$ miles, a little round rocky islet will be seen opening to the southward.

There are some anchorages to the north-east of Cape Providence, according to a plan given in Hawkesworth's Collection of Voyages, but they are too much out of the way, as well as too much exposed to southerly winds, to be of use, or to offer any security to vessels bound through the strait.

CAPE TAMAR is $9\frac{1}{2}$ miles from Cape Providence ; in this space the land bends inwards about 4 miles.  Captain Stokes describes this bay as being formed of two large bights.  In the eastern one there are several islands, of which two are conspicuous; they are of a good height, and well wooded ; and at a distance they appear conical, the eastern one being the lowest.  Between them is a passage to two good anchorages, which

Lieutenant Skyring examined, and considered to be more sheltered than Port Tamar.

**ROUND ISLAND** lies 4 miles to the eastward of Cape Tamar, to the north-west of which there is a well-sheltered anchorage, but with deep water. In standing in, pass midway between Round island and an island to the westward, which lies close to the shore, and haul in round the latter to the mouth of a cove, in the entrance of which, near the south shore, there are 23 fathoms, sand. The shore to the north and north-east of Round island is very rocky.

**PORT TAMAR** lies on the eastern side of Cape Tamar, a useful and excellent anchorage. It is scarcely 2 miles wide, and rather more than half a mile deep. Its entrance is not exactly free from danger, but with attention to the following directions, none need be apprehended.[*]

**Sunk Rock.**—To avoid the sunken rock which lies between a group of rocky islets towards the western side, and a patch of kelp towards the eastern side of the bay, with a westerly wind, give the outermost of the rocky islets a berth of 2 cables' lengths. On this sunk rock there are only 9 feet water, and upon it the *Beagle* struck. An excellent leading mark to clear this danger, is a whitened portion of bare rock, looking like a tombstone, about a third of the way up the green side of the mountainous land that forms the coast of the bay. This white patch bears W. by N. ¼ N from the sunken rock.

The least water found among the kelp on the eastern side of the channel was 4½ fathoms, and near and within the edge towards the rocky islets there are 7 fathoms; so that, with the lead in hand and a look-out for kelp, which should not unnecessarily be entered, there is no real danger. The *Beagle* anchored at about a third of a mile from the northern shore. The plan will show what is further necessary to be known of the anchorage.

**TIDES.**—It is high water, full and change, in Port Tamar at 3h. 5m., and the perpendicular rise and fall is 5 feet. The flood tide on this part of the northern shores of the strait sets to the eastward, and rarely exceeds half a knot.

**TAMAR ISLAND** stands to the westward of Cape Tamar; it is high, and separated from the cape by a deep channel half a mile wide. There is a rock lying half a mile off its south-west end. Abreast of Cape Tamar, Magellan Strait is 7 miles wide; at Cape Phillip, farther to the

---

[*] *See* Admiralty Plan, No. 557; scale, m = 1·5 inches.

westward, the breadth increases to 5 leagues ; and at Cape Parker it narrows again to 4 leagues, which breadth it preserves.

GLACIER AND ICY SOUNDS.—Between Capes Tamar and Phillip a space of 4 leagues, there is a deep bight, with two openings ; the easternmost, in which are Glacier and Icy Sounds, extends to the north-east for 10 miles from the mouth, and the westernmost is the commencement of Smyth channel.

The STRAGGLERS are a long group of rocks extending 6 miles to the south-west, from the point between Cape Phillip and Tamar island ; the point itself is a mass of low rocks and islands.

SKOLL BAY, in which the *Beagle* anchored in 1827, is under the north-east side of Cape Phillip, the western point of entrance to Smyth channel. It is valuable for vessels working through the strait to the westward, inasmuch as, from the discontinuous nature of the northern shore (which here is formed into deep bays), this place will be much more easily recognized than the anchorages on the opposite coast ; besides, the winds hang here, in general, somewhat to the northward of west, hence a better starting-place for the westward is obtained.   There is excellent anchorage in 15 fathoms.

Supplies.—Here, as in every anchorage in the strait, water and fuel are easily procured, but nothing more, unless we except some berries, celery, mussels, and limpets ; the wild goose abounds here, but its nauseous taste renders it uneatable.   No inhabitants, no quadrupeds.

SMYTH CHANNEL, the entrance to which lies between Cape Phillip and the Fairway isles, leads to the northward inside Queen Adelaide, Hanover, and Wellington islands, opening into the Gulf of Peñas, about 330 miles from the Strait of Magellan.   It has been found useful to steamers using the strait bound from the Atlantic to the Pacific, but should not be attempted by a sailing vessel.   For a farther description of these channels, see page 227.

The COAST of the south side of the strait, between Cape Upright and Valentine harbour, the Desolation island of Narborough, was very little examined ; there are several deep bights and spacious bays, which may perhaps contain anchorages, but, in general, they are not found in the large harbours, which are mostly deep, precipitous chasms or ravines in the rock.   The smaller coves, or where the land shelves down to the sea, are more likely to afford anchorages.

ANCHORAGES.—In his Appendix, Cordova describes some which it may be useful to mention here : he says, " In rounding Cape (Ildefonso) Upright we find ourselves in a bay, not very deep, 2 miles across, and divided in its centre by many islets and rocks, the northernmost of which bears west from the extremity of the cape. At a mile N.W. ¼ N. from that northernmost islet lies a rock, which is of dangerous approach.

" To the westward of this bay there is another, 3 miles wide, and about as many deep ; it is full of islets, and narrow channel leads from it to the S.S.E. At the western end of this bay (which seems to have been called by Wallis, the Bay of Islands), commences a third bay, which, with the two preceding, make the great bay, called by the Indians, according to Sarmiento, Alpuilqua." It is contained between Cape Upright and the bold projecting point of Echenique. The country is described by him to be poor, and the vegetation scanty.

**Cuaviguilgua.**—" The eastern point of the third bay has a string of islets extending a mile to the northward ; and on its eastern side lies the bay of Cuaviguilgua ; and a little beyond it, at the bottom of the bay, is Port Uriarte, the mouth of which is 2 cables' lengths across.

" **Port Uriarte** was carefully sounded, but the bottom was generally bad and stony, with 5 to 18 fathoms. It is surrounded by high mountains rising vertically, and with only a few stunted trees on the shores. Its greatest extent, which is from north to south, is half a mile ; the mouth is not visible until close to it; its bearing from Cape Providence is S. by W. ¾ W. There is no danger in entering it but what is visible ; but it is not recommended as a good harbour from the foul ground all over it. A little to the eastward also of Echenique point is Cape Santa Casilda, a low point."

**Puchachailgua.**—" To the westward of Echenique point there is a harbour 2¼ miles wide, the points of entrance bearing N.W. and S.E. An island in the centre forms two channels, but with very deep water, no ground being found with 55 fathoms. From this harbour a channel opens to the south-south-west, but disappears between the mountains. On the eastern side of the island the channel is at first a mile wide, but afterwards narrows gradually : the western channel is scarcely 2 cables' lengths across, the shores are high and precipitous mountains. The Indians, according to Sarmiento, call the place Puchachailgua."

**Port Santa Monica** is another harbour to the westward, supposed to be Sarmiento's Santa Monica. It bears S.S.W. from Cape Tamar, being 14 miles to the westward of Cape Upright, but not more than 3 leagues, according to Sarmiento's account.

**Puerto Churruca,** a deep and spacious bay, lies two-thirds of a mile farther to the westward, round a point with two islets. It is 2 miles wide, the outer points bearing E.S.E. and W.N.W., containing 2 ports and some coves, but with very deep water, and therefore useless, for it would be necessary to make fast to the rocks to secure the vessel.

**Darby Cove**, in which small vessels may obtain good shelter, is a useful anchorage to the westward.   From Darby cove the coast trends to the westward for 7 miles, having in the interval several indentations, but all with deep water; at Felix point the land bends deeply in to the south-west, and forms a bay 5 miles wide and 2½ deep.

**VALENTINE HARBOUR**, in which the *Beagle* anchored, is on its western side.   The plan shows the nature of the anchorage, which seems to be commodious and secure, and of easy approach.   On hauling round the island off Cape Valentine there are some islets extending half a mile from it ;  these must be avoided, but otherwise there seem to be no dangers.   The anchorage, as a stopping-place, has from 20 to 26 fathoms, sand at nearly a quarter of a mile from either shore ; but a more sheltered situation may be obtained farther to the south-west.   The mount on a small island, forming the south point of the harbour, is in lat. 52° 55′ S.′ and long. 74° 18′ 45″ W.[*]

**CAPE CUEVAS**, lying 2 miles to the north-west of Cape Valentine, is the extremity of an island close to the shore.   To the eastward the coast forms a bay with islands ; and to the westward of the cape the coast is broken into several sinuosities.

**TRUXILLO BAY**, 3 miles N.W. ½ W. from Cape Cuevas, was not examined, but the Spanish account describes it to be 1¾ miles wide, in the direction of north-west and south-west, and half a mile deep.   At the head of the bay there is a harbour with an entrance half a mile across, bearing nearly north and south, well sheltered, and trending W.S.W. for 1¼ miles, where it has two small basins.   The depth of water is very great, but close to the western shore there are 8 to 13 fathoms, on sand and coral.   Near the mouth the bottom is generally of stones ; and its several banks and reefs are buoyed by seaweed, but on none was there less than 7 fathoms water.

**TUESDAY COVE.**—There is plenty of wood and water in Truxillo bay, but no one should visit it in preference to Tuesday bay, or, rather, to the more convenient anchorage of Tuesday cove, three-quarters of a mile south of Cape Cortado.   The anchorage is in 12 to 14 fathoms. Tuesday bay is larger, and, therefore, more exposed to the squalls ; but for a ship, perhaps, easier of access.

---

[*] *See* Admiralty Plan, No. 557; scale, m = 1·5 inches.

SKYRING HARBOUR is a league to the westward of Cape Cortado; its entrance is 1¼ miles wide, afterwards narrowing to half a mile, and trending to the S.W. by W. for 1½ miles, where it terminates in a cove extending half a mile to the S.E., with 10 fathoms. There are some islands in it, and anchorage might be obtained in 27 fathoms.

PORT MERCY.—At 3½ miles from the west point of Skyring harbour is the eastern head of this port, one of the best anchorages in the western arm of the strait, and being only 4 miles within Cape Pillar, is very conveniently placed for anchoring, while waiting for a favourable opportunity to run out of the strait. The plan will be a sufficient guide, for there is no danger in entering. The depth is moderate, 12 to 14 fathoms and the holding ground excellent, being a black clay. A ship may select her berth; but abreast of the first bight round the point she will be well sheltered, and find it convenient for many purposes. This is the Puerto della Misericordia of Sarmiento, and the Separation Harbour of Wallis.*

The latitude and longitude were determined upon Observation islet, just north of the port, the summit of which was found to be in 52° 44′ 57″ S., and 70° 39′ 14″ W. Variation 23° 48′ easterly in 1833.

CAPE PILLAR stands 3 miles to the westward of Observation islet. Captain Stokes landed here on 25th February 1827, in order to fix its position, but not without considerable difficulty, owing to the great swell that then, and indeed always, prevails there. Captain Fitz-Roy also landed in a cove under the cape in 1829, with his instruments, to obtain angles from its summit; but the difficulty of the ascent was so great that he did not risk their destruction. The extremity of Cape Pillar is 1,750 feet above the sea.†

CAPE PARKER is a remarkable projection on the north shore of the strait, with three hummocks on the ridge of high land which rises behind it. It lies 11 miles N. by E. off Cape Cortado, and 9 miles W. by S. from Cape Phillip. To the eastward the coast bends deeply into the north, forming Parker bay, in which there appear to be several islands and a narrow opening, perhaps a channel, leading to the northward. Up this inlet the steamer *Fosforo*, commanded by Mr. W. Hall, proceeded N.N.E. for 8 miles.

The western side of the bay is much indented, and affords some anchorages, but the approaches are not clear. The first bay, however, to the eastward of the south-east point of the cape, seems to afford a good

---

* *See* Admiralty Plan, No. 557; scale, m = 1·5 inches.

† *See* View of Cape Pillar and Westminster Hall on Admiralty Chart, No. 554.

stopping place; but it is fronted by a considerable shoal, with two rocky islets; the depth is from 7 to 22 fathoms. The land of Cape Parker will probably prove to be an island. To the westward of it commences a range of islands, rocks, and shoals, fronting a broken coast that should never be approached but for the purpose of discovery or seal-fishery.

**SIR JOHN NARBOROUGH'S ISLANDS.**—This chain consists of 8 or 9 large, and hundreds of small islands, extending 30 miles in a west-south-west direction. To the north-eastward of them there seemed to be a channel, and amongst them are several anchorages, but none to be recommended, especially when on the south coast there are two or three much better, much safer, and much easier of access. Their north-western point is Cape Victory, in lat. 52° 16' 10" S., and long. 74° 54' 39" W., which may be considered as the north-west point of entrance to the Strait of Magellan.

This is all a dangerous coast, as well from the great number of rocks upon which the sea breaks very high, as from the tides, which near the edge of the line of shoals set frequently in amongst them.

**WESTMINSTER HALL,** a high rock, is the south-easternmost island of the chain lying 8 miles W. by N. ¼ N. from Cape Parker; and to the westward there are two or three other conspicuous points, such as the Cupola and Observation Mount, that might be noticed. The *Beagle* ran in amongst the breakers, and anchored near the latter, for the purpose of ascertaining its position, and obtaining bearings for the survey.*

**LOS EVANGELISTAS,** consisting of 4 small islands and some detached rocks and breakers, lie 11 miles S.S.W. ¼ W. from Cape Victory. They were thus named by the early Spanish navigators, but the Direction Isles by Narborough, from their forming an excellent mark for the western mouth of the strait. The islets are very rugged and barren, and suited only to afford a resting place or a breeding-haunt to seals and oceanic birds; but a landing-place may be found on one of them, and anchorage round them, if necessary. The largest and highest may be seen in tolerably clear weather, from a vessel's deck, at the distance of 5 leagues. The southernmost, from its shape called the Sugar Loaf, is in 52° 24' S. and 75° 6' 40" W., and bears from the extremity of Cape Pillar N.W. by W. ¼ W. 23 miles.

**The TIDES** here are very variable, and sometimes set to the E.N.E. towards the rocks that front Cape Victory and the Narborough chain. It is high water, on full and change, at 1h.; and the tide rises about 4 feet.

---

* *See* View of Westminster Hall on Admiralty Chart, No. 554.

# CHAPTER VIII.

## PASSAGES ROUND CAPE HORN AND THROUGH THE STRAIT OF MAGELLAN.

SHIPS bound to the Pacific will find it advantageous to keep within 100 miles of the coast of Eastern Patagonia, as well to avoid the heavy sea that is raised by the westerly gales, which increase in strength according to the distance from the land, as to profit by the variableness of the in-shore winds when from the westward. Near the coast, from April to September, when the sun has north declination, the winds prevail more from the W.N.W. to N.N.W. than from any other quarter. Easterly gales are of very rare occurrence, and even when they do blow, the direction being obliquely upon the coast, it is not hazardous to keep the land on board. And when the sun has south declination, though the winds veer to the southward of west, and frequently blow hard, yet as it is a weather shore, the sea goes down immediately after the gale. In this season the winds are certainly against a ship's making quick progress, yet as they seldom remain fixed in one point, and frequently shift backward and forward 6 or 8 points in as many hours, advantage may be taken of the changes so as to keep close in with the coast.

Having once made the land, which should be done to the southward of Cape Blanco, it will be desirable to keep it topping on the horizon, until the entrance of the strait of Magellan be passed.

STRAIT of LE MAIRE.—With respect to this part of the voyage, whether to pass through Strait Le Maire, or round Staten island, much difference of opinion exists. Captain King approves of the latter: yet he says that he would very reluctantly give up a favourable opportunity of shooting the strait, and therefore of being so much more to windward. With a southerly wind, however, it would not be advisable to attempt that strait; for, with a weather tide, the sea runs very cross and deep, and might severely injure and endanger the safety of a small vessel, and to a large one do much damage. In calm weather it would be still more imprudent in a sailing vessel (unless the western side of the strait can be reached, where a ship might anchor), on account of the tides which set over to the Staten island side; and there, if it becomes necessary to anchor,

it would necessarily be in very deep water, and close to the land. With
a northerly wind the route would seem to be advantageous, and it would
require some resolution to give up the opportunity so invitingly offered;
but it may be doubted whether northerly winds, unless they are very
strong, blow clear through the strait—and if not, a ship is drifted over to
the eastern shores, where, from the tides, she will be quite unmanageable.

Captain Fitz-Roy, whose authority, from the experience he gained
during his survey of the strait, must be good, seems to think that there is
neither difficulty nor risk in passing through it. The chief danger is the
failure of the wind ; and vessels coming from the southward are not so
liable to the failure of a south-westerly wind, unless it be light, and then
it will probably be found from the north-west, at the northern end of the
strait. The anchorage in Good Success bay, on the western shore of the
strait, however, is admirably situated, should the wind or tide fail.

**TIDE RACE.**—In passing to leeward of Staten island, the tide race,
which extends for some distance off Cape St. John, at the north-east end
of the island, must be avoided, and there exist no other dangers.
The anchorage under New Year island, although it is very wild and
the bottom bad, and the tide strong, yet offers good shelter from south-
west winds, and might be occupied with advantage during a gale from
that quarter ; but it would be an unfavourable berth for vessels bound
round Cape Horn.

**NASSAU BAY.**—Ships having to visit the Falkland islands in sailing
from the Atlantic to the Pacific during the summer and most stormy
months, may avoid Cape Horn by making the circuit of Nassau bay, where
there are good anchorages and a convenient stopping place near its southern
entrance, 30 miles to the north-west, and to windward of Cape Horn.
In Nassau bay a ship should wait until the wind has a good deal of
northing, which, if accompanied by a low barometer (29·00), will probably
be the commencement of a semi-revolving gale, and will carry a vessel far
enough to the westward, by the time the wind has veered to that quarter,
to make safe another anchorage under Noir island. This is an easy landfall ; and here a westerly gale may be passed safely in smooth water, and
a good position secured for another start, weather permitting. This
passage by Nassau bay is not recommended.

**BARBARA CHANNEL.**—When not desirous of visiting the Falkland
islands, a passage round Cape Horn may be avoided, even in a frigate, by
going through Magellan strait about half way, or as far as the Magdalen
or Barbara channels. By either of these the vessel may pass into the
Pacific 200 miles to the westward and northward of Cape Horn. But

unless in a steamer, or under some special circumstances, this route would be hardly advisable.

ROUND CAPE HORN.—In proceeding round the Cape, after passing Staten island, if the wind be westerly, the ship should be kept upon the starboard tack (unless it veers to the southward of S.S.W.), until she reaches the latitude of 60° S., in which parallel the wind is thought by some persons to prevail more from the eastward than any other quarter, and then keep upon that tack upon which most westing may be made.

Having reached the meridian of Cape Pillar, or 75° W., it will yet be advisable to take every opportunity of making westing in preference to northing until reaching the longitude of 83° or 84° W., which will enable a ship to lay her course to the northward with the north-westerly winds that prevail between the parallels of 50° and 54° S.

WEATHER off CAPE HORN.—Captain King says, "Never having passed round Cape Horn in the summer season, I may not perhaps be justified in opposing my opinion to that of others, who, having tried both seasons, give the preference to the summer months. The advantage of long days is certainly very great, but from my experience of the winds and weather during those opposite seasons at Port Famine, I preferred the winter passage, and in our subsequent experience of it, found no reason to alter my opinion. Easterly and northerly winds are frequent in the winter off the cape, while southerly and westerly winds are constant during the summer months; and not only are the winds more favourable in the winter, but they are moderate in comparison to the fury of the summer gales."

BAROMETER.—"With respect to the utility of the barometer off Cape Horn, I do not think it can be considered so unfailing a guide as it is in the lower or middle latitudes. Captain Fitz-Roy, however, has a better opinion of the indications shown by this valuable instrument; my opinion is, that although the rise or fall does sometimes precede the change, yet it more frequently accompanies it. The following sketch of the movements of the barometer, and of the weather that we experienced, may not be without its use :—

"Being to the northward of Staten island for three days preceding full moon, which occurred on the 3rd April 1829, we had very foggy weather, with light winds from the eastward and northward, causing a fall of the mercury from 29·90 to 29·56. On the day of full moon the column rose, and we had a beautiful morning, during which the high mountains of Staten island were quite unclouded, as were also those of Tierra del Fuego. At noon, however, a fresh gale from the S.W. set in, and enveloped the land with a dense mist. No sooner had the wind changed, than the

o 2

mercury rose to 29·95, but fell again the next morning ; and with the
descent the wind veered round to N.W., and blew strong with thick
cloudy weather and rain, which continued until the following noon, when
the wind veered to S.W., with the barometer then at 29·54, having
slightly risen ; but after the change it fell and continued to descend
gradually until midnight, when we had a fresh gale from W.S.W.  When
this wind set in, the mercury rose, and continued to rise, while the wind
veered without decreasing in strength to S.S.W., until it reached 29·95,
when it fell again, and the weather moderated, but without any change of
wind.  During the descent of the mercury, the sky with us was dull and
overcast with squalls of wind and rain, but on shore it seemed to be very
fine sunshiny weather.

"The column now fell to 29·23, and during its descent the weather
remained the same, dull and showery ; but as the mercury became
stationary, a fresh breeze set in from the southward, with fine weather.
After this to new moon the weather was very unsettled, the wind
veering between south and W.S.W. ; the barometer rising as it veered to
the former, and falling as it became more westerly ; but on no occasion
did it precede the change.  The mean height of the barometer is about
29·5.  The mercury stands lowest with north-west winds, and highest
with south-east.

"With the wind at N.W. or northerly, the mercury is low ; if it falls
to 29 inches or 28·80, a S.W. gale may be expected, but does not commence until the column has ceased to descend.  It frequently, however,
falls without being followed by this change.  In the month of June, at
Port Famine, the barometer fell to 28·17, and afterwards gradually rose
to 30·5, which was followed by cold weather, in which the thermometer
stood at 12° Fahr.  The following table shows the mean temperature
and pressure of the atmosphere as registered at the Observatory at Fort
Famine in the Strait of Magellan in the year 1828."

| 1828. | Temperature. Fahr. | Barometer. |
|---|---|---|
| | ° | Inches. |
| February - | 51·1 | 29·40 |
| March - | 49·4 | 29·64 |
| April - | 41·2 | 29·57 |
| May - | 35·5 | 29·30 |
| June - | 32·9 | 29·28 |
| July - | 33·0 | 29·57 |
| August - | 33·2 | 29·28 |

**PASSAGE by MAGELLAN STRAIT.**—The difficulties that present themselves to navigators in passing round Cape Horn, as well from adverse winds as from the severe gales and heavy sea to which they are exposed, are so great, that the Strait of Magellan has naturally been anxiously looked to as the route by which they might be avoided. No chart of the strait, however, existed in which any confidence could be placed, till Captain King's laborious and admirable survey was published; but now its navigation, excluding the casualties of wind and weather, has been placed in the power of every well-found vessel; for its shores and its dangers have been delineated, and plans of the ports and anchorages have been given; and in the preceding pages there is a sufficient description to assure the navigator of his place, and to furnish him with advice as to his proceedings. The chief difficulties, therefore, have been removed; but there remain others which should be well considered before a sailing vessel should encounter them, unless detention be an object of no importance.

**SUPPLIES.**—One very great advantage to be derived from the passage through the strait is, the opportunity of obtaining wood and water, without difficulty. At Borja bay, in Crooked reach, watering boats can be hauled to and from the vessel, and a hose may be taken sufficiently high up the stream for the water to run into the boat or casks. Wood, too, can be cut close to the beach. Another advantage is, that by hauling the seine during the summer months, from January to May, at the mouth of the river or along the beach in Port Famine, at the first quarter flood, a plentiful supply of fish may be obtained. Excellent fish are also caught at the anchorage with the hook and line, at all seasons, early in the morning or late in the evening. Fish may also be obtained with the seine at all those places where there are rivers. Freshwater bay and Port Gallant are remarkably productive; and on the outer coast of Tierra del Fuego an excellent fish may be caught in the kelp.

For a square-rigged sailing vessel bound through the strait, the following directions will be useful, in addition to those in Chapter VI. :—

**FIRST NARROWS.**—In the eastern entrance the winds will frequently favour a ship's arrival at the First Narrows; and if she selects a good anchorage near one of the banks which bound the sides of the channel, she may safely wait for an opportunity of passing through them and of reaching Gregory bay. There also a delay may be made in security for the purpose of passing the Second Narrows and getting into the neighbourhood of Cape Negro; at which place the difficulties of the eastern entrance cease.

The dangers being placed on the chart, and sufficiently described in Chapter VI., it is unnecessary to repeat them here; and indeed, much must be left to the judgment and discretion of the navigator.  On leaving the First Narrows care must be taken, in order to avoid the Barranca ledge, which is usually covered with kelp, to keep the north-east entrance of the Narrows one-third open, until 3 miles to the south-west of Barranca point, when Direction hills will open westward of the same point.  From this steer W.S.W. till the Narrows are shut in, which will carry the vessel to the northward of the Triton bank.  From thence a course may be shaped for Cape Gregory, taking care not to go into less than 6 fathoms.*

**TIDES.**—The tides answer best for vessels entering the strait at the period of full and change of the moon, since there are then two westerly tides in the day.  In the winter season, if the morning tide be not sufficient to carry a vessel through the First Narrows, she may return to Possession bay, select an anchorage, and be secured again before night : or, in the summer, if she has passed the Narrows, and enabled to anchor for the tide, there will be sufficient daylight for her to proceed with the following tide to Gregory bay, or at least to a safe anchorage off the peaked hillocks, on the north shore.

**CAUTION.**—Captain King says, that twice, when he attempted to pass the First Narrows, he was obliged to return to the anchorage in Possession bay, and that twice he passed through them against a strong wind blowing directly through, by aid of the tide; which, in the narrower parts, runs with great velocity.  When the tide and wind are opposed to each other, the sea is very heavy, and breaks high over the decks ; it is therefore advisable to close-reef, or lower the top-sails on the cap, and drift through ; for the tide, if at the springs, will generally be sufficient to carry a ship to an anchorage, although not always to one in which it would be safe to pass the night. On this account, it would be prudent to return ; for, although the holding ground is exceedingly good, yet, to part in the night, and drift into the Narrows, could scarcely happen without accident.

**SECOND NARROWS.**—In leaving the anchorage in Gregory bay, attention must again be paid to the tide, which continues to run to the eastward in the Second Narrows, 3 hours after it has commenced to set to the south-west of that anchorage.

**ANCHORAGE.**—With a leading wind through the Second Narrows, a ship will easily reach an anchorage off Laredo bay ; but if the tide fails upon emerging from it, she should seek for a berth in Royal road, inside of Elizabeth island, and as near to it as possible, but always to the westward of its north-east end, so as to be out of the influence of the tide. The depth of water will also be a good guide.

---

* Remarks of Captain C. Barker, R.N., 1857.

In passing round the south side of Elizabeth island, great caution is necessary in a sailing vessel with light winds, in not being set by the tide towards the Santa Marta rocks and the Walker shoal, which latter extends to the south-west of Santa Magdalena islet. Elizabeth island must be kept close on board, as the flood stream, on issuing from the Narrows, sets right across towards those dangers. A full description of those shoals, and directions for avoiding them, have been given at p. 164; and the mariner is cautioned not to attempt the passage without carefully consulting both them and his chart.*

Between Elizabeth island and Port Famine, with a south-westerly wind, it will be advisable to keep close to the weather shore, in order to benefit by the flaws down the valleys; but this must be done with caution, for squalls come suddenly off the high land, with a violence that a stranger cannot well imagine.

**FAMINE REACH.**—The only convenient anchorage between Port Famine and Cape Froward for a ship, is San Nicolas bay, to which, if defeated in passing round the cape, she had better return; for it is easy to reach as well as to leave, and extremely convenient to stop at, to wait for an opportunity of proceeding.

**FROWARD REACH.**—From Cape Froward to the westward, and through English reach, unless favoured by a fair wind, it is necessary to persevere and take advantage of every opportunity of advancing step by step. There are several anchorages that a ship may take up, such as in Snug bay, off Woods bay near Cape Coventry, in Fortescue bay, Elizabeth bay, and York road. To the westward, in Crooked reach, the anchorages are not so good, and, excepting Borja bay, none seem to offer much convenience. Borja bay, landlocked on every side, one of the best stopping places in the strait, is well calculated to supply the deficiency, although for a square-rigged sailing vessel there must be some difficulty in reaching it.†

**LONG REACH** is long and narrow, and ill supplied with anchorages for a ship; such as they are, Swallow harbour, Playa Parda, Marian cove, and Half Port bay, seem to be the best. In the thick weather that often prevails, although the channel is very narrow, yet one side is scarcely visible from the other; it has, however, one advantage over other parts of the strait in the smoothness of the water. In Sea reach there is a heavy rolling swell, besides a short harassing sea, which renders it very difficult to beat to windward.

---

* See Eastern Entrance to Magellan Strait, No. 1316; scale, m = 0·2 of an inch.
† Remarks of Commander F. L. Barnard, R.N., 1852.

SEA REACH.—Port Tamar, Valentine harbour, Tuesday cove, and Port Mercy, are the best anchorages ; and the latter is particularly convenient to those who are preparing for an opportunity of quitting the strait. In the western entrance, the sea is very heavy and irregular during and after a gale ; so that no vessel should leave her anchorage in Port Mercy, without a leading wind to get fairly out into the ocean.

For small vessels, particularly if they be fore-and-aft-rigged, and steamers, most of the local difficulties vanish ; and inlets which a ship would not dare to approach, may be entered with safety, and a snug anchorage easily obtained. A large sailing ship will perhaps be better off in entering and leaving the strait where the open space is frequently accompanied by a heavy sea ; but for the general navigation of the strait, small vessels have considerably the advantage. They have also the opportunity of passing through the Cockburn channel should the wind be north-westerly, which in some cases will much reduce the length of the passage into the Pacific.

FROM the WESTWARD.—In coming from the Pacific to the Atlantic, there must be some great advantage to induce a seaman to entangle his ship with the land by passing through the Strait of Magellan when fair winds, an open sea, and a quicker run invite him to steer at once round Cape Horn. The advantage is this, that to the eastward of the strait, the prevailing winds hang more frequently from the northward than from the southward of west, but are still fair for running up the coast, and the water is comparatively smooth. Whereas, a ship passing round Cape Horn, if the wind be north-west, must go to the eastward of the Falkland islands, and be exposed to a heavy beam sea and strong gales, and to hug the wind in order to make her northing. To a small vessel the advantage is incalculable ; for besides filling her hold with wood and water, she is enabled to escape the severe weather that so constantly reigns in the higher latitudes of the South Atlantic Ocean.

To enter the strait from the northward it will be advisable to keep an offing until the western entrance is well under the lee, so as to avoid being thrown upon the rugged and inhospitable coast to the northward of Cape Victory, which forms as it were a breakwater to the deep rolling swell of the ocean, and is for some miles off fringed by a cross hollow sea almost amounting to breakers.

CAPE VICTORY is high and much broken ; and if the weather be not very thick, will be seen long before the Evangelistas, which are not visible above the horizon from a ship's deck, for more than 4 or 5 leagues. From the *Adventure's* deck, the eye being 13 feet above the water, they were seen on the horizon at the distance of 14 miles. Pass to the southward of these rocks, and steer for Cape Pillar, which make like a high island.* In calm weather do not pass too near the cape, for the current sometimes sets out, and round it to the southward ; but with a strong wind get under its lee and steer along the shore. In the night it will be advisable to keep near the south shore ; and if a patent log be used, which no ship should be without, your distance will be correctly known. The course along shore, by compass, is E. ¾ S. ; and if the weather be thick, by keeping sight of the south shore, there will be no difficulty in proceeding with safety.

We shall here give an example of a passage through the strait from the westward. The *Adventure* entered it on the 1st of April 1830, at sunset ; and after passing within half a mile of the islets, off Port Mercy, steered E. ¾ S. under close-reefed topsails, braced by, the weather being so squally and thick that the land was frequently concealed ; but, it being occasionally seen, the water being quite smooth, and the course steadily steered, with the patent log to mark the distance run, she was allowed to proceed ; though the night was dark and the squalls of wind and rain frequent and violent. When abreast of Cape Tamar, that projection was clearly distinguished, as was also the land of Cape Providence, which served to check the distance shown by the patent log, and both giving the same result proved that the ship had not been subjected to any current ; whereas the account by the common log was very much in error, in consequence of the alternation of squalls and light winds, which rendered it impossible to keep a correct account of the distance. At daybreak she was between Cape Monday and the Gulf of Xaultegua ; and at 8 o'clock abreast of Playa Parda, in which cove, after a calm day, the ship was anchored.

In 1846, the *David Malcolm*, of 600 tons, was piloted through the strait by Captain Fitz-Roy, from Port Mercy to Crooked reach, during a dark and rainy night, without difficulty, the course being so direct and the high land so wall-like, that a comparatively light space was always distinguishable ahead between the walls by which to correct the steerage.

In the summer season there is no occasion to anchor anywhere, unless the weather be very tempestuous, for the nights are short and hardly dark enough to require it, unless as a precautionary measure or for the purpose of procuring wood and water ; the best place for which is Port Famine, where the beaches are strewed with abundance of logs of wood,

* *See* View of Cape Pillar in Admiralty Chart, No. 554.

well-seasoned, and therefore very superior to the green wood that must be
felled.

Notwithstanding that the *Adventure* felt no current on that occasion in
the western part of the strait, there is generally a set there to the eastward.   The direction and strength of the currents are caused by the
duration of the gales.

**DIRECTIONS.**—The chart will be a sufficient guide for vessels bound
through from the westward round Cape Froward and as far as Laredo
bay, at the northern end of Broad reach ; beyond this spot a few directions are necessary.   The land there of Cape Negro and Elizabeth island
should be kept close on board, to avoid the reef off the south-west end of
Santa Magdalena islet.   When abreast of the latter, bear away, keeping
the north-east extremity of Elizabeth island on the starboard bow until
you see Santa Marta rocks in one with, or a little to the southward of, the
south trend of the Second Narrows (Cape St. Vincent) ; this is the leading mark for the fair channel until past the spit of shoal soundings, extending across to Santa Magdalena.   There are also shoal soundings
towards the south-west end of Elizabeth island ; at half a mile off there
are 5 fathoms, Cape St. Vincent being then the breadth of Santa Marta,
open to the northward of that island.   Keeping the cape just in sight to
the northward of Santa Marta, steer on and pass round the low north-east
extremity of Elizabeth island, off which there are several tide eddies ; the
main tide stream here sets across the channel.

**SECOND NARROWS.**—Now steer for the Second Narrows, keeping
Cape Gregory, which will be just discernible as the low projecting extremity of the north side of the Second Narrows, on the starboard bow,
until you are 3 miles past Santa Marta ; the course may then be directed
for Cape Gregory, opening it gradually on the port bow as you approach
it, to avoid the broad shoal that extends off the cape to the southward.

**GREGORY BAY.**—If you anchor in this bay,* which is advisable, in
order to have the whole of the tide for running through the First Narrows,
haul up and keep 1½ miles from the shore.   When the north extremity of
the sandy land of the cape is in a line with the western extremity of the
high table land, you will be near the anchorage ; then shorten sail, and
when the grassy valley begins to open, you will have 14 fathoms : you
may then anchor or keep away to the north-east, and choose a convenient depth, taking care not to approach the shore, so as to bring Cape
Gregory to the southward of S. by W. ¼ W.   The best berth is with the
cape bearing S.S.W.

---

* *See* Admiralty Plan, No. 555.

FIRST NARROWS.—Thence, to the First Narrows, the course is due N.E. by E.  The land at the entrance, being low, will not at first be perceived, but on steering on you will see some hummocky land, making like islands.  These are hills on the eastern or Fuegian side of the Narrows.  Soon afterwards a flat, low sand-hill will be seen to the northward —it stands on the north-west extremity of Barranca point.  On approach-ing the Narrows, when 4 miles off, keep the northern cliff within the eastern side of the Narrows, well open of the trend of Barranca point, by which the shoal that extends off the latter point will be avoided.  You should not go into a less depth than 6 fathoms.  At most times of the tide there are long lines and patches of strong ripplings through which a vessel must pass ; but the shoals are easily distinguished by kelp growing on them.

When the channel through the Narrows bears N. by E. ¾ E., steer through it ; and that or a N.N.E. course will lead through.  On each side the bank extends off for some distance ; but by keeping in mid-channel, there is no danger until the cliffy coast be passed, when reefs extend off either shore for some distance, particularly off Cape Orange to the east-ward.   The N.N.E. course must be preserved until the peak of Cape Orange bears south, and the northern Direction hill W.S.W., or W. by S. ½ S.   Then steer E.N.E. for Cape Possession, taking care not to approach too near to the bank off Cape Orange, or the narrow bank on the north side of Possession bay, for which the chart must be consulted.

For small vessels, the passage through the strait from west to east is not only easy, but to be strongly recommended as the best and safest route.   And it may be here stated that the passage would be quite as ex-peditious, and perhaps safer for small craft to enter the Gulf of Trinidad, and taking Concepcion channel and either the Sarmiento or the Estevan channels, into Smyth channel, and so to enter the Magellan strait at Cape Tamar.  In those channels northerly winds prevail, and there is no want of convenient and well sheltered anchorages for the night.

---

FURTHER REMARKS on the passage between the Southern Oceans by Magellan strait, by Captain J. L. Stokes, R.N., in 1851:—

" The route from the Pacific to the Atlantic is safer for all ships except during winter, viz., from April to September, when easterly winds are frequent, with an occasional gale from that quarter in June and July.

" In these months the passage *westward* is equally easy, and made in five or six days.   Sir Francis Drake, the second navigator who reached the Pacific by this strait, in August 1523, did so in 15 days.

"Between September and May going westward can only be recommended for steamers and small vessels. Early navigators, during these months, have been 60 days on the passage from Port Famine to the western entrance. We were once 30 days (in 1827) attempting it in the Beagle during the months of January and February. On a second occasion it occupied only four days in mid-winter.

"This most difficult part of the western route is generally effected in a week during the winter, and in three weeks during the summer months This estimate is the average of passages made between Valparaiso and Port Famine by a vessel visiting the Chilian settlement formed there in 1843."

H.M. frigate *Fisgard*, Captain Duntze, is the largest sailing vessel that has yet passed westward through the Strait. She, however, was fortunate in having a fair wind from Swallow bay, and thus escaped the most harassing part of the passage, that of beating through Long reach, where it is very desirable more anchorages should be known*, in order that a ship may not lose any of the ground she has thus far with difficulty gained.

LOS EVANGELISTAS.—A ship passing through Magellan strait from the westward should on approaching the coast, be kept near the latitude of these islands, the south-eastern of which, shaped rather like a sugar loaf, is in lat. 52° 24' S., long. 75° 06' W., and bears from Cape Pillar N. 61° W. 23 miles. They consist of four small flat-topped rocky islets, none of which exceed 360 feet in height, and there is a passage 5 miles wide between them and the reefs fronting the main land.

The Evangelistas are visible 15 miles, in clear weather, from a vessel's deck, and on such rare occasions, when the high land about Cape Victory of 2,400 feet elevation is distinctly ˸seen, the name of Direction Isles, as given to them by Narborough, does not appear well bestowed. But in the wild westerly weather common to that locality, when a glimpse through the driving mist of their storm-beaten sea-washed sides, shows the mariner his position, the appellation is at once felt to be admirably applicable.

In the worst gales, however, the winds do not blow directly on the land. The lead is of use, and soundings of less than 50 fathoms may be found for 10 miles on the south-east side of the Evangelistas, and less than 40 fathoms at the same distance on their north-west side.

CAPE PILLAR.—In passing this cape with a scant northerly wind, there is generally found a south-west current setting past it towards the Juezes or Judges and the Apostoles (of the Spanish navigators), dangerous

---

* The attention of officers of H. M. ships and of mariners in general is particularly requested to this point.

clusters of rocks a few miles S.E. of the cape.  The outer and south-eastern of the reefs fronting the Narborough islands is the nearest northern danger to Cape Pillar, from which it bears N. by E. 8 miles.

This passed, a ship may be considered fairly in the strait, and will have before her, in an E. ¼ S. direction, a clear channel passable by night, 55 miles long, and from 5 to 12 wide, with a bold approachable shore to the southward.

The eastern end of this part of the strait (Sea reach) should be gained by early morning, — when with the prevailing westerly winds, Long, Crooked, and English reaches will be passed before dark, and the ship be in a position (between Capes Gallant and Holland,) where the strait is free from islands, and where there is ample space for another night's run.

GREGORY BAY may be reached the next day in good time for anchorage; but should the time of arrival there suit the tide for going through the First narrows, the passage eastward may be continued. Under such favourable circumstances the strait may be passed through without once anchoring.  No ship, however, should leave Gregory bay to pass eastward out of this strait, without having from four to five hours of daylight to clear the Great Orange bank and Possession bay bank.

The narrowest parts of this 350 miles channel, are in the First Narrows and Crooked reach.  In the latter, between the steep helmet-shaped cape El Morion and Cape Quod, it is scarcely 1¾ miles wide.  A rock nearly awash lies 2 miles east of Cape Quod.  This danger (Dolphin or Crooked reach rock) may be avoided by keeping the south shore on board until the largest Ortiz island bears N.E. ¼ N., when a mid-channel course may be again steered.

Crooked reach should not be passed through in the night, without there is good moonlight and clear weather.  Not only is it narrow and winding, but the tides are irregular and set across at the eastern end.

THE TIDES in parts of the strait require particular attention.  Of the entire length of Magellan strait, 125 miles only are subject to tidal streams, 100 miles of which are from the eastern entrance to between Laredo bay and Sandy point, and the remaining 25 miles include English and Crooked reaches.

There is a remarkable difference in the rise and fall in the eastern parts of the strait.  Off Cape Virgins, at the eastern entrance, the rise is from 36 to 42 feet in springs; at the First Narrows it is 36 feet, at the Second 23, and only 9 at Laredo bay.  As these three latter places are scarcely distant 27 miles from each other, it would almost seem that this change of tidal elevation, amounting to 27 feet in a space of 54 miles, was caused by a difference of level in the western and eastern waters of the strait.  But this cannot be; and here, as elsewhere all over the globe, the *mean level* of the sea is the same, or nearly so, at each end of the Strait of Magellan, although the rise and fall may differ from 42 to 6 feet.[*]

The time of high water, full and change, is 9h. at the First Narrows, 10h. at the Second, and 11h. 30m. in Laredo bay.

The changes of stream, as in other straits, differ from the time of either high or low water. The north-easterly stream makes on the full and change days at noon in the First Narrows, and 45 minutes later in the Second, and again about noon at Laredo bay. The stream changing when it is half-tide by the shore is convenient for ships passing the First Narrows, as the formidable banks at its east entrance are then only partly covered and the channel easier to be made out. Sluiced through as this channel is by a sometimes seven knot current, it makes the edges of these banks on either side very steep, and gives them the channel trend of N. 22° E.

Within the Second narrows the stream rarely runs four knots, and scarcely as much as two abreast of Laredo bay. Farther westward in the strait it is only perceptible near projecting points on the shore, until approaching English reach, where, between Rupert and Carlos islands, it again attains a velocity of 3 knots.

In York road, English reach, the tide rises 9 feet at full and change, when the time of high water is 2 o'clock ; the south-eastern stream, which has a strength of 3 knots, makes at 5h. 30m. or 2h. 30m. after it is low water by the shore. In Jerome channel the streams change with the high and low waters, the tides therefore flow and ebb out of it with both western and eastern streams of the strait.

At the junction of English and Crooked reaches, where the streams of tide from St. David's sound on the south and Jerome channel on the north meet, there is a very confused and cross set of tide. Cape Cross Tide, the name given to the north-west end of Carlos island, will draw attention to this tidal irregularity. Westward of Crooked reach there is little or no tide, but almost a constant hourly set to the eastward of from a quarter to 1 knot.

A current of from 1 to 2 miles an hour sweeps round Cape Pillar towards those dangerous clusters of rocks, the Apostoles and Judges, and was found in the *Beagle* to continue along the outer coast to the south-eastward. In the offing, outside the Evangelistas, the currents are trifling, but influenced by the winds. In the passages on the north side of those islands the flood sets to the E.N.E. about a knot ; the time of high water at full and change being 1 o'clock.

---

[*] *See* Voyages of the *Adventure* and *Beagle*, Appendix to Vol. II., page 391.

# CHAPTER IX.

### PATAGONIA WEST COAST, WITH THE SOUNDS AND CHANNELS BETWEEN MAGELLAN STRAIT AND THE GULF OF PEÑAS.

VARIATION 23° to 31° East in 1860.   Annual change inappreciable.

———————

**The WESTERN COAST,** between the Strait of Magellan and the Gulf of Peñas, is formed by a succession of islands of considerable extent, the largest of which, Wellington island, fronts a length of coast of 138 miles.   It is separated from the main by the Messier and Wide channels; and from Madre island by the Gulf of Trinidad.   Madre, which is probably composed of several islands, has for its inner or eastern boundary Concepcion strait.   Hanover island has the Inocentes and Estevan channels on its northern and eastern sides, and on the south is separated from Queen Adelaide Archipelago by Nelson strait, which communicates through Smyth channel with the Strait of Magellan.*

**SMYTH CHANNEL** communicates with the strait through the opening between Cape Phillip and the Fairway isles, and extends 45 miles in a N.N.W. direction from the Strait of Magellan to the Victory pass, and then W. by N. ½ N. 30 miles into Nelson strait.

**Supplies.**—Throughout the whole space between the Strait of Magellan and the Gulf of Peñas, there is abundance of wood and water, fish, shell-fish, celery, and birds.

**DEEP HARBOUR,** the first anchorage on the western shore, is a little more than 6 miles from Cape Phillip; the entrance to it is a quarter of a mile wide, and the anchorage is about half a mile within the first head, off the entrance of a lagoon, in from 30 to 35 fathoms.   North and south of the port are inlets each a mile deep.   In entering, there is a patch of kelp on the starboard hand, and the shore is fronted for a short distance by rocks.†

———————

\* *See* Admiralty Chart of Magellan strait and of the West Coast of South America, Sheets I. and II. Nos. 560, and 561; scale, $d = 7\cdot8$ inches.

† *See* Admiralty Plan of Deep harbour, No. 557; scale, $m = 1\cdot5$ inches.

**GOODS BAY,** 6 miles to the northward on the same shore, is better than the last, the depth being from 20to 25 fathoms. It is convenient for vessels going to the northward, but when bound in the opposite direction, North bay will be better, from the depth being still less ; but it is small, and the entrance is more fronted by rocks than Goods bay. If it is not intended to anchor in either of the above places, the widest and best channel is to the eastward of Renouard island.[*]

**Shoal Island,** on the eastern shore, has a rocky patch off its north-east point, upon which the *Adelaide* struck. The channel for the next 4 miles is rather intricate among the rocks and islets north of Renouard island; but all the dangers are noted in the chart.

**Cape Colworth** is situated on the western shore, to the northward of these dangers. Here the channel widens, and opposite to this cape, on the eastern side, there is the large Inlet of Clapperton, beyond which there is a considerable tract of low country, a rare sight in these regions.

**Hose Harbour,** suitable for a small vessel, lies 2 miles from the entrance of the Inlet; and on the opposite shore of the strait is Retreat bay, fronted by low rocky islets. The depth within is 24 fathoms.

**Oake Bay** is on the western shore, 4 miles from Retreat bay. There is anchorage in this bay in 9 fathoms ; but better is found among the Otter islands in mid-channel, the depth being 6 and 7 fathoms, and the ground clean. The channel is clear between Hose harbour and Oake bay, but for the next 8 miles, becomes more strewed with islands and rocks, and has much shoal water off every low point. The eastern shore is very low as far as the base of Mount Burney, which is 5,800 feet high, and covered with perpetual snow. The best channel lies to the eastward of the Otter islands, and then between the Summer isles and Long island, for which the chart and a good look out for kelp will be sufficient guides. A spit appears to extend off the North point of Long island. The passage to the eastward of Long island is obstructed by rocks, which extend from Rose hill on the main to within a few yards of the beach of the island, and upon which H.M.S. *Vixen* struck in 1857.

**FORTUNE BAY,** opposite the north point of Long island, is at the south-eastern extremity of an island in the entrance of a deep channel, which is, probably, one that Mr. Cutler, the captain of an American sealing vessel, passed through ;[†] and upon the supposition of its leading through into Nelson strait, it has been given by Captain King the name of Cutler channel. Fortune bay is a very convenient and good anchorage, the

---

[*] See Plan of Goods Bay, No. 2,790 ; scale, m = 3·0 inches.

[†] While Captain King was employed upon the survey, he met this intelligent person two or three times, and received much valuable and correct information from him.

depth being moderate, and bottom good ; the best berth is within a low island, in from 8 to 12 fathoms; it is necessary to moor, as the anchorage is too small to allow a vessel to swing at single anchor.  At the head of the bay there is a thickly-wooded valley, with a fresh-water stream. Palmer point, on the opposite side of Cutler channel, is the southern termination of Rennell island, formed by Cutler and Smyth channels.*

WELCOME BAY, on Rennell island, lying 5 miles north of Palmer point, affords good anchorage.  The Hudson Bay Company's steamer *Labouchere* took shelter here from a N.W. gale.  In standing for the 2 islets on the northern shore, she had 16 to 13 fathoms, and then no bottom until right between them, when she brought up in 9 fathoms just south of a patch of kelp, and having veered to 80, found 25 fathoms at the gangway, mud and ooze.  On examining the bay, Captain Trivett found 5 fathoms on the kelp patch, and 17 fathoms mud, in the north-west part of the bay, just inside the second island, and outside the ridge of rocks which form the inner harbour.  This inner harbour appears to be a place where a ship might heave down and refit, but it would require some examination before attempting it.†

Isthmus Bay, a league to the northward of Palmer point, on the main shore, will be found to afford anchorage, but open to the south-west, which, however, is not of much moment, for the channel is only 2 miles wide.  The head of the bay is formed by a very narrow strip of land, separating it from what there can be no doubt is Sarmiento's Oracion bay.

LACE PENINSULA is a long narrow strip of high land, extending 9 miles from Isthmus bay to Victory Pass.  In Sandy bay, about the middle of the peninsula, and off Inlet bay, on the western side of the channel, there are good anchorages ; both have a moderate depth, and are' sheltered from the generally prevailing north-westerly winds.  A rock awash lies about a mile N.E. from Bessel point, the north extreme of the peninsula.

SMYTH CHANNEL is situated at the south point of Hunter island, and leads W.N.W., while Victory pass leads to the north.  On the west side of Hunter island, about 3 miles from the south point, Island bay offers good anchorage both to the northward and southward of some islets which lie off it.  The *Adelaide* anchored in the latter in 17 fathoms.  Piazzi island lies between Smyth channel, Nelson strait, and Sarmiento channel ; at its southern extremity, Hamper bay has anchorage in from 7 to 15 fathoms.  There Smyth channel widens to 3½ miles ; but, at 2 leagues farther on, near Ceres island, under the S.E. end of which the

---

* See Plans of Fortune and Welcome Bays, No. 2,790 ; scale, m = 6·5 and 3·0 inches.
† Remarks of J. F. Trivett, Esq., commanding H.B.C. Steamer *Labouchere*, 1858.

*Adelaide* anchored in 10 fathoms, it narrows to 2 miles. Rocky cove cannot be recommended, and Narrow creek seems confined. Hence to the mouth of the channel, which, during strong north-west winds, sends in a heavy sea, we know of no anchorages; but a small vessel in want will, doubtless, find many, by sending her boat in search. The *Adelaide* anchored among the Diana islands, and in Montague bay, having steered through Heywood pass, at the western extreme of Piazzi island.

The northern point of Piazzi island is Sarmiento's Punta Oeste, or West point, and a league to the south is his Punta Mas Oeste, or Westernmost point. On the eastern shore of Piazzi island, in lat. 51° 40′ 30″ S., is a small anchorage at the entrance of an inlet, useful to vessels unable to reach Relief harbour, at the south end of Vancouver island, before dark. Its position is easily known by 2 small islands, under which is the anchorage in 14 fathoms.*

**TIDES.**—So generally do the northerly winds prevail, that it is troublesome even to working vessels to make a passage to the northward; but it is a safe channel for small craft at any time. The tides are regular; the rise at the southern entrance is 8 or 9 feet, but at the northern only 5 or 6. The flood tide sets to the northward, and the strength of the stream is from half to 1¼ knots; so that a vessel is not so likely to be detained here for any length of time, as she would be in the Strait of Magellan, where there is little or no assistance felt from westerly tides. The channel, besides, is comparatively free from swell, and the winds are not so tempestuous.†

**VICTORY PASS,** in lat. 52° 01′ S., separating Zach peninsula from Hunter island, connects Smyth channel with Sarmiento channel to the north and with Union sound, which, by turning sharp to the south-east, lead to the Ancon Sin Salida (or No Thoroughfare Cove) of Sarmiento.

**DIRECTIONS.**—Vessels running through the Victory pass should keep the south-east coast of Hunter island on board, passing close to the small island off its eastern extreme to avoid the Cloyne reefs. The channel between the island and the reefs is not a mile wide, the reefs only just showing level with the water's edge, and dangerous in thick weather. Having passed this, leave all the small islands lying south-east of Newton island on the port-hand, and steer along the coast of the island, passing between it and an island to the northward, into Sarmiento channel. The passage south of Brinkley island is less direct, but wider, and free of reefs; taking care of the rock off Bessel point.‡

---

* Remarks of Commander A. S. Hamond, R.N., 1843.    † Skyring's MS.
‡ Remarks of J. F. Trivett, Esq., 1858, in Mercantile Marine Magazine for Oct. 1859; a work which contains much useful hydrographic information.

INTERIOR SOUNDS.—As the Sounds within Smyth channel are not very likely to be used for purposes of general navigation, little need be said of them in a work destined solely for the use of shipping frequenting the coast ; although they possess many anchorages for small vessels, affording both shelter and security. The chart will be sufficient to refer to for every purpose of curiosity or information.

Sarmiento, on his third boat-voyage to discover a passage through the land into the Strait of Magellan, gives a detailed and very interesting account of his proceedings. All his descriptions are so good, that the surveyors had no hesitation in assigning positions to the places which he mentions, and to all of which his names have been appended.

CAPE AÑO NUEVO, at the bottom of Union sound, cannot be mistaken, and Sarmiento's description of Ancon Sin Salida is accurate. He says :—" The Morro of Año Nuevo trends round to the south-east and south-south-east for a league to the first water ravine that descends from the summit. In an easterly direction from this appears a large mouth of a channel, about 2 leagues off. We went to it, and found it to be a bay without a thoroughfare, forming a cove to the northward, about a league deep ; so that, finding ourselves embayed, we returned to the entrance, which we had previously reached with great labour and fatigue. This bight has four islets. The bay from the islets to the westward has a sandy beach, backed by a low country for more than a league and a half to the Morro of Año Nuevo."

A comparison of Sarmiento's account with the recent survey cannot fail to claim for that excellent and persevering navigator the admiration of all geographers. Nor should the late Admiral Burney be forgotten ; for the plan formed principally by him, from Sarmiento's Journal and other documents, is an extraordinarily correct delineation of what are now shown to be the true features of the place. Leeward bay, on the east side of Ancon Sin Salida, is an exposed anchorage, and being upon the leeward shore, is not to be recommended.

The CANAL OF THE MOUNTAINS is nearly 40 miles long, by rather more than one mile wide, extending in a N.N.W. direction from Ancon Sin Salida, and bounded on each side by a high snow-capped Cordillera, the western side being by very much the higher land, and having a glacier of 20 miles in extent, running parallel with the canal. Eighteen miles from Cape Earnest, where the canal commences the channel contracts from 2 miles to the breadth of about half a mile. Whale-boat bay is about a mile to the eastward of Cape Grey, the northern point of entrance to the channels leading to Obstruction and Worsley sounds through the Kirke Narrows. There is a small cove on the north shore of Kirke Narrows, about a mile to the eastward of Cape Retford. Fog bay lies 2½ miles to the northward of the east end of Kirke Narrows; and Easter bay is a convenient anchorage within White Narrows, another opening into Worsley sound.

LAST HOPE INLET, an opening on the north side of Worsley sound, is 40 miles in length. Its mouth is 3½ miles wide, but in one place the breadth is narrowed by islands to less than a mile, the channel being from 5 to 14 fathoms deep, beyond this narrow the sound trends to the W.N.W. On the peninsula between Worsley bay and Last Hope inlet is Moore's Monument, a mountain 3,400 feet above the sea. The islets were covered with black-necked swans, and the sound generally is well stocked with birds.

The land round Disappointment bay, on the east side of Worsley sound, is very low, and thickly covered with stunted wood. Mr. Kirke traced its shores, and found them to be formed by a flat stony beach, and the water so shallow, that the boat could seldom approach it within a quarter of a mile. A considerable body of water was noticed by him over the low land; probably a large lagoon, for it communicates with the bay by a rapid stream 50 yards wide. No high land was seen in an easterly direction; so that the country between Disappointment bay and the eastern coast may probably be a continued *pampa*, or plain, like the coast of eastern Patagonia.

OBSTRUCTION SOUND, an opening on the southern side of Worsley sound, extends for 30 miles in a S. by E. direction, and then for 15 more to the W.S.W., where it terminates. It is separated from Skyring Water to the south-east by a ridge of hills, the isthmus being perhaps 12 miles across. Some water was seen from a height, about 6 miles off, in the intervening space, but the shores were so carefully traced that Captain Skyring, who examined it, felt satisfied that no communication exists.

SARMIENTO CHANNEL, communicating with Smyth channel through Victory pass, stretches 70 miles to the northward, running between Piazzi island and Staines peninsula, and then to the eastward of Vancouver and Esperanza islands into Peel inlet, at the mouth of which it unites with Estevan channel, and both merge in the Inocentes channel through the Guia Narrows. Relief harbour, at the south end of Vancouver island, is a convenient anchorage. Between Vancouver and Esperanza islands there is a passage nearly a league wide, but strewed with islands.

**PUERTO BUENO**, on the eastern shore of the channel, 30 miles farther north, is the best anchorage in this locality, and especially noticed by Sarmiento, affording excellent ground and a moderate depth of water; the latter a very unusual occurrence. It is upwards of a mile in length, with 8 and 9 fathoms in the centre, and 4 and 5 close to the shores. A cove round the north point, called Schooner cove, is well adapted for a small vessel, and may be used in preference even to the large harbour. H.M.S. *Gorgon* anchored in Puerto Bueno in 13 fathoms, with three small islands bearing respectively West, W.N. ½ N., and N.E. by E. ½ E. Both the inner and outer anchorages have good holding ground, with the depth varying from 15 to 7 fathoms, and perfectly sheltered. For small vessels, the inner harbour, a completely land-locked basin, is admirable; there are no dangers, and the entrance into it has from 13 to 6 fathoms right across at half-tide. Rise and fall at full and change 6 feet.*

**ESTEVAN CHANNEL** lies on the west side of Vancouver and Esperanza islands, between them and Hanover island. Escape bay, on Vancouver island, although small, is convenient and well-sheltered. Opposite the south end of Esperanza island the deep opening of Ellen bay to the westward may probably be a channel passing through and dividing Hanover island. In the latter island the anchorages of Rejoice harbour and Anchor bay are commodious and useful.

**PEEL INLET**, at the junction of the Estevan and Sarmiento channels, stretches in to the eastward for 21 miles, as far as Pitt channel, which, trending to the north-west, opens through San Andres bay and Andrew sound into Concepcion channel, and thus insulates Chatham island.

**The GUIA NARROWS**, so called from Sarmiento's boat, are between Hanover and Chatham islands, and connect the Sarmiento and Inocentes channels; they are 6 miles long, and from one to half a mile broad, except at the north end, where they are only 2 cables across; but there is no danger in passing, the shores being steep-to on either side. From this the track runs through Inocentes channel for 21 miles, into Concepcion channel, passing Juan island on the starboard, and Inocentes island on the port hand.*

**The TIDES** in the Guia Narrows are not very rapid. It is high water, full and change, at 2h. 8m., the flood running to the southward.

---

* Remarks of Commander J. A. Paynter, R.N., 1848, and Plans of Puerto Bueno and the Guia Narrows, No. 2,790; scale, *m* = 1·5 inches.

At the southern end of Estevan channel, the reverse is the case, of which, for vessels passing through, some advantage may be taken.

GUARD BAY is on the eastern shore of Inocentes channel. The *Adelaide* anchored here. The north-west coast of Chatham island has many bights and coves fronted by islands, but the coast is too much exposed to the sea and to the prevailing winds to offer much convenient or even secure shelter. The north-west points of Hanover and Chatham islands are more than 10 miles apart, and midway between them is the Inocentes island of Sarmiento, from which the channel takes its name.

CONCEPCION CHANNEL separates the Madre islands from the main land. It commences at Inocentes island and joins the Wide channel, or Brazo Ancho of Sarmiento, in lat. 50° 05′ S. On the eastern coast of Madre island there are several convenient anchorages, particularly Walker bay, Michael bay to the northward of Michael point, and Tom bay; all of which, being on the weather shore, afford secure anchorage; but the squalls off the high land are not less felt than in other parts.

ANDREW SOUND, an opening on the east side of Concepcion channel, is 12 miles in length and 5 wide; at 18 miles within, it divides into two arms; the northern one, San Andres bay, is 5 or 6 leagues long, to the foot of the hills; and the southern arm is Pitt channel, passing eastward of Chatham island, and communicating, as before-mentioned, with Peel inlet. The entrance is to the southward of the Canning isles, in the northernmost of which, at the south-west end, is Portland bay, a good anchorage for a small vessel, in 9 fathoms. The other side of the entrance of Andrew sound is the north shore of Chatham island. Expectation bay is a little anchorage 5 leagues within the sound, at the eastern extremity of the Kentish isles, and was used by the *Adelaide* in the examination of these inlets.

WIDE CHANNEL, a continuation of Concepcion channel, commences at Brazo Ancho point, the N.E. cape of the Madre islands. It extends 38 miles to the northward from Topar to Saumarez island, with a breadth of 1¾ to 3½ miles. Open bay is on the eastern shore, opposite the Trinidad channel. The anchorage is sheltered by two islands; but it is too exposed to trust a vessel in, and therefore not to be recommended. Small Craft bight, on the western side of Wide channel, near its commencement, is of small size, but answers every purpose of a stopping-place for the night.

**Sandy Bay,** on the same side, 17 miles to the northward, is a poor anchorage.   There is a rock on the northern side of the bay, on which the H.B. Company's steamer *Labouchere* touched ; and there is no anchorage on the southern shore.   Fury cove, near Red cape, the extremity of Exmouth promontory.   It is very confined, there not being room for more than two small vessels ; but the ground is good, and although open to the south-west, it is a secure haven.

**SAUMAREZ ISLAND** divides Wide channel from Indian reach ; there is a passage on either side of this island, but the eastern is the best, passing close to Bold head, an immense dark mass of rock, rising abruptly from the sea to the height of 1,000 feet.   Abreast of this head is the entrance of Eyre sound, which is 40 miles long, with an average breadth of 4 miles.   Near this entrance on the 'eastern shore, there was a large *rookery* of seals, and another 13 miles farther up, on the same side, in lat. 49° 21' S.

**INDIAN REACH,** between Saumarez island and the English Narrows has many islands and groups of straggling rocks, with deep water close to them ; a vigilant look-out, however, should be kept for kelp, which is an almost certain indication of shoal water.   The track lies on the eastern shore in the first part of the channel ; then to the westward of the low rocks and kelp patches, in lat. 49° 16' S. ; and afterwards on the east shore as far as Level bay.   An exposed anchorage was found by H.M.S. *Salamander* off the islands in front of Level bay.   There is abundance of driftwood on these islands.   Rocky bight, opposite the north-east point of Saumarez island, is a much exposed anchorage, in from 12 to 17 fathoms.

**EDEN HARBOUR,** on the western shore of Indian Reach, just south of the English Narrows, is formed by a group of islands, which are thickly wooded.   This harbour is superior, as a stopping place, to Level bay, there being good anchoring ground in 8 to 12 fathoms, one mile long by a third broad.   Abundance of fish may be caught ; wild fowl are numerous, and the water is good and plentiful.   The main land and islands are thickly wooded with great quantities of driftwood on the north shore of the latter ; wild celery is also to be obtained.*

The entrance is from the southward, steering N.W. between the west shore and the islands, a distance of one mile to the anchorage between Eden island and a rivulet on the main shore.   To the northward the water is deeper, but the holding ground is good all over the harbour.

* *See* Admiralty-Plan of Eden harbour, on No. 516 ; scale, *m* = 1·7 inches.

East of Eden island the space is about half a mile square, with deep water. There is another passage, about 2 cable's length broad, between the second and third islets. In the northern part of the harbour is a convenient cove with a narrow entrance; there is also a similar one just south of the entrance, but south of it again are two rocks, which makes it necessary to keep on the eastern shore of the reach.*

**Level Bay,** on the eastern side of the channel, at the south end of the English Narrows, scarcely deserves the name of Level, as there are 30, 25, 15, and 10 fathoms close to each other on a bottom of sand and clay: this anchorage is confined for anything but a small vessel: the beach was strewn with fine drift timber, suitable for steam fuel. The anchorage in Level bay is on a ridge of sand in 17 fathoms, with the extremes of the bay, N.W. and S.E., the northern point just sheltering the vessel from the tide.

**TIDES.**—It is high water, full and change, in Eden harbour at 1h., the rise and fall being 6 to 7 feet. The flood S.S.E., and the ebb N.N.W.

**ENGLISH NARROWS,** the most intricate passage in this navigation, commences after passing Level bay    It is 12 miles long, and from ¼ to 1¼ miles wide; but in many parts contracted by islands to 400 yards. The southern part, though narrow, is quite apparent to be made out, and entirely free from danger. The channel then becomes an open space for 5 miles, when it appears to close in and forbid further approach; on advancing, however, several small low islands become visible, off one of which on the eastern shore a small rock will be seen, surrounded by kelp. Pass a moderate distance to the westward of this, and to the eastward of a patch of kelp off the western shore, from thence keep the latter shore aboard without fear until a small island is observed, moderately high, apparently blocking up the passage. Keeping the western shore still aboard, haul close round a point which juts out, into a small indentation of the land, which will enable the vessel to pass to the westward of Mid-channel island. On drawing round this island, another, small and rocky, with much kelp about it, comes in view. Pass, southward of this, between it and Mid-channel island.* There is a passage, used by the *Labouchere,* to the eastward of Mid-channel island, but it is not so broad, and has a rock on its eastern shore.†

This is the most intricate part of the English Narrows, but with care, attention, and a vigilant look-out for kelp, there is no difficulty for a

---

* Remarks of Commander A. S. Hamond, and J. Jenkins, Master, R.N., 1843.
† *See* Admiralty Plan of English Narrows, on No. 561; scale, m = 0·8 of an inch; and Remarks of J. S. Trivett, Esq., in Mer. Mar. Mag. for Oct. 1859.

steamer in passing through. The remainder of the Narrows presents an
opening channel, through which the vessel is steered ; and when fairly
passed, the whole extent of the noble Messier channel opens out, extending as far as the eye can reach between lofty mountains covered with snow.

**MESSIER CHANNEL** commences at the north end of the English
Narrows; it is quite open, and free from all impediment, opening into
Tarn bay, in the Gulf of Peñas, a distance of 75 miles. At the end
of English Narrows the Messier channel is something less than 3 miles
wide, but at Iceberg sound, 15 miles to the northward, it increases to
5 or 6, which breadth it retains, with little interruption, to the sea.

**HALT BAY** is on the eastern shore, at the north end of the English
Narrows. On the eastern side of the bay is an island divided from the
main land by a passage only 12 yards wide, with 4 feet water in it, and
having rocks running across it. This island has a commodious circular
harbour in it, fit for small craft, and having a depth of 3 and 4 fathoms
hard bottom. The wood cut at Halt bay was found to be superior for
keeping steam to that found in the other harbours ; it was cedar, burnt
easily, gave out good heat, and did not require to be freshened with
coals ; all the wood we had hitherto cut was saturated with wet, and
required drying in the ash-pits previously to being placed in the fires.*
Here the flood sets to the S.S.E., and the tide, being confined by the
narrow width of the channel, runs with considerable strength. The best
anchorage is with the extremes of the bay, E.N.E. and West, and the
gap in the centre of the island, East.

**DIRECTION ISLETS** lie about mid-channel, 15 miles from the Narrows,
between Iceberg sound on the main, and a deep inlet on the Wellington
island. They are 2 in number, small, but thickly wooded, with quantities
of driftwood lying on their shores. White Kelp cove, 9 miles from Direction islets, on the north side of Lion bay, about a mile within the head,
is confined, and only fit for a small vessel. Middle island lies just to
the northward of this cove. It is high, well wooded, of a conical shape,
and excellently situated to assure a vessel of her position in the channel.
Waterfall bay, 14 miles to the northward of White Kelp cove, is at the
entrance of an inlet on the east side of the channel. About midway
between this bay and Middle island is Caldcleugh inlet.

**ISLAND HARBOUR**, on the eastern shore, 20 miles within the
entrance, is a small but excellent land-locked anchorage, well placed

---

* Remarks of Commander J. A. Paynter, R.N., 1848.

for vessels entering or leaving these channels. It position is marked by an island lying a short mile south of the entrance, in which there is also a small island. A bank or bar of rocky ground stretches across both entrances, having 9 fathoms in mid-channel, and shoaling gradually to 3 fathoms on either side, close to the rocks. At the head of this secure basin there is a large waterfall, from which any quantity of water may be procured with ease. Fish are abundant, and the shores are lined with driftwood.*

The anchorage is in 19 fathoms, good holding ground, a cable from the entrance, with the extremes bearing S.W. ½ S., and S.E. by S. There is also a good anchorage off the harbour's mouth, in from 15 to 20 fathoms.

**Fatal Bay** is open and exposed, lying on the western shore, at the north entrance of the channel, between Millar island and the eastern shore of Wellington island. At this point the Messier channel opens through Tarn bay (see page 240), and the Gulf of Peñas into the Pacific.

Besides the above anchorages, there are many equally convenient, and, perhaps, much better, that might be occupied by vessels navigating these channels. Every bight offers an anchorage, and almost any one of them may be entered with safely. On all occasions the weather shore should be preferred, and a shelving coast is generally fronted by shoal soundings, and more likely to afford moderate depth of water than the steep-sided coasts; for in the great depth of water alone consists the difficulty of navigating these channels.

**TIDES.**—In the northern part of the Messier channel the tides are regular, running 6½ hours each way; and there the flood sets N. by W.

---

**OPEN SEA COAST.**—The following descriptions of the sea-coast to the northward of Magellan strait have been extracted from the manuscript journals of the late Captain Stokes and Skyring, and of Mr. Kirke, Mate of H.M.S. *Beagle*, by Captain King. The coast between Cape Victory and Nelson strait, is much broken and intersected by channels leading between the islands of the Adelaide Archipelago. At 10 miles to the N.N.E. of Cape Victory, there is a remarkable pyramidal hill called Diana peak, which, in clear weather, is visible to ships in the mouth of Magellan strait, at 30 miles off.†

**CAPE ISABEL,** the western point of the Archipelago, is a steep rocky promontory of great height, with a peaked summit, and a sharply

* Remarks of Commander A. S. Hamond, and J. Jenkins, Master, R.N. 1843.
† See Admiralty Chart of West Coast of South America, Sheet I., No. 560; scale, m = 0·13 of an inch.

serrated ridge, having two detached columnar masses of rock. Beagle island, lying off it, is wall sided ; but, although tolerably high, is much lower than the land of the cape.

**CAPE SANTA LUCIA,** the westernmost point of Cambridge island, is high and precipitous. Cape George, at its southern end, is lower, and forms a bluff point. Between Capes George and Isabel is Nelson strait which extends N.E. about 50 miles ; its average breadth being 15 miles. Cutler and Smyth channels open into this strait, and there is an opening into Sarmiento channel between Vancouver and Piazzi islands.

The San Blas channel, Duck and Duncan harbours, the Duncan rock, and other rocks off them, are inserted from the oral information of the master of a schooner, and may be very incorrectly laid down. Augusta island and the White Horse off the west shore of Hanover island, were seen by Captain Skyring.

**CAPE SANTIAGO** is the south-western extremity of Duke of York island, which is separated from Hanover and Madre islands by Concepcion strait, and the West channel. These channels communicate with the inshore passages at Inocentes island to the north-west of the Guia Narrows.

**MADRE ISLANDS,** to the northward of Hanover island, are separated from Wellington island by the Gulf of Trinidad. The most conspicuous summits of these islands, that are seen from seaward, are April peak, Tower rock, and Cape Three Peaks, which latter forms the south-west point of the Gulf of Trinidad.

**Cape Three Peaks,** on the north-west coast, rises to a lofty rocky mountain, nearly 2,000 feet high, the summit having three peaks connected by sharp serrated ridges, with a detached mass of rock of pyramidal form at the base, which shuts in with the land on the bearing of N.N.E. ¾ E.

**Port Henry** is 3 miles to the north-east of Cape Three Peaks ; and the shore between them is filled for nearly a league in the offing with rocks and islets, of which several scores might be counted in the space of a square mile ; but they seem to be of bold approach, and few dangers probably exist that are not above water, or that are not shown by kelp.*

Bound to Port Henry, a vessel should keep on the northern shore of the Madre islands ; for the northern part of the Gulf of Trinidad is strewed with many rocks, some of which seem to be exceedingly dan-

* See Admiralty Plan of Port Henry, No. 1,978; scale, m − 1·5 inches.

gerous. The soundings, also, are very irregular, and the bottom is foul and rocky. From observations made on a rock at the western side of the port, and marked in the plan, the rock was found to be in lat. 50° 00′ 18″ S., long. 75° 18′ 55″ W.

The entrance of Port Henry may be easily distinguished by its being the first sandy beach seen on that shore after entering the gulf; with a lowish sandy cliff at the back, and a round, rocky, and wooded mount at its western end. The Seal rocks are also a good mark for it; they bear nearly N. by W., 5 miles from the west point of the entrance, which is about a mile wide. The channel is bounded on each side by low rocks, and round rocky islets of greater height, and which may all be approached within 1¼ cables' lengths. The soundings are from 20 to 26 fathoms, on a sandy bottom; decreasing pretty gradually to the anchorage, which is in 9 and 10 fathoms.

TIDES.—It is high water, full and change, in Port Henry, at 11h. 45m., and rises 5 feet. The stream of the tide, however, is very inconsiderable, and never exceeds half a knot.

DIRECTIONS.—When the sandy beach bears about S. by E. ¾ E., the fairway of the entrance will be quite open; and a vessel may stand in, keeping the round mount at the western end of the sandy beach on the port bow, until nearly abreast of it; she may then proceed up the harbour as high as convenient, and select her berth; for the ground is quite clear of danger to the line of rock weed, which skirts the shores and islets. The depth of water is between 12 and 8 fathoms, and the bottom generally of sand and mud. In turning in there are some patches of kelp on each side, growing upon rocks that watch at high water; they must be given a good berth. Their positions are given in the plan. As the squalls off the high land are sometimes very strong it will be advisable for a ship to anchor as soon as possible, and warp up to her berth; which, from the smoothness of the water, may be easily effected. Any security may be obtained in this harbour; the plan will show that Aid basin, at the bottom of the harbour, is a complete wet dock. Wood and water at the sandy beach are in abundance.

The GULF of TRINIDAD, leading into the Trinidad channel, separates Wellington and Madre islands. It is nearly 10 leagues long, and from 4 to 9 miles wide. Its southern shore, on Madre island, is very much broken, and probably contains many ports; but none of them were visited unless wanted for night anchorages; under the east side of Divis island lies Port del Morro, which, with Cape Candelaria and Port Rosario, are inserted from Sarmiento's account. Its north side terminates in Cape Primero, the extremity of the mountainous island of Mount

Corso, which may be seen, in clear weather, at the distance of 10 leagues. Viewed from the southward, the summit of the island has a rounded appearance, rising above all the contiguous land, from which a small portion of low coast extends for two degrees beyond it to the westward. The land of the northern shore of the gulf makes in mountainous ridges and peaks, the average height of which Captain Stokes estimated to be about 3,000 feet.

On the northern shore there are two large opening-like channels : the westernmost probably communicates with the Picton opening ; the other, the Brazo de Norte, appeared to reach only to the base of a range of mountains, among which Cathedral mount is a conspicuous object. From the entrance of the strait this mountain resembles the spire and roof of a church, and is visible for more than 20 leagues. Between the two openings is Neesham bay, in which the *Adelaide* found a secure anchorage in 11 fathoms. There is also good anchorage for a small vessel in Windward bay.

**Topar Island.**—Trinidad and Wide channels meet at their junction with Concepcion channel, and the water way is contracted by the Island of Topar, to the breadth of 1½ miles. There are several isles and rocks in the channel, of which the most remarkable are the Seal rocks before mentioned ; the Van isles, opposite the western opening ; and a numerous group extends for a league to the south-eastward of Neesham bay. On the south shore also there are several isles, but they are all near the coast except Medio island, which, with the reef off its south-west end, is well described by Sarmiento.*

**MOUNT CORSO ISLAND** is separated from Cape Brenton on Wellington island by Spartan pass, and some extensive reefs for more than a league off Cape Primero : indeed the whole of the west coast of Madre island is fronted by rocks some of which are 2 leagues from the shore. There are regular soundings in the Gulf of Trnidad, but the water deepens immediately after passing to the eastward of Port Henry, and entering Trinidad channel.

**PICTON OPENING** and Dyneley bay most probably insulate the land that separates them, of which Cape Montague is the western extremity. There are some rocks 8 or 9 miles off the coast to the southward ; and on the west coast of Wellington island, between Cape Montague and Cape Dyer, they are more numerous : several are 10 miles off the shore, and in very deep water ; many are dry, some awash, and others show only by

---

* Sarmiento, page 86.

the breaking of the sea.  The coast to the northward of Dynely bay is very broken.*

**ROCK of DUNDEE** lies 5 miles S. W. by W. ¾ W. from Cape Dyer, the north-west point of Campana island.  This rock is not only a very striking object in itself, but is a good mark for Port Barbara, and bears from its entrance S.W., distant 9 miles, at a mile to the northward of the rock there are 23 fathoms, and the depth gradually decreases on approaching that port.

**PORT BARBARA** lies inside Breaksea island, which bears N. by E. ⅞ E. nearly 5 miles from Cape Dyer.  In the entrance the depth is 4 fathoms, but gradually decreasing to 3 and 2½ fathoms, and deepens again to 6 and 8 fathoms in an inner basin.  This is a very good harbour, and affords the rare opportunity of anchoring in a moderate depth, in a port of easy access.  Dorah peak, on the range running southward from the port, is 2,462 feet above the sea.  Just above high-water mark Captain Stokes found a ship's beam, apparently of English oak, and probably belonging to the *Wager*, one of Lord Anson's squadron.  Wreck point, the west point of entrance, is in lat. 48° 02′ 15″ S., and long. 75° 29′ 15″ W.

**Breaksea Island,** being 2 miles long, and fronting the port, which at its entrance is half a mile wide, effectually shelters it, while between it and the mouth of the port the depth is from 6 to 7 fathoms, good ground, which renders the entrance and exit very easy.  Off the sea coast of the island there are many straggling rocks, so that the *Beagle* having entered the port from the westward, left it by threading the northern rocks, in doing which she had not less than 9 fathoms.  It is high water, full and change, at 0h. 28m., and the tide rises 6 feet at springs.

**DIRECTIONS.** — In steering for Port Barbara, as soon as Cape Dyer bears south, the vessel will be close to some rocks, which should be kept on the port hand.  When passing them at the distance of an eighth of a mile the depth will be 11 fathoms.  The channel here is a mile wide, but gradually narrows on approaching the south-west end of Breaksea island; and at Wreck point, round which the port opens the breadth is not much more than a cable's length.  There are several rocks in this passage, but as the depth is from 6 to 8 fathoms, the anchor may be dropped, and the ship warped clear of them, in case of being becalmed; calms, however, are of rare occurrence there.  Flinn

sound, a deep opening to the eastward of Port Barbara, was not examined.
Bynoe point, with the Bynoe islands, which extend 2 miles, forms the
western head of the Fallos channel.

**FALLOS CHANNEL**, which separates Campana from Wellington
island, was explored for 30 miles, without finding any interesting
feature, by Mr. Kirke, who described it to be clear of rocks, and
abounding in anchorages for small vessels, although the water is deep.
The bottom is generally sandy. Its average breadth from 1½ to 2 miles.
The western side is a ridge of mountains; the eastern side of the entrance
is much lower, and very broken, and formed by many small islands.
At 5 miles within it, on the western side, is Our Lady's bay, of the old
charts. Fallos channel probably communicates with the sea by Dyneley
bay and Picton opening; and even beyond the latter, it is supposed to
enter the Gulf of Trinidad by the opening to the westward of Neesham
bay.

**The GUAIANECO ISLANDS**, off the north end of Wellington island
from which they are divided by the South-west pass, are 20 miles in
extent, and consist of two principal and many smaller ones; the westernmost is called Byron island, and the easternmost Wager island. They
are separated by Rundle pass, the north end of which is Speedwell bay.
Rundle pass is only a quarter of a mile wide, but clear in the
whole extent of its channel, excepting the northern entrance, where
it is guarded by many detached rocks, which render the entrance to
Speedwell bay rather difficult.*
According to Byron's and Bulkeley's Narratives, the situation of the
wreck of the *Wager* is near the western end of the north side of Wager
island. Harvey bay and Good Harbour are mentioned by Bulkeley.
Off the western end of Byron island there are some rocky islets; and its
northern coast is also very much strewed with them, even to a considerable distance from the shore.

**SOUTH-WEST PASS** is a clear but, in some parts, narrow passage,
which at its South-west end is contracted by rocks to 1½ miles broad,
and at the south end of Byron island it is scarcely a mile; though
afterwards widening to 3 miles. The north point of Wellington island is

---

* Machado, the pilot who explored this coast in the year 1769, by order of the
Governor of Chiloe, Don Carlos de Beranger, describes these islands at some length, but
with a little confusion of bearings. His Port Ballenas must be on the south side of
Wager island, for he describes it to be opposite to Cape San Roman; therefore, Port
Eustaquio should be on the north coast, probably in the strait within San Pedro island.—
*Agueros*, p. 211 to 213.

Cape San Roman, round which the Messier channel runs away to the south-eastward.

TARN BAY in which the channels from the Strait of Magellan open into the Gulf of Peñas, is about 5 leagues wide; on its eastern shore lie the Ayautau islands 4 miles from the coast, but the interval is occupied by several rocky reefs, between which Captain Skyring thought there seemed to be a sufficiently clear passage. The pilot, Machado, however, thought differently; and he further describes a small boat haven on the larger island among the rocks. Opposite to the Ayautau islands there is a port, called by the missionary voyagers San Policarpo ; but from its exposure to the westward cannot be thought very inviting. The ports of Tianitau and Asaurituan are also mentioned by the missionary priests in their journals. The former is described to have many islands in its entrance, and to be to the northward of San Policarpo, and the latter to be to the southward of Tianitau, and opposite to Ayantau.

The BOCA DE CANALES of the old Spanish charts is laid down, as well as all this part of the coast, from Machado's account,[*] who describes the opening, and gives its lat. 47° 25′ S., which is only 3 miles in error. It begins in a south-east direction for 11 miles, and then divides into two arms, one turning to the eastward for 15 miles, and the other 11 miles to the southward, where they terminate. They are merely deep and narrow arms of the sea, running between steep-sided ranges of mountains. The shores are rocky, and afford neither coves nor bights, nor even shelter for a boat, and are entirely unproductive ; for no seals nor birds were seen, and the shores were destitute even of shell-fish.

CAPE MACHADO, in lat. 47° 27′ 35″ S., long. 74° 26′ 10″ W., is the northern head of this opening. Two miles off it there are two rocks, which the pilot carefully and correctly describes, as he also does the rocks and breakers that extend from the south head for nearly a league. The *Beagle* twice occupied an anchorage under the Hazard isles, in the entrance, and on both occasions was detained many days from bad weather, with three anchors down.

Excepting this very bad and exposed anchorage, there is no other in the channel ; and Captain Stokes describes it as being extremely perilous. " The anchors," he says, " were in 23 fathoms, on a bad bottom, sand and coral. The squalls were terrifically violent. Astern, at the distance of half a cable's length, were rocks and low islets, upon which a furious

---

* Agueros, p. 210.

surf raged, and on which the ship must have been inevitably driven, if the anchors, of which three were down, had started."

Between the Bocas and Jesuit sound the coast is more unbroken, and lower than usual. In lat. 47° 17' S. lie some reefs which project 2 miles to seaward ; but behind them there was an appearance of a bight which might perhaps afford anchorage.

JESUIT SOUND is quite unfit to be entered by any ship. It terminates in two inlets, Benito and Julian. The former is bounded on either side by high mountains, and terminates in low land, with a rivulet that originates in a large glacier. The latter ends in high mountainous land, with streams of water between the hills ; one part of it is cliffy, and it has on the south-west side a long sandy beach. In its entrance a large island makes the passage on either side very narrow, and both of them are rendered still more so by rocks and islets.

XAVIER ISLAND is separated by Cheape channel from the main, which is 11 miles long, and 4 wide, and is high and thickly wooded with lofty trees. The only two anchorages which it affords are noticed and named by Machado, the northern one Port Xavier, the southern Ignacio bay. The former is by much the better place, being secure from the prevailing winds, with 17 fathoms at 800 yards from the shore. The south end of the bay is a sandy beach, backed by tall beech trees. The shore to the south of Xavier bay, for the first 4 or 5 miles, consists of a high, steep, clay cliff, with a narrow stony beach at its base, rising into mountains of 1,200 or 1,400 feet high, which are covered by large and straight-stemmed trees. The remainder of the coast, to Ignacio bay, is low, and slightly wooded with stunted trees ; and its whole extent is lashed with a furious surf, that totally prevents boats from landing. Ignacio bay affords anchorage in 9 fathoms. The western coast of the island is lined by reefs extending 2 miles off, upon which the sea breaks high.

KELLY HARBOUR lies in the north-east corner of the Gulf of Penas, between San Estevan gulf and Cheape channel. It takes a north-easterly direction for 8 miles. The land about the harbour is high, rugged, and rocky, a peak on the south side being 1,540 feet high, but by no means destitute of verdure. The interior shows many lofty-peaked and craggy ranges of snow-covered mountains. The points of entrance are 2 miles asunder, bearing from each other N.E. ¼ E. and S.W. ¼ W., and are thickly wooded and low, compared with the adjacent land. Between them the channel is from 35 to 40 fathoms deep, over a mud bottom, and to within a cable's length of the rocky islets that fringe

[s. A.]                                                                Q the shore for the breadth of a quarter of a mile. On approaching the harbour the very muddy appearance of the water is rather startling; but the discoloration proceeds only from the freshes of the river, and the streams produced from an extensive glacier that stretches many miles to the northward into the country.

The course in is E.S.E., until in a line between the inner north point, and an inlet on the south shore fronted by five or six wooded islets. Then haul up along the port or north side of the harbour, as close to the shore and as far as requisite to an anchorage. The best berth is when the two points of entrance are locked in with each other, and within 1¼ cables' lengths of the sandy spit which proceeds from the western end of a high and thickly-wooded island. The ground is excellent, and so tenacious, that it was with difficulty that the *Beagle* lifted her anchors. Shelter, wood, and water, however, are the only advantages offered by the harbour. Environed by lofty mountains, from 1,400 to 1,800 feet high, and ice-filled valleys and ravines—it is chill, damp, and dreary. A few birds, and a small number of hair seals, were the only living animals seen. Not a trace of human beings was observed.

For knowing Kelly harbour the glacier is a capital leading-mark, being a large mass of ice lying on the low ground to the northward of the harbour. The water at the anchorage, at half tide, was quite fresh, but too muddy to be fit for immediate use. When in the fair way of the harbour, the Sugar Loaf in Holloway sound in Tres Montes gulf will be seen just in a line with the extremity of the land, to the northward of Purcell island, bearing nearly West. The north point of the harbour is in lat. 46° 59′ S., and long. 74° 05′ W.; and the variation about 20° easterly in 1828, or 20° 40′ E. in 1860.

**SAN ESTEVAN GULF** is 10 miles north of the north-east end of Xavier island, between Cirujano island, and a long sandy beach, which curves round to the north-west, towards the entrance of the River San Tadeo, and on which a furious surf breaks. In the centre of the entrance there is a small islet, called Deadtree island.

San Estevan gulf is one of the best harbours of the coast, being easy of access, and with moderate depth of water all over; good holding-ground and clean bottom. The best anchorage is about 2 miles N. W. by W. of Deadtree island, in from 4 to 6 fathoms, sandy ground. This will be at 2 miles from either shore, but the berth is quite land-locked; and, if necessary, anchorage may be taken up much nearer to the land.

**On Cirujano island,** above-mentioned, the surgeon of the *Wager* was buried. The missionary priests describe a port on the island, called San

Tomas.  The island is separated from the extremity of Forelius peninsula by a strait, from a mile to three-quarters of a mile wide.*

**FORELIUS PENINSULA**, a strip of land 10 miles by 3 wide, forms the southern shore of San Quentin sound; it ends in an isthmus of low sandy land, scarcely a mile wide; the one over which I think it may be inferred, from the "Narrative,"† that the canoes in which Byron and his companions were embarked were carried.  At one day's journey by land to the westward of this isthmus, Byron describes a river, up which the Indian guides attempted to take the *Wager's* barge.‡  This river, if it exists, probably falls into Bad bay.

**SAN QUENTIN SOUND.**—The western branch of the gulf extends for 10 miles between the main and Forelius peninsula; it terminates in continuous low land, with patches of sandy beach, over which, among other lofty mountains, one called the Dome of St. Paul is seen.  The shores are thickly wooded with shapely and well-grown trees; the land near the beach, for the most part, is low, but rises into mountainous peaks; some of which, in the interior, are 1,500 feet high, but not craggy. Aldunate inlet is the north-west branch of the gulf, and is 8 miles in length.

**RIVER SAN TADEO.**—Its mouth is easily distinguished on entering the gulf, by the sand-hills on each side of its entrance, and the eastern trend of Cirujano island bearing S. by W. ¼ W.  A sandy beach extends east and west of it for many miles; the land is low and marshy, and covered with stumps of dead trees.  It has a bar entrance, much of which must be nearly dry at spring tides, a heavy swell breaks upon it for its whole length, so that no opening or swatchway is left, and, except in very fine weather, it is very hazardous to cross.  At the mouth, the breadth is not more than a quarter of a mile; but, within the entrance, it opens to a basin of some extent; and at 3 miles up it is 300 yards wide, after which it gradually narrows.

Nine miles from the entrance, the stream is divided into two arms; the northern, or Black river, takes a northerly, and the other an easterly direction, the former is a strong and rapid stream quite

* This circumstance was obtained from Pedro Osorio, an old soldier at Chiloe, who was one of the old missionary voyagers.  Captain King asked him why it was called El Cirujano, to which he replied; "Porque alli murio el cirujano del Wager."—(Because the surgeon of the Wager died there.)  Pedro Osorio also stated that he well remembered Byron's party, although it was eighty-eight years since they visited the island.—See also *Byron's Narrative*, p. 147.

† *Byron's Narrative*, pp. 119 and 120.          ‡ Ibid., pp. 108 and 111.

uninfluenced by the tide, which, however, extends for a short distance up the eastern arm, but beyond that the current down becomes gradually as strong as in the Black river, its banks are comparatively barren to those of the Black river, where the wood is very thick. The courses of both arms are very tortuous, and the bed of the river was so choked with trunks and branches of trees as to prevent it being properly explored, as well as the discovery of the Desecho, the place where the Indians carry their canoes across the Isthmus of Ofqui.

**PURCELL ISLAND** is separated from Forelius peninsula by a good channel, 2 miles wide; it is moderately high and thickly wooded, and about 6 miles in circuit. About mid-channel, and nearly abreast of the east end of the island, there is a rock only a few feet above the water. The channel between the island and the rock is from 18 to 22 fathoms deep, and the bottom sandy.

**BAD BAY** lies 9 miles W.N.W. from Purcell island. Here the *Beagle* anchored after dark, in 8 fathoms, sandy bottom, and left it at 9 o'clock on the following morning. Of this place, Captain Stokes remarks :—"At daylight, we found that we had anchored in a small bay about half a mile off a shingle beach, on which as well as on every part of the shore, a furious surf raged that effectually prevented our landing to get chronometric observations. The mouth of this bay is N.E. 9 leagues from Cape Tres Montes, which in clear weather may be seen. Like all the shores of the Gulf of Peñas, it is completely open to the south-west, and to a heavy rolling sea. About 9 a.m. we left it, and proceeded to trace the coast to the South-east."

**The GULF of CAPE TRES MONTES** occupies a space of 16 by 12 miles to the westward of Bad bay, and includes Byron's group of the Marine islands,* upon one of which, the Sugar Loaf, a mountain 1,840 feet high, is very conspicuous ; and was noted as being seen from the *Wager* the day before her wreck. Upon the Main, about 6 miles N. ¼ W. from the Sugar Loaf, is another equally remarkable mountain, which was called by him the Dome of St. Paul, and is 2,284 feet high.

At the north-west corner is Hoppner sound, about 5 miles across ; and from its south-west end a deep inlet penetrates 7 miles to the south-west, reaching to within 2 miles of the exterior coast, from which it is separated by an isthmus of low and thickly-wooded land. The *Beagle* anchored in Hoppner sound, near the mouth of the inlet. The Marine islands leave but little room for entering the sound ; but the southern

---

* It was here that four marines voluntarily remained on shore during Byron's perilous boat voyage, after the wreck of the Wager.—*Byron's Narrative*, p. 85.

channel, though narrow, has plenty of water.  On the south-west side of
the Marine islands is Holloway sound, leading to Port Otway, an inlet extending for 5 miles into the land, in a south-west direction.

**Neuman Inlet,** at the north-east corner of Tres Montes gulf, extends for 17 miles into the land, where it terminates ; but it is of no use,
as the water is too deep for anchorage.  It is the resort of large numbers
of hair seals.

Notwithstanding Captain Stokes' conviction that Neuman inlet was
closed to the northward, it is still possible that there may be a communication between it and the Gulf of San Rafael.   It is said that a boat went to
the head of this inlet ; but a boat passing along one side of a sound, or
arm of the sea, 2 or 3 miles wide, may easily overlook, or not discern, the
opening of another arm, which is partly, if not entirely, land-locked to a
distant observer's eye.   Instances of this nature are too numerous to
specify : one instance may be given—that of the Magellan channel, leading from Magellan strait to the Barbara channel, which, after the first
examination, was esteemed to be a sound, closed at its southern entrance.
Where high headlands in a hilly country overlap one another, it is almost
impossible to distinguish all the openings without actually tracing the whole
shore.

**Port Otway.**—The entrance of Port Otway is on the south side of
Holloway sound, 14 or 15 miles from Cape Tres Montes, and may be readily
known by its being the first opening after passing Cape Stokes.   Off the
mouth are the Entrance isles, among which is the Logan rock, having a
strong resemblance to the celebrated rock near the Land's End, in Corn-
wall, the name of which it bears ; it is broad and flat at the top, decreasing
to its base, which is very small, and connected to the rock upon which it
seems to rest.   Immediately within the entrance on the western shore
there is a sandy beach, over which a rivulet discharges itself into the
bay ; and just off that beach anchorage may be had in 9 or 10 fathoms,
the most convenient berth that the port affords.[*]

Nearly 2 miles from the entrance the port opens into a large but deep
inlet, with 2 arms extending to the south-west and south-east 3½ and 2
miles respectively.   This inlet contains anchorage all over it, but the
depth is generally inconveniently great, being from 20 to 30 fathoms ;
an island also lies in the mouth, with a narrow passage about a cable
broad on each side ; the eastern one is shoal.   Wood and water is
plentiful, but H.M.S. *Salamander* hauled the seine here without success.[†]

As a place of refuge, or for any maritime purpose not requiring very
dry weather, few ports on any uninhabited coast can be better adapted than

---

[*] *See* Admiralty Plan of Port Otway, No. 1,325 ; scale, *m* = 1·0 inch.
[†] Remarks of J. Jenkins, Master, R.N. 1843.

Port Otway. It was thoroughly examined by the officers of the *Beagle* under the direction of the late Captain Stokes, in 1828.

CAPE TRES MONTES is a bold and remarkable headland, rising from the sea to the height of 2,000 feet ; free from all outlying dangers, is one of the safest and easiest of landfalls, even in blowing weather and on a lee shore ; there is little or no current, and plenty of drift to leeward, if darkness should prevent anchoring the same day. It lies in lat. 46° 58' 57" S., and long. 75° 27 50" W., and is the southern extremity of the peninsula. Cape Raper lies to the northward, in lat. 46° 48' S., but rocks and breakers extend from it for half a league to seaward.

The CLIMATE of the coast of Western Patagonia, which has been described in the foregoing pages, is cold, damp, and tempestuous. The reigning wind is N.W. ; but if it blows hard from that quarter, the wind is very liable to shift suddenly round to the westward, and to blow still heavier, raising a mountainous and often a cross sea. These westerly gales, however, do not generally last long, but veer round to the southward, when the weather, if the barometer rises, will probably clear up. Should they, however, fly back to the N.W. again, and the barometer remain low, or oscillate, the weather will doubtless be worse. Easterly winds are of rare occurrence—when they do come they are accompanied with fine clear weather ; whereas westerly winds bring with them a constant fall of rain, and a quick succession of hard squalls and showers.

Should a vessel be near the coast during one of these northerly gales, it would be advisable for her to make an offing as quickly as possible, in order to guard against that sudden shift to the westward which is almost certain to ensue. The discovery, however, of the anchorages of Port Henry, Port Barbara, Port Otway, and St. Quentin Sound, has very much reduced the dangers of the lee shore ; and a refuge in either of them will often be preferable to passing a night on this coast during a severe gale.

The barometer falls with northerly and westerly winds, but rises with southerly. It is at its minimum height with N.W. winds, and at its maximum when the wind is S.E. The temperature is rarely so low as 40°, excepting in the winter months. At Port Otway, in the Gulf of Peñas, the maximum and minimum for nineteen days, in the month of June (winter), were respectively 51° and 27°.

TIDES.—High water, on most parts of this coast, takes place within half an hour on either side of noon. The stream is inconsiderable, and the rise and fall rarely more than 6 feet.

# CHAPTER X.

## FROM THE GULF OF PEÑAS TO CHILOE ISLAND.

### VARIATION 21° to 19° East in 1860.  Annual increase about 1'.

FROM Cape Tres Montes, along the coast, by Cape Raper to Cape Gallegos, there is no outlying danger.  The land is high, from 2,000 to 4,000 feet, and the water deep.  *See* view on chart.*

**SAN ANDRES BAY.**—Round Cape Gallegos, a bold promontory, barren to seaward, and rising abruptly from the water, lies San Andres bay, containing no good place for large ships, but a secure anchorage for small vessels in Christmas cove, which is not half a mile across, but sheltered from all winds except those from the northward.  Though apparently much exposed to this quarter, the danger from such winds is more apparent than real, as they do not become very strong till they are some points to the westward of north, and much sea would not be raised till then·  The *Beagle* passed some days there in smooth water, while it was blowing a hard gale from the westward outside.†

**Cone Inlet** is another deep cove, a mile to the southward of Christmas cove, stretching in-shore to the foot of a remarkable cone 1,600 feet high.  It is quite sheltered, but difficult of access, and still more difficult to quit, on account of its narrowness.  Even with a N.W. wind no swell penetrates the interior of Cone inlet; and there is a natural dock at the inner part on the north-east side.  On the northern side of San Andres bay, at the head of a deep bight, there is a large basin which is well called Useless cove, being unfit for any kind of craft.  Cape Pringle forms the north side of San Andres bay ; and between it and Rescue point the land is considerably less high ; there may be anchorage in Stewart bay, and Cliff cove seemed to be a promising little place, but neither was entered.

**PORT SAN ESTEVAN**, 20 miles from Cape Pringle, has a very good anchorage in 10 fathoms water, under Rescue point.  Fresh water may be

---

\* *See* Chart West Coast of America; Sheet II., No. 1,325 ; scale, *m* = 0·13 of an inch.
† *See* Admiralty Plan of San Andres bay, No. 1,207 ; scale, *m* = 1·0 inch.

obtained easily in the stream at the head of the inlet, or from runs near the anchorage. Dark hill, 2,150 feet above the sea, is an excellent mark for this port ; there is no hidden danger, provided that the rocks at the point have a fair berth given to them of a cable's length. A ship should anchor close to the west shore, under shelter of the reef off Rescue point.

The **KELLYER ROCKS** are a cluster of outlying dangerous rocks scarcely above water, on which the sea breaks, lying N. by W. ¾ W. from Rescue point, and 6 miles from the nearest land, Duende island, just outside a line drawn from Rescue point to Cape Taytao. Doubtless there is anchorage under the lee of Usborne islands, or behind Mount Alexander, or within Cornish opening, or Burns inlet ; but the *Beagle's* officers did not explore them.

**CAPE TAYTAO**, bearing N. by W. 25 miles from Port San Estevan, is one of the most remarkable promontories on this coast ; it makes like a large island pointed at the summit ; nearly 3,000 feet in height, and rugged, barren, and steep, several rocks above water lie around it ; none, however, a mile off shore. The coast between this Cape and Rescue point is broken and rugged. *See* view on Chart, No. 1,325.

**ANNA PINK BAY** is round and beyond the headland, within a cove of which one of Lord Anson's squadron, the *Anna Pink*, employed as a victualler, took refuge from westerly gales. She anchored under Ynche-mo island, but drove from thence across the bay, and after slipping or cutting her cables, brought up in Port Refuge in the south-east part of the bay, where she lay some time in security, refitting.*

**Ynche-Mo Island**, 450 feet high, is on the western side of Anna Pink bay. The remains of a large log hut and a number of goats were found on the island. Probably the *Anna* did not in the first instance go near enough to the island of Ynche-mo, for there is good holding-ground in 15 or 20 fathoms on its east side ; Penguin islet bearing North, and the highest part of Ynche-mo S.W.

Cañaveral Cove, at the entrance of Port Refuge, though small, is very convenient for refitting, or for executing any repairs. Patch cove is so small as to be unfit for vessels of any size exceeding 200 tons.

North-east of Ynche-mo, about 6 miles distant, are the Inchin or San Fernando islands, and next to them are the Tenquehuen, Menchuan, and Puyo islands, among which no doubt there are many good anchorages, and abundance of fresh water, wood, wild herbs, and fish, usually found on these coasts. The western extremity of Menchuan island is low and has

* *See* Plan of Anna Pink Bay, No. 1,298 ; scales, m = 0·5 and 1·3 inch.

several rocks near it, therefore a good berth should be allowed in passing. Between Ynche-mo and the Inchin islands are several rocks and islands, lying at the entrance of Wickham or Pulluche inlet. This inlet separates the Chonos archipelago from the peninsula of Taytao.

As a general rule it may be observed that there are no sand-banks little or no current, and, generally speaking, few hidden dangers on the west coast of South America, between the Straits of Magellan and Chiloe. Rocks under water are either buoyed by kelp, or are distinctly visible to an eye aloft, if the sea does not break on them so as to show their position exactly to an eye on deck.

CHONOS ARCHIPELAGO extends from the peninsula of Taytao to the island of Chiloe, between the parallels of latitude 46° and 44° S.; it consists of a large number of barren, rugged, and lofty islands, rising to an elevation of 2,000 to 4,000 feet above the sea. The interior sounds behind these islands have not been surveyed, but they are supposed to contain harbours as numerous as the islands. The inner coasts of Skyring, Clements, Garrido, and Isquiliac islands, are like the outer, high, rugged, and barren, ranging to about 3,000 feet above the sea. In the middle of Darwin bay, that large bight between Tenquehuen and Vallenar islands, is a detached and dangerous islet named Analao. The wide inlet called Darwin channel leads out of Darwin bay into the interior sounds behind the Chonos archipelago.

VALLENAR ROAD, within the Vallenar islands on the north side of Darwin bay, is well pointed out by the mountain of Isquiliac, which is 3,200 feet high, very rugged, and triply peaked. It is an excellent road-stead, easy of access and egress. The best anchorage is in about 12 fathoms water, near the little islet, marked (a) in the plan, which lies off the south-east end of Three Finger island. The *Beagle* lay there during a heavy S.W. gale.[*]

HUAMBLIN or Socorro island, 9½ miles long, lies north-westward about 30 miles from the Vallenar islands ; under it there is good anchorage. It is from 400 to 700 feet in height, comparatively level, and thickly wooded ; generally its shores are sloping, and covered with verdure. Here and there are remarkably cliffy breaks, which show distinctly against the dark woodland. (*See* view on Chart, No. 1,325.) High water at full and change, 12h.

IPUN or Narborough island resembles Huamblin in its character, and therefore differs totally from the rest of the neighbouring islands, which

---

[*] *See* Plan of Vallenar road, No. 1,338 ; scale, m == 1·0 inch.

are high, rugged, and generally barren to seaward; while Huamblin and Ypun are comparatively low, level, and fertile, valuable even now, and likely to become more so, as abundance of vegetables and live stock might be raised on them for the supply of shipping.  They are both easy to approach, or leave, and the rocks which lie around their more exposed points are all distinguished by the sea always breaking on them, and may therefore be easily avoided.

Under Ypun there is good anchorage in 12 to 16 fathoms, over clay and sand; and Scotchwell harbour, at the south-east part of the island, is not only a valuable place of refuge, but a secure and agreeable place for wooding, watering, or refitting.  It should be approached from the north-ward, because, although the passage south of it was examined, and appeared to have no hidden dangers, it is narrow, and there may be undiscovered rocks.  On these islands a considerable number of seals were seen.

ADVENTURE BAY, between the Vallenar islands and Ypun, lies to the eastward of Huamblin; it is encumbered by dangerous outlying rocks, and offers no good anchorage that is easily accessible between Stokes island, with its lofty M'Philip, 2,765 feet high, Rowlett, Williams isle, 2,530 feet, James island, with Sulivan peaks, 4,250 feet high, Kent, Dring, and Lemu islands.  Paz, 1,150 feet, and Liebre islands, in the middle of the bay, are remarkable from their conical form, but they afford no shelter.

The cluster of islands between Narborough and the Guaytecas offers no anchorages so easy of access to a stranger as those previously mentioned.  Coves indeed fit for small craft abound, but to notice each of them would tend to confuse the reader, for whose use these remarks are intended.  A peak in Midhurst, the southern islet of the cluster, rises 1,760 feet, and Mount Mayne, 8 miles to the eastward, 2,080 feet above the sea.  The four northern isles of this cluster are high, about 6 miles in length each, are separated by nearly equal spaces, and preserve a remarkable parallellism in an east and west direction.

GUAYTECAS ISLANDS form a group at the north end of the Chonos archipelago, at the entrance of the Gulf of Corcovado; they consist of one large and several small islands.  The peak of Guayteca Grande is 1,100 feet high, on the north side of it will be found an excellent harbour.*

PORT LOW.—In approaching from the westward, the Guaytecas islands

* See Chart, West Coast of America, Sheet IV., No. 1,289; scale, d = 7·4 inches.

appear in a hummocky ridge ; at their north-east point there is a remarkable flat-topped island, and the south-west part diminishes into low land.
When seen from a considerable distance, such as 20 miles, the flat-topped
island and the hummocky ridge are still conspicuous, this hummocky
ridge appears to be the middle of a group of islands.  On the left, looking
to the south-east, there is a high, single-knobbed hill inland, which looks
as if quite insulated.; and as far again to the left, is the flat-topped island
mentioned above, beyond which there appears to be an opening : the low
land to the westward makes like many islands.  In approaching it a good
berth must be given to the numerous rocks that lie along the north
and north-west shores of those islands, and allowance made for the
stream of tide which is felt off Huacanec islands, and causes a race off
Chayalime point.  The farthest outlying rocks to be guarded against in
approaching Port Low from the south-westward, is rather more than 2
miles N. by E. from Patgui point; but the sea always breaks on it, and
as a precaution against being drifted too near in light winds, it is
advisable to keep to the northward of a line drawn due west from the
north point of Huacanec island : the entrance to the port lies round the
east side of that island.  High water, full and change, 12h. 45m., rise
7 feet.*

**Supplies.**—Port Low will furnish the usual supplies; water of excellent
quality, wood, fish, shell-fish, including oysters, and wild herbs.  Of late
years, potatoes have been planted by the otter-hunting and sealing parties
from Chiloe, and therefore a few may possibly be found there.  This is
a port in which a number of large ships might lie conveniently, it being
one of the best harbours on the coast.

**HUAFO ISLAND** lies distant about 20 miles westward of the Guaytecas islands.  It is without a harbour except for boats ; the highest
part is the north-west head, Weather point, 800 feet above the sea
Reefs extend 3 miles seaward to the north and west.  The island is
composed of indurated clay (*tosca*), which may be cut with a knife like
chocolate.  It is low in the middle, and high again at the south-east
extremity ; it is well wooded, and formerly had many sheep on it, while
the aborigines lived there in peace.  Small and Sheep coves are two landing places on its eastern shore.  *See* view on Chart, No. 1,289.†

**CHILOE.**—This large and fertile island no longer presents to the eye
the wild and rugged features which marked the great islands along the

---

* *See* Plan of Port Low, No. 1,296; scales, m = 0·5 and 1·3 inch.
† *See* Plan of Small and Sheep Coves in Huafo island, No. 1,304 ; scale, m = 2·0
inches.

western coast of Patagonia, and up to the Chonos archipelago. Lower land, softer outline, and continued forest, of the thickest description characterize Chiloe; but the coast of the main land opposite, including the Cordillera, as it is justly called, is as steep and rugged as that to the southward; and in height it exceeds them considerably, though more thickly wooded. In Chiloe, no land exceeds 2,600 feet in height, while its average elevation is not above 500 feet. The small island of San Pedro, however, off the south-east end rises to 3,200 feet.

Chiloe, with the Archipelago, which comprehends that island and the smaller ones in the Gulf of Ancud are divided from the continent by the Corcovado and Ancud gulfs, and at its northern extremity by the narrow strait termed the Chacao narrows. Chiloe extends from north to south about 100 miles, and from east to west 88 miles in its greatest breadth, abreast of Matalqui Paps, but a deep indentation in its centre reduces its breadth at that point to 14 miles; its average breadth may be 25 miles, with an area of 22,500 square geographic miles. The whole island is a mass of rock, covered with earth and clothed with wood, chiefly consisting of a sort of bastard cedar, durable, and affording excellent timber which is largely exported. Population of Chiloe and neighbouring islets may be 45,000.

Among the numerous islands between Chiloe and the main, and along the eastern coast of Chiloe itself, there are many excellent harbours where abundant supplies of provisions may be found, except at its southern end.

SAN PEDRO ISLAND, makes at a distance like a rounded lumpy mountain; when near, it proves to be wooded to the summit, though 3,200 feet in height. It is separated from Chiloe by San Pedro passage or port on the north, and the Huamlad passage on the west. Port San Pedro is a small but secure harbour, which may be known by a white rock lying near the north-east point of entrance. In entering, care should be taken to avoid the 3-fathoms bank, extending two-thirds across the entrance of the harbour, if the tide is low. There are nearly 12 feets rise at springs in this port.*

CADUNUAPI and CANOITAD ROCKS lie off Olleta point, and to the southward of the lofty island of San Pedro. The latter rocks are distant 4½ miles from the nearest land, and as the tide streams set towards them, they are dangerous in the night or during calms.

QUILAN ISLAND, 4 miles long with a hill on each extreme, lies off

* See Plan of Port San Pedro, No. 1,304; scale, m = 2 inches.

the southern end of Chiloe. The roadstead, to the eastward of the island, is wild and unsafe. Off the southern shores of Chiloe there are many out-lying rocks, and it is therefore a coast to be avoided. Round Yemcouma isle, and east of the southern extremity of Quilan, sunken rocks stretch 6 miles from the main land, but the sea breaks heavily on them. In San Pedro and Quilan islands there are no inhabitants.

**CAPE QUILAN**, the south-west point of Chiloe, is wooded; there are cliffs in its vicinity of a light yellowish colour, about 300 feet in height. The land adjoining is less wooded than that on the eastern and more sheltered parts of the island, though some trees are visible everywhere. The outlines of the land hereabouts are rounded, smooth, and rather horizontal; it is a pleasant-looking hill and dale country. The cliffs on the sea-coast are not regular, nor do they extend far. From Cape Quilan to Pirulil head, 35 miles to the north, a similar character of coast line con-tinues; there is no kind of anchorage, scarcely can even a whale-boat find a place of shelter where she could be hauled ashore.

**CUCAO BAY**, 45 miles farther to the northward, is bounded by a low beach, always lashed by a heavy surf. Cucao heights are remarkable, as being the highest and most level lands in the island: they are wooded to their summits, and in height from 2,000 to 3,000 feet. Off all this coast, from Cape Quilan northward, there are no outlying or hidden dangers.

**CAPE MATALQUI** is remarkable; the heights over it rise about 2,000 feet, and make from seaward in three summits. Proceeding northward between Matalqui and Cocotue heads, there is an inlet of the Chepu, behind which low land alone is visible. Cocotue heights do not attain a greater elevation than about 1,000 feet.

**The PENINSULA of LACUY** is joined to the rest of the main island by a low isthmus between Cocotue and Caucahaupi, there is no outlying or hidden danger near them. This part of Chiloe has been thought to have some resemblance to the Isle of Wight, Caucahuapi, Guabun, and Huechucucuy headlands, being bold cliffy promontories; the latter is a high, steep, and bare bluff. The above three headlands are the first seen when making the land near the port of San Carlos.*

**HUAPACHO SHOAL** is a circular reef of rocks about half a mile in diameter, lying 3 miles N.E. by E. ½ E. from Huechucucuy, and 1¼ miles W. by S. ½ S. of Huapacho head. In the night, more especially, this shoal

---

* See Plan of San Carlos and Chacao Narrows, No. 1,313 ; scale, m = 0·5 inches.

should be guarded against, the land behind being a low sandy beach, and not then distinguishable.

**HUAPACHO HEAD,** a light coloured cliff, bare at the top, and broken at the seaward extremity, forms the north point of the peninsula of Lacuy. The low extremity of Huapacho is sometimes called Tenuy point; probably it extended farther seaward, and was more remarkable formerly. The stream of tide is strong hereabouts, and must be allowed for, according to the direction of the ship's course.

**LIGHT.**—Corona point or Huapilacuy light is a *fixed white* light, varied every 2 minutes by a *flash*, at an elevation of 197 feet above high water, visible in clear weather from a distance of from 12 to 18 miles. This light is seen over the land of Huapacho ; vessels therefore approaching Port San Carlos from the southward, after rounding Huechucucuy head, should continue steering to the N.E., until the light bears S.E. by E., when they can haul to the southward. The tower is circular, 32 feet high, and painted white. It stands on the high part of Huapilacuy point, S.E. by E. ½ E. 1¼ miles from Huapacho head, with the west point of Sebastiana islet, bearing N. by E. ¾ E., centre of Cochinos island, E.S.E., and Huechucucuy head, W. by S.*

**PORT SAN CARLOS** lies south of the peninsula of Lacuy. The entrance between Aguy point and Cochinos islet, is about 2 miles across, and from the island the port extends to the westward for nearly 6 miles, with an average breadth of one mile. About 3 cables south from Arena point is a patch with 3 fathoms on it ; the plan is so complete that no further remarks need be made. The town of San Carlos de Ancud, about 180 yards square, with a flag-staff in the centre, stands on Guilmen heights, at the southern entrance of the harbour. On the north *side* there is a strong well built stone store-house, and opposite to it is the church, also of stone.

**SEBASTIANA ISLET,** 160 feet high, lies N.E. ½ E., 4 miles from Huapacho head. A shoal called the Achilles bank extends 4 or 5 miles westward from the islet, over which there is considerably disturbed water, rippling and swelling during a calm, but during a gale breaking in high short seas. About 3 miles from Sebastiana there are 6 fathoms at low water on this ridge, and at 2 miles about 4 fathoms. When on the ridge, Chocoy head, at the entrance of the Chacao Narrows, is hidden by Sebastiana, and on no account ought a vessel to get near this islet, for there the tide runs strongly and with dangerous eddies.

* By the Chilian notice of March 1860, the longitude of this lighthouse is given as 74° 01' W., but on Admiralty Plan, No. 1,313. Huapilacuy point is in 73° 55' 45" W. Corona point should be the Huapilacuy of the Admiralty Chart, not Huapacho.

**Carelmapu Islets** form a rocky ledge about 2 miles long, lying to the northward of Sebastiana island. The north-west island is 140 feet high. These islands should never be approached from the westward within 4 miles. Huapacho head may always be closed, and Sebastiana avoided.

**The Yngles Bank,** lying about 2 miles south of Sebastiana island must also be particularly avoided; it is a very dangerous shoal, 3 miles long, over which the tide runs with great strength; the shallowest spot is on the east end, and is called by the natives the rock of Arenillas, this was found at slack water, and had not a fathom of water on it, the bottom being sand, or sandstone, or of hard tosca. Cochinos Islet with 2 peaks on it, lies nearly a mile from Guilmen heights, it has a shoal extending about a mile off its east point; on it are 2 fathoms water.

**Mutico Point** lies 2 miles south-east from Cochinos islet. There is a patch of rocks lying N.N.W., one mile from this point, and all the bottom thereabouts is very irregular; patches of kelp are seen frequently, but they seem to be attached to large stones as well as to rocks. From Pecheura point, 3 miles north-east of Mutico point, a rocky patch runs out for a quarter of a mile, forming the termination of a bank extending from the shore between it and Mutico point, and is possibly connected with the Yngles bank.

**TIDES.**—It is high water, full and change, in Port San Carlos at 12h. 14m.; the rise of tide being about 11 feet at springs, and 7 to 9 feet at ordinary neaps.

**DIRECTIONS.**—Vessels bound to Port San Carlos should steer for Huapacho head, keeping it to the southward of east, on account of the Huapacho shoal, and thence along shore, at less than half a mile distant round Huapilacuy, and Aguy points, to avoid Achilles and Yngles banks, to an anchorage near Arena point, under Baracura heights. Between Aguy point and Baracura head, a ship should not close nearer than half a mile, as a shoal, called Pechucura, extends to nearly that distance, half way between the two points. Cochinos islet midway between Guilmen heights and Mustico point bearing S.E. by E. forms a good mark from Huapacho head to the entrance of the port. The best anchorage for a large ship is with Aguy point bearing North, and the extremity of Arena point S.W. Trading vessels anchor off the town of San Carlos in 4 fathoms water, with the town bearing about East, but it is an exposed and insecure position.

**CHACAO NARROWS** are entered after passing the Yngles bank; they are about 11 miles long and from one to 2½ miles wide, with from 10 to 40 fathoms in mid-channel. On the north coast, that of the mainland, are Carelmapu and Chocoy heads, steep cliffs, in front of which runs a powerful stream of tide. There are two channels into these narrows, one on each side of Yngles bank. The northern one is the best and has no known dangers. The southern, which leads from Port San Carlos, is a dangerous passage, and seldom used except by the large coasting boats called *periaguas*, which always prefer this channel, and for this reason the mariner is cautioned not to employ the masters of vessels at San Carlos as pilots, if his vessel draw more than 6 feet water. There is water for any ship between Sebastiana and Chocoy head, avoiding the sand-bank half a mile from the east point of the island, as well as to the eastward of the Carelmapu rocks; but so continual is the heavy westerly swell, and so strong the tide, that no vessel ought to attempt those passages unless with a local pilot on board, a commanding breeze, and a favourable time of tide.

**Puñoun Point,** on the south side of the Chacao Narrows, is low, with a sandy beach. In a line between this point and Sebastiana, 2 miles from Puñoun point, there is a knoll with but 4 fathoms. San Gallan point, on the southern shore of the narrows, open to the southward of Coronel point, leads to the northward of this bank, but if touching closely, they lead on it. Near Carelmapu point the passage appears free from danger, but the water is very deep, and to the eastward of the point a long shoal extends nearly half a mile from the shore.

**Periagua Rocks** lie nearly a mile to the westward of Puñoun point, one is awash at low neap tides, and the other lies just to the westward. The easternmost rock shows itself an hour or two before low water at neap tides; it is part of a reef of rocks running N.E. and S.W. about a cable's length. The west end of the reef dries in two places to about 2 feet above low water at neap tides; on the north-west and south-east sides there are 10 fathoms close to the rock, but in the direction of the reef 5 fathoms, at about 30 yards distance. Two cables' lengths to the westward of this rocky reef lies the other rock mentioned above; it runs east and west about a cable's length, and on the east end, which is the shoalest, there is about half a fathom at low water. On the west end there is a patch which has about a fathom on it at the same period. Both rocks are covered with kelp, which, from the strength of the tide, shows only at slack water. During the strength of the flood tide there is heavy tide-rip off Puñoun point, caused by the strong stream running over so very irregular a bottom.

The rock on which the Pacific Company's mail steamer *Prince of Wales* was reported to have been lost, is said to lie 2 miles to the eastward of the Periagua rocks; it was searched for without success by Captain John Williams of the Chilian Navy, no less than 10 fathoms water being found in the locality.

**Lacao Bay.**—Eastward of Puñoun point, between it and Quintraquin point, is the island and the shoal bay of Lacao, where there is no good anchorage, though a ship might hold on there very well for a tide. Quintraquin point is steep-to, a bold cliffy point.

**San Gallan Point,** on the Chiloe shore, 2 miles from Quintaquin point, is steep, with a remarkable clump of bushes on its summit, which is about 500 feet high. The north shore opposite is low, except near Coronel point, where there are cliffs, about 100 feet in height, behind these cliffs the land rises to about 200 feet, and is thickly wooded. Between San Gallan and Santa Teresa points, the distance across is just one mile ; it is the narrowest part of the passage from shore to shore, and half a mile farther eastward the rocks Petucura and Seluian divide the channel into two narrow passages, either of which may be used.

**Petucura Rock,** lying three-quarters of a mile E. by N. ½ N. from San Gallan point, is awash at half tide. A line drawn from the extremity of Coronel point on the northern shore to the extreme point of San Gallan, and a line between the summit of Santa Teresa point and the summit of Chacao head, cross each other at the southern part of Petucura rock.

**Seluian Rock,** more dangerous to large ships than Petucura, lies E. ½ S. from it, distant half a mile ; there are 12 feet on it at low water. Round it, as well as round Petucura, there is deep water, except to the eastward, in which direction a rocky ridge extends a quarter of a mile. The Seluian lies in a line between Remolinos point and the sharp-topped hill near Coronel point ; and also in a line from the summit of a bluff half way between Tres Cruces point to Santa Teresa point. Its line of bearing from Petucura rock is parallel to the trend of the rocky shores on each side. The stream runs very strongly over and past these rocks during the ebb, as well as the flood tide.

**Chacao Bay** is situated on the south side of the east entrance of the Narrows, and there is excellent anchorage in about 10 fathoms, half a mile north of Chacao head. When the Spaniards first settled in Chiloe, their head quarters were at Chacao, and their vessels anchored in this bay. In the eastern entrance between Coronel point and Tres Cruces point there is deep water, about 50 fathoms.

The state of the tide, and there being sufficient wind to keep a vessel under command, are the principal points to consider when about to pass Chacao Narrows ; and, on the whole, as the tide, strong as it is sets to each side of, rather than towards, the Petucura rocks, the passage of these Narrows is not so formidable as it appears to the Chiloe boatmen.

**TIDES.**—It is high water, full and change, in the Chacao Narrows, at 0h. 48m. ; the rise and fall being 11 feet. About a mile south of Tres

Cruces there is a stony point, after passing which, the tide is scarcely felt ; and in Manao bay, just south of the stony point, there is no stream ; abreast of that bay the north and south tide streams usually meet. The nature of the tides around Chiloe will be hereafter described (see page 272). Here it will be sufficient to say, that the tide wave from the ocean sets against Chiloe from the westward. The body of water impelled round the south end of this large island drives the waters of the Corcovado gulf northward into those of the Gulf of Ancud, at the north-west point of which they meet the stream impelled through the Narrows of Chacao. Very little stream is felt in the middle of either of these gulfs, but there is a considerable rise and fall, viz., from 10 to 20 feet, and more or less stream along shore and among the islands.

**CHILEN BLUFF** is a low point of shingle, with a remarkable tree on its extremity ; and half a mile in-shore the land rises suddenly to about 150 feet. Half a mile north-west of the point there is a bluff of the same height, with a rocky point projecting, and between this and the bluff boats may find a good cove.

**LINAO COVE,** south of Chilen bluff, has good anchorage. Off the north-east point of Huapilinao head, on the southern shore of the cove, a reef of rocks extends above a mile from the point. Between Huapilinao and Queniao point, there is a projecting stony beach, which at low waer dries out nearly a mile from the shore. The small village of Lliuco lies about 4 miles from Huapilinao ; the land between them is about 200 feet high, with steep wooded cliffs ; to the eastward of the village it is low, and continues so till near the point of Queniao, when it rises to nearly 200 feet. This point is low and stony, and like Chilen bluff has a remarkable single tree on its extremity. Shoal water extends nearly a mile off the point. About 1½ miles south-west from Queniao point, there is a sandy spit with 12 feet water on it about a quarter of a mile from the shore, when it deepens suddenly to 8 and 12 fathoms near the shingle spit which forms the small but valuable harbour called Oscuro.

**OSCURO PORT,** formed between the mainland of Chiloe and the island of Caucahue, may become of great use, as the tide rises in it to about 20 feet : the water is deep close to the shore, and there is no swell. The entrance is about 3 cables' lengths wide, and the point of the spit steep-to ; the length of the cove is three-quarters of a mile, and its breadth 3 cables' lengths ; there are 7 fathoms water within 50 yards of low-water mark, and from 12 to 16 in the middle, over a bottom of mud and sand. The west side of the entrance is a rocky point, with stones lying off it half a cable's length. Vessels entering the cove should keep close to the other side, under Lobos head, on the island of Caucahue, a steep bluff, above 250 feet high ; behind which the land falls suddenly, and is very low for a short distance, after which it rises again. In this cove a ship might be laid ashore, hove down, or thoroughly repaired, with safety and ease. Any similar place on the west coast of South America is not known. It is high water, full and change, in Oscuro cove at 0h. 55m. The flood tide here runs to the northward and strongly at spring tides, with a rise and fall of 20 feet.

CHOGON POINT is a bluff point about 200 feet in height, and lies a long mile to the southward of Quintergen point, the south point of Caucahue island, which last is low and stony, with a shoal spit of about a quarter of a mile in length. Between them lies Caucahue strait, and in the entrance there is no bottom with 50 fathoms.

QUICAVI BLUFF is the next point to the southward, between this bluff and Chogon point, the coast runs back a little, and in the middle of the bight the River Colu meets the sea ; it appears too small for anything but boats to enter. A flat extends about a quarter of a mile off the latter point, and near the end of it there is a large stone which shows at half tide. A rock is said to exist about the middle of the channel, and to dry at low water, but it was not seen by the *Beagle's* officers. There is a tide race between the Changues islands and Quicavi bluff.

Quicavi Lagoon, lying about a mile to the southward of Quicavi bluff, is an excellent place for boats, for, when inside, they can lie afloat at low water ; but it cannot be entered until the tide has flowed some time. It is formed and may be known by a narrow shingle ridge, carrying a clump of trees, which runs out in a spit to the eastward of Quicavi bluff ; the land at the back rises gradually to about 250 feet, and is thickly wooded, with here and there a cultivated spot.

CHANGUES ISLANDS lie eastward of Quicavi bluff, and at the distance of 3 miles ; they are four in number, and separated by a channel, running nearly north and south, and 1½ miles wide in its narrowest part, in which there are from 48 fathoms to no bottom with 55 fathoms. The Western island is the loftiest, being about 350 feet high, and forming a ridge east and west : the north-east island has a round hill upon it, not as high as the former, and the other parts are much lower ; there are some cleared patches, but they appear thinly inhabited by Indians.

A small island connected to the shore by a reef of rocks lies off the south-east point of the West Changues : at a quarter of a mile from it no bottom was found with 30 fathoms, but on hauling in for the entrance of the channel between the East and West Changues, it shoaled suddenly to 2 fathoms, on a reef, which runs from the rocky point close to the small island. Northward of the reef is the entrance to a narrow channel, fit only for boats, which divides the Western islands. A reef runs off the north-west points of both the Eastern and Western islands ; the latter to the distance of 1½ miles, and at its extreme end there are 10 fathoms water, with the west bluff of Meulin island a little open of the low point under the bluff of West Changues. A small island named Tac, and the Dugoab reef, lie southward of these islands.

**PULMUN REEF** lies N. by W. ½ W., 4 miles from West Changues, and at the same distance from Quicavi point ; it appears to be a long reef running north-west and south-east, and dries in two places, about a quarter of a mile from each other. This reef is always shown by its breakers.

**TENOUN POINT.**—From Quicavi bluff to this point, the shore is flat for half a mile from the beach. A reef runs off the latter point, which dries at low water more than a quarter of a mile from the shore ; it is shallow for nearly half a mile and then deepens suddenly to 10 fathoms. The reef does not run off the extreme point, but from a bluff a little to the north-ward of it ; the point is low and thickly wooded for about a quarter of a mile, when it rises suddenly to a range about 200 feet high : as the south side of the point is steep-to, within a quarter of a mile of it there is no bottom with 20 fathoms, and half a cable's length from the beach there are 7 fathoms.

**TIDES.**—The flood tide sets close round the point of Tenoun, and then across the channel towards the Changues islands ; the ebb tide sets to the south-west close round the point, and at the beginning of the springs at the rate of 2 knots.

**LINLIN**, four miles to the south-westward of Tenoun point, and in the entrance of the Quinchao channel, is low in the centre, gradually rising to a round hill terminated by a bluff, both to the north-ward and southward ; at a mile from its north-west face there is 28 fathoms, sand and mud, but off all the points there are spits of shingle. After passing Linlin, no bottom was found in the centre of the Quinchao channel with 55 fathoms. To the southward of Linlin stands the smaller island of Linna, but it was not visited by the *Beagle's* officers.

**CAHUACHE**, with the isles of Meulin and Quenac, lie south-eastward of Linlin and Linna, and midway between Quinchao and the Changues islands. On Cahuache there is a round hill 250 feet high, which commands a good view of the neighbouring islands ; its north side is low, the south slopes suddenly to the beach, and off the north-east point are the Tenquelil isles, joined to Cahuache by a reef on which there is only sufficient depth for a boat at low water. It is cleared, well cultivated, and many apple-trees are round the houses. It is peopled by Indians, with the exception of one family of Spaniards which occupies four or five houses.

**Tiquia Reef**, from 2 to 3 miles east of Cahuache, is about a league in length north-west and south-east, half a mile broad, and dries at low water. It is said that there is a passage between it and that island, and that a brig once passed through it.

**ALAU, APIAU, and CHAULINEC** lie to the eastward of Quinchao point; reefs extend off the north ends of the two former and from the latter as far as 2½ miles. At the south-west end of Alau, close to the entrance of the channel between it and Chaulinec, there is a small harbour or cove, formed by a low point, which appeared a good place for small vessels; the point is steep-to, and the channel on that side clear.

**The DESERTORES** form a group lying to the south-east of Chaulinec island; in mid-channel between them and Chaulinec, there is 95 fathoms coral and broken shells. Talcan, 9 miles long, and 4 miles broad, is the largest with a deep inlet on its south-east end; the smaller islets, Chulin, Chiut, Nihuel, Ymerquiña, and Nayahue, do not afford any shelter for vessels except at the northern end of the latter, which is divided into two by a narrow channel, with from 2 to 10 fathoms in it, but useless except for boats; some rocks lie half a mile off the south-east point. Just outside the entrance of the inlet, between the points, a bay is formed, in which lie several patches of kelp; and about half a mile beyond the line of the points there is a reef of rocks which dries at low water; a small channel leads to the northward of them into the bay, with 9 and 7 fathoms water till near the entrance of the harbour which is almost blocked up by kelp. The deepest water, a cable's length outside the entrance, is 3 fathoms.

Rocks lie scattered off the south-west and southern part of Talcan to the distance of a mile; and off its north point a shoal extends as far as 1¼ miles, with from 4 to 6 fathoms on it. Two miles from this point there is a rock about 10 feet above the sea, frequented by seals. Vessels seeking anchorage among these islands should be cautious in approaching them, in consequence of these rocks.

**Talcan Inlet** varies in depth from 12 to 7 fathoms, and for 2 miles from the entrance, either shore may be approached within a cable's length: the land on both sides rises gradually to about 200 feet, and is thickly wooded; at the head it is low, and the shore flat and muddy; there are two or three small huts in different parts of the harbour. The inlet is visited by people from the other islands in the season for fishing. The tide runs about 4 knots through the channel at springs.

**MOUNT VILCUN** is a remarkable sugar-loaf hill, on the main land abreast of the south-east point of Talcan ; it rises direct from the water's edge, and is thickly wooded to the summit ; to the southward of it there is a deep inlet with a small islet at its mouth. Midway between the south-east point of Talcan and the main there is 85 fathoms, coral and shells.    Thirteen miles E.N.E. from mount Vilcun is the volcano of Chayapiren, rising to an elevation of 8,100 feet, and to the southward are Corcovado and Yanteles at the respective heights of 7,510 and 6,725 feet above the sea.

**SOLITARIA** consists of small islets, surrounded by a reef lying S.W. ¾ W. 5½ miles from the west point of Nayahue, nearly in mid-channel between Chiloe and the Desertores.

**QUINCHAO** is the largest of the group of islets lying in the bay south of Tenoun point, on the shore of Chiloe.    It is almost 18 miles long running E.S.E. and W.N.W.    Quinchao channel on its northern side narrows gradually to the westward, as far as the north-west point of the island ; it then turns suddenly to the south-west into the Dalcahue channel and is not more than a mile wide.    On the Chiloe shore there is a small village called Dalcahue, where there is a saw mill.

**Dalcahue Channel** has soundings across it in from 4 to 10 fathoms mud ; but the north shore is shallow, and should not be approached nearer than a third of the breadth of the channel : this shoal runs round the bay abreast of Dalcahue, and, off the saw mills of that place, extends half way across the channel ; the best water is close to the shore of Quinchao, where there is 4 fathoms.    The tide runs through the channel about 4 knots at springs.

**RELAN COVE and REEF.**—The Dalcahue channel opens out to the southward into a broad bay on each shore.    On the Chiloe side lies the small cove and village of Relan : in the entrance of the cove there is 18 fathoms.    As far as its eastern point the shore is steep-to, but a flat there commences of shingle and large stones which dries at low water from a quarter to half a mile off, continuing as far as the low shingle point, where it ends in a spit extending above a mile to the south-east ; the north-east side of the spit is shoal to some distance, but on the south-west side it deepens suddenly to 3 fathoms.

**TIDES.**—The ebb stream sets very strongly across Relan reef to the south-east towards the channel, between the islands of Lemuy and Chelin. Between Lemuy and the main the stream was scarcely perceptible, what little was found appeared to set to the eastward ; spring tides rise 18 feet.

**CASTRO INLET.**—South of Relan there is a passage 10 miles long by about 3 broad, with 42 fathoms mud, in mid channel, which leads to the entrance of Castro inlet : the eastern entrance point is low and stony, but a vessel may pass at a quarter of a mile from it in 12 fathoms.  The western side of the entrance is formed by Lintinao islet, which is joined to Chiloe by a sandy spit that dries at low water : on the outer point of this island, a stony point runs off about a cable's length to the eastward, but the south side of the point is steep-to.  At half a mile above the second reach of Castro inlet, the eastern shore may be approached within half a cable, but the other side is flat and shallow for nearly half a mile from the beach, and shoals too suddenly for a vessel to go by the lead : in working up or down, a vessel should keep the former aboard, not going farther across than two-thirds the breadth of the channel.

The eastern shore of Castro inlet is composed of steep wooded slopes, rising to about 150 feet above the level of the sea : the western shore rises gradually forming several level steps, which increase in height to 400 or 500 feet ; behind them, at a distance of 5 miles from the beach, there is a range of hills nearly level, about 1,000 feet high and thickly wooded.  On this shore, 7 miles from the entrance of the inlet, is Castro point, a level piece of land about 100 feet above the sea, running out between the small harbour to the northward, and the River Gamboa to the southward ; it terminates in a low shingle point, which is steep-to on its north side, but to the southward of it a flat commences, which follows the western shore all down that reach of the inlet.  Two miles below Castro there is a small cove where vessels might anchor if necessary ; but there are 20 fathoms between the points, and it shoals suddenly a little inside of them.

The town of Castro stands near the outer part of the point of Castro and consists of two or three short streets of wooden houses : two churches, one of which has been a handsome building, but it is fast falling to decay, and shored up on all sides (1834) ; the other also appears to have been well built, but is now nearly in ruins ; altogether Castro has been much neglected, and the people are poor.  Between San Carlos and Castro, a direct distance of 38 miles, there is a road cut through the forest, 50 feet in width, in the middle of which is a causeway 5 feet broad, formed of logs of wood laid transversely.  This road, however, can only be used in dry weather.

**Castro Harbour.**—The small harbour to the northward of Castro point is half a mile in length and a third of a mile wide ; between the points there is 7 fathoms, but it shoals gradually to 3, about a quarter of a mile farther in ; the best anchorage is nearest to the south point, as the north side is shoal for about a cable's length off.  In running for the harbour a vessel should keep the eastern shore aboard till she is abreast of it, when she may stand across, and will thus avoid the shoal to the southward of Castro point, which extends half a mile off.

QUINCHED is a small harbour lying to the southward of Lintinao islet in which a vessel bound to Castro might wait for a favourable opportunity to go up, in case she found the winds baffling in the two first reaches: this is generally the case with northerly winds, however strong, outside, and no anchorage can be found in either reach until too near the shore for safety. The village of Quinched is about 3 miles to the westward of the harbour: the country is well cultivated and fairly inhabited for about 3 or 4 miles on either side of Castro inlet, and the houses are numerous, and surrounded with apple-trees. After passing Quinched, the land is only cultivated in patches, the rest being thickly wooded; about a mile to the southward of the cove the channel is 1¼ miles wide, with 41 fathoms in the centre. The tide at springs does not run above 1½ knots in the strongest part, and at neaps it is very little felt. Rise 18 feet.

LEMUY ISLAND forms the southern side of the channel leading to Castro inlet. Opposite to the entrance of the inlet, on the north shore of Lemuy, Poqueldon, the principal village on the island, will be seen standing on the east bank of a narrow creek not deep enough to afford shelter for a vessel. There is, however, anchorage in 4 and 5 fathoms water about a quarter of a mile from the east point. The village consists of about twenty houses, forming a square, one side of which is occupied by the church, the largest and the best, between San Carlos and Castro. Although not so large, Poqueldon appeared to be in much better condition than the latter place, and more prosperous.

There is a cove at the north-west point of Lemuy, near which the shore is rocky and steep-to; at its entrance there is no bottom with 20 fathoms, but half way up it there is good anchorage for a small vessel in from 10 to 7 fathoms, mud. The landing is bad; the tide at high water flows close up to the trees; and at low water the shores are very muddy.

Detif Headland, the southern extremity of Lemuy island, terminates in a perpendicular cliff to the westward, about 150 feet above the sea surmounted by a round hill 250 feet high, which falls gradually to a low neck of land about half a mile long, and again rises to the same height. A stony flat extends three quarters of a mile off the point; it is steep-to on the western side, but extends eastward to the next point, about 1½ miles off. Half a cable's length from the end of the shoal there is 7 fathoms, and at 2 cables' lengths no bottom with 30 fathoms.

Apabon Point and Reef.—About a league to the north-east of Detif point the same headland throws out to the eastward Apabon point, with a reef extending therefrom 3½ miles farther; near its outer end there is a rock always dry, and at low water the reef uncovers for about a quarter of a mile on each side of it.   No vessel should attempt to cross this reef although there are 9 feet at low water between the dry rock and the shore, because the tide sets over it strongly and irregularly.

CHILIN and QUEHUY lie between the south-east point of Quinchao and Apabon point ; the north-east extremity of the Quehuy is called Imel, and is connected with it only by a narrow isthmus.   Off Imel, for a mile to the south-east, there is a shingle bank that dries at low water, and which very considerably narrows the channel between it and Chaulinec ; on this bank a French ship struck.

YAL BAY on the Chiloe shore lies on the west side of the bight formed by Lemuy island.   Its northern point has two small low shingle islands off it ; they are connected by a spit, which is covered at high water : between the in-shore island and the point there is said to be a passage for vessels, but it appears very narrow.   Off the north-east end of the outer island a spit extend with only 2 fathoms on it to a quarter of a mile ; its extreme point appeared to reach about a quarter of a mile more out.   A mile to the south-east of Yal point there is a bluff head, and a little in-shore of the point there is a remarkable flat mound covered with trees.   Between the points of the bay, which is 2 miles across, there is no bottom with 55 fathoms.

Yal Cove lies to the northward of the bluff; the points of this little harbour are steep-to on both sides, and between them, in mid-channel, there is no bottom with 20 fathoms, but half way up there is good anchorage in from 5 to 12 fathoms, mud.   There is no anchorage in the outer bay until within a quarter of a mile of the bluff head, where there is 23 fathoms, shoaling gradually to the low water mark ; it is not a fit place for vessels to anchor in, unless obliged to do so.   The tide in the north side of the bay and in the cove is scarcely perceptible.   On the west side there is a flat mound, resembling that on the east side, but a little lower , they both show plainly from the southward, and are excellent marks for knowing the cove : the land at the back is low and thickly wooded.   The south point of the bay, Tebao, is low but steep-to, with 10 fathoms within half a cable of the beach : a little to the southward of it the shore is flat for a quarter of a mile off.

AHONI POINT lies opposite to Detif headland, and from near a rivulet to the eastward of it, a shoal, with 3 fathoms on it, fringes the coast for some miles ; to the southward the shore is rocky, with cliffs about 150 feet high.   Lelbun point lies about 4 miles from Ahoni ; and abreast of it the shoal widens to nearly 1½ miles, deepening to 7 fathoms,

and covered with patches of kelp.  The ebb tide sets to the south-east
about 2 knots at springs.

**CAPE AYTAY** is low and rocky, and about 3 miles to the southward
of Lelbun point.  Some rocks, of a reef which runs out from it, dry about
2 miles from the shore, but there is a passage for boats between and inside
of them.  About a cable's length outside the outer rock there is 5
fathoms ; and from thence the shoal trends in towards a sandy point
with a clump of bushes, about 2 miles to the northward of Quelan point,
After passing that point at a mile from the shore the water deepens to
12 fathoms.

**QUELAN POINT** is a long narrow strip of land, very low and covered
with trees, except in one spot, about 200 yards wide, half a mile from the
point, where the sand runs across.  The beach on the south-east side
slopes gradually, and at a quarter of a mile there is 2 fathoms water.  Off
the point there is 7 fathoms within half a cable's length.  Three miles to
the eastward of the point the small island of Acuy rises from its low
south-west point to a cliff, 200 feet high, on the north-east side ; from
which some rocks stretch off nearly 2 miles ; but the whole island is
surrounded by a shoal of rock and shingle with kelp, part of it drying at
low water, and extending off the west point about a mile.  After rounding
the point of Quelan, by keeping along the inside of the spit, it will lead
to the small harbour or cove of Quelan, the entrance to which is about
half a mile wide.

**TRANQUE Island,** 13 miles long, and about 3 broad, lies south of Quelan
point, and protects Quelan cove and Compu inlet.  The channel between
Quelan point and this island is about a mile wide, and the ebb sets through
it to the westward about 2 knots at neap tides.  A ridge of hills runs
through the island of Tranque, from north-west to south-east ; they are
about 300 feet high in the highest part, which is nearest the north-west
end ; from thence they slope gradually towards the south-east, and terminate in a low point called Centinela.  The north shore slopes gradually,
and is well wooded; the island appears thinly inhabited.  There is a small
bay at the north-west point of the island, where the channel turns
suddenly to the southward.  Off the bay is an islet with 21 fathoms inside
it ; this bay will do well for a vessel to wait the turn of tide.

**QUELAN COVE** is about three-quarters of a mile long, and the same
broad, with 13 fathoms in mid-channel, but a shoal extends from the west
point for a quarter of a mile in the direction of Quelan point, and the shores
on either side should not be approached within a cable's length, at which

distance there are 3 fathoms. In every other part of the cove there is good anchorage in from 5 to 8 fathoms, with 3 fathoms a cable's length off the beach. In its north-west corner there is a narrow creek, but fit only for boats ; and there are three or four houses, with patches of clear land around them. The inhabitants were Indians, and the surrounding country (in 1834) seemed thinly peopled. To the westward the land rises suddenly to about 200 feet, and is thickly wooded.

**Quelan Bay.**—On the north shore, about a mile to the westward of the cove, there is a small bay with an island off it, which affords anchorage for a vessel in from 10 to 13 fathoms ; the island may be approached within a cable's length, where it shoals suddenly from 10 to 3 fathoms. Compu is a deep inlet, on the Chiloe shore, abreast of the north-west end of Tranque island ; a little to the eastward of it is a smaller one, neither of which were examined. Between the points there are deep bays, and on the Chiloe shore about a mile from the turn in the channel there is a small cove, but it appeared unsuited for anything but boats.

**TIDES.**—The flood tide runs close round the points, and then strikes across towards the north shore, outside the small island, within which there is very little tide ; in the narrow channel it runs at least 4 knots at neap tides sweeping round the rocky points.

**CUELLO POINT** is on the Chiloe shore, at the entrance of the south-west channel, between it and Tranque island. About a mile south-east from this point there is a stony reef, extending in a north-west and south-east direction about half a mile, part of it dries at spring tide ; the shallow part has kelp on it, but the shoal extends beyond the kelp about a cable's length each way ; inside, at the distance of a quarter of a mile, there is 4 fathoms, which deepen to 12 about a quarter of a mile from the shore. About 4 miles from Cuello point lies the small island of Chaulin off the entrance of Huildad inlet.

**HUILDAD INLET** lies 5 miles S.E. by S. of Cuello point ; its entrance is only 150 yards wide, but inside the spit on the north shore it opens again to about a third of a mile ; at a mile from the entrance it again narrows to 2 cables' lengths, and then opens into a clear space of water from 1 to 2 miles wide, and 4 miles long. In the outer harbour there is good anchorage in from 5 to 9 fathoms ; the shores are steep-to, except along the bend behind the shingle spit, which is shoal for about 1½ cables' lengths from the beach. In the narrows between the two harbours, there is 5 fathoms water within 40 yards of either shore, and 20 in mid-channel.

On the south shore stands the church, with three or four houses round it ; the remainder (there are about twenty in all) are scattered along the sides of the harbour, chiefly on the south side, with cleared patches of

ground round each of them ; the land rises gradually from the beach for
about a mile, where it joins a ridge of hills 300 feet above the sea.

Should a vessel wait in Huildad for a change of wind or weather, the
outer harbour would be the best, as N.W. gales blow very heavily down
the inner harbour, while in the outer one a vessel would be sheltered from
every wind.

The tides at the entrance run on the ebb at springs nearly 4 knots,
but inside they slacken considerably, except in narrows where the tide
runs nearly as strong as it does in the entrance.

**HUILDAD SHOAL,** to the southward of Huildad, between it and
Chayhuao point, extends above a mile from the shore ; it is nearly covered
with kelp, the tide at the outer edge of it runs about 1½ knots at springs.
The shoal terminates in a long stony reef, which commences half a mile
inside Chayhuao point and runs off to the south-east ; some of the stones
are dry at low water about a mile from the point, and at spring tides the
whole of the reef is dry as far as the outer stones ; there is a channel
between the south end of it and the north-east side of Caylin island, with
deep water close to the reef.

Between Chayhuao point and San Pedro passage there is a deep bay
fronted by the islands Caylin, Laytec, and Colita, with the small cove
of Yalad to the north-west of the latter.  In the channel between Chayhuao point and Caylin island the flood tide sets to the eastward across
the reef at least 3 knots at springs: after passing the reef it meets the
outside tide coming from the southward.

**CAYLIN** is 5 miles long, north-west and south-east, and about a league
broad ; the north shore is steep-to ; in the channel between it and Chiloe
no bottom was found with 40 fathoms.  On the north side of Caylin island,
round a low shingle point, there is an inlet, which runs S.E. for 4 miles,
where it terminates in three small coves ; it is not a good place for vessels,
there being from 22 to 30 fathoms in it, till within a quarter of a mile of
its head, where it shoals suddenly to 12, and at a cable's length inside of
that it dries at low water.  The south-east side of the island is composed
of cliffs about 100 feet high, with a shingle beach at their foot.  A reef
with 4 fathoms on its edge extends 1½ miles from the beach.

We found here an Indian village of 40 houses, containing about 250
inhabitants (in 1834), who were glad to supply us with sheep and
poultry in exchange for tobacco and handkerchiefs ; they seemed anxious
to know when the King of Spain would retake the islands.

**LAYTEC ISLAND** is 6 miles long, and about 3 miles in breadth ;
it is separated from Caylin by a channel 2 miles across, at the southern

entrance of which there are 20 fathoms water : off its south-east end there are a few rocks, but no danger appeared beyond half a mile, where 4 fathoms were found.

COLITA ISLAND is low and thickly wooded, about 4 miles long, and $1\frac{1}{2}$ miles broad ; the channel between it and Chiloe is very narrow, and apparently not fit for a ship ; the land behind rises gradually from the coast, and forms a range of hills above 1,000 feet high. Between Colita and Laytec islands the passage is $1\frac{1}{2}$ miles broad. The tide sets about one knot through the channel north of the islands. San Pedro is described at p. 252.

ABTAO.—Returning to the northward to the Chacao narrows :—Abtao island lies 6 miles to the eastward of Coronel point. It is $2\frac{1}{2}$ miles long, and one mile broad ; the north-west point is the highest, and ends in a bluff, 80 feet above the sea, off which a stony flat runs a distance of 2 cables ; close to the flat there is 12 fathoms, and a quarter of a mile to the north-east 30 fathoms ; the shoal from the main runs off nearly a mile. From the south-east end of Abtao a shoal extends $1\frac{1}{4}$ miles, with 5 fathoms near the extremity. The shore of the main land rises gradually from the beach to about 200 feet, and is thickly wooded ; at $1\frac{1}{4}$ miles from the beach there is 35 fathoms.

CARVA lies north-east of Abtao ; it is a round hummock, about 200 yards long, surrounded by a bed of shingle, which is covered at high water, except at the north point, where a narrow spit remains dry ; a shoal extends a mile off its south-east end. Lami Bank has its north-west edge 2 miles east of Carva, and is always dry in several places ; the north side is about 2 miles long, and runs parallel to the shore, at the distance of about $1\frac{1}{2}$ miles ; in mid-channel there are 35 fathoms.

TABON ISLAND is composed of a number of detached hummocks of land joined together by low shingle ridges, some of which are overflowed at high water ; the land is clear, except the apple trees round the houses, and its greatest height does not exceed 150 feet. Half a mile to the north-east of its western extremity a stony reef runs to the northward, in the direction of the banks of Lami, and is dry at low water three-quarters of a mile from the shore. The channel between it and the south end of Lami bank is about three-quarters of a mile wide, and at 2 cables lengths from the end of the reef there is 7 fathoms. Another reef runs off more to the westward, and to the distance of a mile.

CALBUCO, QUENU, and CHIDHUAPI lie to the northward of Tabon, and are of a similar nature. Chidhuapi is low and nearly all cultivated, between its south end and Tabon there was no bottom found with 55 fathoms. Quenu passage between the islands of Quenu and Calbuco, is about three-quarters of a mile wide, with 21 fathoms in mid-channel; the points of both islands are low. A rocky flat runs off the point of Quenu, but it does not quite bar the channel.

The town of Calbuco or El Fuerte, near the north-east end of the island, on a steep slope, is only a third of the size of San Carlos, but superior to any of the other settlements; the church is a large wooden building, though not equal to either of those at Castro, and the land about Castro is better cleared and cultivated. The beach off El Fuerte dries at low water about a cable's length, and close outside there is 6 fathoms, and a very little farther 17 fathoms near it; the channel then deepens to 24 fathoms. The best anchorage is abreast of the town, about a third of a mile distant, and in from 20 to 22 fathoms, muddy bottom.

**PU LUQUI**, 7 miles in length, is the largest and most eastern of the islands between the Chacao narrows and Reloncavi Sound. It is thickly wooded; on the eastern side the patches of clear land are very few, but on the other side, where the land is lower and swampy, they are more numerous. The south point is low, and rises gradually to a ridge about 300 feet, which runs through the island from north to south near the eastern shore.

**Puluqui Channel** between the island and the main shoals gradually, having 8 fathoms in the narrowest part, between the north point of Puluqui and the islet of Tantil, where it is about a third of a mile wide. In hauling round the point of Puluqui to the south-east it shoals to 5, and then deepens to 8 and 16 fathoms; the passage between Tantil and the main appears to be shallow.

**Soldado Point**, the southern point of Puluqui, is low, shingly, and thickly wooded; the high land rises about 200 yards in shore, and a flat extends a cable's length from the point: it runs nearly east and west for 3 miles, and then turns to the north-west. After rounding this point, about a mile to the northward there is a small cove, the entrance of which is very narrow, and too shallow for a boat after half tide; but inside it is about half a mile across with 8 fathoms in one part. Cullin islet lies off Aulen point, on the eastern or main shore. San Jose shoal lies 1½ miles to the northward of Cullin, with the clear space of a league between it and Puluqui island.

**RELONCAVI SOUND.**—The strait between Cullin and Puluqui, only 2 miles wide, forms the entrance to Reloncavi sound, which extends 20 miles to the northward, and is about 12 miles across, from east to west. In this strait there is no bottom with 60 fathoms, and as far as information could be obtained, there is no bottom with 120 fathoms throughout the sound, except in the neighbourhood of the islands and shoals; anchorage may be found under both the former, and doubtless close along the shores on either side, according to the prevailing wind.

**Huar Island** lies on the west side of the sound, separated from the mainland by the Huar passage, round the north point of which there is the small settlement of Ilque. To the south-east of Huar are two shoar patches, Pucari and Rosario ; the eastern side of the latter lies 3 miles from the island.

**MAYLLEN** is on the same side of the sound, and 5 miles to the north-ward of Huar. There is a passage between the main and this island, but shoal patches off its south and west points.

**PORT MONTT.**—Three miles north of Mayllen is the islet of Tenglo separated from the coast by a narrow passage, at the north end of which is Port Montt, a prosperous town established by the Chilian Government in 1853. The anchorage is good but open to the southward. There is a rise and fall of 15 feet. This port is only 15 miles south from the German colony on the banks of Lake Llanquihue, to which it has become the sea-port. The road lies through a forest of Alerse, a species of pine much in request for building purposes, upwards of 1,200 men are employed near Port Montt in felling this timber, and over a million deals are annually exported.[*] The Pacific Mail Company's steamers call here every month.

**RELONCAVI INLET.**—There is a deep inlet on the eastern side of Reloncavi sound, by way of which and the River Petrohue, through Todos los Santos lake, and up the Peulla, a communication was formerly kept up with the Spanish Missionaries' settlement, on an island in the great lake of Nahuelhuapi ; but this mission was abandoned towards the close of the last century. To the south-west of the lake named by the Spaniards Todos los Santos, and about 23 miles north-east of Port Montt, is the volcano of Calbuco, 18 miles north of which, between the lakes Llanquihue and Todos los Santos, is the lofty volcano of Osorno or Purraraque, called also Huañauca, rising 7,750 feet above the level of the sea.

**The COAST** of the continent in the Gulfs of Ancud and Corcovado, together with the shores of the interior sounds of the Chonos archipelago, have not been surveyed. They were explored by Don Jose de Moraleda in 1795, from whose charts the coast line on the Admiralty sheets Nos. 1325, and 1289 is drawn in a faint outline, the mountains being fixed by Captain now Rear Admiral Fitz Roy. The plans of Reloncavi and Comau inlets in the Gulf of Ancud, as also those of Tictoc bay, Piti Palena, and Port of St. Domingo at the entrance of the Gulf of Corcovado, are from the same authority.[†]

---

[*] *See* Mittheilungen, &c., Gotha, by Dr. A. Petermann, Part IV., 1860; page 133.

[†] *See* Plans, Nos. 563, 564, 565, 566, and 567 ; scale, m = 0·7 of an inch.

The **TIDES** on the east coast of Chiloe are very irregular, being much influenced by the winds, as appears by the following table. The time of high water at Castro, and other places, is earlier in going to the south-ward; yet at Huildad, which is more than 13 miles south of Castro, it was high water three-quarters of an hour later than at Castro; but at the time it was blowing a heavy north-west gale at Huildad. The average time of high water in the north part of the archipelago is pro-bably about 1 o'clock on full and change days, which decreases gradually to about 12h. 15m. near the south end. It appears to be seldom regular, and was found to vary half an hour in two following tides. The rise was also very irregular, as the tides often rose higher when they were taking off. The night tides were always higher than the day, during the *Beagle's* visits.

In Port Oscuro, the rise and fall at one time, at dead neap tides, was 18 feet, and the next springs it only rose 16; by the marks on the shore, the rise and fall at some high tides had been above 24 feet. The greatest rise and fall is at this place, and it is the best for heaving down in the gulf, or for cutting docks, if they should ever be required. The only other place that would answer well for that purpose is the outer part of Huildad inlet, on the west side of which there is 9 fathoms close to the shore, and the coast is composed of rock, which would answer better than the sand and shingle of Port Oscuro; but the rise and fall is only 15 feet at spring tides, which would be too small for large ships. Port Oscuro may therefore be considered preferable.

TIDE TABLE for the GULF of ANCUD.

| Place. | High Water at Full and Change. | | Rise. |
|---|---|---|---|
| | H. | M. | |
| Carelmapu - - - | 0 | 50 | 10 |
| Chacao narrows - - | 0 | 48 | 11 |
| Abtao island - - | 1 | 06 | 14 |
| Calbuco beach - - | 1 | 15 | 16 |
| Chacao bay - - - | 0 | 40 | 14 |
| Huapilinao - - - | 1 | 25 | 16 |
| Oscuro cove - - | 0 | 55 | 22 |
| Quicavi lagoon - - | 0 | 50 | 20 |
| Changues islands - - | 0 | 31 | .. |
| Castro - - - - | 0 | 11 | 18 |
| Alau island - - - | 0 | 31 | 18 |
| Poqueldon harbour - | 0 | 54 | .. |
| Talcan island - - | 1 | 03 | 15 |
| Quelan cove - - - | 0 | 26 | .. |
| Huildad inlet - - - | 0 | 56 | 15 |

# CHAPTER XI.

## CHILOE ISLAND TO COQUIMBO BAY.

VARIATION 19° to 15° East in 1860.   Annual increase about 1'.

CARELMAPU ISLETS.—Continuing along the coast from Chiloe, northwards, the islets of Sebastiana and Carelmapu require to be again noticed, in order that their vicinity may be widely avoided.   The tide sets strongly at times in races near them ; and when there is a swell from seaward with an ebb-tide running, the short high sea north westward of these islets is very straining to a ship, as well as dangerous to boats or even to small vessels.*

GODOY POINT is low, with two small islets off it ; it lies 5½ miles north of the outer and largest Carelmapu islet.   Maullin inlet, 6 miles to the eastward, is a shallow, wild place, exposed to a heavy breaking sea, and unfit for vessels.   It forms, however, the mouth of the river Maullin, which communicates with the German colony on Lake Llanquihue.   Small deposits of coal have been found near this inlet, and also at Punta Pargu, 20 miles to the northward.

The COAST from Godoy point trends N.W. 8 miles to Quillahua point, thence N.N.W. 17 miles to Estaguillas point, and 9 miles beyond this to Cape Quedal, a projecting and bold promontory ; under a height which is very conspicuous (a part of the range called Pargo Cuesta) is a point called Capitanes.   Both Quillahua and Estaguillas, as well indeed as most of the projecting points on the coast between Godoy and Galera point in latitude 40°, have many detached rocks about them, but all close to the shore, and the greater part above water.   This part of the coast may be described in a few words.   The land is high and bold without any outlying danger ; but at the same time without a safe anchorage between San Carlos and Valdivia, a distance of 120 miles.   Soundings extend some miles into the offing, though the water is deep.   At 2

* See Charts of West Coast of South America, Sheets IV. and V., Nos. 1,289 and 1,374 ; scale, m = 0·12 of an inch.

[A. S.]

S

miles off shore there is usually about 40 fathoms water, at 3 miles about 60, and at 5 miles from 70 to 90 fathoms, over a soft sandy and muddy bottom.

**CAPE SAN ANTONIO,** 9 miles from Cape Quedal, is a high, bold headland, dark-coloured, and partly wooded ; the land hereabouts ranges from 1,000 to more than 2,000 feet in height. San Pedro bay lies between these capes. In it there is the mouth of a small river, with a shallow bar. Manzano Cove, 23 miles to the northward, may afford temporary shelter for small coasters : a river of no consequence flows into it. Milagro cove is of a similar character. The river Bueno is navigable within, and flows through a valuable tract of country, but there is a bar at its mouth which excludes all but the smallest craft.

**GALERA** is a prominent point of land with a low hill backed by the remarkable heights called the Valdivia hills, three in number, very conspicuous, pointed at their summits, and about 1500 feet in height.* Two miles and a half N.N.E. from Galera point is Cape Falsa, a low projection, with rocks half a mile off it, but above water ; it is in a line with the ridge of Valdivia hills, which are excellent marks for this part of the coast. Hence the shore trends north-eastward 13 miles to Gonzales head, a wooded bluff cliff, immediately to the eastward of which is the Port of Valdivia.

**VALDIVIA.**—N.E., 2¾ miles from Gonzales head, is Mill point, off which some rocks lie about 3 cables' lengths. Mill point is rather steep and covered with wood ; between these is the entrance to Valdivia, a port apparently spacious and really secure, but the portion affording sheltered anchorage for large ships is somewhat confined. Two river-like inlets open into the port on the south, and from the north-eastward come the rivers Calla-calla and Cruces, winding, and full of banks, and navigable only for small vessels assisted by a local pilot. About 9 miles up the former river, on the east bank, is the town, still called the city, of Valdivia, founded by Pedro de Valdivia, a follower of Pizarro, about the year 1540. Woods clothe every hill about the town, and all the adjacent country is hilly, the land about Valdivia ranging to 1,000 feet in height. Water is plentiful, the climate being almost as rainy as that of Chiloe. Provisions are cheap but not abundant. There is a German Colony at this port in communication with the one on Lake Llanquihue. Population of Valdivia is said to be 7,000.†

---

* *See* View on Sheet No. 1,374.
† *See* Admiralty Plan of Port Valdivia, No. 1,318 ; scale, *m* = 2·0 inches.

Fort San Carlos, which may be closely passed, stands on 'the second point from Gonzales head; on the opposite shore, nearly east of San Carlos, is Niebla castle, off which there is 8 fathoms at 2 cables' lengths, and in mid-channel 7 fathoms water, from which the depth gradually increases seaward. Amargos Point, on the western shore of the port, rather less than a mile from Niebla castle, is low, and has a small battery on it, close to which there is deep water. About a mile to the southward of Amargos point, at the farther side of a well-sheltered cove, 3 or 4 cables' lengths square, is the Corral fort. The best anchorage is Calvary cove, near Fort Corral, which is also the watering-place.

Manzera Bank lies midway and in a line from Corral fort to Piojo point, to which the water gradually shoals; it is, however, dangerous to a stranger, because there is nothing to indicate its situation from the appearance of the water, which is always discoloured during the ebb-tide, by the silt brought down the river. This bank, which extends nearly across to Corral fort, with some very shallow spots, detracts very materially from the goodness of Valdivia harbour. Manzera island, 300 feet high, lies to the south-east. From this island three openings are seen, the north-east being the river Calla calla, leading to the town of Valdivia, the south-east an unexplored inlet, and the southern the shoal bay of San Juan. Rocks and banks extend for one-third of a mile off the southern end of Manzera island. Although the plan of the port and river made by the *Beagle's* officers was correct in 1835, it ought not to be trusted either for the river banks or for the limits of the Manzera shoal for more than a few years. It is high water, full and change, in Port Valdivia at 10h. 35m.; the rise and fall being 5 feet.

BONIFACIO HEAD is about 8 miles north of Gonzales head; it is bold, and has deep water near: 2 miles off it are 20 fathoms. Thence the coast trends north, about 20 miles to Chanchan cove, at the mouth of the river Mehuin; a tolerably good anchorage, in from 5 to 7 fathoms, sand or mud, for coasters in summer only, sheltered from the west by islets and a reef of rocks extending a mile to the northward, but quite open to the north-west.

At GOCALE HEAD, 8 miles north of Chanchan point, the coast changes its character, becoming low and sandy, but with occasional cliffs; the high lands which to the southward of this point bordered the ocean, here retreat 5 or 6 miles, leaving a level and apparently very fertile country, as far as abreast of Mocha island. This piece of coast lies about N.W. by N., and extends nearly 60 miles. Off its whole extent there are comparatively shoal soundings, 10 fathoms at 2 miles distance, 20 at 4 miles, and everywhere a sandy bottom: it is therefore dangerous to approach at night without the lead going. A heavy surf breaks everywhere, even in fine weather.

**TOLTEN and CAUTEN RIVERS**, though said to have been navigable formerly, are scarcely distinguishable at 2 miles distance from the shore, and are both closed by bars. Ranges of cliffs extend for several miles, at intervals, along this shore; and on their level summits may often be seen troops of the unconquered Araucanian Indians riding, lance in hand, watching the passing ship. The summits of the Andes are visible for a great distance northward and southward, whenever the weather is clear, and the active volcano of Villa Rica, 100 miles east of Tolten, is said to visible at sea at 60 miles off the land.

**CAUTEN HEAD**, is a bold, cliffy headland, standing nearly 8 miles N.W. of the Cauten river. It is about 300 feet in height, with 20 fathoms 2 miles off-shore, and apparently steep-to. From thence cliffs, more or less broken, extend 10 miles to Manoel point, bearing east from Mocha, distant 20 miles: 8 miles N.N.W. from this point is Cape Tirua, the point of the main land nearest to Mocha.

**CAPE TIRUA** has a small islet close to it, and in a little bay just to the northward is the mouth of the river Tirua, whence a communication used to be kept up by the Indians of the main land, with those who lived on Mocha island, by means of rafts, *balsas*,* and large canoes. The tide runs about a knot during springs, the flood to the northward. There is no sheltered anchorage on the part of the coast that has been described; but 9 miles north of Cape Tirua the small cove of Nena may afford temporary protection, and possibly a good landing place.

**MOCHA ISLAND** is lofty, and therefore a prominent land-fall for seamen, its summit being 1,250 feet above the sea: it should not, however, be approached too freely, as dangerous rocks lie off its west and south sides: the most out-lying are 3 miles south of the island. During the flood-tide these rocks are particularly dangerous, as it sets towards them from the south-westward. Sometimes the ebb stream is scarcely felt for days together, and then the flood stream has the effect and appearance of a continuous northerly current.†

---

* Inflated skins supporting a slight frame work.
† *See* Plan of Mocha island, No. 1,305; scale, m = 0·5 of an inch, and View on No. 1,374.

Previous to the eighteenth century it was inhabited by Araucanian Indians, but they were driven away by the Spaniards, and since that time a few stray animals have been the only permanent tenants.

Mocha is about 7 miles long, by 3 in breadth, and lies about 18 miles off the coast, with which it is connected by a bank, having on it less than 20 fathoms. To the westward the water is deep, a matter which might seem of minor consequence, as this island is so good a landmark, but occasionally there is thick weather for days together on this coast. The channel between Mocha and the main land is free from danger ; the depth varying regularly from 10 to 20 fathoms, over a sandy bottom.

The Anchorages are indifferent ; one on the north-east side, the other, near the south-east point, called by the Spaniards Anegadiza. The landing is bad, and there are now no supplies to be obtained except wood, and with considerable difficulty water, but it is of excellent quality. The anchorage near Anegadiza point is good, during northers, just in front of the first little hills, and in 6 or 7 fathoms sand : the other anchorage is off English creek, in 13 to 20 fathoms, over a sandy bottom; nearer the shore it is rocky. Were there any adequate object in view, a good landing-place might easily be made, and there is abundant space on the island for growing vegetables, as well as for pasturing animals.

The COAST from Cape Tirua to Tucapel point is wild and exposed, totally unfit to be approached : it is lashed by the south-west swell, and has no kind of shelter. At the north-west end of a long low beach, on which there is always a heavy surf, is Molguilla point, on which H.M.S. *Challenger* was wrecked in 1835. Eight miles N.W. of Molguilla point is Tucapel point, a low, projecting, rocky point flat-topped, and dark coloured. The interior country hereabouts is very fertile and beautiful. Hill and dale, wood land and pasture, are everywhere interspersed, while numerous streams plentifully irrigate the soil. Tucapel head is a high bold hill, 7 miles N.N.W. of Tucapel point; from the top of the hill, which overlooks the mouth of Leübu river, and which is a commanding position likely to be of future consequence, the high land slopes gradually to the southward ; the summit is about 600 feet above the sea.*

RIVER LEUBU.—Between Tucapel head and Millon point, a rocky projection 2 miles farther north, there is a cove, into which the Leübu runs. Coasters may find shelter there if the wind does not blow too strongly from N.W., but it has no defence from that quarter. Boats can enter

---

* *See* West Coast of South America, Sheet VI., and River Leübu, and Santa Maria Island, Nos. 1,286 and 1,303 ; scales, *m* = 0·12 and 0·5 of an inch.

Letibu at half-tide, when there is not much swell on the bar.  In former days there was a settlement called Tucapel Viejo, at the mouth of this river.   The pirate Benavides at one time resorted to a cave near this point ; and in 1835, the *Challenger's* crew encamped under the heights, till taken off by the *Blonde* frigate, which anchored in 27 fathoms, a mile N.W. of the head.   At 150 miles E.N.E. of the Letibu is the lofty volcano of Antuco.

CARNERO BAY is a wild exposed bight unfit for shipping, but Yanass cove, at its northern end, affords anchorage for coasting vessels of small size, a bank however is said to exist to the southward of the cove.  Carnero head is a cliffy bluff.   From thence to Cape Rumena and Lavapie point the shore is bold and cliffy, and backed by high land, well wooded.

CAUTION.—The coast between Santa Maria island and Carnero bay is reported to be foul.  A dangerous rock has lately been discovered lying 2½ miles N.W. from Cape Rumena, in the direct way of vessels from the southward going through the channel between Santa Maria island and the main ;  it is not seen except at low water, or when there is a heavy swell.*

SANTA MARIA ISLAND is comparatively low and dangerous on account of numerous outlying rocks.  It has a cliffy coast, and somewhat irregular currents.  Between it and Lavapie point there are two dangerous rocks under water, on which the sea does not always break.  The first, named the Hector, requires especial care, as it lies North 1½ miles of the east side of that point near mid-channel, and exactly in the track that most vessels would incline to take.  The other also lies North, and *is* distant half a mile from the same point, around which there are several other rocks ;  but the sea always breaks on them.

COCKATRICE and RENWICK ROCKS.—Besides the above two rocks and a cluster of others half a mile South and S.W. of Cochinos point, there is another danger which vessels making the land should be careful of.  It was discovered in 1849, by Mr. James Rundle, commanding the *Cockatrice* schooner, who was searching for the rock, pretended to be in the offing, on which the ship *John Renwick* was lost, and which proved to be the Dormido, to the northward of Santa Maria island.   This Cockatrice rock lies S. by W. of Cadenas point, the western extremity of

the island, and W.S.W. of its south point, Cochinos, from which it therefore appears to be upwards of 3 miles distant.

**Santa Maria Road.**—South-east of Santa Maria island there is a
tolerable roadstead, with from 4 to 8 fathoms water, over good ground;
but the only place really sheltered is quite close to the eastward of Cochinos
point. Formerly there was good anchorage between it and Delicada point,
but the earthquake of 1835 raised the bottom nearly 9 feet; so that where
there was a depth of 5 fathoms in 1834, the *Beagle* found only 3½ in 1835.

In passing round Santa Maria to the eastward, a wide berth must be
given to the shoal which now runs off towards the south-east; it is not
prudent to go a cable's length to the northward of a line drawn E. by
S ½ S. from Cochinos point until 3 miles to the eastward of it, where
there are but 4 fathoms at low water. From thence the shoal turns to
the northward round Delicada point, off which the water deepens
suddenly to 10 and 20 fathoms. On the north-east side of the island
there is anchorage during southerly winds. Water is good and abundant,
there is also plenty of wood and vegetables, and but little else.

**Dormido and Vogelborg Rocks.**—Off the N.W. end of Santa Maria
island there are many rocks, the principal being the Dormido, lying
3 miles N.W. ¼ W., and the Vogelborg (which are 2 in number), 4 miles
N. by W. ½ W. from the northern point of the island. They are sometimes
undistinguishable by breakers, and it is not safe to pass between them and
the island; neither is it prudent to approach the western side of Santa
Maria nearer than 3 miles.

**ARAUCO BAY** is an extensive bay inside Santa Maria island about
15 miles broad and 18 deep. In southerly winds there is good anchorage
throughout Arauco bay, but except in Luco bay it is everywhere exposed
to northerly winds and sea. In the bight of the bay lies Arauco, once so
renowned, but now only a small square fort, or rather enclosure of earth,
about 200 yards square; it stands a short distance from the sea.

**Luco Bay** under Lavapie point has a tolerably good anchorage, but is
not quite sheltered from N.N.W., and liable to heavy squalls off the heights
over Cape Rumena when it blows strong from the south-westward: there
is 5 fathoms water over good ground.

**Tubul River.**—For 3 or 4 miles on each side of this river the coast
is steep and cliffy, with high down-like hills. Tubul was formerly capable
of receiving vessels of considerable burden, but the earthquake of 1835
raised its bar so much as to prevent access to more than boats: it is supposed that the bar will not remain; the neighbouring country is very
beautiful and fertile. Off the outer point of the long cliff west of Tubul

river, and a mile from the land, there is a rock called El Frayle, on which the sea always breaks unless the water is unusually still.

**Laraquete Beach** extends 10 miles to the N.E. from Tubul cliffs; and 2 miles off it are from 8 to 10 fathoms water, over a sandy bottom. The river Carampangue is not navigable at its mouth, though deep and rather wide 2 miles inland : its exit is choked by sand-banks.

From the river Laraquete to Coronel point the coast runs N. by W. ¼ W., high and bold, free from outlying dangers, and affording temporary anchorage for small vessels, or at least shelter for boats, in three or four coves ; and the mouth of the little river Chivilingo affords shelter for small craft except during south-west gales. Fuerte Viejo cove, immediately south of Colcura, is exposed to both the south-west and north-west.

**Lota,** a little cove just to the northward of Colcura, is the best of the three, but it also is open to the south-west. This port has lately become of much importance in consequence of its considerable export of coal. It may be known by two long white houses on the hill above the cove, and a long iron jetty, with a wheel and drop on the outer end of it for coaling purposes. Fresh meat and vegetables are to be obtained at a moderate price.

**Coals.**—Good steaming coals were obtained here in 1858, at 5 dollars per ton, it is however quick consuming fuel ; 250 tons can be put on board from the drop in one day, under which there is 20 feet water at L.W. spring tides. In 1858, 20 vessels were loading at Lota and Coronel point. The port charges are 4 dollars. A steamer calling for coals should visit Lota in preference to Coronel point, for the former is the only place where they can be obtained from the drop, at other places they are put on board by lighters ; the mine is only a few fathoms from the jetty. Good fire bricks are also to be got at Lota.*

**TIDES.**—It is high water, full and change, in Arauco bay at 10h. 20m.; the rise and fall being 6 feet.

**BIO BIO RIVER.**—This great river is not accessible on account of its sand-banks, and of the south-west swell : its situation, together with that of Port San Vicente and Concepcion bay, are well pointed out by the remarkable pointed hills about 800 feet high, called the Paps of Bio Bio, which lie 11 miles N. by W. ¼ W. from Coronel point ; there is no danger near them except a few rocks close to the shore.

---

* *See* Mercantile Marine Magazine ; October 1859, page 301.

**PORT SAN VICENTE** on the north side of the Paps, is an exposed bad anchorage, entirely open to the N.W. winds and to the western swell and has rocks on its north side, nearly half a mile from the shore.

Close to Tumbes, the peninsula which forms the bay of Concepcion, there are many straggling rocks, some under, some above water, from Lobo point to the end of Tumbes promontory; this piece of coast trends north 6 miles from Port San Vicente. Quiebra olla, or Break-pot rock, is above water, lying N.W. three-quarters of a mile from Tumbes point; between it and the point no vessel should pass, as there are several rocks under water there; outside of it there is no danger.

**CONCEPCION BAY.**—Between Tumbes point and Loberia head, 6¼ miles to the N.E., lie the entrances to the Bay of Concepcion, the finest port on this coast; being 6 miles long, and 4 miles wide, with anchorage ground everywhere, abundant space, and all well sheltered. Mount Neuke, about 5 miles to the eastward of Loberia head, the highest land in the vicinity, is 1,790 feet above the sea.*

**Quiriquina Island** lying north and south, 3 miles long by nearly a mile wide, is situated in this entrance, giving shelter from northerly winds; and ships may freely anchor near Arena point, at its south-east extremity. This island and the shoals south of it protect the anchorage off the town of Talcahuano from the northers. The peak of the island is 395 feet high.

There is a passage into Concepcion bay on either side of Quiriquina island. The western one, called the Quiriquina channel, is a mile wide between Tumbes and Quiriquina, with deep water on the island side, but the Buey rocks, which project from the north-east shoulder of Tumbes reduce the available passage to the breadth of half a mile. The Great Channel is 2 miles in width, with no dangers at a reasonable distance from either of the points of Loberia or of Pajaros Niños; and there is also less tide in this wide passage. The best entrance, therefore, for those not locally acquainted, is to the eastward of Quiriquina.

**PORT TALCAHUANO** is at the south-west angle of Concepcion bay. Until lately this was the principal port in the bay, but Penco and Tomé are fast rising into importance; the former was the ancient port of Concepcion, and bids fair to become so again. The town of Concepcion, with a population of 10,000, lies on the right bank of the Biobio, at 7½ miles from its mouth.

**Supplies.**—The country around amply rewards cultivation. Some of the valleys at the back of the town of Concepcion are very fertile, pro-

* *See* Plan of Concepcion bay, No. 1,319; scale, m = 1·0 inch, and View on Chart, No.1,286.

ducing grain of all kinds in abundance. Beef and mutton are cheap and
good, the former about twopence per pound, the latter one dollar the
carcase ; pork and fowls rather dearer. Vegetables of all kinds cheap
and plentiful, as well as fruits in their season ; wood plentiful, at two
dollars per 1,000 billets. Good water can be procured from a tank which
carries 30 tons, at 1½ dollars per ton.

The Coal of this country, though easily worked and covering a large
space, is not a very good coal for steaming purposes, burns rapidly, and
throws out a great heat with much flame, but does not last, consuming
nearly as quickly as wood, so that the funnels are either red hot or steam
cannot be kept ; used with Welsh or other coal which requires much
draught, it is, however, very serviceable, and when the high price of
English coal is considered the value of such a supply can scarcely be
over-rated; it was to be had in Valparaiso for from six to eight dollars a ton, but if in regular demand could be brought to market at a
much smaller price. The communication with Sydney now, however,
opens a source of supply independent of England, the emigrant ships
being glad to take anything on freight or even for ballast on their return
voyage, out of the route of which Valparaiso does not lie many miles.*

Belen Bank.—Near the principal anchorage, in Port Talcahuano, are
the Belen, the Choros, and the Manzano banks, but their positions are
so clearly shown on the large plan of this port, that it would be a
loss of time to do more here than draw attention to them. On the Belen
there is generally a red buoy lying with the south-east end of the cliffs
on Tumbes peninsula, called Talcahuano head, just open of Mount
Espinosa in Port Vicente, bearing S. by W. ; and on the Choros there is
a rock which shows at low water, on which a perch has been lately fixed.

Vessels bound to Talcahuano should keep Point Arena on with the north
point of Quiriquina till the flag-staff on the town fort bears S. 40° W. when
they may haul in for the anchorage.

Beechey Rock.—About 1½ miles W.S.W. from Lirquen point, at the
S.E. part of the bay, Captain Beechey found a rock, or rocky shoal, with
only 15 feet on it. The Beagle's boats searched for it in every direction
near the place indicated by him, but could not succeed in finding less
water than 9 fathoms. Nevertheless such authority as that of Captain
Beechey is not to be doubted, and ships should avoid that part of the
bay till the exact situation of this danger is decided : it is not at all
necessary to stand over so far towards the eastern shore when working
up to Talcahuano.

* Remarks of Lieut. J. Wood, R.N., 1845.

**COLIUMO BAY.**—Three and a quarter miles to the northward of this, head, which is dark-coloured, and has some straggling rocks close to it, the coast turns short round to the eastward, to Cullin point and Coliumo head, and then again to the southward, where it forms the Bay of Coliumo. Coasters may anchor there in security, but there is not much shelter for large ships during northerly winds. The best anchorage is in Rare cove, just round Coliumo head, offering good landing for boats, and a convenient watering-place. It has always been the scene of smuggling transactions.*

**The COAST** from Coliumo bay, 16 miles north, to Boquita point, and thence 40 miles farther, in a similar direction, to Cape Carranza, assumes an unbroken line, without any place for shipping. It is a deep water shore; and the land, which rises to a considerable height, is partially wooded. Among the rocks at Cape Carranza boats find shelter occasionally; it is a projecting though rather low part of the coast, and therefore to be avoided; for about 10 miles on each side of the cape there is a sandy or shingle beach. Cape Humos lies 17 miles N. by E. nearly from Carranza; it is a remakable headland projecting westward, and higher than any other land near that part of the coast; it is bold-to, and there are no outlying dangers in the vicinity.

**CHURCH ROCK,** so named from its appearance, lies 4 miles N.N.E. from Cape Humos, and one mile S.W. from the entrance of the river Maule. There is no mistaking the entrance, for on the south side the land is high and the shore rocky; while on the north side a long low sandy beach extends beyond eye-sight. Not far from Church rock a remarkable bare space of grey sand may be seen on the side of a hill, but generally the heights between Cape Humos and the Maule are covered with vegetation and partially wooded. The highest hills in the vicinity range from 1,000 to 1,300 feet; those actually on the coast between Humos and the river, from 500 to 900 feet. A ship may anchor in fine weather, in from 10 to 15 fathoms, sandy ground, from 2 to 3 miles N.W. of the Church rock. There is no hidden danger, but an extensive sand-bank north of the river shelves out to seaward, and should have a wide berth.

**MAULE HEAD** is the south point of entrance to the river Maule, it forms a steep cliff with a beach on each side; a sandy point called Entrance point, extends about a quarter of a mile from it to the N.N.W. Boats sometimes land on the outer beach, under this head; but there is great

* See Plans of Coliumo bay, and River Maule, No. 1,312; scales, m = 2·0 and 0·5 inches.

difficulty and risk in doing so, for the surf is always high and treacherous, and the sandy beach is so soft and steep, that it is extremely difficult to haul up even a whale-boat. A better outside landing is close to the Church rock, but even there it is bad enough. *Balsas* should be provided, and the boats anchored near the beach, safe from the surf. High water at full and change at 10h.

**CONSTITUCION.**—This little town, on the south bank of the river, a mile from its mouth, may flourish hereafter, by the help of small steamers, and some engineering assistance at the bar, which shuts up the river Maule. A most productive country surrounds it, abounding with internal and external wealth, and a fine river, that communicates with the interior is navigable far inland; besides which the best pass through the Andes (discovered in 1805) is not far from the latitude of the Maule ; it is said to be nearly level, and even fit for waggons; the only pass of such a description between the Isthmus of Darien and Patagonia. Small steamers now run (1859) between Valparaiso and this port.

**FALSE MAULE.**—To the northward of the river is an extensive sand-bank evidently formed by the detritus brought down the river ; behind this sand, there is a flat, several miles in extent ; this flat in front of the high ground, reaches to within 5 miles of a very remarkable valley, called the False Maule, from its having been taken for the entrance of that river.

**TOPOCALMA POINT.**—From False Maule the coast trends north to Lora point, and thence nearly N by W. to Topocalma point, a distance of 55 miles, without an anchorage or any outlying danger ; the shore is high and bold, and deep water everywhere. Near Topocalma point coasters sometimes anchor for a few hours, but there is no place fit for a vessel of 200 tons.*

**PAPUYA COVE** lying 13 miles north of Topocalma point, is fit for small vessels only during southerly winds. North-eastward about 4 miles is Natividad bay, it also offers no good anchorage, and is much exposed. Rapel point is on the north side of the bay, close to the river Rapel. Three miles N.W. of Rapel point stands Bucalemo head, a bold cliff 200 feet high ; and 2 miles west of this head lies the Rapel shoal, sometimes, but erroneously, called Topocalma shoal.

* *See* West Coast of South America, Sheet VII., No. 1,282 ; scale, m = 0·12 of an inch.

**RAPEL SHOAL** extends nearly a mile, and shows three rocks above water, on which the sea breaks in all weathers, at nearly 2 miles off shore. Plenty of water exists all round, the soundings increasing from 10 to 50 fathoms gradually. Vessels should by no means approach the land in the neighbourhood of this shoal, as the heavy S.W. swell sets upon this dangerous part of the coast, as well as the prevailing current, which sometimes runs upwards of a knot an hour round Topocalma point towards the reefs.*

**TORO POINT** lies north from Bucalemo head ; close off it are a few rocks ; the water here is less deep than on the coast further southward, there being 15 fathoms at a mile from the shore. To the eastward of Toro point there is a bight, which contains a sand-bank lying a mile off shore. Possibly there may be some depth of water, if not a sheltered anchorage, between this sand-bank and the shore, but it was not examined by the *Beagle* in 1835.

**MAIPU RIVER.**—The coast trends north-eastward from Toro point 18 miles to White Rock point forming a slight bay, about half way the Maipu falls into it, having a bar across its mouth, and a dry sand stretches for nearly 2 miles to the northward parallel to the shore. Three miles farther north is San Antonio cove, a small place affording indifferent shelter to a few coasters, and immediately under a pointed hill. Two miles north of this hill there is a diminutive cove called La Bodega, which large boats frequent occasionally. At 13 miles to the eastward the land rises to a height of 3,280 feet. Cartagena beach, northward of La Bodega, is quite exposed to south-west winds. Tres Cruces point is low and rocky ; and 5 miles from it to the north-west is White Rock point, so called from the remarkable appearance of the white rock which forms a good land-mark.†

**In ALGARROBA COVE,** 4 miles north of White rock, small coasters find temporary shelter during southerly winds. About Algarroba point the coast is cliffy, but the cliffs are dark-coloured ; the land in the neighbourhood is high, and rather barren, of a dark colour, generally a brownish hue. In the distance, at about 90 miles, the Andes, stretching from north to south, show their majestic height, and appear much nearer than they are in reality.

**GALLO POINT** is a steep cliff, 7 miles north of Algarroba point ; between them are two sandy bights divided by a rocky point. At the

---

* *See* Nautical Magazine, June 1847.   † *See* View on Chart, No. 1,282.

corner of the northern bight, called Tunquen, and close under Gallo point, a boat might find shelter in a northerly wind, but there is no place for a sailing vessel. Steep cliffs extend 6 miles north of Gallo point. Quintay cove affords no good anchorage, though boats might take refuge in one corner.

CURAUMA HEAD, 3 miles farther north, is a remarkable promontory, and one that demands special notice, because it is generally the first land made out distinctly by ships approaching Valparaiso from the southward. The head itself is a high cliff; and above it the land rises steeply to the two high ranges of Curauma, the higher one being 1,830 feet above the sea, about 2 miles inland and north-east of the head. Usually, when first made out from seaward, the high part of the range of Curauma appears directly over the Head, and then, if tolerably clear weather, the Bell or Campana de Quillota, 6,200 feet high, is seen 30 miles off in the distance. If the Andes should be also visible, the volcano of Aconcagua, 60 miles beyond, will at a glance be distinguishable by its superior height, said by Captain Fitz Roy to be 23,200 feet.

CURAUMILLA POINT, projecting 4 miles W.N.W. from the high land over Curauma head, low by comparison with the neighbouring land, though not so really, is rugged and rocky; two or three islets lie close off it. Angeles point, the north-west extremity of the land forming Valparaiso bay, bears N.E. by N., distant 7 miles from Curaumilla point; between them is a deep angular indentation of the coast, bordered by scattered rocks on the west side, and steep cliffs on the east. A light on the hummock of Curaumilla point would be valuable by night, and the lighthouse would be a distinctive object to make out in thick weather on closing the land.

LIGHT.—Valparaiso light is a *fixed white* light, varied once a minute by a *flash*, shown at an elevation of 197 feet above the sea, and should be visible from the deck of a ship in clear weather at a distance of 20 miles. The lighthouse stands on Angeles point, the western point of the bay, at the end of a plain called the Playa Ancha, and is a white circular tower 50 feet high. This light is eclipsed by Caraumilla point until it bears to the eastward of N.N.E. ½ E.

VALPARAISO BAY, as will be seen by the plan, is of a semi-circular form and capable of accommodating a large fleet. It is well sheltered in the east, south, and west, but is entirely open towards the north; and during the prevalence of winds from this quarter in the winter season, accompanied, as they always are, by a heavy rolling sea, the shipping is much exposed, and serious accidents often take place. The town of Valparaiso lies at the foot of a range of *cuestas* or hills from 1,000 to 1,400

feet in height, on one of which there is a signal-staff, to give notice of
the approach of shipping. In 1832 the town consisted of one long
straggling street, not far from, and parallel to the beach. It has since
been enlarged and greatly improved; buildings have been erected, the
streets have been paved, and other improvements are still in progress.* A
pile of bonded warehouses has been constructed on San Antonio point on
the west side of the bay, and a battery and barracks placed on the heights
over it; a new and handsome custom-house faces the quay, which has
been carried farther out; repairs to vessels' bottoms are effected by heaving
down, which can be done with safety during the fine weather from September to May, and by a lift dock capable of taking up vessels of 1,000
tons burthen. A railway from Valparaiso, by the rich valleys of Quillota
and Aconcagua to Santiago, is in progress.

There is a bi-monthly steamer belonging to the Pacific Steam Navigation Company, which takes the mails to Panama in transit to
England, and calls at the *Puertos Intermedios*, as Coquimbo, Huasco,
Caldera, Cobija, Iquique, Arica, Islay, Pisco, Callao, Huacho, Casma,
Huanchaco, Lambayeque, Payta, Guayaquil, and Panama. The whole
voyage, including stoppages, is accomplished in rather more than 17
days, and the post to England is from 42 to 50 days.

There is also a communication once a month by a small steamer belonging to the same company with the ports to the south as far as San Pedro,
in the island of Chiloe; the principal places touched at are,—Constitucion, on the River Maule; Concepcion, Valdivia, Port San Carlos, and
Port Montt.

**Supplies.**—Water is supplied from tanks with force pumps, and is of
good quality. It is the best port for supplies and repairs on the west side
of South America; beef, vegetables, and stores of all kinds plentiful, and
comparatively cheap. In 1845, it was remarked that where the surf was
breaking in 1830 houses are now standing, and further observation confirms
the fact that the beach is gaining on the sea. In 1845, the town contained
about 30,000 inhabitants, but so rapidly has it increased since, that 50,000
is probably nearer the number in 1853. In 1845 the number of vessels
that entered this port was 859; in 1851 it had increased to 1,561, and in
1855 it amounted to 2,757. In 1845 the customs dues were 321,600*l.*;
in 1851 they had risen to 487,400*l.*; and in 1856 to 831,459*l.* The
value of the imports in 1851 was 3,176,800*l.*, and of exports 1,821,000*l.*;
and in 1856, respectively, 3,960,808*l.*, and 3,631,904*l.*, the great increase
being due to the discovery of gold in California, Australia, and British
Columbia.

* *See* Admiralty Plan of Valparaiso bay, No. 1,314; scale, m = 2·4 inches.

**Baja Rock,** a small rock above water, with deep water close to it, lies 300 yards E.N.E. from Angeles point. A ship may pass as near this rock as may be wished, and then steer into the bay. It is high water, full and change, in Valparaiso bay at 11h. 32m. ; the rise and fall being 5 feet.

**DIRECTIONS.**—All vessels bound to Valparaiso should endeavour to make the land about Curaumilla point, and in thick weather or in approaching the land at night, the greatest attention should be paid to the deep sea lead, as soundings may be obtained at from 2 to 6, and even in some place 12 miles off the land.*

During the morning and forenoon, although a vessel may have a fine breeze outside, she will generally lose it on opening the bay ; in this case the best course is to make the most use of the breeze by passing close round the Baja rock, and then steering direct for the shipping, shortening sail if the wind heads her and trusting to her way to gain the anchorage.†
After noon with southerly winds care must be taken to reef in time, for, however moderate and steady the southerly winds may be in the offing, squalls, which are not to be disregarded, blow down from the high land in the bay. When outside, the wind from the southward requires only a single reef in the topsails on a wind, probably treble-reefed topsails and foresail will be quite enough in the bay ; and when it is blowing strong in the offing from the same quarter, close-reefed topsails, over reefed courses, or over reefed foresail only, will be as much sail as can be carried. Should a ship find it blowing too hard to work up to an anchorage, she had better stand out and remain under easy sail off Angeles point till it moderates, which it does generally in a few hours.

In the event of a ship approaching with a northerly wind, likely to blow strong, she should keep an offing till the wind has shifted to the westward of north-west, which it always does after some hours of strong northerly winds. The best anchorage is close off Fort San Antonio, or in the south-west corner of the bay ; but occupied as that part always is, a ship must take as good a berth near the part as she can find. During summer, the closer in shore the better ; but during the winter, and if outside of other vessels, if it can be managed, so as to be safe from their driving during a northerly gale, which sends a heavy sea into the bay.

A norther as it is called, often passes over without doing damage, but at intervals the effects are most disastrous, and all the ill-secured or ill-placed vessels are driven ashore. Some prefer riding near the shore, on account of the undertow ; but in such a position there is more risk

* *See* Remarks of George Peacock, Esq., in Nautical Magazine for June, 1847.
† Remark book, H.M.S. *Havannah*, Captain Harvey, 1858.

of being fouled by driving vessels, besides feeling the sea considerably.
In the summer, southerly gales blow in furious squalls off the heights.
Clear weather and a high glass presage strong southerly winds; cloudy
weather, with a low glass, and distant land being remarkably visible, such
as the hill over Port Papudo, and the heights over Pichidanque, are sure
indications of northerly winds.

**CONCON ROCKS.**—The north-east side of Valparaiso bay is formed
by alternate beaches and rocks, as far as Concon Point, behind which
there is a cove where boats can land in moderate weather. The rocks
lie 3 miles N.N.W. from that point, and, though always above water,
should have a wide berth given to them during light winds, as there is
usually a swell and a northerly current setting towards them from the
southward.

**QUINTERO BAY,** distant 7 miles from the Concon rocks, is situated
round Liles point, which may be closely passed. It is roomy, and during
southerly winds sheltered; but quite open to the N.W., although some
shelter during northerly winds, may be found at the north-east corner
of the bay, under Ventanilla point, and also fresh water, when the
season is not very dry. This bay affords spacious and good anchorage in
the summer months; some even prefer it to Valparaiso; the best anchorage is in 13 fathoms, half a mile east of Liles point.[*]

**Tortuga Rocks.**—There is a little shoal or rocky patch at the south-
west corner, nearly 2 cables off shore, and 4 cables from the junction of
the cliff and sandy beach. This shoal, called Tortuga, does not dry, and
requires caution if hauling in near the shore. The land between this
bay and Concon is rather high and rugged, and all this coast has rather a
barren and weather-beaten aspect, here and there only any trees being
visible. During the winter and spring alone is there verdure near the
sea-coast.

**QUINTERO ROCKS** lie N. by W., 4 miles from Liles point and 1¼
miles west of Horcon head; they are above water, but low, straggling,
and dangerous; they are of a dark colour, and spread over half a mile of
space. Horcon head has a remarkable hole in the extreme point of the
cliff; the cliffs are dark-coloured, about 80 or 100 feet high, and the land
immediately behind them is higher and level. Inland are considerable
heights, and in the distance, at about 70 miles, the Cordillera of the Andes.

**HORCON BAY** is a landing-place between projecting rocks, situated
E.N.E. a mile from Horcon head; good water and plenty of fish may be

---

[*] *See* Admiralty Plan of Horcon and Quintero bays, No. 1,300; scale, m = 1·0 inch.

procured as well as fire-wood, and fresh provisions in small quantities. The roadstead is good during southerly winds, that is, in effect, during nine months out of the twelve; and there are 10 to 15 fathoms water half a mile north of the landing-place, over a clean sandy bottom. This bay was somewhat unaccountably omitted in all the Spanish charts of this coast.*

**PORT PAPUDO** is 13 miles from Horcon; between them there is no anchorage, the shore is steep, and free from out-lying dangers. The high pointed hill over Papudo, called El Gobernador, or the Cerro Verde, and 1,287 feet in height, is an unfailing land-mark for this small open bay. (*See* view on Sheet No. 1,282.) In the south-west part of the bay a small pier is run out, called the Muelle Frances, to facilitate the embarkation of corn, wood, and copper, which are shipped here for Valparaiso. Fish are to be had by the seine. There is a fresh-water stream close to the landing-place: wood and small quantities of fresh provisions may be obtained, but not cheaply.*

**Zapallar Point,** at the western extremity of the bay, is low, and must have a berth of nearly half a mile, in order to avoid the straggling rocks lying about the little isles off it called the Litis; the usual anchorage is 3 cables' lengths, N. by W. of the landing-place, at the southern part of the bay. It is safe during nine months of the year, but quite the reverse during the other three. The north-east point of the shore forming this bay is called Lilen; half a mile N.W. of it is Lobos island, to the eastward of which no vessel ought to pass.

**LIGUA BAY** lies 5 miles to the northward of Papudo, and behind a low rocky point, is the mouth of the River Ligua, not navigable; nor does the bay afford anchorage for any but the smallest craft, chiefly on account of the usually heavy swell.

Advancing northward, the projecting point named La Cruz de la Ballena, and Muelles point, require only a passing notice: they are steep and bold-to. The trend of the coast from Ligua river to the former point is W.N.W. for 5 miles, it then bends to the northward, and finally, to the westward towards Muelles point, which is low, dark-coloured, and rocky. The shore round Muelles bay is sandy, with low rocky points backed, as all the coast is, by high land. From Muelles point to Salinas point the southern extreme of Pichidanque or Herradura bay, the broken, dark-coloured, rocky shore runs nearly N. by W. ¼ W. for 8 miles. In

---

* See Plans of Horcon and Quintero bays, and Port Papudo, No. 1,300 ; scale, m = 1·6 inch.

sailing along this part of the coast, care should be taken to avoid a few outlying rocks which may be seen by day close to Salinas point, and those which lie half a mile off shore, 3 miles N. by W. ¼ W. from Locos islet at the entrance of Pichidanque.

PICHIDANQUE BAY is a blind place to make, but the Cerro de St. Ynez, a saddle-topped, conical, and conspicuous hill, overlooking the bay, is an excellent mark : it is 2,260 feet in height, and only 2 miles from the harbour.  This hill brought to bear S.E. by E. ¼ E. will lead right up to the entrance, which shows plainly on approaching the land.  The best anchorage here is close to the little island of Locos, on the eastern side, in about 5 fathoms water.*  See view on Chart No. 1,282.

Provisions.—Large quantities of corn are said to be produced here. Water could be had, but with some trouble.  Sheep and cattle plentiful, and 1½ dollar each for the latter.  Pichidanque is used occasionally for loading copper ore, or for smuggling affairs; there are only a few fishermen's huts near the harbour, but at the village of Quilimari, behind the nearest hills, supplies can be obtained.

Casualidad Rock is 60 feet long, by about 18 broad, has only 10 feet water on it, and lies N.E. 1¾ cables from the north end of Locos island. Care must be taken to avoid this dangerous rocky patch, as there is neither ripple nor weed upon it in fine weather, though it breaks when a swell sets in.  This danger, on which there is said to be only 9 feet at low water, is in a line between the north end of Locos island, and a gully at the north-east part of the bay, through which a river runs from the neighbouring village of Quilimari ; the southern end of the patch is within 2 cables' lengths of the islet.  As the point of Locos is bold-to on the north side, by keeping it close on board, the rock is avoided, and with the sea-breeze a good berth can be fetched by hauling sharp up.  It is high water, full and change, in Pichidanque bay at 9h.  The tide rises 5 feet at springs.

BALLENA POINT is 9 miles N. by W. ¼ W. from the same islet, dark-coloured, broken, and with an islet close to it ; it is the extreme point of a ridge of land, extending southward.  Two miles to the eastward there is a bight under Negro valley ; but it is rocky, and has no shelter for vessels larger than boats.  From Ballena point, the coast trends N. by W. ¼ W. to the low rocky point of Penitente ; between which and Cape Tabla lies Conchali bay, an exposed roadstead, seldom used, but by

* See Plan of Pichidanque and Ligua, No. 1,307 ; scale, m = 0·4 of an inch.

smugglers; the anchorage and landing are both bad except in one little cove at the north part of the bay.

**PENITENTE** is an out-lying rock awash at about 1¼ miles south of Cape Tabla, and N.N.W. ¼ W., nearly 3 miles from Penitente point; there are also two islets, and some rocks above water in the bay; and it is a very wild place, exposed to much swell.

**CAPE TABLA** is a projecting and dangerous point 4 miles to the north-west of Penitente. All the coast, except a few corners, is steep high, and barren, but picturesque in the outline. North of Cape Tabla there is an indentation of the coast, at the north-east corner of which there is a cove called Chigua Loco: in this bight there are two or three detached rocks above water: but it is not a place for any large vessel. From Chigua Loco to the River Chuapa, is a nearly straight piece of cliffy coast, extending N.N.W. 7 miles; and thence to Maytencillo is 22 miles in a similar direction without a break.

**MAYTENCILLO** is a little cove, fit only for *balsas*; at certain times a boat may land, but there are many hidden rocks. Its situation is pointed out by a large triangular patch of white sand having an artificial appearance, on the face of the steep cliffs which here line the coast; this mark is made by the sand that is drifted by the eddy winds, against the north side of the cove.*

Ten miles to the northward of Maytencillo, the two points of Vano throw several rocks off from them, and the whole interval of coast between them and that cove is composed of blue rocky cliffs about 150 feet high; the land above the cliffs rises to between 300 and 400 feet, and about 3 miles farther in-shore the range of hills runs from 3,000 to 5,000 feet in height. Sixteen miles to the northward of the cove lies the deep valley of Arenal with a sand-hill on its northern side close to the shore; and at the mouth of the valley there is a small sandy beach.

**MOUNT TALINAY** is a remarkable hill 2,300 feet high, about 3 miles in shore, in lat. 30° 51′ S.; it is thickly wooded on the top, but the sides are quite bare. The coast from Maytencillo cove may be said to extend in an unbroken line 33 miles N.N.W. to the next opening, which is that of the River Limari; and which looks large from seaward, but it was found to be inaccessible. The coast near the Limari is steep and rocky; and 2 miles from its mouth there is a low rocky point with a small beach, on

---

* *See* Plan of Maytencillo, and West Coast of South America, Sheet VIII., Nos. 1,307 and 1,287; scales, m = 0·4 and 0·12 of an inch.

which boats sometimes land, but through a heavy surf. The land rises suddenly to a range of hills about 1,000 feet high, which runs parallel to the coast, and extends 2 or 3 miles north and south of the river : the summits of the hills to the northward are covered with wood. The north entrance point is low and rocky, the south point is a steep slope with a conspicuous white sandy patch on its side. The river at its mouth is about a quarter of a mile wide, but the surf breaks heavily right across ; inside it turns a little to the north-east, and then again to the eastward through a deep gully in the range of hills before mentioned.

TORTORAL COVE.—About 14 miles to the northward of Limari, there is a small bay, with a sandy beach in the north corner, but a heavy surf. From this bay to the northward the east coast is rocky and much broken ; and about 8 or 9 miles farther we come to a small rocky penin- sula, with a high sharp rock rising from its centre. To the southward of it lies a small deep cove, with a sandy beach at the head, but the entrance is so blocked up by small islets and rocks, both above and below water, that it is impracticable for the smallest vessel, though in fine weather boats can get in and land in the cove. The outer breaker is not more than 2 cables' lengths from the shore, but when calm the swell sets directly on it. This cove is called the Tortoral della Lengua de Vaca.

The LENGUA DE VACA, 8 miles farther, is a very low rocky point, rising gradually in-shore to a round hummock 850 feet high about a mile to the southward of the point. There are rocks nearly awash, about a cable's length from the point, and at 2 cables' lengths there are but 5 feet water.

TONGOY BAY.—After rounding the Lengua, the coast turns short to the south-east, into Tongoy bay, and is rocky and steep for about 2 miles from the point, where there are 15 fathoms about half a mile from the shore. About 3 miles from the point a long sandy beach commences, which extends the whole length of that large bay as far as the Peninsula of Tongoy ; the southern part of the beach is called Playa de Tanque, and its eastern part Playa de Tongoy.

TANQUE.—Off the south-west end of the beach, near Tanque, there is anchorage about half a mile from the shore, in from 5 to 7 fathoms ; the bottom is a soft muddy sand in some places, but in others it is hard. With a southerly wind the bay is smooth, and the landing good, but a heavy sea sets in with a northerly breeze. This anchorage was once frequented by American and other whalers. The village, which is called the Rincon de Tanque, consists of about a dozen *ranchos*. The only water to be had is brackish; and about 2½ miles to the E.N.E., where there is good water; it is at some distance from the beach, and the landing there is generally very bad.

All the way from Tanque to the peninsula of Tongoy, there is anchorage in any part of the bay within 2 miles of the shore, in from 7 to 10 fathoms, sandy bottom. There is also good anchorage with a northerly wind, for small vessels, to the south-west of the peninsula, abreast of the small village on the point, with the Lengua bearing W.N.W. in 4 fathoms, sandy bottom, with clay underneath; but no vessel, however small, should go into less than 4 fathoms, as the sea breaks inside of that depth when blowing hard from the northward. Even large vessels might find a little shelter there with the wind to the northward of N.W. With a strong south-westerly breeze, the sea across the bay would render any vessel unable to remain at anchor in this berth south of the peninsula; but there is a small bay on its north side which is completely sheltered from southerly winds. In the south-west corner of this latter bay there is a small creek, into which when smooth, boats can go: it runs about a mile inland, and near its head there is fresh water, for which the whalers sometimes send their boats.

TONGOY VILLAGE consists of half-a-dozen small houses, built on a high point on the south side of the peninsula. The Mexican and South American Company had here a smelting work and large ore stations. The furnaces when at work can at night be distinguished at sea. The company have embanked part of the shore with copper slag and have made a mole alongside of which small coasters load and discharge. The company's copper ships are laden and unladen by launches. A screw steamer of the company runs from here to Herradura, and can be employed for towing. A trade in the shipment of ores to Herradura, America, England, and Hamburgh is carried on here by the company. There are two or three small stores, and ready access to Herradura. In summer time the place is frequented for bathing.

HUANAQUERO, 2 miles from the sea on the east side of Tongoy bay, is treble peaked, 1,850 feet above the sea, and forms a conspicuous mark on this part of the coast. The coast to the westward is broken and rocky, affording no shelter for anything but a boat; to the northward there is a deep bay, well sheltered from southerly and westerly winds, but open to the northward; between this point and Port Herradura, there is no place fit for a vessel.

HERRADURA.—From Huanaquero point it is 13 miles to the narrow entrance of Herradura de Coquimbo, a small land-locked harbour that is separated from Coquimbo bay by an isthmus of about a mile in breadth.

Vessels, however, of any size may freely enter with a leading wind by keeping the southern shore on board in order to avoid a rock off Miedo point; and when in, may anchor in any depth they please on a bottom of sand covering very tenacious marly clay. In the south-west angle they will find shelter from all winds, and the water so smooth that they may carry on any repairs with the utmost security. The *Beagle* lay there some weeks refitting, the crew being encamped on the beach.*†

A singular phenomenon was observed in this port, by R. E. Alison, Esq., H. M. Consul, on April 24th, 1858. At 7.30 a.m. a smart shock of an earthquake was felt, instantly followed by a sudden rise of the sea of 15 feet. The sea continued ebbing and flowing for 1 hour and a half, at intervals of 3 to 5 minutes, every flow being gradually lower than the previous one. The moon wanted 4 days of being full, and the rise took place one hour and a half before high water. Although the rise of the sea was 15 feet at the time, it was not more than 9½ feet above the usual high-water mark. On this occasion the Bay of Coquimbo was affected in a similar manner, the sea rising suddenly and covering the wharves and ground floor of the houses. Mr. Alison has witnessed several of these sudden convulsions; on one occasion a 1,400 ton ship, lying in 4½ fathoms, was left nearly dry on the first receding of the wave. No serious damage was done at either time to vessels lying in the port.‡

COAL.—The South Chile coal is used here in large quantities; it is shipped at 5 dollars and landed on the wharves at 9 dollars per ton, including all expenses. This land-locked and convenient bay would be of much importance if farther from Coquimbo, which has the advantage in size,

---

* *See* Admiralty Plan of the Bay of Coquimbo, No. 574; scale, m = 1·5 inches.

† In 1848 the Mexican and South American Company formed a large establishment at Herradura for smelting copper ores. A town of about 1000 inhabitants, English and Chilians, has been formed. A boat mole has been run out on the Whale islands, and a long mole, at which vessels of 300 tons can discharge, has been constructed to the northward. For large vessels discharging, launches and peons can be hired. An iron screw steamer belonging to the company is occasionally employed in towing ships in and out. The furnaces are constantly alight, so that the harbour is entered in the night by the company's ships. Herradura is a second-class port, and vessels have, therefore, first to enter Port Coquimbo for a pass. A large trade is carried on by the company coastwise for ores and Chile coals. The foreign trade consists of the import by them of English coals, bricks, clay, iron, &c., and the export of bar-copper, copper regulus, and copper and silver ores, to England, the United States, and Hamburgh. About 20,000 tons of shipping yearly frequent the bay. The banking operations are conducted by the company, who issue bills, on the coast, and on England. The water is brackish, but fresh water is regularly brought from Coquimbo bay. A large stock of coals is kept in the company's works.—*M & S. A. Co.*, 1856.

‡ Remarks of Captain T. Harvey, R.N., 1859.

but not in shelter; for during the winter months the northers are often severe, and send a heavy sea into the anchorage of that place, whilst in Herradura there is good shelter. The only disadvantage is that sailing ships sometimes find difficulty in getting out, as the entrance is narrow and the wind draws through it into the bay, which, with the heavy swell that rolls on the coast and the deep water outside, makes it rather difficult for a deep-loaded ship, unless with a leading wind, which seldom prevails more than a few hours in the forenoon and is mostly light and uncertain.*

COQUIMBO lies to the northward of the peninsula which forms the Port of Herradura. The Spanish survey of Coquimbo bay, as republished by the Admiralty, is correct, and if the lead be kept briskly going when approaching either the eastern shore or the bottom of the bay, it will be a sufficient guide, as the water shoals gradually towards the beach, which is low and sandy. In approaching this port, vessels must guard against being swept to the northward by the prevailing swell, current, and wind, which almost always come from the southward. The land is remarkable, and easily recognized by the views in the chart; and Signal hill, being upwards of 500 feet, can be easily made out at a moderate offing.

This is a much frequented port, though one great inconvenience attends it, which is, that the fresh water is not only not good, but difficult to procure, the watering-place being at a lagoon on the eastern side of the bay, wood is also scarce, and far from the anchorage. An aqueduct, however, is in course of erection, which will bring fresh water into Coquimbo from Serena. Plenty of fish may be caught with the seine; and fresh provisions are cheap and plentiful. There is no landing at La Serena, the town of Coquimbo, in consequence of the heavy surf, except in *balsas*; but its distance is only 6 to 7 miles from the landing-place under Signal hill, where horses and conveyances are readily obtained, and where a small town has sprung up, with a convenient mole, to the westward of which is a smelting establishment, and a wharf for loading and unloading copper, in which metal the hills in the neighbourhood are very rich. The mail steamer calls twice monthly, and the coasting steamer thrice monthly. A steamer of the Mexican and South American Company occasionally tows ships in and out.

LA SERENA, the town of which Coquimbo is the port, is situated on the north-east side of the bay, the road lying along the beach. The houses are mostly of sun-dried bricks, and only one story in height; so built in consequence of the earthquakes to which all Chile is subject. The town

---

* Remarks of Commander J. Wood, R.N.

and its gardens are supplied with water by canals cut from the river on its
northern side ; it is remarkable for its extreme quietness. Population 7,000.

The WEATHER at Coquimbo is so uniformly fine, the climate so
charming, and the atmosphere so clear, as to have given the city the name
of La Serena. The country, however, frequently suffers from want of rain.
In 1853 (June) they had been 15 months without a drop. Some notion
of the value of a shower under such circumstances may be gathered from
the fact that in the small valley of Huasco (the next port to the northward) a night's rain was estimated as worth a million of dollars for that
district alone ; indeed the effect is almost miraculous. Before rain the
whole country is a barren sandy desert, looking as hopelessly arid as the
African Sahara ; in a week or 10 days after, the ground is covered with
verdure and flowers.

There is one point that should be noticed, which is the prevalence
of thick weather here, and all along this coast as far as the river Guayaquil,
particularly during the winter months. These fogs are frequent and
sometimes very dense ; they are a great source of anxiety to the navigator.
This is especially the case on the Peruvian coast, where clear weather is
an exception to the general rule.

Pelicanos Rock.—The western shore of the bay is high and bold, particularly at its northern end, off which lies this insulated rock, having 4½
fathoms within a boat's length of it. About 60 yards to the northward
is a patch with 22 feet water on it, and 6 fathoms between it and the
rock. On the point there is a platform with 2 guns, and a hut that
answers the purpose of a guardhouse, but they are scarcely visible,
having much the appearance of the rock on which they stand.

Pajaros Niños, are 2 rocky islets with reefs and outlying rocks
around them, situated to the north-west of the Pelican rock ; the
outer one being about 1¼ miles off shore. It is necessary in going
in to give these islets and rocks a good berth in case of falling calm
lest the vessel should be obliged to anchor, the ground near them being
rocky ; and for those reasons all vessels are advised to pass outside of these
rocks.

A conical rock with only 8 feet water on it, on which two vessels, the
P. S. N. Co. steamer *New Granada* and British barque *Chelydra* have
struck, lies in a line between the Pelicanos rock and the city of Serena ;
it was searched for by Captain T. Harvey, R.N., without success : in
sounding, a patch with 6 fathoms was found, on which the lead would not
rest, with 9 and 10 fathoms around it, and 12 fathoms between it and the
shore. This patch lies N. by E. 1¼ cables from the Pelicanos rock. It
is high water, full and change, in Port Coquimbo at 9h. 15m. ; the
rise and fall being 5 feet.

**DIRECTIONS.**—Vessels bound to Coquimbo after rounding the Pajaros Niños, should keep off the rocky ground between them and the Pelicanos rock by steering E. by S. till a house on the isthmus opens of the cliff under the signal hill ; then steer in for the anchorage. This port may easily be entered at night in consequence of the constant light shown by the furnaces. Coming from the southward run along within 3 miles of the land until the lights at the town of Serena bear E. by N., steer for them on this bearing, which will clear the Pajaros Niños, and when the furnace lights open of the Pelicanos rock the vessel can haul into the bay and anchor in 8 fathoms.*

The usual anchorage for strangers is in 8 fathoms, with the extreme north point of the western shore N.W. ¼ W., the church at La Serena (or town of Coquimbo) N.E., and the houses near the landing place S.W. ¼ W. The best anchorage is in 6 fathoms, in the south-west angle of the bay, and the holding-ground excellent ; but a swell usually rolls in and produces such a surf along the beach that landing is difficult, except in a few sheltered spots. The winds at Coquimbo are in general moderate and southerly, or chiefly off shore during the greatest part of the year, and are interrupted for short intervals only in winter by strong breezes from the north-west.

**VARIATION.**—The variation of the compass along the extent of coast comprised in this chapter, namely, from Chiloe to Coquimbo, or from 42° to 30° S. lat., is as follows :—About the middle of Chiloe 19° ; at Valdivia 18° ; at Concepcion 17°, at the river Maipu 16°, and at Tortoral Cove 15°. The annual increase is small, being about 1'. The curves of equal variation strike the coast and extend across South America in about a south-east direction, and the degrees of variation are about 180 miles apart.

---

* Remarks of Captain T. Harvey, and Mr. Thomas A. Hull, Master, R.N. 1859.

# CHAPTER XII.

## COQUIMBO BAY TO LAVATA BAY.

VARIATION 15° to 13° East in 1860.   Annual increase about 1'.

---

**TEATINOS POINT**, the northern extreme of Coquimbo bay, is bold and rugged, the land behind it rising in ridges, which gradually become higher as they recede from the coast to Cobre hill, which is 6,400 feet high.   The point which makes the north extremity of the bay in coming from the northward, is a low rocky point, called Poroto; about 4 miles to the northward of which is the cove of Arrayan, or Juan Soldado, but it does not deserve any name, it being merely a small bight, behind a rocky point, scarcely affording shelter for a boat from southerly winds, and entirely open to northerly.

**MOUNT JUAN SOLDADO.**—A little to the northward of Cobre hill is another mountain in the same range 3,900 feet high, called Juan Soldado; its northern side is steep, and at its foot lies the small bay of Osorno, which is about half a mile long, but it would not afford any shelter for the smallest vessel: about half a mile to the northward of the bay there is a hamlet, consisting of a few small houses, called Yerba Buena.

**The PAJAROS** are two low rocky islets, lying about 12 miles from the coast; the northernmost is much smaller than the other, and as well as could be seen there is no danger round or between them.   A little to the northward of Yerba Buena, there is a small island, called Tilgo, separated from the shore by a channel about a cable's length broad, but it is only fit for boats.   The island, except when very close, appears to be only a projecting point; there is a large white rock on its west point.

**TORTORALILLO BAY** lies about 3 miles to the northward of Tilgo island, it is formed by a small bay facing the north-west, with three small islands off the west point.   In coming from the southward, the best entrance for small vessels is between the southernmost island and the point, where there is a channel about a cable's length wide, with from 8 to 12 fathoms water; the dry rock off the point on the main land should not be approached nearer than half a cable's length, as a sunken rock lies nearly that distance from it.   There is no channel between the islets, as the space is blocked by breakers. Temblador is a small cove to the north-east of Tortoralillo, but the landing there is worse than on the other beach, and it is not so well sheltered.

A vessel may anchor about half a mile from any part of the beach, in from 6 to 8 fathoms, sandy bottom ; the landing is not good ; the best is on the rocks near the entrance, but nothing could be embarked from thence ; the east end of the beach is the best for that purpose. From the land about Choros running so far to the westward, it is not likely that a heavy sea would be caused by a northerly gale.*

The village of Tortoralillo is chiefly dependent on the Mexican and South American Company, who own a large copper ore station here and small smelting works, and carry on a trade in the shipment of ores for Herradura, Caldera, England, Hamburgh, and the United States ; a small steamer belonging to the company runs occasionally to Herradura, and one of their officers resides here. They have a small mole for the shipment of ores, and launches and peons can be hired.

CHUNGUNGA ISLAND, lies about 4½ miles to the northward of Tor-toralillo at about a mile from the shore, and it is a good mark for knowing the little cove of the same name : there is a rocky point abreast of it, and a little way in-shore a remarkable Saddle hill, with a nipple in the middle, which to a person coming from the southward appears as the end of the high range that runs thence to the eastward of Tortoralillo, and is from 2,000 to 3,000 feet high. A little to the northward of Chungunga, there is a large white sand patch, which is seen distinctly from the westward ; it is at the south end of the Choros beach, which runs for 7 or 8 miles to the north-west to Cape Choros ; a heavy surf always breaks upon it.

OF CAPE CHOROS there are three islands ; the inner one is low, and so nearly joins the shore that nothing but a boat can pass. The channel is clear of danger between this island and the other two Choros islands ; the southernmost of them is the largest, being about 2 miles long ; the top is very much broken, and the south-west end very much resembles a castle ; there is a small pyramid off the south point, and rocks break about a quarter of a mile from the shore. The channel between the two outer islands is also clear of danger ; but about half a mile to the west-ward of the northern island there is a rock nearly awash.

TORO REEF lies 5 miles S.S.E. of the southern Choros island ; it is dangerous, being only a little above water. Cape Carrisal is low and rocky, about 7 miles to the northward of Cape Choros, with a remarkable round hummock ; to the southward of it is the small cove of Polillao, where there is shelter for small vessels, but the landing is bad ; there are two small rocky islets off the south point of the cove.

**Carrisal Bay** lies to the northward of the cape, but it is not fit for sea-going vessels : a heavy surf breaks about half a mile from the shore. The north side of the bay is formed by a rocky point, with outlying rocks and breakers all round it ; there is a landing-place in the bay near the south-east corner, where the rocky coast joins the beach, but in bad weather the surf breaks outside of it.

**CHAÑERAL BAY** is situated nearly a mile to the northward of the north point of Carrisal Bay ; it is well sheltered from northerly and southerly winds, but the swell sets in heavily from the south-west, which makes the landing bad. The best landing is in a small cove on the north side near the beach ; there is also a landing-place on the north side of the bay, but it is bad when there is any swell. On the beach, in the bight of the port, there is always too much surf to land, except after very fine weather.*

**CHAÑERAL ISLAND** lies about 4 miles to the westward ; it is nearly level, except on the south end, near which there is a remarkable mound, with a nipple in its centre. There are rocks nearly half a mile from the south point of the island, and one about the same distance off the north-west point. On the north side there is a small cove, where boats can land with the wind from the southward; and there is anchorage close off it, but the water is deep. An American sealing schooner was lost in this cove from a norther coming on while she was at anchor.

The land round Chañeral bay is low, with ridges of low hills rising from the points ; their tops are very rugged and rocky, and the land is sandy and very barren. A range of high hills will be seen several miles from the shore, but between them and the coast there are several smaller hills springing out of the low land. The village of Chañeral is about 3 miles from the port, and is said to consist of about twenty houses ; there are none near the port.

**LEONES CAPE** is about 3½ miles N.W. by N. from the north point of Chañeral bay ; the coast between is low, and, falling back, forms a small bay. Cape Leones has several rocks and reefs extending from it to the

---

* *See* Plan of Chañeral on Sheet VIII., scale m = 20·7 of an inch.

distance of a mile ; there is also a reef which projects nearly a mile from
the shore a little to the northward of Chañeral bay.

CAPE VASCUÑAN.—From Cape Leones the coast projects N. by
W. ¼ W., 4½ miles to Pajaros point, and from thence about north
4 miles to Cape Vascuñan ; this cape has a small rocky islet off it about
2 cables' lengths from the shore.  The land in-shore rises gradually to a
low ridge about half a mile from the sea ; the high range is about 3 miles
in-shore.  From Cape Vascuñan the coast runs to the north-eastward,
forming a small bay, open to the northward, but well sheltered from
southerly winds : there is anchorage in from 8 to 12 fathoms about a third
of a mile from the shore, but the landing is bad.

Sareo Bay, in which there is some shelter from southerly winds, lies to
the eastward of Cape Vascuñan : a deep gully runs inland from the south-
east corner of the bay, at the mouth of which there is a sandy beach, with
anchorage a third of a mile off, in from 8 to 12 fathoms, but the landing is
not good.  There are two or three small huts close to it.  To the northward
of Sareo, the high land comes close to the coast ; the sides of the hills are
covered with yellow sand ; the summits are rocky, and the whole coast has
a miserable, barren appearance.  To the northward of the deep gully, about
4 miles, there is a projecting rocky point at the foot of the high range of
hills, with a very remarkable black sharp peak near its termination ; the
coast to the northward of this runs nearly north and south, and is very
rocky for about 8 miles, when it turns to the westward, forming a deep
bay, in the north-east corner of which is a small beach called Tontado.

ALCALDE POINT is a rocky promontory forming the seaward termination of a projecting spur of the coast range ; the point is rocky, with
small detached rocks close to it ; in-shore it rises a little, and there are
several small rocky lumps peeping out of the sand, one of which from the
southward shows very distinctly ; it is higher than the rest, and forms a
sharp peak, a little in-shore of which the land rises suddenly to the
break of the high range.

HUASCO POINT, forming the south-western extreme of the bay or port
of the same name, is low and rugged, with several small islands between
it and the port, one only of which is of any size, and it is separated from
the shore by a very narrow channel, so as to appear from seaward to be
the point of the mainland ; it shows distinctly coming from the southward, but from the northward it is mixed with the other rocks behind
it : to the south-west of the island there are several other small rocky
islets.  Round this point to the eastward, is the outer roadstead, in which

there is no good anchorage, the water being very deep, and the bottom in most parts of it rocky.

A little in-shore of the extreme point, there is a short range of low hills, forming four rugged peaks, which show very distinctly from the southward and westward ; the land falls again inside of them for a short distance, and then rises suddenly to a high range, running east and west, and directly to the southward, of the anchorage. The top of the range forms three round summits, the easternmost of which, being 1,900 feet, is a little higher, and the middle one a little lower than the other ; they all form part of the Cerro del Huasco.*

The country around presents a more barren appearance than any part of this coast ; the ground being everywhere covered with small stones, mixed with sand, out of which project masses of craggy rocks. A little in-shore, the stony ground changes into a loose yellow sand, which covers the sides and bases of nearly all the hills round ; the summits are stony, without any appearance of vegetation ; but in the low grounds a few stunted bushes grow among the stones, and after rain (which is rare), they look fresher than might be expected in such soil, and the valley through which the river runs, then also appearing green, forms a striking contrast to the country around.

HUASCO PORT, is to the eastward of a second and inner point lying 2 miles from Huasco point, having 2 large rocks off it in an N.W. direction. It is a blind and inconvenient anchorage, it is, however, one of the places at which the P. S. N. Company's steamers call, as it is the port of Ballenar, a considerable town in the interior. There is a custom-house, 20 or 30 houses, and a smelting establishment belonging to Mr. Hardy, half a mile from the landing place, where a good deal of copper is manufactured, and with certain kinds of ore sent to England. The valley, after the rains, is as green and fertile as any spot in England, and Mr. Hardy's hacienda at Huasco bay, two miles from the port, may boast of as pretty a garden and as cool shady lanes as are to be found anywhere in the south of Chile.

On the night of the steamer's arrival a light is shown above the pier, which should be brought to bear S.E., but care must be taken not to mistake for this the flame from the chimneys of the smelting works which show a redder and more uncertain light, as they would lead a vessel too near to the beach and on to the rocky ground which lies abreast them.†

Nearly 3 miles to the N.E. of the outer port, there is another range of hills about 1,400 feet high, on the south slope of which there is a sharp

* See Plan of Port Huasco, No. 575 ; scale, m = 2·0 inches.
† Remarks of Commander J. Wood, R.N., 1853.

peak, rising immediately above a valley that conveys a small stream of excellent water to the sea ; a heavy surf breaks on its bar.   There is a narrow lagoon, or small streamlet in the valley, nearer the port, but the water is very brackish.   The anchorage is much exposed to northerly winds, and a heavy sea then rolls in, but a mischievous norther does not occur more than once in two or three years.

**LOBO POINT,** about 10 miles to the northward of Huasco, is rugged, and carries several small hummocks; to the southward of it there are some small sandy beaches, with rocky points between them, but a tremendous surf that breaks throughout allows no shelter even for boats.   A little in-shore of the point there two low hills, and within them the land rises suddenly to a range of about 1,000 feet in height.   In the bay to the northward of Lobo point there are several small rocks, and about 6 miles from it there is a reef which runs perhaps half a mile off a low rocky point ; the outer rock of which is high and detached from the others.

**HERRADURA DE CARRISAL.**—About 11 miles to the northward of Lobo point is another rugged point with several sharp peaks on it, the highest reaching 3,050 feet, and half a mile beyond them lies the small bay of Herradura, which can hardly be made out till quite close. Between it and Herradura, (which is distinguished from the other Herraduras on the coast of South America by the additional name of de Carrisal,) there are breakers a quarter of a mile from the shore.*

Off Herradura point there is a patch of low rocks, which, in coming from the southward, appears to extend right across the mouth of the bay; but the entrance faces the north-west, and lies between that low patch and a small islet to the north-east of it, and there is no danger within half a cable's length of either of them.   The bay curves in about three-quarters of a mile to the eastward of the islet, and is sheltered from both northerly and southerly winds, but with a strong northerly breeze a swell rolls in round the islet ; it is rather small for large vessels, and they would not be able to lie at single anchor in the inner part of the cove, but there is room enough to moor across at about a quarter of a mile above the islet in 4 fathoms, fine sand.

In this place, an American ship, the *Nile*, of 420 tons, was moored during a northerly gale, which blew very heavily, and she was quite sheltered ; the landing is better than in any place between it and Coquimbo, but the want of water is a very serious inconvenience.   There

* *See* Chart, West Coast of South America, Sheet IX., No. 1,726 ; and Plan of Herradura de Carrisal, No. 1,315 ; scales, m = 0·12 and 4·0 inches.

is a small lagoon about a mile off, in the valley at the head of Carrisal cove, but it is worse than brackish, yet the peons, who work at shipping the ore, make use of it. A deep valley which runs in from the head of the cove, and separates the high ranges of hills, is a good mark to know the place. The range to the southward of the valley is the highest near the coast, and is distinctly seen, both from the northward and southward: there is a small nipple on the highest part of it.

CARRISAL COVE is small, about a mile to the north-east of Herradura, well sheltered from southerly winds; but as it is so close to Herradura, which is very much superior, it is not likely to be of much use. To the northward of Carrisal, the coast is bold and rugged, with outlying rocks a cable's length off most of the points. About 7 miles farther north there is a high point, with a round hummock on it, and several rugged hummocks a little in-shore. To the northward of it, a cove was found, sheltered from the southward, where small vessels might anchor, but not fit for large vessels: and another cove similar to it about a mile farther to the northward.

MATAMORES.—A little to the northward of the second cove, a high rocky point terminates the high part of the coast; and north of that point there is a small port, which, from the natives, appears to be called Matamores; it is well sheltered from southerly winds, and the landing is good. In the inner part of it, a vessel not drawing more than 10 or 12 feet might moor, sheltered from northerly winds, in 3 or 4 fathoms, but with a northerly wind there would be a heavy swell. There is anchorage farther out under the point, in from 8 to 10 fathoms, but a vessel should not go nearer the shore there than 8 fathoms, as the bottom inside of that depth is rocky; during the summer months this would be a good port for small merchant vessels, but there is no appearance of fresh water. Abreast of Matamores the high range of hills recedes to four miles inland, where a high point reaches 2,450 feet; the coast is low, with some moderately high rocky hills a little in-shore.

TORTORAL BAXO is a low rocky point about 2 miles farther north, a little to the northward of which there is a small deep bay, at the mouth of the valley Tortoral Baxo or Low (which distinguishes it from Tortoral cove, in latitude 30° 20' S.) In this bay apparently there is anchorage for a vessel, but from the heavy surf on the beach, and the bad landing, it was not examined. To the northward of this, the low hills are not rocky, but covered with yellow sand, except near the summits. White rock, about 6 miles farther north, is a remarkable rocky point, with a detached white rock off it, and a hump with a nipple on it a little in-shore.

[S. A.]　　　　　　　　　　　　　　　U

**PAJONAL COVE** lies about 1½ miles to the northward; in coming from the southward, it may be easily known by the above nipple, and by a small island, with a square-topped hillock in its centre, off the point to the northward of the cove. A range of hills, higher than any near this, rises directly from the north side of the cove; and in the valley, about a mile from it, there is a range of small and very rugged hills, rising out of the low land.*

**The ANCHORAGE** is better sheltered from southerly winds than any to the southward, except Herradura, and there ought not to be much swell, as the point and island to the northward project considerably to the westward. The southerly swell rolls into the mouth of the cove, but along the south shore it is smooth, and the landing pretty good. There is a dangerous breaker about a quarter of a mile W. by S. of the south extreme point, but it shows only when there is much swell. The best anchorage is about half way up the cove, near the south shore, in 5 fathoms : near the head it is shallow.

In 1835 the officers of the survey found a cargo of copper ore there ready to be shipped, but no vessel had yet been in the cove ; no water was to be had within 2 miles, and there it was very bad. The name of Pajonal was given to it by a young man who was getting the ore down ; but he appeared to know scarcely anything of the coast, and there were no inhabitants near the place.

**SALADO BAY.**— At Cachos point, about 1½ miles farther north, the coast turns to the eastward, forming the spacious bay of Salado, and close round the point the large cove of Chasco, which, at a distance, looks very inviting, but a mile from its head there are only 3 fathoms, with rocks all round, some above water and others sunk, which, from the bay being well sheltered from the southward, do not show. There are two patches off the north point that are always uncovered. A vessel must not attempt to pass in shore of the islet off Cachos point, nor of Square-topped island.

**Middle Bay** is another recess, a mile to the northward, and is quite clear of danger ; in the south corner is a small cove, where there is good anchorage in 7 fathoms, well sheltered from southerly winds, but very open to northerly : the water is quite smooth with a southerly wind, and no swell could ever reach it unless it blew from the northward. There is a small bay, half a mile to the northward of this, where a vessel may anchor, but it is not so well sheltered; there are no signs of inhabitants, nor the least appearance of fresh water in the valleys.

---

* *See* Pan of Pajonal, No 1,315 ; scale, m = 2·0 inches.

**SALADO POINT.**—The land at the back of the bay is low, but to the northward of the north bay it rises to a ridge of sand hills, running east and west, and terminating at this steep rocky point, with a cluster of steep rocky islets off it.  To the northward of this point the coast is rocky and broken, with rocks a short distance from the shore for about 4 miles ; then, a rugged point, with a high sharp-topped hill a little in-shore, which, from the southward, shows a double peak.

**BARRANQUILLA.**—Directly to the northward of this point there is a deep rocky bay, with a small cove close to the point ;  the *Beagle* anchored there in 5 fathoms, but only half a cable's length from the shore on either side.  It appears unfit for any vessel, for though the bay is partly sheltered from northerly winds the northerly swell rolls in.  From an old fisherman, who was living there in a hut, it appears that the name of the place is Barranquilla de Copiapó, and it was surprising to see a cargo of copper prepared for shipping in such a place : but he also told us that a cargo had been shipped from the same spot about a year before ; yet the cove is far too small for any vessel to anchor in it with safety, and outside the water deepens very suddenly.  There is no anchorage in the cove at the head of the bay, and the landing there is very bad ; but in the small cove the landing is good.  Middle bay is much superior to Barranquilla, and might be a much better place to embark the ore.  There is no fresh water nearer than the river of Copiapó, which is about 12 miles off.

**DALLAS POINT** is all black rock, with a hummock on its extremity, and coming from the southward it appears to be an island ; the land rises to a range of low sandy hills, with rocky summits.  From Barranquilla to this point the coast is rocky and broken, without any place sufficient to shelter the smallest vessels.  Detached reefs extend from Dallas point for nearly 4 miles in a N.N.E. ¼ E. direction, a small sharp-topped rock called the Caxa Chica (small chest) being the only one that shows above water.  The patch near Dallas point is awash ; and the channel between it and the point appears to be wide enough for any vessel, though the reef off the point projects so far as to show in a high sea a breaker above a quarter of a mile out ; but at a quarter of a mile farther there are 11 fathoms.  When the swell is not high the breakers off the point would not show ; they appeared to be detached from the reef which joins the point.  *See* view on Chart, No. 1,276.

**COPIAPÓ,** inside of these ledges, is a bad roadstead, the swell rolls in heavily, and the landing is worse than in any port to the southward.  It may easily be known by the Morro, a hill 850 feet high, 10 miles to the northward, which is very remarkable, nearly level at the top, but near its eastern, extremity there are two small hummocks.  The east fall is very steep ; the end of another range of hills shows to the northward.  To the south-west, apparently forming part of the same range, stands another hill, the west side of which forms a steep bluff.  In coming from the southward these hills will be seen in clear weather before the land about the port can be made out.

This port, rightly termed by Captain Fitz Roy a very bad one, is now deserted ; the large and daily increasing value of the Copiapó silver mines, and the trade thus created, called for better accommodation, which fortunately existed as near to the mines as the old port, the only wonder being that with two such good ports as Caldera and Port Ingles within reach, such a dangerous anchorage should have been used at all, although the difficulty of making the place has been much exaggerated.*†

**Anacachi Rock.**—The chief dangers to be avoided in entering the harbour of Copiapó are the Caxa Grande and Caxa Chica shoals ; and between these and Dallas point several other small but dangerous patches of rock, on one of which, the *Anacachi*, a Chilian brig, was wrecked. It lies about half a mile N.W. ½ W. from the Caxa Chica, and W. ½ S., distant 3 miles from Copiapó flagstaff, and carries only 10 feet at low water.

**Caxa Grande,** the outer of the reefs extending from Dallas point, is the northernmost of the first two mentioned, is a bed of rocks under water, about three quarters of a mile long, and a third of a mile broad, and lying nearly in a north and south direction ; its situation is apparent from the heavy breakers on it whenever a swell sets into the bay.

**Caxa Chica,** 2¼ miles from Dallas point, is a small rocky shoal, having in its centre one large rock always above water.  It lies south of the Caxa Grande, with a passage between them of nearly a mile in breadth, though appearing much less, from the rollers which extend sometimes across it on the Caxa Grande side.

**Isla Grande.**—The island to the northward of Copiapó bay, called Isla Grande, is very remarkable, having a small nipple on each extremity ; that on the eastern end is the highest ; and just to the westward of the middle of the island there is another small round nipple.

The channel between Isla Grande and the Main is clear of danger in the middle ; but such a heavy swell rolls through that it is scarcely fit for any vessel.  Off the north end of the island there is a reef

---

\* *See* Admiralty Plan of Copiapó, No. 1,315 ; scale, m = 1·0 inches.
† Remarks of Commander J. Wood, R.N., 1853.

projecting under water two cables' lengths to the eastward ; but at one cable's length from that reef there are 8 fathoms. The main land abreast of the island appeared to have no danger off its points ; and the rocks to the southward of it are inside the line of the northern points. The swell in the channel was by far the worst the *Beagle* had experienced on this coast. To the northward of the island there are several small rocks ; one of which is high, but there is no danger within a quarter of a mile of them. It is high water, full and change, in Copiapo, at 8h. 30m., with a rise and fall of 5 feet.

**DIRECTIONS.**—The passage between Caxa Chica and Caxa Grande, though often used, is very dangerous on account of the Anacachi. The Caxa Chica should be given a berth from 4 to 6 cables' lengths, but unless the wind is steady, and to be depended on, it should not be taken on any account. The flag-staff above the town of Copiapó, bearing East, leads through the passage. Neither should the passage between Dallas point and the southern shoals ever be taken by a sailing vessel ; as, should the wind fail, which when so near the high cliffs in the vicinity is a common occurrence, a ship would be placed in a very dangerous position.

The obvious and best passage is to the northward of the Caxa Grande ; and to avoid those rocks when coming from the southward bring Isla Grande to bear N.E., and steer for it on that bearing till the northern end of the sandstone rocks, to the northward of the town of Copiapó, bears at least E. by S. ; then haul in for that mark, and when the flag-staff above the town of Copiapó bears S.E. ½ E., steer towards it, and anchor where convenient. Should the flag-staff, which is small, not be quickly seen, a large house in the town, remarkable from its bright green roof, which is of copper, and always visible if brought on the same bearing, will be an equally good mark.

Coming from the northward, vessels will most probably have to work in ; in which case the shore may be approached to half a mile, and Isla Grande to within that distance ; and when approaching the Caxa Grande stand no nearer to it, or any of the shoals, than to bring the western end of Isla Grande to bear N.N.W., or the bluff part of Dallas point S.S.E. Should the wind be from the northward, the flag-staff on a S.E. ½ S. bearing will lead up to the anchorage in from 12 to 6 fathoms. A large scope of cable should always be given in this road, and it might be prudent to drop another anchor under foot, as the rollers often set in with very little warning, and the bottom is bad holding-ground. The soundings are very regular, from 12 fathoms to 3 close up to the beach, but the bottom is chiefly a hard yellow sand, with occasional patches of yellow sandstone rock. Several vessels have been driven on shore here from their anchors by the rollers setting in suddenly.

When the Morro de Copiapó, which is so high as to be seen 30 or 35 miles in clear weather, is open of Isla Grande, the vessel will be well to the westward of all the dangers off Copiapó.

**ANCHORAGE.**—The in-shore anchorage for a large vessel is in 5 fathoms, with the following bearing, viz., the Caxa Chica W.S.W., western extremity of Isla Grande, N.W. ¾ N., the jetty (or landing-place) S. ¼ E., and the flag-staff over the town S.S.W. ¼ W.

**MEDIO POINT**, on the main to the northward of the island, is very rocky : on the S.W. point there are two rugged hummocks, and several rocks and islets close to the shore, but no danger outside of them. From this to the Morro the shore is steep and cliffy, with remarkable patches of white rock in the cliffs to the southward of the point, which is steep, with rugged lumps on its summit. The Morro rises suddenly a little in-shore.

**YNGLES BAY** is a deep bay opening to the south-east, after passing Morro point; there are several small rocky patches in it, and at the north end of the long sandy beach there is a piece of rocky coast, off the extreme point of which there is a small island.

**PORT YNGLES** is just to the northward of that island, round the peninsula of Caldereta, off which there is a rock awash at high water about a cable's length, but it always shows : after passing this rock the land is steep-to, and may be approached within a cable's length. The harbour inside forms several coves, in the first of which, on the starboard hand going in, there is anchorage for small vessels, but the bottom is stony and bad : there is a low island to the eastward of this cove, and about half-way between it and East point will be found the best anchorage with southerly winds. Small vessels may go much closer into the bight to the south-east of the island, where the landing is very good.*

The bay in the north-east corner is well sheltered from northerly winds, and no sea could ever get up in it, but the landing is not good; the best there is at a rocky point at the south end of the north-east beach, where in a small cove among the rocks the water is perfectly smooth. This north-east cove is by far the best in the harbour, but it has no fresh water. The South cove is too shoal for a vessel of any size to go higher up than abreast of East point, where she will have 4 and 5 fathoms in mid-channel. The bottom in this bay is hard sand, and may be seen in 12 fathoms water, which makes it appear shallow. In the entrance there are 18 fathoms close to the shore on both sides.

---

* *See* Admiralty Plan of Ports Caldera and Yngles, No. 1,302; scale, m = 2·0 inches

**PORT CALDERA** is close to the northward of Port Yngles, and is directly round the island off Caldera point : it is a fine bay, pretty well sheltered, but more open than Port Yngles.   The land is entirely covered with loose sand, except a few rocks on the points : the bottom of the bay is low, but the hills rise a little inland, and the ranges become higher as they recede from the coast.   To the eastward stands a very remarkable sharp-topped hill, the sides of which are covered with sand, with two low paps near it.   The northers the great enemies of all the harbours and roadsteads of Chile, blow into the bay, throwing a good deal of sea into the southern angle ; but as Caldera is near their northern limit they are seldom of sufficient strength to be dangerous to a properly found vessel ; the point of Cabeza de Vaca, though 12 miles distant, is also some protection, and in the north-east corner, which they call the Calderillo, it is always smooth.   There are fish to be had in the bay, but only with a net ; and in none of the ports visited by the *Beagle* were any caught alongside. Near the outer points of these two ports rock-fish are to be caught, but there is always a heavy swell in such places.

**Light.**—The P. S. N. Companies have managed to get a small wooden beacon carrying a light erected on the top of the hill of Caldera point ; a larger and better one was to be put up as soon as ready upon the point itself, where it would be more clear of the fogs and a better guide to vessels entering at night.

The port of Caldera is one of the most rapidly improving places in South America.   A fine mole is run out from some rocks at the south-east angle of the bay into water deep enough to admit of four ships of 1,000 tons and upwards lying alongside discharging and loading their cargoes (a further extension is contemplated) ; for this purpose the railway which is continued to the pier end offers the greatest convenience.   This rail goes up to the city of Copiapó (about 60 miles distant) and crosses the track of the mules from the silver mines ; it is supplied with excellent American locomotives and cars, and is under very able and energetic management ; the communication is twice a day.   The great difficulty to be overcome was the want of water, but as no lack of enterprise or means existed, this was obviated by a large distilling apparatus erected at the station, which also partly supplies the wants of the town now rising, as if by magic, from the desert of rock and sand which forms the shores of the bay ; it is laid out in squares, and many of the houses already built are large and handsome.

Captain Fitz Roy ascribes, with justice, the chief merit of these improvements to Mr. Wheelwright, a gentleman to whose energy, enterprise, and ability the west coast of South America is indebted for most valuable plans and suggestions, in a great majority of cases successfully carried out ; but the unexpected and extraordinary yield of the silver mines seems at present the principal source of prosperity in this place.*

The Mexican and South American Company have a large copper and silver smelting work at Caldera.  The company carry on a large trade in the import of copper and silver ores from the coasts of Chile, Bolivia, and Peru ; and in the export of copper and copper ores, and of silver regulus and silver ores to England, Hamburgh, and the United States. Fresh beef, &c., can be had from the interior by rail, and in a short time water also, as it was proposed to bring it down by pipes to the end of the pier.  Coal can be procured at from 12 to 15 dollars per ton.

**The CABEZA DE VACA** is a remarkable point, about 12 miles to the northward of Caldera : it has two small hummocks near its extremity; inside of them the land is nearly level for some distance, and then rises into several low hills, which form the extremity of a long range.  The coast between Caldera and the Cabeza forms several small bays with rocky points between them, off all of which there are rocks a short distance: there is no danger within a quarter of a mile of this point.   To the northward of it there is a small rocky bay called Tortoralillo,  and off its north entrance point a reef,  which extends about a quarter of a mile from the shore, with a high rock at its extremity : about half a mile to the northwest of this,  some  heavy  breakers  were  seen  whenever  there was much swell.

**OBISPITO and OBISPO COVES.**—To the northward the coast is steep and rocky for 3 or 4 miles, with a high range of hills running close to the shore ; then, a small cove called Obispito, with a white rock on its south point ; and to the northward of this the land is low and very rocky, with breakers about a quarter of a mile from the shore.   About 2 miles from that cove there is a point with a small white islet off it ; to the northward of which the coast trends to the eastward, and forms the small cove of Obispo, in which there is a high sand-hill, with a stony summit.  H.M.S. *Beagle* anchored in this cove, but it is not fit for any vessel, and the landing is bad.  A little in shore of the cove, and to the northward, a higher range of stony hills runs close to the coast for about 7 miles, where it terminates in low rugged hills in-shore of a brown rough point off which is an islet, appearing from the sea like a white patch on the point.

**PORT FLAMENCO,** lying to the northward of this patch point, is a good port, well sheltered from southerly winds, and protected from the

---

* Remarks of Commander J. Wood, R.N., 1853.

northward, as the point projects far enough to prevent a heavy sea running in.

On the north side of the bay the land is very low ; the north point is low and rocky, with a detached hill rising out of the lowland a little in-shore.   To the northward there is another hill very much like it ; in the depth of the bay the land is very low, and a deep valley runs back between two ranges of rugged hills.   The hills are all covered with yellow sand from their bases, to about half-way up their sides ; the tops are stony, with a few stunted bushes.

The landing is good in the south-east corner of the bay, either on the rocks, or on the beach of a small cove in the middle of a patch of rocks a little more to the northward, where, in 1835, there were a few huts in which two brothers with their families were living in a miserable way, in huts made of seal and guanaco skins, much worse than a Patagonian *toldo :* the only water was half salt, and lay some distance from the shore.   Their chief employment was catching and salting large conger eels, and drying them to supply Copiapó ; in one day they had caught 400.   Occasionally they catch guanacos, running them down with dogs, of which they have a great number.*

The only vessel they had ever seen there was a ship which anchored one night on her way to Las Animas for copper ore, six years before ; they described Las Animas as a very bad place, not fit for any vessel, and in consequence no cargo had ever been shipped there again, but taken to Chañeral, which was better, but not so good as Flamenco : there are no mines so near Flamenco as to Chañeral.

**LAS ANIMAS.**—In the bay to the northward of Flamenco, in which Las Animas was said to be, no place could be seen in 1835 even fit for a boat to land ; the whole bay is rocky, with a few little patches of sand, and a heavy surf always breaking on the shore.   The north point of this bay is low, but a little in-shore there is a high range of hills, the outside of which is very steep; and to the northward of this point there is a small rocky bay, which appears to answer better to the description of Las Animas than the other ; it did not, however, appear to be a fit place for vessels, and the landing was bad.   The north point of this bay is a steep rock, with a round hill rising directly from the water's edge ; the sides of the hills are crossed by dark veins running in different directions, and are very remarkable.

**CHAÑERAL,** from description, must be the deeper bay to the northward ; the south side of it is rocky, with small coves, but the

* *See* Plan of Port Flamenco, No. 1,302 ; scale, m = 1·0 inch.

landing appeared to be bad; the east and north shores are low and sandy, and a heavy surf was breaking on the beach. There were no signs of people, nor any heaps of ore along the coast; and as it did not appear a good place for vessels, it was not thought worth bestowing any time on a particular examination; the north point of the bay is low and rocky, with a high range a little in-shore. To the northward of this point the hills and coast are both composed of brown and red rocks, with a few bushes on the summits of some of the hills; the sandy appearance that the hills have to the southward ceases, and the prospect is, if possible, still more barren.

**SUGAR LOAF ISLAND**, 600 feet high, stands nearly 9 miles to the northward of the bay of Chañeral, about half a mile from the shore. In coming from the southward, there is a similarly shaped hill on the Main, a little to the southward of the island, for which it may be mistaken; but the island is not so high and the summit is sharper. Between Sugar Loaf island and Chañeral the coast is rocky and affords no shelter; but there is a small bay to the southward of the island, which might afford some shelter from northerly winds, though with southerly it would be exposed, and the landing is very bad.*

**ANCHORAGE.**—In the middle of the passage between the island and the Main there are 5 fathoms in the shallowest part; the water in its northern end is smooth, and a vessel might anchor off the point of the island, sheltered from southerly winds, in 6 or 7 fathoms; but outside of 8 fathoms it deepens suddenly to 13 and 20 fathoms about half a mile from the island. There is a small bay on the Main to the northward of the channel, where a vessel might apparently be sheltered from southerly winds.

**BALLENA POINT.**—About 19 miles to the northward of Sugar Loaf island there is a projecting point, with some small rocky islets; it was supposed to be Ballena point, from the description given to the surveyors at Port Caldera. Between it and Sugar Loaf island the rocky coast falls back a little, with a range more than 2,000 feet high, running close to the shore. A little to the northward of Ballena point there is a small bay with a rocky islet, about half a mile off the south point of it. The top of the islet is white, and answers the description given of a port called Ballenita, but it is not worthy of the name of a port; two or three small patches of sandy beach, on which a heavy surf was breaking, appeared on the rocky shore, and the hills, which come close to the water, have a very rugged appearance.

* *See* Plan of Sugar Loaf isle, No. 1,302; scale, m = 1·0 inch.

**LAVATA BAY** is a little to the northward of this; the south point has several low rugged prongs from it, and in-shore the hills rise very steeply. There is a small cove, in which the *Beagle* anchored, with excellent landing, directly behind this point; and there was a still better-looking port inside, but it was far from the outer line of coast, and her time would not allow of more than a hasty glance. The outer cove in which she anchored, appeared to afford good shelter from southerly winds, and the water was very smooth.*

**TORTOLAS ISLETS.**—A little to the northward of Lavata there is a point which, till close, appears to be an island; but it is joined to the shore by a low shingle spit; its summit is very rugged, with several steep peaks from it; and several rocky islets that lie scattered off the point are named the Tortolas.

**POINT SAN PEDRO** lies nearly 3½ miles to the northward of them, very rugged, and with a high round hummock a little way in-shore. To the eastward of this point there is a deep bay, in which it was expected to find the town of Paposo, according to its position given in the old charts, but there was no appearance of any houses or inhabitants.

**YSLA BLANCA BAY** is very rocky, and does not afford good anchorage; several rocks lie off San Pedro point, and inside of it there is a reef projecting half a mile from the shore; in the bottom of the bay there are several small white islets; and two or three small sandy coves, none of which are large enough to afford shelter for a vessel. About three miles from Taltal point there is a white islet, with some rugged hummocks upon it; and a little way in-shore there is a hill of much higher colour than any in the neighbourhood.

**HUESO PARADO.**—Between Taltal and Grande points the shore recedes, forming a long bay, in which it was thought that Paposo would certainly be found, and as the *Beagle* was becalmed, a boat was sent to search for it. Two fishermen were met with who stated that the place was called Hueso Parado, and that Paposo was round another point about 17 miles to the northward. On inquiring for water, they brought some, which was better than that tasted in some other places to the southward, but still it was scarcely fit for use; they said it was as good as that at Paposo, and *they* thought both good. In the south corner of the bay there appeared to be anchorage for vessels and the landing good, but very open to northerly winds. No vessel had ever been there in the recollection of those men, neither had they heard of any.

---

* *See* Plan of Lavata bay, No. 1,302; scale, m = 1·0 inch.

On this uninhabited coast it was impossible to ascertain the exact spot where the republic of Chile terminates and that of Bolivia begins ; but according to the best information, afterwards obtained, the line of limitation comes down to the shore in the bight of Hueso Parado, or somewhere between it and the point of San Pedro.

**TIDES and CURRENTS.**—The only place on this part of the coast at which the time of high water was satisfactorily determined was at Huasco, where it is high water, full and change, at 8h. 30m. ; the rise 4 feet at neap tides, and at springs about 2 feet more. From the swell on all this coast it is very difficult so get the time of high water at all near the truth ; the rise and fall everywhere appeared to be 5 or 6 feet. The tide at Huasco was very carefully observed in a cove without swell, yet from the small rise the exact time of high water could not be ascertained, the water remaining at the same level above half an hour.

The only perceptible current was in the channel between Sugar Loaf island and the Main, where a very slight stream sets to the northward, but not more than a quarter of a mile in an hour ; and this was after a fresh breeze from the southward for several days. It is said, however, by coasters, that there is usually a set towards the north of about half a mile an hour.

**WINDS.**—Very few words will suffice to give strangers to the coast of Chile a clear idea of the winds and weather that they may expect to find there, for it is one of the least uncertain climates on the face of the globe.

From the parallel of 35° S., or thereabouts, to near 25° S., the wind is southerly or south-easterly during nine months out of the twelve ; in the other three there are some calms and light variable breezes, but the remainder is really bad weather,—northerly gales and heavy rains prevailing not only on the coast, but far across the ocean in parallel latitudes.

From September to May is the fine season, during which the skies of Chile are generally clear, and, comparatively speaking, but little rain falls. It is not, however, meant that there are not occasional exceptions to this general rule : strong northers have been known (though rarely) in summer ; and two or three days of heavy rain, even with little intermission, now and then disturb the equanimity of those who have made arrangements with implicit confidence in the serenity of a summer sky. These unwelcome interruptions are more rare, and of less consequence, to the northward of lat. 31° than south of that parallel ; and indeed so nearly uniform is the climate of Coquimbo that the city, as already mentioned, is called La Serena.

In settled weather a fresh southerly wind springs up a little before noon (an hour sooner or later) and blows till about sunset, occasionally till midnight. This wind is sometimes quite furious in the height of summer, so very strong that ships are often prevented from working into their anchorages, such as Valparaiso bay, although they may have taken the previous precautions of sending down topgallant-yards, striking topgallant-masts, and reefing their sails. But the usual strength of this sea breeze (as it is called, though it blows along the land) is such as a good ship could carry no more than double-reefed topsails to it when working to windward.

This is also nearly the average strength of a southerly wind in the open sea, between the parallels above mentioned; but there it is neither so strong by day, nor does it die away at night. Within sight of land a ship finds the wind freshen and decrease nearly as much as in the ports, where the nights are generally calm till a land-breeze from the eastward springs up; but this light message from the Cordilleras is never troublesome, neither does it last many hours. With these winds the sky is almost always clear; indeed, when the sky becomes cloudy, in summer, it is a sure sign of little or no sea-breeze, and probably a fall of rain: in the winter it foretells an approaching northerly wind with rain.

In summer ships anchor close to the land, to avoid being driven out to sea by those strong southerly winds; but as the winter approaches a more roomy berth is advisable, though not too far out, because near the shore there is always an undertow, and the wind is less powerful. Seamen should bear in mind that the course of the winds on this coast, as in all the southern hemisphere, is from the north round by the west; that the winds which blow the hardest, and bring the most sea, come from the westward of north; and that therefore they should get as much as possible under the shelter of rocks or land lying to the westward, rather than of those which only defend them from north winds. Northers, as they are called, give good warning: an overcast sky, little or no wind unless easterly, a swell from the northward, water higher than usual, distant land remarkably visible, being raised by refraction, and a falling barometer, are their sure indications. All northers, however, are not gales; some years pass without one that can be so termed, though few years pass in succession without ships being driven on shore on Valparaiso beach. Thunder and lighting are rare. Wind of any considerable strength from the east is unknown. Westerly winds are only felt while a norther is shifting round, previous to the sky clearing and the wind moderating. The violence of southerly winds lasts but a few hours; and even a northerly gale seldom continues beyond a day and a night, generally not so long.

Some persons say that the strength of northerly winds is not felt to the northward of Coquimbo, but there is good evidence of many gales, with heavy seas, at Copiapó ; and Captain Eden states that he had a very heavy gale of wind in H.M.S. *Conway*, in lat. 25° S., and long. 90° W., where such an interruption to the usual southerly winds was little expected.

**PASSAGES.**—There are but two ways on the coast of Chile to make passages. When going to the northward steer direct to the place, or as nearly so as is consistent with making use of the steady winds which prevail in the offing ; and if bound to the southward, steer also direct to the place, if fortunate enough to have a wind which admits of it ; but if not, stand out to sea by the wind, keeping every sail clean full, the object being to get through the adverse southerly winds as soon as possible, and to reach a latitude from which the ship will be sure of reaching her port on a direct course. Every experienced seaman knows that, in the regions of periodic winds, no method is more inconsistent with quick passages than that of "hugging the wind." Juan Fernandez island may be made with safety, being high and bold.

Ships coming from Peru, southward, during the northers, should be careful not to get to the southward of their port. Vessels standing in for the shore during the summer months, when about 100 miles from the coast, will often find the wind heading them, in which case they need not tack as the wind will haul to the westward on approaching the land.

In concluding this chapter on the coast of Chile, it may be convenient to the seaman to have a brief description of the islands of Juan Fernandez and Mas a fuera, which belong to that republic.

**JUAN FERNANDEZ ISLAND** lies about 360 miles to the westward of Valparaiso, and is generally called by the Spaniards Mas a tierra, or nearer the mainland. It is of an irregular form, 3,000 feet high and about 12 miles in length, but scarcely 4 miles across in its widest part. Its north-eastern half is composed of alternate craggy ridges and fertile valleys, and mostly covered with wood, while the southern division, which is comparatively flat and low, is nearly barren.* When seen from a distance the mountain El Yunque (the anvil), so called from its shape, appears conspicuously placed in the midst of a range of precipitous mountains ; it is wooded nearly from the summit to the base, whence a fertile valley extends to the shore.

**Supplies.**—Fresh water is good and easily procured ; wood can be purchased ; beef of excellent quality, pigs, poultry, vegetables, and fruit

---

* *See* Admiralty Plans of Juan Fernandez and Cumberland bay, Nos. 1,383 and 1,344 ; scales, m = 0·7, m = 2·9.

can be obtained at a moderate cost. Peaches grow wild in large quantities; wild goats are numerous, and the bay has fish in great abundance. In 1856 a German was superintending the island for the Chilian Government, several families, amounting in all to 50 persons, residing under his care, cultivating the land, stacking wood, and attending to the cattle, which are fine and fast increasing. During the last 3 years some 25 American whalers have called annually for supplies.*

**Cumberland Bay** is on its northern shore, about 3 miles from its eastern extremity; it affords safe anchorage for vessels of any size. A vessel approaching this roadstead from the southward may pass round either end of the island according to circumstances; but she should not approach the shore nearer than a mile, to avoid the eddy winds down the valleys. In the event of taking the eastern end, which is the best, when off Bacalao point, Cumberland bay will open out, and also some caverns which will be seen on the face of the first high land rising from the beach; and when the western part of West bay shuts in with the western point of Cumberland bay, deep soundings will be obtained. The ship should anchor with the next cast in about 25 fathoms, on a bottom of fine clear sand, about a quarter of a mile from the beach.

The marks for the best berth are, a small rock off the west point of the bay, which just shows at high water, N.N.W. ¾ W., the easternmost of the caverns, of which there are six in number, about S.E. ¾ E., or the flagstaff on the fort W.S.W. In approaching the bay from the westward, keep about a mile from the shore, and when the caverns over Cumberland bay are well open, run in towards them, and anchor as above directed.

All vessels visiting this bay should moor, placing the in-shore anchor in about 16 fathoms, and the off-shore anchor in 35 fathoms, the ship will then be in a good berth with a southerly wind, and have room to veer should the wind come in from the northward. In the summer season, southerly winds prevail, and at times heavy gusts rush down from the valleys; consequently the bottom being of sand, a good scope of cable on the southernmost anchor is necessary. A kedge astern will serve to keep the hawse clear.†

**TIDES.**—It is high water, full and change, in Cumberland bay at 9h. 30m.; rise of tide about 4 feet.

**MAS-A FUERA ISLAND** lies about 92 miles to the westward of Juan Fernandez. It is covered with trees, and has several falls of water pouring down its side into the sea, but offers no convenient anchorage. There

---

* Remarks of Captain T. Harvey, R.N., 1856.
† Remarks of Mr. J. Penn, Master R.N., 1845.

is a sort of bank on its north side, but it is steep-to, and the water on it so deep, that if a vessel did anchor, she must necessarily lie close to the shore and be exposed to all winds but those from the southward.

**ST. AMBROSE and ST. FELIX ISLES** lie off Copiapó, from which they are distant nearly 500 miles. In former years immense herds of seals frequented their shores, but these have almost all disappeared; their pursuit is no longer profitable; guano has been collected here to a small extent.[*]

St. Ambrose, the easternmost island, is about 4 miles in circumference, and 1,500 feet high. On the north side near the centre is a snug little cove for a boat, good landing at all seasons with a southerly wind, and fresh water of an excellent quality. A remarkable rock, resembling the Bass of the Frith of Forth, lies off the east end of this island, with small rugged rocks to the eastward of it again; there is a pinnacle also off the west end.

St. Felix lies 11 miles west from St. Ambrose, and consists of two islands connected by a reef, presenting from the southward the appearance of a double-headed shot. The west and south-west sides of the northern island are steep cliffs, sloping down to beaches on the north-east side; there is a place for landing just to the eastward of the north-west bluff. The southern island is inaccessible, and about 600 feet high. About 1½ miles W. by N. from the north point of St. Felix is a remarkable islet, which has been named Peterborough cathedral from its form; it lies in lat. 26 16′ 12″ S., and long. 80° 11′ 43″ W.

**VARIATION.**—The variation at the beginning of the year 1860 at Juan Fernandez was about 17° E.; off Tongoy and at St. Felix and St. Ambrose 15° E.; off Port Huasco 14° E.; off Port Caldera 13° E.; and off Hueso Parado 12° E. The curves of equal variation on this part of the coast assumé a S.E. by E. direction (true), and the degrees are about 150 miles apart. The variation is increasing at the rate of about 1′ annually.

---

[*] *See* Admiralty Plan on Sheet IX. West Coast of South America.

# CHAPTER XIII.

## HUESO PARADO TO CALLAO BAY.

VARIATION 13° to 11° East in 1860.   Annual increase about 1.'

NUESTRA SEÑORA BAY.—At Taltal point, near Hueso Parado, is said to be the southern limit of Bolivia and of the desert of Atacama, and from thence the coast sweeps to the northward round the Bay of Nuestra Señora to Grande point, a distance of 17 miles.   This point, which rises 1,570 feet above the sea, when seen from the south-west, appears high and rounded, terminating in a low rugged ridge, with several hummocks on it, and surrounded by rocks and breakers to the distance of a quarter of a mile. At 9½ miles farther north, lies Rincon point, having off it a large white rock, and between those two points, in the lat. of 25° 20′ S., lies the village of Paposo.   It is a poor place, containing about 200 inhabitants, under an Alcalde; the huts are scattered, and difficult to distinguish, from their being the same colour as the hills behind them.*

Vessels touch at Paposo occasionally for dried fish and copper ore; the former plentiful, but the latter scarce.   The mines lie in a south-east direction, 7 or 8 leagues distant; but are very little worked.   Wood and water may be obtained on reasonable terms; the water is brought from wells 2 miles off, but owing to the swell which constantly sets in on the coast, it is difficult to embark.

DIRECTIONS.—Vessels bound for Paposo should run in on a parallel of 25° 5′, and when at the distance of 2 or 3 leagues the White rock off Rincon point will appear, and shortly after the low white head of Paposo. The course should be immediately shaped for the latter; and on the bearing of S.S.E., distant half a mile, they should anchor in from 14 to 20 fathoms, sand and broken shells.   Should the weather be clear (which is rare) a round hill, higher than the surrounding hills, and immediately over the village, is also a good guide.   *See* view on Chart No. 1,277.

PLATA POINT, lying N.N.W., 23 miles from Grande point, is similar to it in every respect, being 1,670 feet high, and terminating in a low spit, off which lie several small rocks, forming a bay on the northern side,

---

* *See* West Coast of South America, Sheet X., No. 1,277; scale, m = 0·12 of an inch.

[S. A.]                                                                              X

with from 17 to 7 fathoms water; rocky, uneven ground. From this point to Jara head, which lies N. ¾ W., 52 miles, the coast runs in nearly a direct line; a steep rocky shore, surmounted with hills from 2,000 to 2,500 feet high, and without any visible shelter, even for a boat.

JARA HEAD is a steep rock, with a rounded summit, and has on its northern side a snug cove for small craft; it is visited occasionally by sealing vessels, who leave their boats to seal in the vicinity; water is left with them, and for fuel they use kelp, which grows there in great quantity, as neither of these necessaries of life is to be had within 25 leagues on either side. Mount Jaron, 4 miles to the eastward of Jara head, is 3,990 feet above the sea.

MORENO BAY, or La Playa Brava, commences nearly 4 miles N. ¼ E from Jara head; the intermediate land being high and rocky, with a black rock lying off it. The coast of this bay runs North for 14 miles to Bolfin point, terminating in Jorge bay under mount Moreno.

MOUNT MORENO, formerly called Monte Jorge, stands on a peninsula forming the western side of Jorge bay and is the most conspicuous object on this part of the coast; its summit is 4,160 feet above the level of the sea, inclined on its southern side, but to the northward ending abruptly over the barren plain from which it rises. It is of a light brown (*moreno*) colour, without the slightest sign of vegetation, and split by a deep ravine on its western side. Tetas Point, the south-west point of Moreno peninsula, slopes gradually from the summit of Mount Moreno, and terminates in two nipples, whence its Spanish name.

CONSTITUCION HARBOUR, a small but snug anchorage, formed by the mainland on one side, and by Forsyth island on the other, lies immediately under Mount Moreno, about 5 miles north of Tetas point. Here a vessel might haul in to the land, and careen without being exposed to the heavy rolling swell which sets into most of the ports on this coast. The landing is excellent; and the best anchorage is off a sandy spit at the north-east end of the island, in 6 fathoms water, muddy bottom; but it would be advisable to moor securely, as the sea breeze is sometimes very strong. Farther out the holding ground is bad; and when running in, the island or weather side should not be hugged too closely, as a number of sunken rocks lie off the low cliffy points, some only being buoyed by kelp. A mid-channel course would be the best, provided the wind allows a vessel to reach the anchorage before mentioned.* N. by W., 12 miles, Monte Jorgino, a steep

* *See* Admiralty Plan of Constitucion Harbour, No. 1,301; scale, m = 1·0 inches.

bluff, terminates the range of table-land which extends from Monte
Moreno ; on the northern side of this headland lies the Bay of Herradura
de Mexillones, a narrow inlet, running in to the eastward, without
affording any shelter.

**LEADING BLUFF.**—North 9 miles from Monte Jorgino, is Low point
surrounded with sunken rocks ; and 5 miles farther north is Leading
Bluff, a very remarkable headland, which, with the hill of Mexillones,
a few miles to the south, is an excellent guide for the port of Cobija.
The bluff is about 1,000 feet high, and faces the north, and being entirely
covered with guano, it has the appearance of a chalky cliff.   There is an
islet about half a mile to the north-west of the bluff, and attached to it
by a reef ; but there is no danger of any description outside.

**MOUNT MEXILLONES** is 2,650 feet high ; it has the appearance of
a cone with the top cut off, and stands conspicuously above the surrounding heights.   In clear weather this is undoubtedly the better of the two
marks ; but as the tops of hills on the coast of Bolivia are frequently
covered with heavy clouds, Leading Bluff is a surer mark, as it cannot
be mistaken ; for, besides its chalky appearance, it is the northern extremity of the peninsula, and to the eastward of it the land suddenly falls
back.

**Mexillones Bay** is the spacious bay to the eastward of the bluff.
The shore is steep-to ; but there is anchorage on the western side, 2
miles inside the bluff, and a cable's length off a sandy spit in 7 fathoms,
sandy bottom ; at the distance of 8 cables' lengths there are 30 fathoms.

**COBIJA BAY.**—From Mexillones the coast runs nearly north and
south, without anything worthy of remark, as far as the Bay of Cobija,
or as generally called Puerto la Mar, which lies N. by E. 30 miles from
Leading Bluff.   It is the only port of the Bolivian Republic, and contains
about 1,400 inhabitants.   Vessels call occasionally to take in copper ore
and cotton ; but the trade was very small in 1835, in consequence of the
recent revolution in Peru.*   There has, however, been a slight improvement of late years, a quay, barracks, and custom-house having been built,
and the number of vessels frequenting the port being apparently on the
increase.

Good water is very scarce as rain never falls here ; an occasional rill
(caused by condensed fog) runs down a ravine to the northward of the
town, but so small that a musket barrel is sufficient to convey it to the
reservoir of the inhabitants ; that generally in use is distilled.   There

* *See* Admiralty Plan of Cobija Bay, No. 1,301 ; scale, m = 2·0 inches.

are wells, but the water from them is very brackish, and will not keep
in casks.  Fresh meat may be procured, but at a high price ; fruit and
vegetables, even for their own consumption, are brought from Valparaiso, a distance of 700 miles.  There is a mud-built fort of 5 or 6 guns,
on the summit of the point.

The only means of transport with the interior being by pack mules,
prevents any extensive exports of ore or wool.  From the top of the
ridge, which rises directly over the coast to a height of 3,000 feet, there
is a desert extending to the eastward for 135 miles, without water or
refreshment of any kind.  The muleteers cross this in 2 days.  To Potosi,
distant 540 miles, it takes 14 days, but the Indians, on foot, by relays of
men, carry the mail in 6 days.

This place was the only port assigned to Bolivia, when the boundaries
of the several states forming the Columbian republic were settled at the
time of their separation.  Geographically viewed, both Iquique and Arica
appear to belong of right to Bolivia, as the line of demarcation which
crosses the lake of Titicaca nearly east and west, would, if continued in
the same line, have included the valley and city of Tacna, with its
accompanying port of Arica ; instead of which, to preserve these two
important places to Peru, it was made to turn at right angles along the
ridge of the Andes, and run south for many miles, when it joins the river
Loa, and again bends to the coast, thus giving Peru a strip of coast but
a few miles deep, and depriving Bolivia of two ports, one of them only
second to Callao in capacity and importance.

**DIRECTIONS.**—A white stone lies on the slope of Cobija point, which
shows very plainly in relief against the black rocks in the back ground ;
and a white flag is usually hoisted at the fort when a vessel appears in the
offing, which is also a good guide.  In going in there is no danger, the
point being steep-to, may be rounded at a cable's length, and the anchorage is good in 8 or 9 fathoms, sand and shells.  In the bay there are a
number of straggling rocks, but they are all well pointed out by kelp.
Landing at all times is indifferent, and owing to the heavy swell it requires some skill to wind the boats in through the narrow channel formed
by rocks on each side.  The long strong kelp is also dangerous if a boat
should be swamped, in 1854 five men belonging to the French brig of
war *Obligado* were drowned here.

It requires care, however, to make this place, the hills rise directly from
the coast and form an almost unbroken ridge of from 2,000 to 3,000 feet
high, having no sufficiently marked feature to point out the position of
the town at their base ; the white-topped rock would be a good mark were
there not now a precisely similar rock some miles to the northward.
One of the best marks is a kind of gully, or indentation in the hills, a

little to the northward of the town, and another is the road winding up the valley from the port, but neither of them is easily distinguished.

If coming from the southward towards this bay, after having passed Leading Bluff (which should always be made), it would be advisable to shape a course so as to close the land about 9 miles to southward of the port, and then coast along until two white-topped islets, off False point, are seen ; 1¼ miles to the northward of them is the port.* *See* view on Sheet No. 1,277. It is high water, full and change, in Cobija bay at 10 o'clock, and the tide rises 4 feet.

COPPER COVE is 2 miles to the N.N.E., a convenient place for taking in the ore ; there being anchorage in 12 fathoms, a short distance from the shore. After passing the north point of Cobija bay, off which lies a number of straggling rocks at a short distance, the coast takes about a N. ¼ W. direction : generally shallow sandy bays with rocky points, and hills from 2,000 to 3,000 feet high close to the coast, but no anchorage or place fit for shipping until you reach Algodon or Cotton bay, 28 miles from Cobija.

ALGODON BAY, is small, and the water deep : the *Beagle* anchored a quarter of a mile from the shore in 11 fathoms, sand and broken shells, over a rocky bottom. One use of this anchorage is that you may send from it for water to the Gully of Mamilla, 7 miles to the northward. The spring there is 1½ miles from the beach ; and the usual method of bringing it is in bladders made of seal skin, holding 7 or 8 gallons each, with which most of the coasters are provided, the only vessels that profit by a knowledge of these places.† Algodon bay may be distinguished by a gully leading down to it, and by that of Mamilla to the northward, which has two paps 4,020 feet high over its north side. There is also a white islet off Algodon point.

CAPE SAN FRANCISCO, or PAQUIQUI, is a projecting cape, 10 miles farther north, having on its northern side, and near its extremity, a large bed of guano. A brig of 170 tons was loading with it for Islay in 1835, she was moored head and stern within a cable's length of the rocks, on which a considerable surf was breaking, and the guano was brought off in a *balsa* to a launch just outside the surf. There is better anchorage farther in the bay ; but the other is chosen for convenience. Several vessels were loading here in 1859.

---

* Remarks of Commander James Wood, R.N. 1853.

† *See* Plan of Algodon bay and West Coast of South America, Sheet XL., Nos. 1,301 and 1,278 ; scales, m = 2·0 and 0·12 inches.

**ARENA POINT,** low and sandy, lies N. ¼W., 16 miles from Cape Paqui-que, with a rocky outline; between the two is a small fishing village, near a remarkable hummock.    Anchorage may be obtained under Arena point, in 10 fathoms, fine sandy bottom.

**LOA RIVER** and gully, forming the boundary line between Bolivia and Peru, is N. ¾ E., 12 miles from Arena point.   It is the principal river on this part of the coast ; but its water is extremely bad, in consequence of running through a bed of saltpetre, as well as from the hills surrounding it containing copper-ore.   It is said that the ashes of a volcano also fall into it, which may add to its unwholesomeness ; but bad as it is, the people residing on its banks have no other.   In the summer season it is about 15 feet broad and a foot deep, and runs with considerable strength to within a quarter of a mile of the sea, where it spreads, and flows over, or filters through the beach ; but does not make a channel, or throw up any banks.   A chapel on the north bank, half a mile from the sea, is the only remains of a once populous village.   People from the interior visit it occasionally for guano, which is abundant.

The best distinguishing mark for the Loa is the gully through which it runs : and that may easily be known from its being in the deepest part of the bay formed by Arena point on the south and Lobo point on the north ; as well as from the hills on the south side being nearly level, while those on the north are much higher and irregular.   *See* view on Chart No. 1,278. There is good anchorage, but rather exposed to the sea-breeze, with the chapel bearing north, half a mile from the shore, in from 8 to 12 fathoms, muddy bottom ; and landing may be effected under Chileno point, 3 miles to the southward.

**CHIPANA BAY,** 6 miles N.N.W. from Loa river is a better anchorage, and there is a snug cove for landing near the tail of the point ; but at full and change of the moon a heavy swell sets in and a boat would scarcely be able to land with goods at those times.*

After making the land in the latitude of the Loa, a large white double patch may be seen on the side of a hill near the beach, and a similar one a little to the northward : on discovering these marks (which are visible), a course should be shaped directly for the southern point, where lies the anchorage in 7 fathoms, sand and broken shells, sheltered by low level ground.   No danger need be feared in entering ; for though the land is low, it may be approached within half a mile, in from 10 to 6 fathoms.   The anchorage inside the long kelp-covered reef might perhaps be preferred ; but the landing is not so good there.

---

* *See* Admiralty Plan on Sheet XI., scale, *m* = 2·0 inches.

**LOBO** or **BLANCA POINT** 18 miles farther to the N.N.W. is a bold point 3,090 feet high, and on its extremity there are several hillocks.  In the interval there is a small fishing village, called Chomache, under a point, with a long reef, on the outer part of which a cluster of rocks show themselves a few feet above the water.  The people of this village get their water from the Loa, a passage requiring, on a *balsa*, four days or more.  Patache Point low, rugged, and projecting, lies N.N.W., 14 miles from Lobo point, with an islet a quarter of a mile in the offing ; and all clear outside.

**PABELLON de PICA**, a remarkable tent-shaped hillock said to be all guano, lies half-way between these two points ; its appearance being in strong contrast with the barren, sunburnt brown of the surrounding hills.  This also is a place of resort for the guano vessels, as they find pretty good anchorage close to the northward of the Pabellon.  East, a little southerly, a few miles in-shore, stands the bell-shaped mountain named Carrasco, 5,520 feet high.

**OYARVIDE COAST.**——From Patache point to Grueso point, N. ¾ W. 28 miles, the coast is low and rocky, the termination of a long range of table-land, called the heights of Oyarvide, or the Barrancas (ravines), from its cliffy appearance : it has numerous rocks and shoals off it, and should not be approached on any account within a league, for the frequent calms and heavy swell peculiar to this coast render it unsafe for nearer approach.  Inland of these heights, 28 miles to the northward of Mount Carrasco, is Mount Oyarvide 5,800 feet high.

**GRUESO POINT**, at the north end of the Barrancas, is low but cliffy, with three white patches on its northern side, round which lies the bay of Cheurañatta.  There is so much sameness in the aspect of the land on this part of the coast, the whole presenting the appearance of a rugged wall rising from the sea, that, unless a vessel is close inshore, it is difficult to distinguish the various points, especially in the morning, when the sun is over the land.

**IQUIQUE** anchorage, formed by an islet and reef, in all about a mile long, which jut out from the land, lies 11 miles farther north.  The town affords scarcely sufficient provisions for its inhabitants, about 500 ; with no water nearer than Pisagua (40 miles to the north), from whence it is brought by boats built for the purpose, and therefore dear.  Yet with these disadvantages, it is a place of considerable trade, from the quantity of saltpetre, and the silver-mines of Huantacayhua, in its neighbourhood ; the latter are but little worked, as the saltpetre is a

surer profit, large cargoes of it being taken in English vessels, about 50
of which load annually.   Silver and copper ore, and bars are also
shipped, though in small quantities.   There are no imports.   All the
property belongs to merchants in Lima, where the vessels are chartered,
and have only to call here and take in their cargoes.*

The French Government has a large contract for saltpetre, which,
although almost inexhaustible, yet the amount obtained is inconsiderable
compared with the demand, from the difficulty of procuring labour.  Vessels
calculating on return freights are often obliged to seek them elsewhere,
in 1856 the demand was for 120,000 tons but not more than 30,000 could
be procured.†

Iquique, however, still holds its ground as a mercantile place ; large
vessels from 600 to 1,000 tons are often seen at anchor here.   As it is a
port that is made at night by the P. S. N. steamers a small light is on that
night displayed from the reef, but so badly managed as to be worse than
useless ; no dependence can be placed as to what part of the reef it will
be shown from, and vessels have been found nearly in the breakers,
from the light being shown from the light-keeper's house, nearly
a mile from its proper spot, the end of the reef.   A launch comes out
into the bay to the southward of the reef, and shows a light with
flashes, which is the only way communication could be had with such a
place at night.

Landing is bad, and the approach to the shore hazardous, owing to the
number of blind breakers with which it abounds ; and at the full and
change of the moon, a heavy swell sets in.   Balsas are employed to bring
the cargoes to launches at anchor outside the danger, as is the case at
most places on this coast.   It is high water, full and change, at 8h. 45m.,
the rise and fall being 5 feet.

DIRECTIONS.—Vessels bound for this place should run in on the
parallel of Grueso point, until the white patches on that point are dis-
cerned, when a course should be shaped for the northern of three large
sand-hills : stand boldly in on this course till the church steeple appears,
when, or shortly after, the town will be seen, and a low island, 50 feet
high, under which is the anchorage ; care must be taken in rounding
this island to give it a good berth, a reef extending off it to the
westward, to the distance of 2 cables' lengths.   The anchorage is good
in 11 fathoms, with Piedras point bearing N. by W.; the outer point
of the island S. W. by W.; and the church steeple S. by E. ¼ E.

* See Admiralty Plan of Iquique, No. 1,340 ; scale, m = 0·7 inches.
† Remarks of Captain T. Harvey, R.N., 1859.  Remarks of Commander J. Wood,
R.N., 1853.

Vessels have attempted the crooked passage between the island and the main, and thereby got into danger, from which they were extricated with some difficulty ; it is only fit for boats or very small vessels.

OF PIEDRAS POINT, which is 2 miles north of Iquique island, there is a cluster of rocks ; and N. by W. 18 miles from it the small low black island of Mexillon, with a white rock lying off it. It may be known by the Gully of Aurora, a little to the southward, and a road apparently well trodden on the side of the hills, leading to the mines. And 33 miles farther north, is Pichalo point, a projecting ridge at right angles to the general trend of the coast, with a number of hummocks on it.

PISAGUA roadstead and village lie to the northward of Pichalo point : this place, as well as Mexillon, is connected with Iquique in the saltpetre trade, and is resorted to by vessels for that article. In rounding the point, a sunken rock lies about half a cable's length off, and should be looked out for, as it is necessary to hug the land closely, in order to ensure fetching the anchorage off the village ; for baffling winds are frequent, and may throw you near the shore, but that does not signify, as the water is smooth, and the shore steep-to. The best anchorage is with the extreme of Pisagua point N. ¾ W., a quarter of a mile off the village, in 8 fathoms ; by which you will avoid a rock with 4 feet water on it, lying off the sandy cove at the distance of 2 cables' lengths.*

The RIVER PISAGUA makes a conspicuous break in the shore, 2¼ miles north of Pisagua point ; its water supplies all the neighbouring inhabitants. For a few months during the winter season, when this river attains the greatest strength, it appears to be about 10 feet in width, but even then has not sufficient force to make an exit for itself into the sea ; like the Loa to the south, it merely filters through the beack, or is lost in the parched-up soil around. During nine months of the year no water is found in its bed ; though a scanty supply may always be had from the wells dug near it, yet no vessel should trust to renewing her stock at this place, for, besides its unwholesomeness, the difficulty and expense attending its embarkation would very great.

GORDA POINT is a low jutting spur, where a long line of cliff, from 2,000 to 4,000 feet high, commences ; and continues, with only two breaks or interruptions (quebradas in Spanish), as far northward as Arica. From Pichalo point to Gorda point, 18 miles, the coast is in

* See Plan of Pisagua bay on Sheet XI., scale, m = 2·0 inches.

low broken cliffs, with a few scattered rocks off it, and ranges of high hills near.

CAMARONES GULLY.—These breaks in the cliffs, or gullies, as they are called by the sailors, are remarkable, and very useful in making Arica from the southward. The first is the Quebrada de Camarones, which lies 7 miles north of Gorda point, and is about a mile in width, lying at right angles to the coast with a stream of water running down it, and a quantity of brushwood on its banks ; it forms a slight sandy bay, but not sufficient to shelter a vessel from the heavy swell.

The GULLY of VICTOR is the other; it lies 29 miles to the northward of Camarones, and 16 miles to the southward of Arica ; it is about three-quarters of a mile in breadth, and traverses the country in a similar manner to that of Camarones, and has likewise a small stream passing through, with verdure on its banks.* Cape Lobos a high bold point, projecting to the south-westward, forms a tolerably good anchorage for small vessels.

ARICA, the seaport of Tacna, is situated north of Arica head, at the extreme of the long line of cliff before mentioned, is one of the best anchorages on the coast, although there is frequently a considerable swell, which makes it desirable to lay a kedge out astern. The roadstead is protected by the low island and reef of Alacran. The town is composed of houses chiefly constructed of canes and reeds covered with mats from fear of earthquakes. Of late Arica has been the seat of civil war, from which it has severely suffered, as well as from the earthquake of 1833. It was in contemplation, in the latter end of 1836, to make it the port of the Bolivian territory ; and had that taken place, it would perhaps have become next in importance to the harbour of Callao, the principal port of Peru: its present exports are bark, cotton, and wool, for which is received in return merchandize, chiefly British.† Customs revenue in 1846 amounted to 64,578l., population may be 3,500.

A new and much more convenient mole, has lately been constructed close to the old one, and a large space enclosed to the southward of it, to form the terminus of a railroad to Tacna, considered one of the most salubrious towns in this part of South America.

On a clear day Tacna can be plainly seen from the sea, though more than 20 miles off, the valley rising gradually to it ; to the eastward also lie some of the highest peaks of the Andes : indeed, few scenes can vie in magnificence with the view on a clear day from between Sama head and

---

* See West Coast of South America, Sheet XII., No. 1,283; scale, m = 0·12 of an inch.

† See Plan of Arica, No. 578 ; scales, m = 1·0 inches, and m = 3·0 inches.

Arica, where the coast being low and the country sweeping back in one unbroken ascent to the stupendous snowy masses of the Andes, displays the endless variety of their outlines and the grandeur of their vast proportions to great advantage; even the volcano of Arequipa, more than 90 miles off, is often seen.

**Supplies.**—Fresh provisions and vegetables, with all kinds of tropical fruit, may be had in abundance, and upon reasonable terms; the water also is excellent, and may be obtained with little difficulty, as a mole is built out into the sea, which enables boats to lie quietly while loading and discharging; the only inconvenience is having to carry or roll it through the town. Fever and ague are said to be prevalent; this in all probability arises from the bad situation which has been chosen for the town, the high head to the southward excluding the benefit of the refreshing sea-breeze, which generally sets in about noon.

**DIRECTIONS.**—Vessels bound to Arica should endeavour to make the Gully of Victor, and when within 3 or 4 leagues of it they will see Arica head, which is 500 feet high, and appears as a steep white bluff, with a round hill in-shore, 880 feet above the sea, called Monte Gordo. Upon nearer approach the island of Alacran will be observed, joined to the head by a reef of rocks. To the northward of this island, is the roadstead, in entering which there is no danger whatever: the low island may be rounded at the distance of a cable's length in 7 or 8 fathoms, and anchorage chosen where convenient.

**MORRO DE SAMA.**—From Arica the coast of south America, which for some hundred miles to the southward extends nearly due north and south, takes a sudden turn to the north-westward; and as far as the river Juan Diaz, it is a low sandy beach, with regular soundings; it then gradually becomes more rocky, and increases in height till it reaches the Point and Morro de Sama, where it attains the elevation of 3,890 feet. This is the highest and most conspicuous land near the sea about this part of the coast, and at a distance appears, from its boldness, to project beyond the neighbouring coast line. On the western side of the Morro there is a cove formed by Sama point (45 miles from Arica), where coasting vessels occasionally anchor for guano; and there are three or four miserable looking huts, the residence of those who collect the guano. It would be impossible to land, except in a balsa, and even then with difficulty. Should a vessel be drifted down here by baffling winds and heavy swell, which has been the case, she should endeavour to pass the Head (as a number of rocks surround it) about a mile to the westward; and there anchorage may be obtained in 15 fathoms.

**CUMBA RIVER.**—At 9 miles from Sama point, is a low rocky point, called Tyke, and between those points issues the small stream Cumba, with low cliffs on each side : like most of the streams on the coast, it has not strength to make an outlet, but is lost in the shingle beach at the foot of the cliffs. Regular soundings which continue gradually as far as Coles point, may be obtained at the distance of 2 miles, in from 15 to 20 fathoms.

**COLES POINT** lies 31 miles W.N.W., from Sama point ; the shore between is alternately sandy beach, with low cliff, and moderately high table-land a short distance from the coast. It is doubtful if a landing could be effected anywhere between Arica and Coles point, as a high swell sets directly.on shore, and appears to break with unceasing violence.

Coles point is very remarkable : it is a low sandy spit, running out from an abrupt termination of high table land. Near its extremity there is a cluster of small hummocks : and at a distance it appears like one island. Off the point, to the south-west, there is a group of rocks or islets, but no hidden dangers. The rebound of the sea beating against both sides of the point, causes a ripple, and much froth, which leads one to suspect a reef in the vicinity.

Coles point should be carefully avoided by sailing vessels or steamers running along the coast, as at night it is difficult to see, and if inside they may be amongst the foul ground to the northward of it before they can haul out. This ground extends nearly two miles from the shore, and should be avoided, as it is shoal, and full of rocks on which the sea breaks heavily at times.*

**YLO ROAD** and village is N.E., 5½ miles from Coles point. This is a poor place, containing about three hundred inhabitants, with a governor and captain of the port. But little trade is carried on, and that chiefly in guano : a mine of copper has lately been discovered, which may add to its importance. The inhabitants have full occupation in collecting the necessaries of life, and do not care therefore to trouble themselves about luxuries. Water is scarce, and wood is brought from the interior, so that it is not on any account a suitable place for shipping.

**Pacocha Road,** off the village of Pacocha, about 1½ miles to the southward of Ylo, is the best anchorage, in 13 fathoms, and the landing place is in Huano cove : but great care must be taken lest the boat be swamped or hurled with violence against the rocks. In going into Ylo, the shore should not be approached nearer than half a mile (as many sharp rocks and blind breakers exist), until three small rocks, called the Brothers,

---

* *See* Admiralty Plan of Ylo Road, No. 1,340 ; scale, m = 0·7 inches.

which are always visible under the Table End, bear East, when the village of Pachocha may be steered for, and anchorage taken abreast of it, as convenient. English cove affords the best landing, but boats are forbidden that cove, to prevent the contraband trade carried on there.

**TAMBO VALLEY.**—From Ylo the coast trends to the north-westward for 40 miles, with a cliffy outline, from 200 to 400 feet in height, and with one or two coves, useful only to small coasters, as far as the valley of Tambo, which is of considerable extent, and may be easily distinguished by its fertile appearance, contrasting strongly with the barren and desolate cliffs on either side; those to the eastward maintaining their regularity for several miles, while to the westward they are broken, and from the near approach of the hills the aspect is bolder. Mexico point is the outer extreme of the low land of the Tambo valley; it is covered with brushwood to the water's edge, and projects considerably beyond the general trend of the coast. At the distance of 2 miles to the southward, soundings may be obtained in 10 fathoms, muddy bottom; from that depth in the same direction, it increases to 20 fathoms; but on each side of the bank there are 50 fathoms.

**MOLLENDO COVE,** 16 miles farther to the westward, was once the port of Arequipa; but of late years the bottom has been so much altered, that it is only capable of affording shelter to a boat or very small vessel: in consequence of which it has been thrown into disuse, and the bay of Islay now receives the vessels that bring goods to the Arequipa market.

**ISLAY BAY,** the port of Arequipa, formed by a few straggling islets and by Flat Rock point, which extends to the north-west, is capable of containing 20 or 25 sail. The town is built on the west side of a gradually declining hill, sloping towards the anchorage, and is said to contain 1,500 inhabitants, chiefly employed by the merchants of Arequipa. As in all the seaports of Peru, a governor and a captain of the port are the authorities; and it is also the residence of a British vice-consul. Trade was in a more flourishing condition here, even during the civil war, than at any place that was visited by the *Beagle;* there were generally four or five, and often double that number of vessels discharging or taking in cargoes. The principal exports were wool, bark, and specie. In 1858 there entered the port 49 vessels of 19,835 tons; value of cargoes £258,125, while the exports were valued at £461,594. Attempts have been made to develop the resources of the intermediate country by sending a steamer once a month to Ylo, but have failed.[*]

* *See* Admiralty Plan of Islay Bay, No. 1,340; scale, m = 5·5 inches.

The space for anchorage inside the islands is much circumscribed, and the water very deep : the P. S. N. Companies have a buoy laid down in about 30 fathoms.  Being the port of so considerable a town as Arequipa, it has a good sized custom-house, beside the pier, but the road up the cliffs from one to the other is so steep that most of the goods landed have to be carried up and down on men's backs, almost incredible loads being in this way transported to the plateau at the summit, where the custom-house and town are built.

**Supplies.**—A fountain supplied with very good water from the hills, is placed in the square or plaza near the custom-house, and from which shipping can always be supplied, the pipes reaching down to the mole. Last year an iron mole was erected, thus improving the landing of passengers and goods.  Fresh provisions may be had on reasonable terms.  Mollendito Cove, 3 miles from Islay, is the residence of a few fishermen ; there is a similar cove, named Sacetano, 9 miles farther east. The coast between Islay and Cornejo point is an irregular black cliff, from 50 to 200 feet high, bounded by scattered rocks to the distance of a cable.

**CORNEJO POINT,** 14 miles farther west, is about 200 feet high, with the appearance of a fort of two tiers of guns, and quite white.  The coast to the westward is dark, and forms a bay ; and to the eastward there are low black cliffs, with ashes on the top, extending half-way up the hills.  If the weather be clear, the valley of Quilca may be seen, which is the first green spot west of Tambo.

**TIDES.**—It is high water, full and change, in Islay, at 8h. 50m.; the rise and fall being 7 feet.  Ships have frequently been in sight, to the westward of the port, yet from the set of the current, which runs to the westward, from one to half a knot an hour, have been 'prevented from anchoring for several days.

**DIRECTIONS.**—Vessels bound to Islay, from the southward, should make the land abreast of Tambo, Mexico point, which is low and covered with brushwood, being conspicuous.  That place ascertained, which (according to the state of the weather) may be seen from the distance of 3 to 6 leagues : the course should then be shaped towards a gap in the mountain to the westward ; through this gap lies the road which leads to Arequipa, and which winds along the foot of the hill from Islay.

The best mark for making Islay, are the white rocks which form the port ; those to the northward off Cornejo point, however, are very similar when first seen, but the dark bay to the northward of that point, and the road to Arequipa, which shows plainly when approaching Islay, will be a sufficient guide to distinguish them.  There is also a rather remarkable bell-topped mountain 3,840 feet high, to the north-east of the town, which is useful in making the port when the fog hangs on the low land.*

As the coast is approached, the foot of the hills will be seen to be covered with white ashes (said to have been ejected from the volcano of Arequipa, 50 miles distant), not found on any other part of the coast. This peculiarity commences a little to the westward of Tambo, and continues as far as Cornejo point, and when within 3 leagues of Islay point, the White islets forming the bay will be plainly observed, and should be steered for.

Islay point must be closed with caution, as a rock, barely covered, lies a quarter of amile to the southward of the cluster of islets off that point. It is the custom to go to the westward of the White islets; but with a commanding breeze it would be better to run between the third outer and next islets,† which enables a vessel to choose her berth at once ; for the wind heads on passing the outer island, and obliges a vessel to bring up and use warps, or endangers her being thrown by the swell too near the main shore.

The mark to run between the third and fourth islets is, the Flat rock just open of the point north of the town. Pass close to the rock, or you will get off the bank, and anchor directly the town is well open, with Flat rock S. by E. and point north of town N.E. The mail steamers anchor with the fourth island shut in by the Flat rock. A hawser is necessary to keep the bow to the swell, to prevent rolling heavily even in the most sheltered part. Vessels from the eastward should observe the same directions, allowing for the north-west current of one mile an hour.‡

If coming from the westward, run in on the parallel of 17° 5' which will lead about a league to the southward of Islay point; and if the longitude cannot be trusted, Cornejo point, being the most remarkable land, and easily seen from that parallel, should be recognized in passing, and when abreast of it, Islay point will be seen, topping to the eastward, like two islands off a sloping point. The bell-topped mountain before named in the near range will also be seen, if favourable weather ; and shortly after the town will appear like black spots in strong relief against the white ground, when a course may be shaped for the anchorage under the White islets, as before.

AREQUIPA town lies 50 miles N.E. by N. of Islay, on the elevated plain of Quilca at a height of 7,850 feet above the sea, and has a population of

---

* Remarks of Commander J. Wood, R.N., 1853.

† Her Majesty's ships *Menai, Challenger* and *Havannah* passed in between these islands.

‡ Remarks of Captain Thomas Harvey and Thomas A. Hull, Master, R.N., 1859.

35,000. About 12 miles to the eastward of the town are several snowy peaks, the volcano of Arequipa towering over all, and rising to a height of 20,200 feet, and when the weather is clear about sunrise, visible at sea fully 100 miles off shore.

QUILCA.—Westward of Cornejo point the coast retires and forms a shallow bay, in which are three small coves—Noratos, Guata, and Aranta; and 13 miles to the north-west is the valley and river of Quilca, off which vessels occasionally anchor, under the Seal rock to the south-eastward of Quilca point. This anchorage is much exposed; but landing is good in the cove west of the valley. Watering is sometimes attempted, by filling at the river and rafting off, but it must always be attended with difficulty and danger. The valley is about three-quarters of a mile in breadth, and, differing from the others, which are level, has a rapid descent. From the regularity of the cliffs by which it is bounded, it has almost the appearance of a work of art. *See* view on Chart No. 1,283.

CAMANA VALLEY lies W. ½ N., 6 leagues from Quilca; the coast between is nearly straight, with alternate sandy beach and low broken cliff, the termination of the barren hills immediately above. The valley is from 2 to 8 miles broad, near the sea, and apparently well cultivated: the village stands about a mile from the beach; but being small, and surrounded with thick brushwood, is scarcely perceptible from seaward.

Monte Camana is a remarkable cliff, resembling a fort, forming the eastern slope of the Camana valley; it will be seen near the sea, on approaching from the eastward; this is an excellent guide till the valley becomes open. There is anchorage in 10 to 12 fathoms, muddy bottom, due south about a mile; but landing would be dangerous. Ocoña valley, the next remarkable place, 23 miles beyond, is smaller and less conspicuous than the former, but similar in other respects. An islet lies at the southern extremity, and several rocks near the end of the cliff, on its eastern side.

PESCADORES POINT is a projecting bluff point, 12 miles to the westward; it has a cove on its eastern side, surrounded by islets, and off the point, at the distance of three-quarters of a mile in a southerly direction, lies a rock barely covered. To the westward of the point there is a bay, but no anchorage; and the coast then runs in a direct line W. ¾ N. 26 miles, as far as Atico point,* which is a rugged peninsula, with a number of irregular hillocks on it, barely connected with the coast by a sandy

---

* *See* West Coast of South America, Sheet XIII., and Atico road, No. 1,279 and 1,340; scales, m = 0·12 and 0·5 of an inch.

isthmus, and at a distance appears like an island, the isthmus not being visible far off; there is a tolerable anchorage in 19 or 20 fathoms on its western side, and excellent landing in a snug cove at the inner end of the peninsula. By keeping a cable's length off shore, no danger need be feared in running into this road. The valley of Atico lies 1¼ leagues to the eastward, where there are about thirty houses, scattered among trees, which grow to the height of some 20 feet.* From Atico point the coast continues its westerly direction (low and broken cliff, with hills immediately above) to the foot of Capa point; it then forms a curve towards Chala point; and in these two intervals several sandy coves were observed, but none that appeared serviceable for shipping.

CHALA POINT bears from Atico point W. by N. ¾ N., distant 51 miles; it is a high rocky point, the termination of the Morro, or mount of that name. This mount shows prominently, and has several summits, the highest being 3,740 feet high: on the east side there is a valley separating it from another but lower hill, with two remarkable paps; and on the west it slopes suddenly to a sandy plain; the nearest range of hills to the northward is considerably in-shore, making Morro Chala still more conspicuous. About 9 miles to the south-east of Chala point, is a small cove protected by some off-lying rocks, where the P.S.N. Companies tried to establish a port of call for their steamers, but it is a difficult place even for a boat to land, and no vessel should go within the rocks as the heavy rollers will sometimes come in without warning, and then it is all broken water. A merchant brig was enticed in by the natives, but was nearly lost before she got out again. There are only one or two ranchos on the beach, but the district at the back looks fertile and is said to be worth attention.†

CHAVINI POINT appears like a rock on the beach, lying W.N.W. 18 miles from Chala point: between them there is a sandy beach, with little green hillocks and sand-hills; and two rivulets, running from the valleys of Atequipa and Lomas. These valleys are seen at a considerable distance. Half a mile to the westward of Chavini there is a small white islet and a cluster of rocks level with the water's edge; hence to the roadstead of Lomas a sandy beach continues, with regular soundings off it, at 2 miles from the shore.

LOMAS POINT projects at right angles to the general trend of the coast, and, like Atico, is all but an island; it may easily be distinguished,

---

* See West Coast of South America, Sheet XIII., and Atico Road, No. 1,279 and 1,340; scales, m = 0·12 and 0·5 of an inch.

† Remarks of Commander James Wood. R.N., 1853.

[S. A.]　　　　　　　　　　　　　　　　　　Y

although low, from the adjacent coast, by its marked difference in colour, being a black rock. Lomas road is the port of Acari, and affords good anchorage in from 5 to 15 fathoms, and tolerable landing; it is the residence of a few fishermen, and used as a bathing-place by the inhabitants of Acari, which, from the information obtained, is a populous town several leagues inland. All supplies, even water, are brought here by those who visit it : the fishermen have a well of brackish water, scarcely fit for use. Boats occasionally call here for otters, which are plentiful at particular seasons.*

**PORT SAN JUAN,** 23 miles to the westward, offers a fit place for a vessel to undergo any repairs, or to heave down in case of necessity, without being inconvenienced by a swell; but all materials must be brought, as well as wood and water, none being found there. The shore is composed of irregular broken cliffs, and the head of the bay is a sandy plain ; the harbour is good, indeed much better than any other on the south-west coast of Peru, and might be an excellent place to run for, if in distress.

**Steep Point and Reef.**—S.W. three quarters of a mile from Steep point (the southern point of Port San Juan) lies a small black rock, always visible, with a reef of rocks extending a quarter of a mile to the northward ; and nearly 2 miles to the S.E. there is an islet which also shows distinctly. A passage may exist between the reef and Steep point, but prudence would forbid its being attempted ; the sure plan is to pass to the southward, giving it a berth of a cable's length, and not to close the shore until well within Juan point, off which lies a sunken rock. Then haul the wind and work up to the anchorage at the head of the bay, and come to in any depth from 5 to 15 fathoms, muddy bottom. In working up, the northern shore may be approached boldly ; it is steep-to, and has no outlying dangers.

**Morro de Acari,** a remarkable sugar-loaf hill, 1,650 feet high, rising very steeply from the cliff, on the north side of the bay, forms a good mark for this port ; and 9 miles to the eastward, a short distance from the coast, a high bluff head forms the termination of a range of table land, and is well called Direction bluff. Between this bluff and the harbour the land is low and level, with few exceptions, and has a number of rocks lying off it to the distance of half a mile. *See* view on Sheet No. 1,279.

**PORT SAN NICOLAS** lies N.W. ¼ N., 8 miles from San Juan ; it is quite as commodious and free from danger as the latter, but the landing

---

* *See* Admiralty Plan of Lomas road, with Ports of San Nicolas and San Juan, No. 1,369; scale, *m* = 0·5 inch.

is not so good.   Harmless point which forms the south side of the port,
may be rounded within a cable's length ; there are a number of scattered
rocks to the southward of it, but as they all appear, there is no danger
to be feared.   There are no inhabitants at either of these ports, so that
vessels wanting repairs may proceed uninterruptedly with their employment.   High water at full change 5h. 30m., rise 3 feet.   Beware point
N.W. by W. 8½ miles from Harmless point, is high and cliffy, with a
number of small rocks and blind breakers in its immediate vicinity.   From
this point the coast is alternately cliffs and small sandy bays, for 14 miles
to Nasca point, round which lies Caballos road.

CAPE NASCA may be readily distinguished ; it has a bluff head, of a
dark brown colour, 1,020 feet in height, with two sharp-topped hummocks,
moderately high, at its foot : the coast to the westward falls back to the
distance of 2 miles, and is composed of white sand-hills : in the depth of
this bight is Caballos road, a rocky, shallow spot, and should only be
known to be avoided.   The *Beagle* lay at anchor in 7 fathoms, as far in
as it was thought prudent to go, for twenty-four hours, without being able
to effect a landing : the wind came round Cape Nasca in heavy gusts,
which, with the long ground-swell, made it doubtful if two anchors would
hold till the surveying observations were concluded.   The only traces of
there ever having been any inhabitants at this dreary place, was a pole
sticking upon the top of a mound near the head of the bay.

SANTA MARIA POINT, 28 miles beyond, has lying off it a rock
called Infiernillo.   The point is low and rugged, surrounded by rocks and
breakers.   At the distance of a league and a half inland, to the eastward, is a remarkable flat-topped mountain 2,160 feet in height, called
La Mesa de Doña Maria ; which may be seen in clear weather at a
considerable distance from seaward, and from its height and peculiar
shape is a good mark for this part of the coast.   The Infiernillo rock lies
due west from the northern end of Santa Maria, at the distance of a
mile ; it is about 50 feet high, quite black, and in the form of a sugar-
loaf ; no dangers exist near it, and there are 54 fathoms at 2 miles
distance.   Between this rock and Caballos road, the coast to a short
distance west of the small river Yca, is a sandy beach, with ranges of
moderately high sand-hills.   From thence to the Infiernillo it is rocky,
with grassy cliffs immediately over it, and some small white rocks off
the shore.

AZUA POINT, a high bluff with a low rocky point off it, lies N.N.W.
¼ W., 10 miles from Sta. Maria point : there is a sandy beach between,
interrupted by rocky projections, and a small stream running from the

hills. N.W. by W. from Azua point, and at the distance of 21 miles, is Dardo head, forming the northern entrance to the bay of Independencia.*

INDEPENDENCIA BAY is 15 miles in length in a north-west and south-east direction, and 3½ miles broad. This bay was till lately unknown, or at least unnoticed, no mention being made of it in the Spanish charts, and it was not till the year 1825 that the authorities at Lima became aware of its existence; and then only by an accidental discovery. The *Dardo* and *Trujillana*, two vessels that were conveying troops to Pisco, ran in, mistaking it for that place, and were wrecked; and many of the people on board perished.

The bay is bounded on the west by the islands of Viejas and of Santa Rosa, and on the east by the main land, which is moderately high, cliffy, and broken by a sandy beach, at the end of which is the small fishing village of Tungo. The people of this village are residents of Yca, the principal town in the province, which is about 42 miles distant; they come here occasionally to fish, and remain a few days, bringing with them all their supplies, even to water, as that necessary of life is not to be obtained in the neighbourhood. Serrate channel, the southern entrance to the bay, takes its name from the master of the vessel by whom it was discovered, and is formed by the islets of Santa Rosa on the north, and Quemado point on the south: it is three-quarters of a mile wide, and free from danger.

Trujillana or Northern Entrance, is named after one of the wrecked ships, and is formed by Caretas head on the north, and to the southward by Dardo head, so called after her consort; it is 4¾ miles in breadth, and clear in all parts. Approaching this part of the coast from seaward, it may be distinguished by the three mountains, Quemado, 2,070 feet, Viejas island, and Carretas 1,430 feet; they are at equal distances from one another, and nearly of the same height. The south-west sides of Mount Carretas and the island of Viejas are steep dark cliffs; but Mount Quemado slopes gradually to the water's edge, and is of a much lighter colour. A few miles farther inland are Carrasco heights, 3,000 feet high.

At the southern extremity of Viejas island there is a remarkable black lump of land in the shape of a sugar-loaf; off which lies the white level island of Santa Rosa, the south-west side of which is studded with rocks and breakers, but there is no danger a mile from the shore. It is high water, full and change, in Independencia bay, at 4h. 50m.; with a rise and fall of 4 feet.

ANCHORAGE.—There is anchorage in any part of this spacious bay; the bottom is quite regular, about 20 fathoms all over, excepting off the

---

* *See* Admiralty Plan of Independencia bay, No. 1,295; scale, m = 0·5 inches.

shingle spit on the north-east side of Viejas island, where a bank runs off that spit to the northward, on which there are 5 and 6 fathoms ; and this is the best place to anchor, for on the weather shore, near Quemado point, there are such sudden gusts off the high land, that great difficulty would be found in landing ; whereas at the spit a vessel is not annoyed by the wind, and there is a snug cove or basin within it, where boats may land, or lie in safety at any time.

**BOQUERON de PISCO,** or the entrance to that bay, formed by the peninsula of Paracas on the east and the island of San Gallan on the west, lies N.W. $\frac{3}{4}$ N. 20 miles from Carretas head ; the shore between that head and Huacas point, the south-west extreme of the Paracas peninsula, forming a deep angular bay, with the island of Zarate near its centre.*

**San Gallan Island** is $2\frac{1}{3}$ miles long, in a N.N.W. and S.S.E. direction and one mile in breadth; it is high, with a bold cliff outline. There is a deep valley dividing the hills, which, when seen from the south-west, gives it the appearance of a saddle ; the south end terminating abruptly, while its northern end slopes more gradually, and carries several peaks. Off this point there are some detached rocks, the northernmost of which has the appearance of a nine-pin, and shows distinctly.

**Pinero Rock,** a very dangerous rock, lies S. $\frac{1}{4}$ E., at the distance of a mile from San Gallan, and is much in the way of vessels bound to Pisco from the southward ; it is just level with the water's edge, and in fine weather can always be seen, but when it blows hard and the weather tide is running, there is such a confused cross sea that the whole space is covered with foam, rendering it difficult to distinguish the rock ; at such a time the shore should be kept well aboard on either side, until the ship is in a line between the south point of the island and the white rock off Huacas point, when she will be within the rock, and may steer for Cape Paraca, on rounding which the bay of Pisco will open.†

**DIRECTIONS.**—Vessels bound from the southward for Pisco bay or the Chinchas, should endeavour to make the land about Carretas head, as a westerly current of from 12 to 20 miles a day will generally be found setting off this part of the coast. From thence a course should be shaped for the Boqueron, which though narrow is bold-to on both sides, and is the best passage into the bay, as it is usually free from fog, and the wind blows right through it. The description of the Pinero rock should attended to, as from the broken sea before mentioned it is so difficult to make out, that in a fast steamer there would be hardly time to clear it ; by keeping close to the Paracas side, however, it will be avoided.

---

* *See* West Coast of America, Sheet XIV., No. 1,323 ; scale, m = 0·12 of an inch.
† *See* Admiralty Plan of Pisco Bay, No. 1,291 ; scale, m = 0·5 inches.

**PISCO BAY.**—This extensive bay, formed by the peninsula of Paracas on the south, and the Ballista and Chincha islands on the west, is the principal port of the province of Yca. The town of Pisco is built on the eastern side, about a mile from the sea; and is said to contain 3,000 inhabitants, who derive considerable profit from a spirit they distil, known by the name of Pisco or Italia, great quantities of which are annually exported to different parts of the coast; sugar is also an article of trade, but the pisco is the staple commodity. Refreshments may be obtained on reasonable terms: wood is scarce; excellent water may be had at the head of Pisco bay, under the cluster of trees, 2 miles south from the fishing village of Paraca; the landing there is very good, and the wells are near the beach.

The frequent calls of the P. S. N. Companies' steamers and the enormous trade in guano, procured from the Chincha islands, have given it an impulse which is evidenced by its increased exports, for in addition to the supply of several hundred large merchant vessels lying off these islands who draw most of their fresh provisions from hence, large quantities of fruit, Italia, olive oil, &c., &c., are ready for the steamers as often as they call, which is at least twice a month, and frequently from 40 to 50 passengers also. The landing is always bad, sometimes impracticable; though managed by launches built for the purpose.*

There may be said to be four entrances to this capacious bay: the Boqueron already mentioned; between San Gallan and the Ballista islands, the southern rocks of which lie 2½ miles from Cape Paraca; between the large Ballista island and the Chincha islands, a distance of 3½ miles; and, lastly, the northern entrance between the Chinchas and the main; all of which, from appearances, may be safely used; but of those between the islands, time would not allow a full examination, and therefore there may be dangers that were unseen.

**Salcedo Rock**, among the Ballista isles, has about 4 feet on it at low water, and there are 4 fathoms close to. It bears S.E. by S. distant about 7 cables from the south-eastern end of the second island. Between San Gallan and the southern rocks there are from 4 to 16 fathoms, and from 10 to 30 fathoms between them and the main Ballistas.

**Paracas Shoal.**—When hauling into Pisco bay, round Cape Paraca, care must be taken to avoid some shoal ground to the north-east of the head which looks dangerous. Steamers have had to haul to the northward in less than 4 fathoms, from turning the point too sharp; the safest plan is to run on till a short mile past the head when a course may be steered for Blanca island, in the middle of Pisco bay, which may be passed close to on its southern side.

**Paracas Bay** in the south-east corner of Pisco bay, formed between the main land and the peninsula of Paracas, is now used as a watering-place for vessels from the Chincha islands ; there is good anchorage off the watering-place in 4 fathoms ; no surf is found in this bay, and a plentiful supply of fish can be obtained, by hauling the seine on any of its beaches. It is high water, full and change, in Pisco bay at 4h. 50m. ; the rise and fall being 4 feet.

**DIRECTIONS.**—In coming from the southward, after passing Cape Paraca, a course may be shaped rather outside of Blanca island, in order to give a berth to the Paracas shoal, about a mile to the northward of the peninsula, where 4 fathoms are marked in the plan ; and then towards the church of Pisco, which will lead directly to the anchorage. Abreast of Blanca island there is 12 fathoms, muddy bottom ; and from this depth it decreases gradually to the anchorage.

In coming from the northward it is all plain sailing. After passing the Chincha islands, stand in boldly to the anchorage ; the water shoals quickly on that side of Blanca island, but there is no danger whatever. Vessels having to ballast in Paracas bay should work up and anchor under Shingle point, on the north-east side of the peninsula ; they can lie close to the shore, and boats may land with expedition.

In coming from seaward, this part of the coast may easily be known by the island of San Gallan and the peninsula of Paracas, on which is mount Lechura, 1,300 feet high, making like large islands, the land on each side being considerably lower, and falling back to the eastward so as not to be visible at a moderate distance. As the shore is approached, the Chincha and Ballista islands will be seen ; which will confirm the position, there being no other islands lying off the coast near this parallel.

**ANCHORAGE.**—The best anchorage off the town of Pisco is with the church open of the road, bearing W. by N. ¼ N., in 4 fathoms, muddy bottom, three quarters of a mile from shore. A heavy surf beats on the beach with rollers to the distance of a quarter of a mile off, rendering it dangerous to land in ships' boats. Launches built for the purpose are used in loading and discharging vessels ; but at times even these cannot stand it, and all communication is cut off for two or three days together.

**The CHINCHAS** are 3 small islands situated 10 miles W. b. N. from the town of Pisco. They have become of the utmost importance to the Republic of Peru ; the immense quantity of guano deposited on them insuring it a vast addition to the public revenues whilst the demand for the article lasts, or until it is all removed ; the latter will take some years to accomplish, as there must be many millions of tons on the two principal islands. In June 1850 there were about 40 vessels (principally English) of large tonnage waiting their turns to load, whilst departures and arrivals were of daily occurrence. In August 1853 there were 180 sail waiting for cargo ; the amount shipped weekly being 50,000 tons.   In 1858, 346 vessels of 266,709 tons cleared from Peru with cargoes of guano.*

**Supplies.**—Not the slightest trace of vegetation is to be found on the islands, or a drop of water ; the latter necessary article is supplied by the shipping and so managed, that each vessel shall contribute an equal proportion, for instance, the vessel whose next turn it is to haul under the hose, provides enough for the use of the island, and so on in succession.   All the supplies for the ships are drawn from Pisco, where good beef, poultry, fruit and vegetables are plenty, and at reasonable prices.   Fish, such as herrings, mackerel, rock cod, &c., are caught in abundance about the islands, and numbers of seals may be seen sporting about in pursuit of them at all hours of the day.

**The North Island** is about 1,600 yards long, and from 700 to 800 yards broad, the highest part about 200 feet above the level of the sea ; but as the guano at that point is certainly 90 feet deep, the island itself (terra firma) is in no part more than 110 feet above high water mark. On the north-east side of this island are 2 sunken rocks with 6 and 7 feet water on them lying nearly 2 cables' from the shore ; they were both buoyed in 1858.   Should the buoys be gone, the east point of the middle Chincha kept open of the south point of North island until the north-west rock opens of that island will clear these dangers.

The northern end of the North island was, in June 1850, the principal loading-place, vessels hauling alongside a perpendicular cliff 100 feet high to receive their cargo, which was conveyed into their holds through canvas hoses ; and when a sufficient number of labourers were at hand, 400 or 500 tons could be shipped daily ; a little engineering skill would greatly facilitate the loading of vessels, and decrease the expense of shipments. There were about 150 labourers on the island, principally convicts, a few natives and Chinese, a number far too few for the duties required ; and although inducements are held out to volunteers, a feeling of disgust exists among the poorest classes against such disagreeable employment, though not considered an unhealthy one.

**Anchorage.**—Although the depth of water is great, the anchorage may be considered secure, as it never blows strong, except from the southward ; the breeze from that quarter is termed the "Paraca" (coming from the peninsula of that name), but as the merchant ships usually anchor to the northward of the island, they are consequently sheltered from its violence; the best anchorage is between the North and Middle Chincha, in 18 or 19 fathoms (white sand and shells), taking care to be a little to the eastward or westward of the guano hoses, to avoid the dust that is blown from them (when the southerly wind sets in), which is disagreeable and offensive.

---

* *See* Admiralty Plan on Sheet XIV ; scale, m = 2·0 inches.

In proceeding from Callao to the Chincha islands, it is recommended to stand off the land at night, and towards it during the day until to the southward of lat. 13°, when it is advisable to keep within 4 or 5 miles of the shore down to Pisco. The currents are uncertain at the Chincha islands, but generally set to the northward about 1½ knots per hour.*

**CERRO AZUL.**—From Pisco the coast, a low sandy beach with regular soundings off it, runs in a northerly direction, as far as the River Chincha, and from thence to the River Cañete it is a line of clay cliffs, from 430 to 540 feet high. From this river to Frayle point a beautiful and fertile valley fringes the shore, and to the north-eastward of Frayle point stands the town of Cerro Azul. The valley of Cañete produces rum, sugar, and *chancaca,* a sort of treacle, for which it is resorted to by coasters. The anchorage is W.N.W. from the bluff that forms the cove, three-quarters of a mile distant, in 7 fathoms ; nearer the shore the water is shoal, which causes a long swell. The landing-place is on the northern side of the point, on a stony beach, where a heavy surf, however, constantly breaks.

**ASIA ISLAND** lies N.W. ½ N., 17 miles from Cerro Azul ; it is round, white, and about a mile in circumference, with some rocks extending from it to the shore, forming a bay, but scarcely affording anchorage. The coast-line is partly a rocky, and partly a sandy beach ; in-shore are heights of about 1,400 feet, declining gradually towards the coast.

**CHILCA PORT.**—Chilca point, lying N.W. 20 miles from Asia island, is about 300 feet in its highest part, has several rises on it, and terminates in a steep cliff, with a small flat rock close off its pitch. The valley of Chilca lies 3 miles to the eastward of the point, and the snug little port of Chilca 1¼ miles to the northward. Port Chilca is safe, but very confined ; anchorage is good in any part of it, and landing tolerable ; there is a small village at the head of the bay.†

**PACHACAMAC ISLANDS.**—From Chilca the coast forms a bend to the valley of Lurin, off which lie these islands distant 10 miles from Port Chilca. The northern is the largest, half a mile in length, about a cable's length broad and 400 feet high ; San Francisco is the most remarkable, being quite like a sugar loaf, rounded at the top; the others are mere rock, and not visible at any distance. At the northern end of these islands lies a small reef even with the water's edge ; the group runs nearly parallel to the coast, in a north-west and south-east direction, and is about a league

---

* Remarks of Mr. W. Dillon, Master R.N., 1850.
† *See* Plan of Port Chilca, No. 1,710 ; scale, *m* = 4·0 inches.

in extent. There is no danger on their outer side, but towards the shore the water is shoal, which causes a long swell to break there heavily. To the north of the Pachacamac islands, the river Lurin brings its small stream from the interior, but without sufficient force to make its way into the sea; the valley, however, which it waters, appears fertile and well cultivated when seen from the offing. From thence to the Morro Solar is a sandy beach, with moderately high land a short distance from the sea.

**SOLAR POINT** is 10 miles from San Francisco island, and forms the south-west point of the Morro Solar. Off Solar point there is an insignificant islet with some rocks lying about it, and off Chorillos point a reef of rocks projects about 2 cables' lengths; round this reef, on the north side of the Morro, lies the town and road of Chorillos. The former is built on a cliff, at the foot of one of the slopes of that mountain, and is used chiefly as a bathing-place for the inhabitants of Lima. The Morro Solar is a remarkable cluster of hills, 860 feet high, standing on a sandy plain; when seen from the southward it has the appearance of an island in the shape of a gunner's quoin, sloping to the westward and falling very abruptly in-shore; its sea face, however, terminates in a steep cliff, named Codo point, with a sandy bay on each side.

**CHORILLOS BAY** lies round the point of the same name on the north side of the Morro Solar. When political revolutions render the road of Callao a dangerous berth, the vessels resort to Chorillos bay, though in every other respect an unfit place for anchoring, as the bottom is a hard sand, with patches of stones and clay, mixed together, called *tosca*; and the heavy swell that sets round the point, causing almost a roller, brings a vessel up to her anchor and throws her back again with a sudden jerk, which endangers dragging the anchor or snapping the cable.[*]

Vessels having to anchor there should keep Solar point open of Codo point: by so doing they will ride in 8 or 9 fathoms, and not have so much swell as there is farther in. The landing in the bay is bad; canoes built purposely, and dexterously managed, are the usual means of communication; for though, no doubt, there are times when a ship's boat may land without danger, yet very seldom without the crew being thoroughly drenched. From Chorillos the coast runs in a steady sweep, with cliffs diminishing in height, till it reaches the point of Callao, which is a shingle bank stretching out towards the island of San Lorenzo, and which with it forms the extensive and commodious bay of Callao.

---

[*] *See* Admiralty Plan of the Boqueron of Callao, No. 1,853; scale, *m* = 1·5 inches.

# CHAPTER XIV.

## CALLAO TO THE RIVER TUMBEZ.

VARIATION 11° to 9° E. in 1860. Annual increase 1.'

---

**CALLAO** is well known as the sea-port of the City of Lima, which stands at the foot of the mountains 7 miles to the eastward, but lying 500 feet above the level of the sea, and the ground having a gradual ascent, it is well seen from the anchorage in clear weather, and has a very imposing appearance. The town, castle, and forts of Callao extend about a mile along the beach that fronts the bay. Formerly the houses were chiefly of mud, of one story and flat roofs, but recently the principal street, which runs parallel to the bay, has been widened, and contains some well built houses of two stories. The forts of San Sebastian and San Rafael still exist, but the castle, which was considered the key of Lima, has been partly dismantled as a fortress and is part used as a custom-house. A good mole has been erected, but is too small for the increased demand for space by shipping, and a railway with its terminus close to the mole head connects the port with the capital. In the year 1858 there entered at the port 1,296 vessels of 649,909 tons, and nearly as many cleared. In 1857 the value of imports was £1,750,387, and of exports £441,434. Population 8,435 in 1852.

**CALLAO ROAD**, distant 11 miles from the Morro Solar, is formed by Callao point and the long spit that stretches off from it towards San Lorenzo island; a part of this spit, termed the Whales Back, just shows at the water's edge, the sea breaking violently along its ridge. Callao point is low, and consists of a bank of small round stones. This roadstead, assisted by the climate and prevalent southerly winds, becomes a fine harbour, the island of San Lorenzo protecting it from the long swell from the ocean.

The port of Callao, as well as the Government, and indeed the whole country of Peru, has profited much by the vast quantities of guano exported from the Chinchas. Of the hundreds of large vessels that come here for this purpose all are obliged to clear at Callao on their arrival and departure; as much as 33,000 tons a month were shipped in the early part of 1853, so that some idea may be formed of the crowded state of the bay, extensive as it is, and of the vast increase of business in the port.

**Supplies** of all sorts may be obtained for shipping ; fresh provisions as well as vegetables, with an abundance of fruit ; watering is also extremely convenient, a well-constructed mole being run out into the sea, at which boats can fill from the pipes that project from its side.

Water is brought off in tanks and pumped on board at 2 dollars per ton. Wood is scarce, mostly imported, and of course dear. All kinds of stores, &c., are to be had, though far from reasonable. Repairs to ships' bottoms are effected by heaving down, which, however, is sometimes interrupted by the rollers coming in, which is mostly the case at the same periods, viz., full and change of the moon, and during the equinoxes. The P. S. N. Companies had their head-quarters here till lately, and besides two store ships, had mooring buoys laid down for two of their steamers, and a factory and coal depôt on the beach under the castle walls.*

**LIMA,** the capital city of Peru, founded by Pizarro in 1535, at 7 miles east of Callao, lies on a plain at the foot of some granite hills chiefly on the southern bank of the Rimac, which river falls into the sea 2 miles north of Callao. In the range of mountains which extends north and south-east of the city, two conspicuous peaks, one 3 miles to the north and the other 8 miles to the south-east, rise respectively 3,000 and 3,420 feet high and are visible from the anchorage in Callaó bay. The city is about 6 miles in circumference and the south side is surrounded by a wall having 9 gates. The streets, of a fair width, intersect each other at right angles, and those lying east and west or parallel to the river have each a stream of water running down the centre. The Plaza Mayor and other squares are spacious, and the cathedral and convent of San Francisco with some other churches and convents are handsome buildings. The Alameda del Acho, on the banks of the Rimac, is a spacious and beautiful promenade. Lima is the seat of an university having a national library of 20,000 volumes and a museum ; there are also well conducted hospitals.

The climate is very agreeable, the range of the thermometer remarkably small, varying from 73° Fahr. in winter to 87° in summer. From April to October, that is in the winter, a heavy mist overhangs the city in the morning and evenings. Rain is of exceedingly rare occurrence ; thunder and lightning are unknown. Lima is very subject to earthquakes ; the

---

* A nauseous smell, usually called the Painter or Barber, is frequently experienced by vessels in this port. It deposits on white paint and whitewash a thick slime of a chocolate colour, washing off from the former, but spoiling its after appearance ; it is supposed to proceed from the mud at the bottom of the sea.—*Remarks of Captain Harvey, R.N.*, 1859.

most destructive on record occurred in October 1746. On that occasion the port of Callao was suddenly submerged by a huge wave and it is said that only 200 persons escaped out of 4,000. The population of Lima has fluctuated greatly, but it is supposed to be now about 70,000.

**SAN LORENZO ISLAND** distant 11 miles from Solar point, is 4¼ miles long in a north-west and south-east direction, and one mile broad, rising to an elevation of 1,284 feet above the sea. Off its south-east end lies a small but bold-looking isle, called Fronton, almost connected with San Lorenzo by a reef; and to the south-west are the Palominos rocks. Its northern point, or Cape San Lorenzo, on which stands the lighthouse, is clear, and round it is the usual passage to the anchorage of Callao. An American Company (1859), are building a Sectional Lift Dock on San Lorenzo, calculated to be equal to taking vessels from 1,200 to 1,400 tons. The parts and machinery connected with it were brought from the United States.

**LIGHT.**—The lighthouse on Cape San Lorenzo is a wooden octagonal tower, 60 feet high. It stands on the summit of the cape, and exhibits, at an elevation of 980 feet above high water, a *fixed white* light, visible in clear weather from a distance of 18 miles. Between the bearings of N.W. ¾ N., and W. by N. ¾ N., it is hidden by the peak of the island; and when just open on the latter bearing leads through the Boqueron channel in 4½ fathoms. From its lofty position, however, this light is not always seen, being often enveloped in the thick fog or haze which hangs over the high land, causing it at night to appear like a star only.

**DIRECTIONS.**—Vessels bound to Callao road in rounding Cape San Lorenzo should not close the land nearer than half a mile, for within that distance there are light baffling airs caused by the eddy winds round the island, by getting among which the vessel would be more delayed than if she gave the island a good berth, and had to make an additional tack to fetch the anchorage.

Should there be occasion to work to windward to reach the anchorage, the Whales Back, with another rock, said to lie off Galera point on the island of San Lorenzo, are so far to the southward that the vessel need scarcely apprehend borrowing on them. Run or work up close to the shipping, and anchor in from 7 to 5 fathoms; with the pier-head bearing about S.E. and Cape San Lorenzo W. by S. Although the above mark is given, for the most convenient anchorage, yet ships may lie with the greatest safety in any part of Callao Road, and in any depth of water, on clear ground and gradual soundings from 20 to 3½ fathoms up to the mole-head and landing-place.

**BOQUERON CHANNEL.**—The above is the obvious route to Callao ; but there is another which, with common precaution, may be used with great advantage to vessels coming from the southward, by passing through the Boqueron channel between the island of San Lorenzo and Callao point.

**TIDES.**—It is high water, full and change, in the Boqueron channel at 5h. 47m.; the rise and fall being 4 feet.   There is no regular tide in this passage, yet a little drain of current is always felt, sometimes to the north-west and at others the contrary : should the stream be adverse, and it falls calm while in the channel, there is good anchorage in 8 or 9 fathoms with the leading marks in one.   Var. 10·50 E. in 1860.

**DIRECTIONS.**—After making Fronton island steer so as to keep its southern end about a point open on the port bow ; continue on this course until Callao castle is seen, which, with its two martello towers, stands on the inner part of the shingle bank that forms the point ; then steer for that castle till Horadada rock, which has a hole through it, lying between Fronton isle and Chorillos bay, comes in one with the middle of Solar bay, on the bearing of E.S.E. ; with these in one, and there-fore steering about W.N.W. for the farthest point of Lorenzo that can be seen, the vessel will be clear of all danger ; and when the western martello tower in the castle comes in one with the northern part of Callao point she may haul gradually round to the northward till that tower opens clear of the breakers on the spit, when a direct course may be shaped for the anchorage, taking care not to come nearer the sand called Whales Back than 6 fathoms.

These marks will also lead clear of the bank that extends three-quarters of a mile to the northward of Fronton ; and as soon as the round islet between Fronton and San Lorenzo bears S. by W. the Fronton shoal will have been passed, and San Lorenzo may be approached as above directed.   H.M.S. *Collingwood*, drawing 25 feet, came through with the above mark, in not less than 5¼ fathoms on the port side, and 5¾ on the starboard.

The lighthouse if seen affords a better mark for running through the Boqueron.   Haul to the north-west directly it opens, and run through with the south base just touching the high land ; when the round islet between Fronton and Lorenzo is shut in, steer N.N.W. till the cliff of Cape San Lorenzo is well open, bearing W. ½ N., then steer a direct course for the anchorage.   These marks make a vessel independent of the castles, which are bad objects and frequently enveloped in the mist.

**Working through.**—In March 1859, Captain Thomas Harvey, R.N., worked through this channel in H.M.S. *Havannah*.   The following are his remarks :—" Finding that the guano laden ships were not allowed to run through the Boqueron, in consequence of reported errors in the charts, and that the loss of sundry vessels was attributed to these errors, or shifting sands, the passage was sounded and the *Beagle* survey of 1835 verified, so far as the channel portion of it was concerned and found to be correct ; Mr. Hull, master, selecting such turning marks as would enable ships drawing 24 feet to work out through the Boqueron.

"On weighing from Callao road steer for the north peak of San Lorenzo till the Horadada rocks open of Callao point ; then S.W. ¼ W. for a sandy bay, in which is a lift-dock ; and when the cliffs of Fronton and San Lorenzo touch haul to the wind, and tack when the hole in the Round islet is shut in by San Lorenzo. When standing towards the Whales Back, tack directly the cliffs of Fronton and San Lorenzo open; vessels drawing 24 feet and over should tack a little before, when a conspicuous dark mark on Fronton opens. When the cliff of the Red bluff on San Lorenzo is in line with a saddle on the summit of the island, bearing W. ¼ S., a vessel will be clear of the Callao shoal, and may stand to the eastward. Very little current was found in the channel, the set is generally to the northward, it is therefore necessary to watch the last named mark in standing away to the eastward.

"This northerly set is, at times, considerable outside Fronton islet, near which the winds, when light, are treacherous and baffling, it is consequently advisable to give it a berth of at least half a mile. In the narrows, when standing over to the Whales Back, the first shoal cast gives timely warning to tack. The bold appearance of San Lorenzo has led strangers to suppose themselves nearer to it than they were, and the fear of approaching it has been the cause of getting on the Callao shoal."

**The HORMIGAS DE AFUERA** are a small cluster of rocks lying due west from the north end of San Lorenzo, at the distance of 30 miles ; the largest is about three-quarters of a mile in circumference, 25 feet high, and covered with guano ; no sign of vegetation was observed : it is merely a resting-place for birds and seals ; landing may be effected, if requisite, on its north side, but with difficulty. Being somewhat in the way of vessels bound to Callao from the northward, and of those leaving that port for the westward, care should be taken not to approach those rocks too closely, for fear of being overtaken by one of those dense fogs which are so frequent on the Peruvian coast. The water is deep close-to all around and no warning would be given by the lead.

**PORT ANCON.**—From Callao, the coast is a sandy beach, lying in a northerly direction, until it reaches Pancha point ; it there becomes higher, and breaks into cliffs and maintains this character as far as Cape Mulatas, round to the eastward of which is the little bay of Ancon.

**THE PESCADORES**, a small group of nine or ten rocks, lie to the west and south-west of Ancon, above 3 miles off shore, the outer and largest of which bears N.N.W. ¾ W. from Callao castle, and at a distance of 18 miles. A sunken rock lies to the eastward of the large islet, otherwise there appears to be no danger among the group; they are steep-to, with 10 to 30 fathoms near them. The Hormigas de Tierra, or the in shore Hormigas or Ants, are two rocks above water, about 1 mile to the N.N.W. of the Great Pescador. Caution is required in approaching Ancon from the westward, as the lead would give no warning of these rocks.

**CHANCAY BAY** lies N.W. by N. from Cape Mulatas, 12 miles distant; this bay may be known by the bluff head that forms the point, and has three hills on it, in an easterly direction; it is a confined place and fit only for small coasters. From Chancay, the coast runs in a more westerly direction, as far as Salinas point, a shingle beach, with a few broken, cliffy points; the hills are near the coast, and from 400 to 500 feet high, but 13 miles inland to the eastward Mount Stokes rises to 4,000 feet.*

**SALINAS POINT** is 27 miles N.W. by W. ½ W. from Chancay head; it is the south-western extremity of the broad promontory of Salinas, the sea face of which has a north and south direction of 5 miles in length. Off its southern point there is a reef of rocks, a quarter of a mile from the shore; and at its north-western angle, called Bajas point, an islet at a cable's length distant. There are two coves in the sea-face, but they are fit for boats only. Salinas Bay lies round Bajas point; it is of large dimensions, and affording roomy anchorage. The remarkable round hill of Salinas rises at a short distance from the coast, from a level sandy plain to 930 feet in height; and at the south side of the plain lie the salinas, or salt-lakes, which give the promontory its name. These lakes are visited occasionally by people from Huacho.

**HUAURA ISLETS** lie off Salinas point, in a south-west direction, the largest of which is called Mazorque. It is 200 feet in height, three-quarters of a mile long, and quite white: sealers occasionally frequent this island; as there is a landing-place on its north side.

The next in size is Pelado, it lies S.W. ½ W. 6½ miles from Mazorque, is about 150 feet high, and apparently quite round; and between these two islands a safe passage exists, and may be used without fear in working up to Callao. Between Mazorque and Salinas stand several other islands, which, from their appearance, may be approached without danger;

* See Admiralty Plan of Chancay bay, No. 1,347; scale, m = 2·0 inches.

but as no advantage could be gained, it would not be prudent to risk going between them. Vessels, in working up, do sometimes pass between the inner one and the point ; but what they gain thereby does not appear, for when the current sets to the southward, it runs equally as strong near the shore as it does between Mazorque and Pelado.

There are two passages between these islands used by the steamers; the inner one is narrow and lies between the main and Tambillo island, which is rather a blind object at night ; but the outer one, between the Mazorque and Pelado is much used, being six miles wide and quite free from danger. Pelado is a high conical rock, as seen from the south-eastward, and is generally the first seen at night.

**HUACHO BAY.**—From Salinas bay the coast is moderately high and cliffy, without any break as far as the bay of Huacho, which lies round a bluff head, and is small ; the anchorage, however, is good, in 5 fathoms, just within the two rocks off the northern part of the head.*

The town of Huacho is built about a mile from the coast, in a fertile plain, and in approaching from seaward has a pleasant appearance ; it is not a place of much trade, but whale ships find it useful for watering and re-freshing their crews. Fresh provisions, vegetables, and fruit are abundant, and on reasonable terms ; wood is also plentiful, and a stream of fresh water runs down the side of the cliff into the sea. Landing is tolerably good ; yet rafting seems to be the best method of watering.

**BEAGLE MOUNTAINS.**—In coming from seaward, the best distinguish-ing marks for this place are the three Beagle mountains, 4,000 feet high, in the near range, having each of them two separate peaks. They lie inland about eight miles, and on closing the land, Mount Salinas to the southward, as well as the island of Don Martin, to the northward, will be seen ; about midway between them is the bay of Huacho, under a light brown cliff, the top of which is covered with brushwood. To the south-ward the coast is a dark rocky cliff. *See* View on Sheet No. 1,323.

**CARQUIN BAY** lies N.N.W. ½ W. 3½ miles from Huacho, scarcely as large, and apparently shoal, and useless to shipping. Off Carquin head, which is a steep cliff with a sharp-topped hill over it, there are some rocks above water, and an islet a short mile distant. N.N.W. ¾ W. 3 miles from this islet stands the island of Don Martin, and round to the northward of the point, abreast of it, is the bay of Begueta.

**BEGUETA BAY** is no place for a vessel, being full of rocks and breakers, and having nothing to induce one to go there. From this bay

---

* *See* Admiralty Plan of Huacho bay, No. 1,347; scale, *m* = 1·8 inches.

the coast is moderately high, with sandy outline all the way to Atahuanquí point, distant 8 miles N.N.W. ¼ W.   This is a steep point, with two mounds on it, and is partly white on its south side ; there is a small bay on its north side, fit only for boats.   Between this point and the south part of Cape Thomas the coast forms a sandy bay, low and shrubby ; with the town of Supé, about a mile from the sea.†  Point Thomas is similar in appearance to Atahuanque point, without the white on the south side. To the northward of the point there is a snug little bay, capable of containing four or five sail ; it is called the Bay of Supé, and is the port of that place and of Barranca.‡

**SUPÉ BAY** has been a forbidden port by the government : in consequence of which it is little known, and has had few opportunities of exchanging its produce for the goods of other countries.   Very little information could be gained there as to the size of the neighbouring towns, or the number of inhabitants they contain, but from their appearance it was thought they might be of considerable extent.   These places produce chiefly sugar and corn, cargoes of which are taken in the various little vessels that trade along the coast.   Refreshments may be obtained ; but water is scarce, the greater part of which is brought from Supé town, for the use of the inhabitants of the village.   There is a fishing village at the south end of the bay, which is used by the inhabitants of Barranca during the bathing season.

The best anchorage is in 4 fathoms, with Point Thomas shut in by Patillo point, about a cable's length from the rocks off that point, and rather more than a quarter of a mile from the village.   Good anchorage may be obtained farther out, in 6 or 7 fathoms, though but little sheltered from the swell.   In entering, no danger need be apprehended ; Point Thomas is bold, with regular soundings from 10 to 15 fathoms, extending three quarters of a mile.   Off Patillo point there are a few rocks, but there is no necessity for hugging the shore very closely, as you can always fetch the anchorage by keeping at a moderate distance when standing in.

**MOUNT USBORNE**, 8,060 feet above the sea, the highest and most remarkable mountain in the second range, is the best guide at a distance to recognize Supé bay.   It bears from the anchorage N.E. ¼ E. ; it has something of the shape of a bell, and has three distinct rises on its summit —the highest at the north end.   On that side it shows very distinctly,

* See West Coast of South America, Sheet XV., No. 1,285 ; scale, m = 2·12 of an inch.
† See Admiralty Plan of Supé bay, No. 1,347 ; scale, m = 2·0 inches.

there being no other peak within a considerable distance. Mount Dar-win, 20 miles to the north-west and 8 miles from the shore, is next in height and reaches 5,800 feet. Supé anchorage has a white rock off its northern point, and cannot be mistaken, for there is no other like it near this part of the coast. From Supé the coast is a clay cliff, about 100 feet in height, to the distance of 1½ leagues ; it then becomes low and covered with brushwood to the foot of Horca hill, already mentioned ; here it again becomes hilly near the sea, with alternate rocky points and small sandy bays, which continue for the distance of six leagues to Jaguey point and the bay called Gramadel.

**GRAMADEL BAY** is a wild-looking place, with a heavy swell rolling in ; but it is visited occasionally for the hair-seal, with which it abounds ; there is anchorage in 6 or 7 fathoms, sandy bottom, with the bluff that forms the bay bearing S.S.E. about half a mile from the shore ; landing is scarcely practicable. The coast to the northward maintains its rocky character, with deep water off it as far as the Bufadero, a high steep cliff, with a hill having two paps on it, 1,620 feet high, a little in-shore.

**GUARMEY BAY.**—From this bluff a rocky cliff from 200 to 300 feet high, with a more level country, extends for 14 miles, as far as Legarto head, a steep cliff, with the land falling immediately inside it, and rising again to about the same height. Round this head is Guarmey bay, considered a tolerable harbour, in comparison with other places, having good anchorage everywhere in from 3½ to 10 fathoms, over a fine sandy bottom.*

Firewood is the principal commodity, for which it is the best and cheapest place on the whole coast. Vessels of considerable burthen touch here for that article, which they carry up to Callao, and derive great profit from its sale. There are also some saltpetre works established by a Frenchman, but little business is done in that line. The town lies in a north-easterly direction, about 2 miles from the anchorage, but it is hidden by the surrounding trees, which grow to the height of 30 feet. It has only one street, and cannot contain more than 500 or 600 inhabitants. At the anchorage there is a small house used for transacting business, but no other building, which is unusual, as at most of these places a small village has been established near the sea. Large stacks of wood are piled up on the beach ready for embarkation.

**Supplies.**—Fresh provisions, vegetables, and fruit, are plentiful and moderate ; but water is not to be depended on. It is true there is a river, and for several months after March a plentiful supply may be obtained ;

---

* *See* Plan of Casma and Guarmey bays, No. 1,368 ; scale, m — 2·0 inches.

but in the summer season great drought is sometimes experienced. In 1836 a whale-ship put in to supply her wants, and had to remain several days waiting for the water to come down from the mountains. The rise and fall of tide is very irregular, and the time of high water uncertain, but near 6 h. full and change ; 3 feet may be considered about the extent to which it ranges.

**DIRECTIONS.**—In coming from seaward, the best way to make this bay is to stand in on the parallel of 10° 6', and when within a few leagues of the coast, a sharped peaked hill with a large white mark on it, will be seen standing alone a little north of the bay ; the break in the hills through which the river runs, is high and cliffy on each side. The land is also much lower to the northward of Legarto head, and there is a large white islet at the north end of Guarmey bay.

In sailing in, after having passed the head a small white islet will be seen towards the middle of the bay ; steer for it, that the vessel may not border on the southern shore, for there are many straggling rocks running off the points ; and when sufficiently far to the northward to shape a mid-channel course between this Harbour islet and the point opposite it, to the south-ward, do so, and it will lead to the anchorage. In standing in, in this direction, the water shoals gradually to the beach, but the southern shore must on no account be approached nearer than a quarter of a mile.

The anchorage is in 4 fathoms, with Harbour islet bearing N.W. ¼ W. and the ruins of a fort on a hill inshore E. ¼ N., about a quarter of a mile from the landing-place on the beach. This landing-place does not seem to be so good as at a steep rock on the outer side of the bluff, where the sandy beach commences ; but probably it is the most convenient for loading boats. The sea-breeze sets in so strongly occasionally, that it is difficult for boats to pull against it ; this is particularly the case under the high land, whence it comes in sudden gusts and squalls.

**CULEBRAS POINT** is N.W. by N. 7½ miles from the white islet at the north end of Guarmey bay, level and projecting, and when seen from the northward, similar in appearance to Legarto head ; the intervening coast is a mass of broken cliffs and innumerable detached rocks, with moderately high land near the shore. Culebras cove is an anchorage off the valley of that name, on the north side of Culebras point. From that point the coast is broken into small sandy bays; and the Erizos rocks lie three-quarters of a mile off-shore.

**CORNEJOS ISLET** is white and cliffy, lying 5 miles to the northward of Culebras, and from thence the coast takes a bend inwards, forming a bay, which terminates at Mongoncilla point. A straight shore of 10 miles of length then leads towards the Colina Redonda, a promontory with two hummocks ; when seen from the southward it appears like an island. On

its north side is a caleta or cove, but it is fit for boats only, and immediately from the shore rises the Cerro or Mount Mongon.

The CERRO MONGON, 3,900 feet high, is the highest and most conspicuous object on this part of the coast; when seen from the westward it
has a rounded appearance, though with rather a sharp summit; but from
the southward it shows as a long mount with a peak at each end. It is said
there is a lake of fresh water on the range between these peaks, and that
its valleys abound with deer; but the truth of this depends on report only.

From the Mongon a range of hills runs parallel to the coast, which is
high and rocky, with some white islets lying off it as far as Casma, where
it terminates in Calvario point, a steep rocky bluff that forms the southern
head of that bay. *See* View on Sheet No. 1,285.

The BAY of CASMA is a snug anchorage, something in the form of a
horse-shoe; between the two entrance points it is 1¾ miles in a north-west
and south-east direction, and 1½ miles wide from the outer line of the
cheeks, with regular soundings from 15 to 3 fathoms near the beach.
The best distinguishing mark for Casma is the sandy beach in the bay,
with the sand-hills in-shore of it, contrasting strongly with the hard dark
rocks, of which the heads at the entrance are formed; .there is also a small
black islet lying a little to the westward of the North Cheek.*

The anchorage is with the inner part of the South Cheek about S.S.E.
a quarter of a mile off-shore, in 7 fathoms; for by not going farther in,
you escape in a great measure the sudden gusts of wind that come down
the valley with great violence. Captain Fergusson, of H.M.S. *Mersey*,
mentions a rock on the south side of the bay, half a mile from the shore,
carrying only 9 feet water, and sometimes breaking; it was not seen by
the *Beagle*, but, as doubtless it exists, it has been inserted in the plan.

From Casma the coast takes rather a more westerly direction, but continues bold and rocky. N.W. ¼ N., 14 miles from Casma, is the great bay
of Samanco or Guambacho; and midway between them the shore recedes
into a deep bight, with the two islands in front of Tortuga and Viuda;
but, from circumstances which need not be stated here, neither the bay
nor the islands were examined as to their capabilities.

The BAY of SAMANCO is the most extensive on the coast of Peru
to the northward of Callao; being 6 miles in length in a north-west and
south-east direction, and 3 miles wide; the entrance is 2 miles across between Samanco head on the south and Seal island on the north, and there
are regular soundings all over the bay.†

---

* *See* Admiralty Plan of Casma bay, No. 1,368; scale, m = 2·0 inches.
† *See* Plan of Port Samanco or Guambacho, No. 1,311; scale, m = 1·0 inch.

At the south-east corner, in a sandy bay, stands a small village (the residence of some fishermen), at the termination of the River Nepeña. This river, like most others on the coast, has not sufficient strength to force a passage for itself through the beach, but terminates in a lagoon within a few yards of the sea.   The town of Guambacho is about a league distant, at the eastern extremity of the valley ; and Nepeña, which is the principal town, lies to the north-eastward, about 5 leagues.   There is very little trade at this place ; small coasting vessels from Payta sometimes call here with a mixed cargo, and they get in exchange sugar and a little grain.

**Supplies.** — Refreshments may be obtained from the neighbouring towns, but wood is scarce.   The water of the river is brackish and unfit for use, but there are wells on the left bank, a short distance from the huts.   When taken on board, this water is not good, but, contrary to the general rule, after it has been some time confined on board, it becomes wholesome and pleasant tasted.

**Samanco Head,** the south point of Samanco bay, is a steep bluff, with some rocks lying a cable off it ; on opening the bay Leading bluff will be seen, a large mass of rock, on the sandy beach at the north-east side, and looking like an island.   In going in, give Samanco head a fair berth in passing ; a vessel may then stand in as close as convenient to the weather shore, and anchor off the village in 4, 5, or 6 fathoms, sandy bottom.   When rounding the inner points, take care of the small spars, for the wind comes off the Bell Mount, on the southern shore of the bay, in sudden and variable puffs.

**Mount Division,** with three sharp peaks 1,880 feet high, rising from the peninsula between Samanco and the bay of Ferrol, is, at a distance, the best mark to distinguish these bays.   Mount Tortuga, a short distance inland to the eastward, will also be seen; it is higher, but similar in appearance to the Bell mount.

**FERROL BAY** opens at a distance of 9 miles from Samanco ; it is nearly equal in size to Samanco, and separated from it by a low sandy isthmus.   It is an excellent place for a vessel to careen, being entirely free from the swell that sets into most of these parts.   On its north-east side is the Indian village of Chimbote, where it is said that refreshments of any kind may be had, but no water.   The entrance is clear ; but there is a reef of rocks off Blanca island, 1½ miles to the northward, which must be avoided.*   Santa Island, N.W. ½ N., 6 miles from the entrance of Ferrol, is about 1½ miles in length, lying N.N.E. and S.S.W., and very

_____

* *See* Plan on Sheet XV.; scale, m = 0·4 of an inch.

white ; just without it there are two sharp-pointed rocks, 20 feet above the sea.

**SANTA BAY.**—Two miles N.N.E. from the island, Santa head forms the south side of the bay of that name.  Although small, it is a tolerable port, without danger : the soundings are regular from some distance outside, and vessels may anchor anywhere in a moderate depth of water, but of course exposed to the swell.  To the north-west of the harbour, is Corcovado, a small but remarkably white island.  The best anchorage is in 4 or 5 fathoms, with the north-west extremity of the head bearing S.W. Fresh provisions and vegetables may be obtained on moderate terms.  It is also a tolerable place for watering.*

The town lies about two miles east from the anchorage, and the mouth of the river Santa 1¼ miles north of it.  This is the largest and most rapid river on the coast of Peru, from Santa head it is seen to wind its way along the valley, with several islets interrupting its course ; but at it termination it branches off and becomes shallow, with only sufficient strength to make a narrow outlet through the sandy beach, on which there is such a heavy and dangerous surf that no boat could enter the river with any degree of safety.  This part of the coast may be known by the wide-spreading valley through which the river runs, bounded on each side by ranges of sharp topped hills ; and in approaching, Santa island and head will be plainly seen.

**CHAO ISLANDS,** 1¼ miles off Chao point, lie N.W. ½ N. 15 miles from Santa.  The southernmost and largest is a mile in circumference, about 120 feet high, and, like most of these islands, quite white ; there are regular soundings from 10 to 20 fathoms, at the distance of a mile outside of them.  Between Santa and Chao the coast is a low sandy beach, with moderately high land a few miles in-shore.  The same charac- ter of shore continues through some shallow bays, as far as the hill of Guañape.

**GUANAPE ISLANDS,** lying N.W. by W. 16 miles from Chao point consist of two large islands, with some islets and rocks lying about them ; the southern one, 540 feet above the sea, is the highest and most con- spicuous.  There is a safe passage between them and the shore, from which they are distant between 6 and 7 miles.

**GUANAPE HILL,** 20 miles from Chao point, is about 700 feet high rather sharp at its summit and when seen from the southward appears

* *See* Plan of the Bay of Santa, No. 1,311 ; scale, m = 1·0 inch.

like an island.  On its north side there is a small cove, with tolerable
landing just inside the rock, that lies off the point.

From the hill of Guañape the coast continues to be a sandy beach with
regular soundings, and ranges of high sharp-topped hills, between 4,000
and 5,000 feet high, about 6 miles from the sea, till interrupted by the
little hill of Carretas, which stands on the beach, with the Morro de
Garita, which is 3,720 feet above the sea, overlooking it.  Here commences the valley of Chimu, in which Pizarro built the city of Truxillo ;
and 5 miles farther north we find the village and road of Huanchaco.
This is a bad place for shipping, and seems to have been unwisely chosen ;
for the north side of Carretas hill would appear to have been a better
place for landing and embarking goods.*

**HUANCHACO ROAD** is on the north side of a few rocks, which
run out from a cliffy point, and which shelter the beach in a slight
degree, but afford no protection to shipping.  The village is under the
cliff, and not distinguishable till to the northward of the point ; but the
church, which is on the rising ground, shows very distinctly, and is a good
guide when near the coast.  The usual anchorage is with the church and
a tree that stands in the village in one, bearing about East 1¼ miles from
the shore, in 7 fathoms, dark sand and mud.  Vessels often have to
weigh, or slip, and stand off, owing to the heavy swell that sets in ; it is
also customary to sight the anchor once in twenty-four hours, to prevent
its being embedded so firmly as to require much time to weigh it when
required.

Landing cannot be effected in ships' boats ; but there are launches
constructed for the purpose, manned by Indians of the village, who are
skilful in their management ; they come off on every arrival, and will
land passengers safely, for which they charge six dollars, equal to twenty-
four shillings sterling ; no more is charged for a cargo of goods, the risk
of the surf being that for which you pay.  Fresh provisions may be had
from Truxillo, but the surf renders watering nearly impracticable.

**TRUXILLO** city lies 9 miles to the south-east, in the rich valley of
Chimu.  It was founded by Pizarro, who gave it the name of his native
city.  Around it are numerous tumuli and other ancient Peruvian remains.
Population said to be 8,000.  Rice is the principal production of the valley;
and it is for that article and spice, and for the supply of Truxillo, that
vessels call at Huanchaco.

**DIRECTIONS.**—If bound for Huanchaco road, stand in on the parallel
of 8° 7' (which is a mile to windward), and Mount Campana, a bell-shaped
hill, 3,450 feet high, which is 7 miles to the northward of the road, will be
seen standing alone.  To the southward of this mount, on the north side

---

* See Plan of Peruvian ports, No. 1,294 ; scale, m = 1·0 inch.

of the valley of Chimu, is the sharp peak of Huanchaco, and shortly after the church and the shipping in the roads will come in sight.   The coast is cliffy for a few miles to the northward of Huanchaco ; the low, sandy soil with bushes on it then commences, with regular soundings in the offing, and continues as far as Malabrigo road.   Macabi island lies S. ½ W. 6½ miles from Malabrigo, with a safe channel of 10 fathoms between it and the main.

**MALABRIGO ROAD,** is formed by a cluster of hills, 790 feet high, projecting beyond the general trend of the coast, which at a distance appears like an island.   Although badly sheltered (as its name denotes), it is considerably preferable to Huanchaco.   There is a fishing village at the south-east side, but no trade is carried on.   The town of Paysan lies some leagues to the south east, and, by the account they gave of it at Malabrigo, must be of considerable extent.*

The best anchorage in Malabrigo road is with the village bearing about E.S.E. three quarters of a mile from the shore, in 4 fathoms sandy bottom. Landing is bad, but the fishermen have what they call *caballitos,* bunches of reed tied together, and turned up at the bow like a *balsa* of Chile, but much higher.   They are so light that they are thrown from the top of the surf to the beach, when the people jump off and carry them to the huts. It seems that each different bay or road has wisely its own peculiarly constructed caballito, adapted to the surf which it has to go through.

**PACASMAYO ROAD** bears N.W. by N., 20 miles from Malabrigo.   The coast between them is low and cliffy, with a sandy beach at the foot of the cliff, and soundings of 10 fathoms 2 miles off shore.   It is a tolerably good roadstead, under a projecting sandy point, with a flat running off it to the distance of a quarter of a mile.   The best anchorage is with the point bearing about S. by E., and the village east ; you will there have 5 fathoms sand and mud.   There is no danger in standing in ; the soundings are regular, shoaling gradually towards the shore.   Landing is difficult. Such launches are used as at Huanchaco.   The principal export is rice, which is brought from the town of San Pedro de Yoco, 2 leagues inland.   Fresh provisions may also be obtained from the same place ; wood and water may be had at the village on the beach, which is principally inhabited by Indians employed by the merchants of San Pedro.

**DIRECTIONS.**—To distinguish Pacasmayo road from seaward, the best guide is to stand in on the parallel of 7° 25′, and when within 6 leagues the hill of Malabrigo will be seen, like an island sloping gradually on each side ; and a little to the northward Arcana hill, rugged, with sharp

---

* *See* West Coast of South America, Sheet XVI., No. 1,335, scale, m.= 0·12 of an inch ; and Plans of Malabrigo and Pacasmayo roads, No. 1,294.

peaks, lying 7 miles to the southward of Pacasmayo point, and if clear Mount Sulivan, rising 5,000 feet above the sea, will be seen 18 miles inland. As the vessel approaches, the low yellow cliffs will appear (those north of the road the highest), on the summit of which, on the north side of the point, there is a dark square building that shows very distinctly. The best mark for the anchorage is the shipping, when any are there. *See* View on Sheet 1,335.

**ETEN POINT**, 11 leagues N.W. $\frac{1}{2}$ N. is a double hill (the southern one the highest, being 640 feet), with a steep cliff facing the sea. The north side of this cliff is white, and shows conspicuously. The coast between Pacasmayo and Eten continues low, with broken cliffs. Inland at 7 and 14 miles distance are two peaks respectively 1,900 and 2,450 feet above the sea.

**LAMBAYEQUE ROAD** lies N.W. $\frac{1}{2}$ N., a little more than 12 miles from Eten point. It is one of the worst anchorages on the coast of Peru. There is a small village on the rising ground, with a white church, off which vessels anchor in 5 fathoms 1$\frac{1}{4}$ miles from the shore. The bottom is a hard sand, and bad holding-ground : it is always necessary to have two anchors ready, for the heavy swell that sets on this beach renders it almost impossible to bring up with one, particularly after the sea breeze sets in.*

Rice is the chief commodity for which vessels touch at Lambayeque ; the only method of discharging or taking in a cargo (or in fact landing at all) is by means of the *balsa* of that country, which is a raft of nine logs of the cabbage palm, secured together by lashings, with a platform raised about 2 feet, on which the goods are placed. They have a large lug-sail which is used in landing ; the wind being along the shore enables them to run through the surf and on the beach with ease and safety, and it seldom happens that any damage is sustained by their peculiarly apt mode of proceeding. Supplies of fresh provisions, fruit, and vegetables may be obtained, but neither wood nor water.

**LOBOS DE AFUERA**, or off-shore Seal rocks, are a small group of islets 45 miles W.S.W. from Lambayeque. These islets are 3 miles in length north and south, 1$\frac{1}{4}$ miles broad, and about 100 feet high, of a mixed brown and white colours, may be seen several leagues. They are not quite barren, but afford neither wood nor water. The guano, with which the islands are more or less covered, is not so good in quality nor so plentiful as that found in the Chinchas, and apart from it the islands were not worth the trouble of taking care of.

There is a cove on the north side formed by the two principal islets but with deep water and rocky bottom ; within this cove there are

---

* *See* Plan of Peruvian ports, No. 1,294.

some little nooks, in which a small vessel might careen, without being much interrupted by the swell. It has, however, a rock in the centre with only a few feet water on it; by keeping close to the western point, where there is nothing to be avoided that cannot be seen, this danger will be cleared. The best anchorage is well up at the head of the bay. Should the guano come into much demand this bay will afford great facilities for shipping it.

These islets are resorted to by fishermen from Lambayeque on their balsas: they carry all their necessaries with them, and remain about a month salting fish, which fetch a high price on the mainland. There is no danger round the islets, at the distance of a mile; and regular soundings will be found between them and the shore, with which they are connected by a bank of less than 50 fathoms in depth. *See* View on Chart No. 1,335.

**LOBOS DE TIERRA,** consisting of one island and three or four islets or rocks, lie N.N.W. ½ W., 10 leagues from Lobos de Afuera, and only 10 miles from the nearest beach of the main. The main island is 5½ miles long in a north and south direction, and on an average one mile wide. When seen from seaward, the chief islet has a similar appearance to the former islets, and many rocks and blind breakers lie round it, particularly to the westward. There is tolerable anchorage on the north-east side, in 11 or 12 fathoms, sand and broken shells. A safe passage is said to exist between this island and the main, distant 10 miles; but as no advantage can be gained by using it, it was not thoroughly examined.

**CURRENTS.**—Navigating in the neighbourhood of these groups, and especially to the northward off Aguja point, great attention must be paid to the longitude, as the currents seem to be both irregular and strong. Vessels have been set as much as 36 miles to the westward of their courses in 24 hours, and at other times as much to the eastward; as a general rule after a strong breeze from the southward a westerly set may be expected, and therefore a set off the land when passing from Aguja to the northward.

**AGUJA POINT.**—The coast continues low and sandy, similar in appearance to that of Lambayeque, to the distance of 25 leagues W.N.W.: an extensive range of table-land of considerable height, with broken rocky points, then commences, and continues to Aguja or Needle point, which is long and level, terminating in a steep bluff 150 feet high, and has a finger rock a short distance off it, with several detached rocks round the point.* *See* View on Chart No. 1,335.

**NONURA POINT** is 3½ miles beyond, and 5 miles farther, in a N.E. by N. direction, is Pisura point, the south point of the Bay of Sechura:

---

* It was at Punta de la Aguja that Captain D'Aignan and M. de la Pinelas, of the French steamer of war, Mégère, observed the Total Eclipse of the Sun, on the 7th September, 1858.

between Aguja and Pisura points there are two small bays, where anchorage might be obtained if required.  The land about this part of the coast is much higher, and has deeper water off it than either to the north or south of it ; and may be known by its regularity and table-top, at about 1,000 feet.

**SECHURA BAY** is 12 leagues in length, from Pisura point to Foca island, in a N.N.W. direction, and is 5 leagues deep.  On the south-east side the coast shows low sand-hills ; but as it curves round to the northward it becomes cliffy and considerably higher.  Near the centre of the bay is the entrance to the river Piura, and the town of Sechura is situated on its banks.  This town is inhabited chiefly by Indians, who carry on a considerable trade in salt, which they take to Payta on their balsas, and sell to the shipping.  The river is small, but of sufficient size to admit the balsas when laden.  There is anchorage anywhere off the river in from 12 to 5 fathoms, coarse sand ; the latter depth being better than a mile from the shore.

This place may easily be distinguished by Sechura church, which has two high steeples, and shows conspicuously above the surrounding sand-hills ; one of these steeples has a considerable inclination to the northward, which, at a distance, gives it more the appearance of a tree than of a stone building.  From Foca point the coast is cliffy, about 20 feet high, and continues so as far as Payta point, which is 3 leagues distant N. ¼ E.

**SADDLE or SILLA de PAYTA** is a peculiar range of three insulated hills, 1,300 feet high, and 1¼ miles from the coast, situated between Foca and Payta points.  They vary in colour from bright yellow sandy, to black, according to the position of the observer and the sun ; and they only show the Saddle when seen on a north-easterly bearing.  They are an excellent land mark, and the Port of Payta may readily be found by knowing that their peaks lie 6½ miles south of it.  *See* View on Chart No. 1,335.

**PORT PAYTA** is the best open port on the coast.  A considerable trade is carried on, vessels of all nations touching there for cargoes, principally of cotton, bark, hides, and drugs, in return for which they bring the manufactures of their several countries.  In the year 1835 upwards of 40,000 tons of shipping anchored at this port.*  In common with all the ports before mentioned Payta has a bi-monthly communication with Europe across the isthmus of Panama.  The trade of Payta is rapidly growing in importance from its large exports of cotton and grain, which

* *See* Plan of Port Payta, No. 1, 293 ; scale, m = 2·0 inches.

are abundantly raised in the interior; and the place is considered extremely healthy, though without water.

The town is built on the slope and at the foot of the hill, on the south-east side of the bay; at a distance it is scarcely visible, the houses being of the same colour with the surrounding cliff, which is 200 feet high. It is said to contain 5,000 inhabitants, and is the sea-port of the province of Piura, the population of which is estimated at 75,000.  Fresh provisions may be had at Payta on reasonable terms, but neither wood nor water, except at a high price, most of that now used (1854) being brought on mules a distance of 18 to 20 miles.*

DIRECTIONS.—There is no danger in entering the bay or port: after rounding Payta point, which has a signal station on its ridge, False bay, which is rocky, will open; this must be passed, as the true bay (the beach of which is sandy) is round Inner point, the eastern point of False bay.  That point ought not to be hugged closely, for there are some rocks at the distance of a cable's length, and the wind baffles often. After having passed Inner point and opened the town, stand in towards its centre, and anchor in from 10 to 7 fathoms mud, tough holding-ground. The landing-place is at the mole, about the middle of the town.

CAUTION.—When leaving Payta for the southward at night time, care must be taken not to mistake Rocky point for Foca island, as they are then much alike, the black rocks of the point relieved by the sandy bays on each side, showing at that time precisely like an island, and should the course be then altered for Aguja point, it would most likely lead amongst the rocks and foul ground to the northward of the island before the mistake is found out.†

PARINA POINT, the western extremity of South America, is 27 miles from Payta, and rises to a bluff about 80 feet high, with a reef out to the distance of half a mile on its western side; between this point and Payta the coast is low and sandy, with table-land of a moderate height at a short distance from the beach, at 16 miles east of Pariña point, and thence extending 40 miles to the north-east is the range of the Amatape mountains, rising from 3,000 to 4,000 feet above the sea.‡

TALARA POINT.—After rounding Parina point, the coast trends abruptly to the northward for 24 miles, and becomes higher and more cliffy in approaching Talara point.  This is a double point, the southern

---

* It was near Olmos, about 100 miles S.E. of Payta, that Lieut. J. M. Gilliss of the U.S. navy, observed the Total Eclipse of the Sun on the 7th September, 1858.  He determined the position of his station to be 6° 0′ 2″ S., long. by chronometers 79° 42′ 53″ west of Greenwich.

† Remarks of Commander James Wood, R.N., 1854.

‡ See West Coast of South America, Sheet XVII., No. 1,813; scale, m=0·12 of an inch.

part of which is cliffy, about 80 feet high, with a small black rock lying off it ; the northern part is much lower, and has a few breakers near it. On the north side of this point there is a shallow bay, in the bight of which the high cliff coast again commences, and continues in a line towards Cape Blanco.

CAPE BLANCO, apparently the angle of a long range of table-land, is 900 feet high, bold, and sloping gradually towards the sea. Near the extremity of the cape there are two sharp hillocks ; and midway between them and the commencement of the table-land is another rise with a sharp top. There are some rocks that show themselves about a quarter of a mile off, but no danger exists without that distance. From Cape Blanco the general trend of the coast is more easterly, in nearly a direct line to Malpelo point, which is 66 miles distant.

SAL POINT, 22 miles from Cape Blanco, is a brown cliff, 120 feet high. The coast between is a sandy beach, with high cliffs as far as the valley of Mancora, where it is low, with brushwood near the sea ; the hills being at a distance inland. Northward of Sal point the coast is cliffy to about midway between it and Picos point ; it then becomes lower and similar to Mancora.

PICOS POINT is a sloping bluff, with a sandy beach outside of it, and another very similar point a little to the northward ; behind there is a cluster of hills with sharp peaks, 710 feet above the sea ; from whence arises probably the Spanish name of that point. From Picos the coast is a sandy beach, with a mixture of hill and cliff of a light brown colour, and well wooded. There are several small bays between it and Malpelo point, which bears N.E. ¼ N. 24 miles distant.

MALPELO POINT forms the southern side of the entrance of Guayaquil river, and may be readily known by the marked difference between it and the coast to the southward. It is very low, and covered with bushes, and a short distance in-shore there is a clump of bushes higher and more conspicuous than the rest, which shows plainly on approaching.

TUMBEZ RIVER issues at the extremity of the point, and a reef stretches out from thence for a quarter of a mile. This place is much frequented by whalers for fresh water, as they can fill their boats from alongside, about a mile up the river ; but great care is necessary in crossing the bar, as a heavy and dangerous surf beats over it, and renders that operation at all times difficult. The entrance to the river may be distinguished by a hut on the east side in going in ; it may be perceived immediately on rounding the point of Malpelo. About two leagues up the river stood the old town of Tumbez, now scarcely more than a few huts, and barely sufficient to supply the whalers with fruit and vegetables. They anchor anywhere off the point in 6 or 7 fathoms. This river is said to be the boundary between the States of Peru and Ecuador.

**WINDS.**—The prevailing winds on the shores of Peru blow from S.S.E. to S.W.; seldom stronger than a fresh breeze, and not often on certain parts of the coast more than sufficient to enable shipping to make a passage from one port to another. This is especially the case in the district between Cobija and Callao. Sometimes during the summer, for three or four successive days there is not a breath of wind; the sky beautifully clear, and with a nearly vertical sun.

On the days that the sea breeze sets in, it generally commences about ten in the morning; light and variable, at first, but gradually increasing till one or two in the afternoon; from that time a steady breeze prevails till near sunset, when it begins to die away; and soon after the sun is down all is dead calm. About eight or nine in the evening light winds begin to come off the land, and continue till sunrise; when it again falls calm until the sea-breeze after mid-day.

During the winter (from April to August) light northerly winds may be expected frequently, and are generally accompanied by thick fogs, or dark lowering weather; but this seldom occurs in the summer months, although even then the tops of the hills are frequently enveloped in mist. To the northward of Callao the winds are more to be depended on; the sea-breeze sets in with greater regularity and more vigour than on the southern parts of the coast; and near the limit of the Peruvian territory (about Payta and off Cape Blanco), a double-reefed top-sail breeze is not uncommon. It is to be remarked, that although these moderate wind are the general rule on the coast of Peru, yet that sudden and heavy gusts do often come over high land after the sea-breeze sets in, and from the smallness of the ports are attended with some inconvenience, if precautions are not taken in duly shortening sail previous to entering them.

**WEATHER.**—The only difference between winter and summer, as far as regards the winds, is the frequency of light northerly airs during the former months; but in the winter the difference in climate is far greater than one would imagine in so low a latitude. In the summer the weather is delightfully fine, with the thermometer (Fahr.) seldom below 70°, and often as high as 80°, in a vessel's cabin; but during winter, the air is raw and damp, with thick fogs and a cloudy overcast sky. Cloth clothing is then necessary for the security of health; whereas in summer, the lighter one is clad the more conducive to comfort and health.

**CURRENTS.**—The waters of the South Pacific ocean form a current on the west coast of South America, which extends as a river of cooler water from the latitude of Chiloe to the equator, along which it sets to the westward. From its becoming more evident in the warmer latitude of Peru, it has been denominated the Peruvian current. Its westerly set is felt on the coast between Arica and Pisco, especially to the southward of the latter port. Its greatest force on the American coast is between Payta and the Galápogos, where vessels have occasionally been drifted 50 miles to the W.N.W. in 24 hours. The general set of the current on the coast of Peru is along shore to the northward, from a half to one knot; and yet occcasionally it sets to the southward, with equal or greater strength.

The periods at which this southerly movement takes place cannot be foreseen with any degree of certainty. Neither the seasons, the age of the moon, nor any of those causes to which we so freely ascribe the currents of other coasts, seem to have any influence here. The oldest navigators in the coasting trade can neither predict these changes by their experience, nor connect them by the closest observation—they only know that they will suddenly take place, and endeavour to profit by them accordingly. These irregular currents may be connected with the causes of the remarkable meeting of oceanic currents about the Galapagos.

During the continuance of the *Beagle's* survey, these southerly sets were frequently experienced immediately preceding or during northerly winds; but this was far from being always the case, and no general rule could be found to hold; however, it appeared natural that there was some connecting link between them, for at times the current was found to change; and to set to the southward after a fresh wind had for several days been blowing from that quarter. Again, no inequalities or irregularities in the coast line seemed to have any effect on the main body of the current, and every fresh observation that was made served to awaken fresh curiosity, without helping to elicit the source of these singular but interesting anomalies.

**PASSAGES.**—With regard to making passages in sailing vessels along this coast, little difficulty is found in going to the northward; a fair offing is all that is requisite to ensure any vessel making a certain port in a given number of days; but in working to the southward, some degree of skill and constant attention are necessary.

Much difference of opinion exists as to whether the in-shore or off-shore route should be preferred; but Captain Fitz-Roy's experience, added to the information he obtained from those who were said to understand the navigation of that coast, led him to propose the following rules :—

On leaving Guayaquil or Payta, if bound to Callao, work close in-shore to about the islands of Lobos de Afuera. All agree in this. Endeavour always to be in with the land soon after the sun has set, so that advantage may be taken of the land wind, which, however light, usually begins about that time; this will frequently enable a ship to make her way nearly along shore throughout the night, and will place her in a good situation for the first of the sea-breeze.

After having passed the before-mentioned islands, it would be advisable to work upon their meridian until the latitude of Callao is approached; then stand in and if it be not fetched work up along shore, as above directed, remembering that the wind hauls to the eastward on leaving the coast. Some people attempt to make this passage by standing off for several days, hoping to fetch in well on the other tack; but this will generally be found a fruitless effort, owing to a northerly current which often is found on approaching the equator.

Sailing vessels bound from Guayaquil to Valparaiso should stretch out to sea, crossing the Peruvian current before passing the meridian of 92° west. From this they should push to southward, not caring about being driven to the westward if southing can be made, as they will have no difficulty in making their easting in the parallel of Valparaiso. This passage is generally made in 37 days.

For a sailing vessel bound from Callao to Valparaiso, there is no question but that by running off with a full sail the passage will be made in much less time than by working in-shore, for she may run quite through the trade, and fall in with the westerly winds which are always found beyond it. But for the intermediate ports (excepting Coquimbo) the case is different, as they lie considerably within the trade-wind, and must be attained through that medium alone. A very dull sailer might indeed do better by running through the trade, and making southing in the offing, so as to return to the northward along the coast, than by attempting to work to windward against a trade wind which never varies more than a few points.

For the Puertos Intermedios, or Intermediate ports it may be recommended to work along shore as above directed, as far as the island of San Gallan, from whence the coast trends more to the eastward, so that a long leg and a short one may be made (with the land just in sight) to Arica, or to any of the ports between it and Pisco.

From Arica the coast being nearly north and south, vessels bound to the southward should make an offing of not more than 15 or 20 leagues (so as to ensure keeping the sea-breeze), and work upon that meridian till in the parallel of the place to which they are bound. But on no account is it advisable to make a long stretch off; for as the limit of the trade wind is approached, it gradually hauls to the eastward, and great difficulty will be found in even fetching the port from which they started.

The average passage in a well-conditioned merchant vessel from Guayaquil to Callao is from fifteen to twenty days ; and from Callao to Valparaiso
about three weeks.   Fast sailing schooners have made these passages in
much less time ; and there is an instance of two men-of-war, sailing in
company, having gone from Callao to Valparaiso, remained there two days,
and re-anchored at Callao on the twenty-first day.   But these are rare
occurrences, and only to be done under most favourable circumstances,
such as meeting with a *norther* soon after leaving Callao.

**VARIATION.**—The variation on this part of the coasts of Bolivia and
Peru at the beginning of the year 1860 was as follows :—Off Hueso
Parado 13° E. ; off Cobija 12 E. ; at Callao 11° E ; off Sechura 10° E. ;
and at the Rio Tumbez 9°.   The curves of equal variation in this district assume nearly an E.S.E. direction, and each degree is separated by
an interval of about 100 miles.   The variation is slightly increasing at
the rate of about 1' annually.*

---

\* *See* Admiralty Chart of the World, showing the curves of equal variation for the
epoch 1858, by Mr. Fred. J. Evans, Master, R.N.

## CHAPTER XV.

RIVER TUMBEZ TO CAPE CORRIENTES AND THE GALAPAGOS
ISLANDS.

VARIATION 9° to 7° E. in 1860.   Annual increase about 1'.

THE GULF of GUAYAQUIL, into which the river of the same name
empties itself, is the largest inlet on the west coast of South America
north of Chiloe island; it extends inland upwards of 50 miles, and is
17 miles wide at the entrance between the Payana shoals and the island
of Puna.   It is encumbered with extensive shoals, which, however, are
buoyed, so that with the assistance of a pilot, vessels drawing 18 feet may
cross the bar north of Puna at high water and proceed up the river to the
city, a distance of 80 miles from the outer entrance, with tolerable
facility.   The southern shore of the gulf is low, thickly wooded, and
edged with shoals, which extend about 3 miles from the coast.   The
northern or Puna shore is also low and well wooded, but except at Salinas
point, the bank does not extend more than $1\frac{1}{2}$ miles from the shore, and
there is a range of hills named Zampo Palo, about 4 miles inland.*

SANTA CLARA or AMORTAJADA (shrouded) island is the best mark
for making the Gulf of Guayaquil, being high, remarkable, and lying
near the centre of the entrance, 14 miles from Payana point and 12 miles
from Salinas point, the south-west extreme of Puna island.   It is narrow
and about one mile and a half long, lying N.E. by N. and S.W. by S. ;
a spit extends for nearly half a mile off its north-east end, and the bottom
is foul for about a mile in the direction of Puna.   This island, on many
bearings, assumes the appearance of a gigantic shrouded corpse, the
resemblance being more complete when the centre bears W. $\frac{1}{4}$ S.

LIGHT.—The lighthouse on Santa Clara island is a wooden tower,
about 30 feet high.   It stands on the breast of the island, one-third from
the southern head, and exhibits, at an elevation of 250 feet above high
water, a *fixed white* light, visible in every direction except from S. $\frac{1}{4}$ E.
to S. by W. ; and, should a vessel approach too near the island in a
southerly direction, it will be shut in by the edge of the cliff.   Its most

---

* See Admiralty Charts, West Coast of South America, Sheet XVII., No. 1,813 ;
scale, m = 0·12 of an inch, and Guayaquil River, No. 586 ; scale, m = 0·2 of an inch.

brilliant face is seen from E.S.E. to W.S.W. by the north, in which direction it is visible in clear weather a distance of about 16 miles.*

The **AMORTAJADA SHOALS** lie off the south end of Santa Clara, at a distance of 2 miles in a south-west, and one in a west direction ; they consist of rocky patches, some awash, and others with 12 and 15 feet water, with 4 to 6 fathoms among them and 7 on their outer edge. Vessels approaching the island should keep the lead going, and come no nearer than 12 fathoms.

**TUMBEZ BAY.**—The south coast of the gulf north-east of Malpelo point recedes, and forms the shallow bay of Tumbez, the northern point of which is called Payana, on the Tembleque islands, distant 16 miles from Malpelo point, and forming a portion of the low land at one mouth of the Rio Tumbez. Here in the year 1527 the first Spanish colony in Peru was formed at a city named Tumbez, distinguished at that time for its stately temple and palace of the Incas, and the country described as fully peopled and cultivated with an appearance of regular industry.†

**PAYANA SHOALS** extend 5½ miles to the north-east of Payana point, and dry at low water to a distance of 2½ miles from the shore. A *black* buoy with *staff* and *ball* is moored in 4½ fathoms water, about two cables from the nearest breaker at low water, and about 2½ miles from the nearest land, with the lighthouse on Santa Clara bearing W. by N. ¾ N. 14 miles, and Payana point S.S.W. ¾ W. 4½ miles. There are 12 and 15 fathoms immediately outside this buoy, and the whole space is clear between it and the Amortajada shoals.

**BOCA JAMBELI,** or mouth of the Santa Rosa river, is 14 miles N.E. by E. ¼ E. from this buoy; there are 8 fathoms in the entrance, which is nearly 2 cables broad, between shoal patches of 16 and 18 feet. The river trends to the southward with 6 and 7 fathoms in the channel as far as Machala creek, nearly 5 miles from the entrance, beyond which it was not examined. To the northward of this river the soundings become irregular, and the patches of shoal water more numerous. The land is low with several creeks, and runs N. by E. ¼ E. a distance of 35 miles as far as Valao chico at the entrance of the Mondragon channel.

---

* Numerous complaints have been received of the light not having been shown for several nights together, from the keeper deserting it to procure provisions, oil, and other necessaries.

† Robertson's History of America, Book VI.

**SALINAS POINT** on Puna island, forming the north point of entrance to the Gulf of Guayaquil is low and woody ; mud banks, dry at low water, extend 2 miles on its western, and nearly one mile on its southern side. It is shoal for 5 miles in the direction of Santa Clara island, and to the north-west along the western coast of Puna island for 10 miles ; the latter lie out of the general track of shipping, and have not been thoroughly examined.

**ARENAS POINT**, 9 miles to the eastward, is the pilot station, and does not appear sandy as its name implies, but is wooded like the rest of the coast ; it has sometimes been called Salinas point ; the shore between them is edged with shoals which extend about two miles from the land, and about the same distance west of the point is the entrance of a creek called Salinas river. Arena point is 12 miles from the Boca Jambeli on the opposite shore, and here the difficulties of the navigation may be said to commence. Vessels not acquainted with the place should anchor off this point, and wait for a pilot.

**MALA HILL.**—At Arenas point the coast trends to the N.N.E. for 16 miles to the foot of Mala hill, which is useful as a leading mark and visible 16 miles to the southward, making like a moderately high island. The coast on this side of Puna is of a similar nature to the southern shore ; a large creek, called Puna Vieja, extends some distance inland, but it is shoal at the mouth, which is nearly 5 miles to the northward of Arenas point. Between Mala hill and the range of Zampo Palo the island of Puna is nearly all mangrove swamps intersected by several inlets or creeks, of which Puna vieja is the largest, and is said nearly to separate the island. Zampo Palo begins on the south shore of this creek and extends across the island in the direction of Salinas point ; on the eastern part of the range are several peaks, while to the westward it forms a long flat table-land.*

**ESPAÑOLA POINT**, 3½ miles E.N.E. from Mala hill, is a remarkable cliffy point, and forms a useful leading mark to vessels from the south-ward to clear the Mala bank. To the westward is the house and planta-tion of the English consulate, and about three-quarters of a mile south of the point is a 10-feet patch with 6 and 4½ fathoms close to. From this the coast again trends to the north to Mandinga point, a bold bluff, forming the north-east extreme of Puna island ; just to westward is the village of Puna, a place famous in the annals of the Buccaneers, by whom it was occasionally sacked, and the inhabitants at last retreated to Guayaquil. It

---

* Remark book of Lieutenant R. Collinson, R.N., 1836.

is now frequently the resort of the principal persons of the latter city during the rainy season.

PUNA consists of about 50 houses and a church, the former like most in this country are raised on piles, about 10 feet from the ground. This style of building has several advantages; the houses are in a great measure freed from the noxious exhalation of the earth, better ventilated, and ants and other insects cannot make their inroads so easily. The island of Puna appears to have been a place of some importance, and well inhabited under the Incas. Here Pizarro met the first check on his march along the coast, the islanders defending themselves with such valour that their reduction cost him six months. The coast from Mala hill to this village is much higher than any other on the island, presenting a line of cliffs fronted with a beach, forming a pleasant contrast to the swampy mangrove shore to the southward. At Puna the river Guayaquil commences, and no ship should attempt the passage without a pilot, which is easily procured at that place.

MALA BANK is an extensive shoal consisting of a chain of banks, 16 miles long by 1 broad, with from 2 to 15 feet water on them. It is nearly parallel to the east shore of Puna, at a distance of about 3½ miles, and is marked by a *white* buoy on each extreme; that on the southern lies in 4 fathoms, on the south end of a patch, known as the shoal of Punta Arenas, with the termination of the trees on Arenas point bearing W. by S. ¼ S. 4½ miles, and Mala hill N. ¼ E. 14 miles. The north buoy also lies in 4 fathoms, with the English Consul's house on the hill, just shut in with the sandy bluff of Española point bearing W. by N., the west point of Mondragon island a ship's length open of Puna bluff N. by W., and Mala hill W. ¾ S. The shoalest part is about 5 miles to the southward of this buoy, 1½ miles to the eastward of which are 3 small patches with 18 and 16 feet on them. There is a channel on each side of the Mala bank; the eastern is broader and deeper, yet the western is the one generally preferred by the pilots.

Puna Patch is a small bank 4 miles N. by E. ¼ E. from the north buoy of the Mala bank, and 2 miles E. by N. from Mandinga bluff, lying in the route of vessels using the eastern channel bound to Puna for a pilot. It is about half a mile in circumference, with 1 foot on it in the shoalest part, and 4 to 5 fathoms close to.

TIDES.—It is high water, full and change, at Santa Clara island at 4h., and at Puna village at 6h.; the rise and fall at each place being 11 feet. In the gulf the tide stream appears to set in the same direction as the trend of the shores, from 3 knots at springs to 1 at neaps, the ebb stream being the strongest. Off Santa Clara the ebb sets to South and the flood to the East. In all parts of the gulf, in a sailing vessel, it will be well to anchor with light winds and an ebb-tide. Off Puna the flood sets to the N.W., and the ebb S.S.E. at about 2½ miles an hour at springs.*

**DIRECTIONS.**—Vessels bound to Guayaquil from the southward should make the land about Picos point, which is remarkable, having many small sandy peaks ; a few miles farther north is Malpelo point, low and covered with trees. Soundings of 41 fathoms sand and ooze will be obtained when 10 miles to the north-west of this point. Coming from the northward Santa Clara island may be made, which is visible about 16 miles, and at first appears like three hummocks, and Zampo Palo, the high range on Puna island, will generally be seen at the same time. Santa Clara should not be approached nearer than 2 miles or within the depth of 12 fathoms, the best track being about 5 miles to the southward of it in from 20 to 15 fathoms water, from whence a N.E. ½ E. course for 25 miles will lead towards Arenas point, between it and the South buoy of the Mala bank. A pilot can generally be obtained at Arenas point, but with common care a vessel may proceed as far as Puna. The channel west of the Mala bank is preferred, Mala hill forming a good leading mark.

After passing Arenas point continue on the same course N.E. ½ E., keeping a good look out for the south buoy of the Mala bank, and when Mala hill bears N. by E. steer for it. On this course you will have from 20 to 9 fathoms water, but northward of Puna vieja it shoals to 4 fathoms, deepening again after passing Española point. The water shoals gradually towards the island, so by keeping that on board as near as the vessel's draught will permit, and not going in more than 7 fathoms, which will keep you clear of the Mala bank, there is no danger, should the hill be hidden. When Centinela point is shut in by Española bluff, bearing N.E. ½ N., haul out N.E., passing about 1 mile south of the bluff, in not less than 7 fathoms, and looking out for the North buoy. When Mandinga bluff opens of Centinela point, bearing N. ½ E., steer along the land until the village of Puna is seen, when you may stand towards it and come to about half a mile to the northward, with Mandinga bluff bearing S. by W.

The Eastern channel is wider and about the same depth, but has no good leading marks, and a ship must trust to her lead. After passing the South buoy of the Mala bank steer N.E. by N. 10 miles, and then N. by E. 8 miles to the North buoy, taking care not to go into less than

* Remarks of Captain Basil Hall, R.N., 1821 ; and Lieut. H. Trollope, R.N., 1846.

4 fathoms, and when the houses south of Española point are shut in by the same bluff, bearing W. by S., you will be clear of Mala bank, and may steer for Mandinga bluff, which may be rounded at the distance of one-third of a mile. This channel may be used by sailing vessels working down the gulf, taking care not to go into less than 4 fathoms, and not to cross the Mala bank.

The **RIVER GUAYAQUIL**, the largest on the west coast of South America, is navigable 75 miles above Puna, and for vessels drawing 18 feet water as far as the city. The lower part of the river is from 1 to 1½ miles wide, being similar to all tropical rivers, bordered by low mangrove islands, upon which it is scarcely ever practicable to obtain a footing ; large mud-banks, which dry at low water, extend from these islands, covered with cranes and herons in great variety, dotted with numerous alligators, and infested with mosquitoes.

The **CITY of GUAYAQUIL**, 35 miles from Puna, is the principal seaport of the republic of Ecuador. It stands on the west bank of the river, to the southward of three remarkable hills called Los Cerros de la Cruz, the centre one being 320 feet high, from the foot of which it extends about 1 mile. It is well laid out, the houses chiefly being built of wood, with several churches and a custom-house, but, from its being on a dead level, the drainage is bad, and in the wet season it is unhealthy ; this is increased by an extensive marsh situated to the westward. There is a fine quay, or marina, which extends along the bank of the river, the whole length of the town ; it is 60 feet broad, coped with stone, and lined with a row of respectable houses, which gives the city a fine appearance from the river, especially in the evening when the rooms are lighted up.*

A patch, with only 13 feet water on it, caused by the sinking of a ship, lies directly off the centre of the town, 1½ cables from the quay. The marks for it are the clock-tower on the quay, midway between the cathedral and church of San Augustin, bearing W. by S., and the western Cerro de la Cruz open left of the church of San Merced, N.W. ½ W.

**Supplies.**—Bullocks can be obtained here, and the water for the use of the town and shipping is brought from a considerable distance up the river in earthen jars, from 100 to 150 of which are packed on a balsa or raft formed of logs of a very light wood lashed together with vine, and floated down with the tide. The water of the river off the town is fresh at the last of the ebb, but is considered unfit for drinking, passing as it does through a mass of mangroves. Most of the produce of the

* Voyage of H. M. S. *Herald*, Vol. 1., page 209.

interior, consisting of every variety of tropical fruits and vegetables, is brought down in the same manner, whole families living on the balsas, which are met with in all parts of the river above the bar. The population was about 20,000 in 1847. In 1858, 190 vessels, of 42,913 tons, entered inwards, and the same number cleared outwards; the value of imports was 506,456*l.*, and of exports 474,324*l.*, the principal being cocoa. Opposite to the town, on the island of Santay, was a building slip in 1847, where several vessels had been constructed, and there is a gridiron to the northward of the city on the eastern side of Cerros de la Cruz.

CHIMBORAZO mountain, rising 21,424 feet above the level of the sea, is visible in clear weather from Guayaquil, from which it is distant 76 miles, presenting the imposing appearance of a vast double peaked mass covered with perpetual snow. Chimborazo is the culminating point of the Colombian Andes, at about 90 miles south by west of Quito. Its elevation was measured by Humboldt and Bonpland in the year 1802, who ascended it to a height of 19,286 feet, and Boussingault in 1831 reached to 19,689 feet. This mountain was long supposed to be the highest of the Cordillera of the Andes, but it is now known that it is only the sixth in height, being surpassed by four peaks in Peru, and by Aconcagua in Chile. Its whole height as above stated is 21,424 feet above the sea level, but above the valley of Quito, which forms its base, it only rises 11,958 feet.

The BAR of Guayaquil river lies 6½ miles to the northward of Puna, between the south end of Mondragon and Green island; it is 2½ miles across, the least depth at low water springs, being 12 feet. A long mud flat, dry at low water, extends from the south part of the bar for 4 miles in the direction of the Puna patch. Green island, on the western side of the river, also has a bank to the southward in the same direction, and it is between these banks that the main channel lies.

MONDRAGON and MATORILLOS islands are at the entrance of the river, and, although termed islands, are nothing but large banks covered with mangrove trees, actually growing in the water, forming a grove of innumerable pillars, at a distance quite ornamental, but from their monotony soon becoming to a stranger as wearisome as a barren desert, this being the general appearance of the banks of the river. Both to the northward and southward of Mondragon and Matorillos are extensive mud flats, covered at half tide. There is a channel to the east of Mondragon, and also of Matorillos, but both are barred at their northern extremes, where they join the main river. The rivers Naranjal and Taura open into the Mondragon channel, down which there is considerable trade to Puna.

**Piedras point.**—The western or main bank of the river is of a similar nature to that of the islands, possessing occasionally small cleared spots, on which landing may be effected at high water. The first of these spots is called Puerto de Balsas, just over the bar, about 9 miles north of Puna. Piedras point, 9 miles farther to the northward, is the most considerable, and in the time of the Spaniards had a fort ; there is a small hill over it, which is remarkable amid the low land. Two miles and a half to the southward of the point, at the entrance of the Mondragon channel, is a small rock called the Baja, awash at low water.

The next landing is on the eastern bank of the river, 6 miles above Piedras point, called Estero de Tiramano, and there is also one on the opposite shore, at the northern point of Masa island. At Casa de Josefina, 3 miles farther on the western shore, a bank commences which continues to the city. At this point the river divides, having a passage on either side of Santay island ; the western being the only available one for ships.

**TIDES.**—It is high water, full and change, off the city of Guayaquil at 7h. ; the rise being 11 feet. The stream runs from 3 to 4 knots, following the trend of the shores.

**DIRECTIONS.**—No vessel bound to Guayaquil should leave Puna without a pilot ; the following remarks, however, may be useful. On weighing from Puna with the flood, a N.N.W. ¼ W. course may be steered just open of Green point, at the east end of Green island, and when the south point of Green island is on with Cascajal point, or the north end of Puna, bearing W.S.W., cross the bar on a N. by E. ¾ E. course for the extreme point of Mondragon island, having Puerto de Balsas, the cleared spot before mentioned, just open on the port bow ; from thence the course lies along the western bank as far as Piedras point, from which a shoal dry at low water runs for 7 miles, having a channel on each side of it, but the better is on the eastern shore, until the vessel is abreast of the Estero de Tiramano opposite to Masa island. From the north extreme of this island steer for a similar spot on the opposite shore, along which the course lies to the town. It is necessary to moor off the city, taking care of the shoal off the quay.

With a sailing vessel the passage up the river is made easy by the prevailing fair wind, the return is generally accomplished by kedging.

**MORRO CHANNEL** is another approach to the river Guayaquil to the westward and northward of Puna island, but the entrance is encumbered with shoals, and the land so low that no good leading marks could be obtained, so that, with the good channels to the eastward of Puna island, there is but little occasion to use this dangerous passage.

The western coast of Puna island is fronted by shoals for a distance of

10 miles ; northward of these is the entrance of the Morro channel, the
mark for which is the northern hills on Puna in line with the Morro point
bearing E. ¾ N., this course may be used until a red cliff on Puna bears
N.E. ½ E.   Then haul up N.E. by E. for the small bay to the northward,
keeping the cliff on the starboard bow ; on this course the channel will
open between the Morro and Trinchera points.   In this channel some distance to the northward will be seen Mangles island, which should be steered
for, when bearing N. by E. ½ E., until Isla Nueva, a small mangrove islet to
the eastward, just touches the main island of Puna bearing E. by N. ¼ N.
From this point the vessel must steer along the northern shore of Puna,
by the lead, there being a 3 fathom channel ; the chart however, is the
only guide that can be given to this dangerous and intricate passage.

ESTERO SALADO is a singular creek running parallel to the river
Guayaquil, the head of it being only about 2 miles distant from the south
end of the city ; it is deep and free from shoals, but with an extensive bar
only partly examined on which no greater depth than 7 feet could be
found.   The entrance is 6 miles wide, lying northward of Puna island and
the Morro channel ; it is only a cable wide at the head near the city, but
deep, having 7 and 8 fathoms water.   Like the Morro channel, this creek
is of little value, and has only been used as a point from which the city
may be attacked.

SHOALS of CHANDUY.—The coast westward of the entrance of the
Morro channel, as far as Carnero point, a distance of 48 miles, has never
been examined.   In the Spanish chart of Malaspina of 1791, it is represented as fronted with shoals, extending 7 miles from the shore, probably of
the same nature as those off the western coast of Puna island ; this piece of
coast, therefore, should be carefully avoided.   The heights of Chanduy
are a table land, situated midway between Puna island and Carnero point,
distant about 11 miles from the sea coast.   Leaving Guayaquil for the
northward, these heights form a conspicuous object.   The indraught of the
Morro channel must be guarded against.   A heavy swell and strong tides
will be found between Santa Clara and Carnero point, which latter is low
and flat.

POINT SANTA ELENA, 11 miles N.W. by N. from Carnero point,
terminates in an abrupt hill, 424 feet high, flat at the top, and appears
like an island when made either from the northward or southward, the
isthmus connecting it with the mainland being low.   The southern shore
of the point for a distance of 8 miles is a line of beach and sand hills,
ending in a small creek, on the south shore of which is a rocky bluff ; one

mile to the eastward of this is a rock awash, the only danger on this side
of the point.*

**SANTA ELENA BAY** is to the northward of the point, forming a good
anchorage, with convenient depth of water. One mile N. by W. of the
hill is a spot that should be avoided, where the sea lifts heavily ; no less
than 4½ fathoms was found on it, with 6 and 8 between it and the shore.
The land is barren, producing only stunted bushes ; but there are several
natural salt pans, and about 4 miles from the point a kind of pitch is found.
The town of Santa Elena is 8 miles to the eastward, and 1¼ miles from
the shore, lying at the foot of a hill about 400 feet high ; landing is easy
at the custom-house, and the chief trade is salt, obtained from the pans
before mentioned. It is high water, full and change, in Santa Elena bay
at 1h. 18m., the rise and fall being 8 feet.

**AYANGUI POINT** is a rocky peninsula, lying 18 miles to the north-
east of Santa Elena point, the coast between forming a deep bay, with
alternate beaches and rocky bluffs. Three miles south of Ayangui point,
at the north extreme of a long beach, is a small creek called Estero Balsa,
on the south bank of which is a large village, well marked by a table hill
258 feet high, lying to the southward of it, and the hill of Colonche
6 miles inland. Pelado, a small rocky islet, 72 feet above the sea, is 3 miles
to the north-west of the point, with a reef extending about a cable on the
northern side. Northward of the point is another village, called Valdivia,
at which bullocks can be obtained, the landing however is bad. The
coast from this runs in a N.N.W. direction for 24 miles to Salango island
clear of shoals, with the two projecting points of Montanita and Jampa.
Two miles north-west of the latter, with a clear passage between, are
the Ahorcados, a small group of rocky islets. Inland is a high ridge,
running parallel to the shore, rising to an elevation of 2,400 feet.†

**SALANGO ISLAND,** 524 feet high, covered with luxuriant vegetation,
and about 3 miles in circumference, forms an anchorage much resorted to
by whalers who come for wood, water, and fresh provisions, all of which
are to be obtained from an extensive plantation in the neighbourhood.
The anchorage is to the northward of the island in 15 to 20 fathoms
water on a line N.N.E. and S.S.W., from the north point of the bay to the
east point of Salango island, and about half a mile from a rivulet on the
main shore, which forms the watering place. There are a few rocks off the

---

* See Plan of Santa Elena Bay; scale, m ⌐ 1·5 of an inch.
  † See Chart of the West Coast of South America, Sheet XVIII., No. 1,814 ; scale,
m = 0·12 of an inch.

west point of the island, but they are steep-to; the passage between the island and the main should not be used.  Large bamboos are found here, and fish are plentiful; the greatest surf prevails with a rising tide. The coast to the northward is much indented, with several houses on the beach, but the landing is bad.*

**CALLO POINT,** a bluff in the bight of the bay, 14 miles to the north-east, with a small islet both north and south of it, is the next point worth remarking; from this spot the land trends out to the north-west, a clean coast having a narrow beach in front of a steep bank.

**LA PLATA ISLAND,** 769 feet high (said to be the spot where Drake divided the spoils of the *Cacafuego* in 1579), is W. by N. 18½ miles from Callo point, the channel being clear and generally used by steamers bound to Panama.  It is of a moderate height covered with low trees, the western side forming a precipitous cliff with a few small islets ; there is landing on the beach on the eastern side.  Wood, and frequently turtle, can be obtained; the anchorage is deep, 18 and 20 fathoms, 2 cables from the beach, the bay of Salango is far preferable.

**CAPE SAN LORENZO,** 16 miles north-east of La Plata, forming the outer part of a high projection of the coast, ends in a tongue of land about half a mile long, off which are three detached rocks, one of them resembling a marlingspike; from this cape the land trends more to the north-east towards San Matéo point ; to the eastward of the cape are some hills, the principal of which is Monte Christo, 1,430 feet high.  The water deepens off this cape, the 100 fathoms line extending only 5 miles from the land, with 40 fathoms at a distance of one mile.

**PORT MANTA,** a village 6 miles to the eastward of San Matéo point, is the seaport of the town of Monte Christo, situated to the eastward of the hill of the same name, and said to contain a population of 3,000 in 1847.  The anchorage is to the northward of the houses, in 6 fathoms. Care must be taken of the shoal patches, which extend about three-quarters of a mile from the shore ; there is no danger if attention is paid to the soundings.  The landing is good, but no water can be obtained, the surrounding country being like Santa Elena, a mere desert ; bullocks may be procured by communicating with Monte Christo.

There is some little trade here, principally in hats and hammocks. These hats, which form a considerable item in the commerce of this coast, are worn nearly throughout the whole American continent, and vary in

---

* See Plan of Salango Island ; scale, m = 2·0 inches.

price from 2 to 150 dollars. They are distinguished from other straw hats
by their lightness and flexibility, consisting only of a single piece. From
Manta the coast runs east for 4 miles to Garamigo bluff, from which it
trends to the north for 12 miles as far as the river Charapota which
flows through an extensive plain bounded by white cliffs, 789 feet above
the sea; the mouth of the river is barred and landing difficult. It is
high water, full and change, in Port Manta at 3h. 4m., the rise and fall
being 6 feet.

CARACAS RIVER, to the eastward of Point Bellacas, 13 miles to the
northward of the river Charapota, is shoal and difficult of access, but
much used by the coasters, there being a considerable trade in cocoa.
Santa Marta bank, a small rocky ledge, lies about 1½ miles N.N.W. from
Bellacas point with a depth of 5 fathoms in mid-channel between it
and the main. The entrance to the river is to the eastward of this bank
round Punta Playa, which is low and grassy with a small rock off it;
northward of this are two banks dry at low water. Between the small
rock and the southern bank vessels drawing under 12 feet may enter at
high water, but the channel is narrow about a cable broad, and should
not be attempted by a stranger. There is another channel along the
northern shore, but its entrance requires more examination. The village
is about half a mile to the southward of Punta Playa, from which the
river runs in a south-east direction for a distance of 6 miles; beyond
that distance it was not examined. Fresh water may be procured by
sending boats about 3 miles up the river. Bullocks can be obtained
at the town, and fish is plentiful. It is high water, full and change, at
about 3h. 30m., the rise being 10 feet.

CAPE PASADO, 14 miles to the north-west of Caracas, is a high
round point apparently split, the land on both sides covered with short
trees, that to the southward being bounded by white cliffs. A reef of
rocks extends half a mile to the northward of the cape, and in the small
bay formed by them is a fair anchorage in 6 fathoms about 1 mile
N.N.E. from the reef. Fresh water may be obtained in the bay. Between
this and Borrachos point, a distance of 11 miles, the beach is studded
with rocks, with wooded cliffs of a uniform height a short distance inshore.
Near Point Borrachos the bare white cliffs again make their appearance,
fronted with long sandy beaches.*

At JAMA POINT 5 miles to the north-east, the coast takes a sudden
turn to the southward, forming a wide but shoal bay, in the bight of

---

* See Plans of Caracas River and of Cape Pasado; scale, m = 1·5 of an inch.

which is an extensive plain with two rivers running through it. The northern shore is bordered by the high land of Quaques, a white patch in which is remarkable from the sea.

**PEDERNALES POINT,** nearly 20 miles to the north-east of Jama point, is a narrow cliffy ridge with a few rocky islets off it. In the bay to the eastward, about 1 mile from the point, is the mouth of a small river, on which is a village very conspicuous from the sea, but the landing is difficult and supplies indifferent. From this spot low cliffs extend in a straight line for 8 miles, terminating in a sandy beach, with thick under-wood and occasional tall trees.

**The COGIMIES SHOALS** commence 18 miles northward from Peder-nales point, and appear to be a bar formed by the mouth of a large river. Their outer edge extends 4 miles from the shore, and assumes the form of a semicircle over 5 miles in diameter, with the depth of 6 fathoms about 1 mile from the edge, deepening suddenly to 40 fathoms at a distance of 3 miles. No rocks were seen, and, although attempted in several places, no passage could be found through the breakers. The land in the vicinity is low, and mangroves make their appearance on Point Mangles at the northern extreme of the shoal. From Cape Pasado to these shoals a bank extends with 5 fathoms, and under from 1 to 3 miles off the coast.

**PORTETE RIVER,** a stream of some magnitude but with a dangerous bar, is 2 miles to the northward of Mangles point; here the cliffs begin again, and there was in the year 1836 a remarkable clump of cocoa-nut trees making like a lighthouse or church steeple. Two miles farther is a small village to the eastward of Munpiche point, which in some measure protects the landing, where water, cattle, and especially pigs may be obtained in great abundance.

**CAPE SAN FRANCISCO,** 9 miles from Munpiche point, forms the southern end of a large round bluff, which trends in a curve to the northward for 11 miles to Point Galera, a low shelving point forming its northern extreme. This bluff is composed of abrupt cliffs of a moderate height, clothed with tall trees and generally steep-to, though in some places the rocks lie a little distance off the coast, but the heavy breakers always show their position; the country inland is mountainous and wooded. Cape San Francisco and Munpiche point form the horns of a bay, into which flow several rivers, the largest being the Muisne, about 5 miles from the cape, tracing its course for some distance along the beach, but a bar at the entrance renders its navigation precarious even for boats. The Bunche river, 3 miles to the north-west, may be known by four remarkable rocks.

The town of Cape San Francisco is in a small bay on the northern side of the four rocks ; on the western shore of the bay is another little river, that may be entered at high water, and the landing at all times is generally good.   This bay having only 2½ and 3 fathoms in it will not permit a large vessel to go nearer than a mile, but by anchoring a quarter of a mile S.W. of the four rocks she would command both the river Bunche and the town of San Francisco.

**Supplies.**—Water and refreshments of all kinds may be obtained on this part of the coast, the river Bunche perhaps offers the most convenient place for procuring bullocks, pigs, or vegetables.   The small bay under the cape is advantageous for watering, the water being very good.   Most of the cattle supplied to whalers at Atacames come from these places.

**The COAST.**—There is a small bay with convenient landing just round Galera point.   From this point the coast trends to the north-east, consisting, as far as Sua and Aguada points, a distance of about 11 miles, of low white cliffs, crowned with trees and fronted by beaches studded occasionally with black rocks.   Sua and Aguada points are remarkable, being small cliffy peninsulas, each connected with the main by a sandy isthmus.

The most remarkable features of the coast of Ecuador are the sudden changes in the aspect and climate of the country ; places separated but a few miles differ widely from each other.   At Guayaquil there are mangrove swamps and impenetrable thickets; at Santa Elena, aridity and a scanty vegetation ; at Salango, a moist atmosphere, and a soil densely covered with plants ; at Manta, a desert ; and in the bay of Atacames, again, thick forests and plenty of rain.

**ATACAMES BAY** lies eastward of Aguada point, which forms a good mark for the bay.   Water is obtained in the smaller bay just round Aguada point in a river accessible at all times of tide ; on its eastern bank is a small village and plantation.   Atacames river, on the east bank of which the town is situated, nearly 2 miles to the eastward of Aguada point, can only be entered at high water, and the landing on the beach is sometimes dangerous.   The town is small, containing about 500 persons, the houses, like those of Puna, are built on piles.   The soil is fertile and yields two crops a year, so that vegetables and fruit are always abundant. It has been a great resort of whalers for water, fresh provisions, and yams, the latter being particularly good.   At the village of Sua, about a mile inland, there are extensive plantations of sugar cane, for manufacturing aguardiente.   Tobacco is grown on a large scale, and fetches a high price.

**ATACAMES LEDGE.**—At Galera point a small bank, similar to that north of Cape Pasado, commences, extending nearly 2 miles from the shore. Off Atacames it juts out to the northward, ending in a dangerous ledge of coral, distant 7 miles from the land, lying N.W. by W. and S.E. by E., nearly 1½ miles long by half a mile broad, having an average depth of 12 feet and only 6 feet water on its shoalest part, with Point Galera bearing S.W. ¼ S., Aguada point South 7 miles, and Gorda point West. It is high-water, full and change, in Atacames bay at 3h. 37m., the rise and fall being 13 feet.*

**DIRECTIONS.**—This ledge is dangerous to sailing vessels working to the southward out of Panama bay, as both wind and current generally cause them to make the land about Atacames; care must be taken not to bring Galera point to the westward of S.W. by S. or not to go into less than 10 fathoms. Vessels bound to Atacames should make Galera point and run along the land with Aguada open of Sua point, bearing E. by S. ½ S. anchoring about 1½ miles north of the former in 5 fathoms with the town bearing S.E. ½ E.

**GORDA POINT** is a steep bluff with a small reef off it about 8 miles from Atacames, the coast between being low land and beach with an occasional cliff near the point; the shoal water of the Atacames ledge apparently stretches into this part of the coast, and therefore, vessels from Atacames to Esmeraldas, drawing more than 12 feet water, should pass outside the ledge.

**ESMERALDAS RIVER** is a rapid stream formed by the junction of numerous rivers about 40 miles S.E. from the mouth, said to have their rise in the neighbourhood of Cotopaxi. The entrance is about 6 miles from Gorda point, the coast between forming a succession of cliffs and small valleys; it is well marked by the narrow and precipitous gorge through which it runs, and a small round peak over the eastern point of entrance. The bottom off the bar of this river is deep and singularly uneven, 84 fathoms being found alongside 7, and the stream runs out with great velocity, the water being fresh 2 miles from the mouth. There is a safe anchorage for small vessels just inside the bar which may be reached with common care. There is only a village at the mouth, the town being situated 13 miles up the river, and it was said to contain 4,000 inhabitants, in 1847, mostly negroes.

**SUPPLIES.**—The principal trade at this port is cocoa and tobacco, the latter said to be the best on the coast; a considerable quantity of grain is raised, cotton is plentiful, and there is a small export of caoutchouc. There are mines of emeralds in the vicinity, from which the place takes its

---

* See plan of Atacames Ledge; scale, m = 2·0 inches.

name ; these were formerly worked, but latterly this branch of industry
has been neglected.

**VERDE POINT,** the next remarkable place on the coast, is a cliffy
bluff with a hill over it, 13 miles to the eastward of Esmeraldas, the bight
between forming the bay of San Matéo, the land reached by Pizarro
on his first attempt on Peru 1526, and where he again landed in 1531 and
marched from thence to the famous conquest.   His experience shows,
under a primitive system of government, the considerable resources of
this country, as on his march along the coast from this place to Puna island,
he collected gold and silver to the amount of 30,000 dollars, with other
booty of such value as dispelled all the doubts of the invaders and inspired
the most desponding with sanguine hopes.   There is a river one mile to
the westward with a bar navigable at high water and a small town on the
eastern bank.   From this point the shore is lower, with fewer cliffs, but
having several huts and cultivated spots near them, giving the coast a
more civilized appearance. The cliffs cease altogether at the river Majaqual,
13 miles to the eastward, at which point the low river land commences,
continuing with only two breaks as far as  Cape Corrientes a distance of
300 miles.

**The COAST** between the river Majaqual and Point Mangles is a shallow
bight in which are three large openings, apparently the mouths of rivers,
forming the entrances to a considerable inland navigation, and leading
to the Pailon, a deep basin of some extent.   The land in this vicinity is
owned by the Ecuador Land Company.   Santiago, the first of these, 11 miles
from Majaqual, is of considerable width, and in December 1836 there
were three passages through the breakers, the southern one being the best,
but they probably alter with the freshes.   There are several houses on the
southern bank and numerous cattle were seen.   About one mile from the
mouth is the village of Tola, from which there is an inland communication by a 3-fathom channel for 30 miles.

**ROSA HARBOUR,** 4 miles from Santiago, is the second of these openings, the coast between being fringed with shoals, extending 4 miles from
the shore, on which the sea breaks continually.   The passage into the
harbour is through these, about one mile wide, with a depth of 12 feet at
low water ; when inside, the harbour is capacious and secure.   The
breakers extend in the same way to San Pedro, the best of the three esteros,
into which the channel is broader, with about the same depth of water.
No vessel bound to any of these ports should attempt to enter without a
pilot, which can easily be obtained at Tola ; the coast being so low and

---

* See plans of Santiago, San Lorenzo, and Port Tumaco ; scale, m = 2·0 inches.

similar that no leading marks can be given distinguishable by a stranger.
There are apparently two more openings on the northern shore of the
bay, but they were not examined. No vessel should approach this coast
within the depth of 10 fathoms, which is generally found at a distance of
4 miles from the land. It is high water, full and change, in San Pedro, at
3h. 30m. ; the rise being 13 feet.

**POINT MANGLES,** the northern point of the bay of Panguapi, is low
and sandy, forming the south-west extreme of a low narrow island ; the
water is deep off it, there being 38 fathoms within half a mile of the point,
and 100 fathoms at a distance of 3¼ miles. Bullocks and fresh water can
be obtained from a small village close to the point. From this the coast
runs in a north-easterly direction for 19 miles, low and intersected with
esteros or creeks, said to be the mouths of the river Mira, as far as Boca
Grande, which is the largest, and rendered conspicuous by a considerable
village surrounded with cocoa-nut trees.

**PORT TUMACO,** the boundary town of the state of Nueva Granada, lies
to the eastward of Boca Grande. The port is formed by the three islands of
Tumaco, Viciosa, and El Morro, lying at the mouth of an estero, and may
be recognized by the white cliff on the north-east end of the latter island.
There was but little commerce at this port in 1847, although the convenience for transporting merchandise to the interior of the country is well
worthy of further attention. The town is at present a mere village of
bamboo huts ; it is well supplied with fruit, and exports timber, chiefly
mangrove and cedar.

**La Viuda** is a small rock, lying about 1¼ miles to the north-east of
El Morro, useful as indicating the position of the entrance to Tumaco.
Farallon de Castillo is a similar rock, off the north point of El Morro, to
which it is connected at low water. There is a small channel, with only
6 feet water in it, between Viciosa and El Morro, and also a boat channel
from the Boca Grande. It is high water, full and change, in Port Tumaco,
at 2h. 23m.; the rise and fall being 12 feet. The current in the offing will
generally be found setting to the north-eastward.

**DIRECTIONS.**—Vessels bound to Tumaco should make Point Mangles,
and then run along the land on a N.E. by N. course for about 25 miles,
until the cliff on El Morro is seen ; and when it bears S.E. by E. ¼ E.
shape a course so as to pass midway between the Farallon de Castillo and
La Viuda rock, keeping, if anything, nearer the former. This will lead
through a channel with not less than 12 feet at low water; when Vernicita
point, the east extreme of El Morro, bears South, steer for it, from which
you may run along the south shore of that island for the town. This, how-

ever, should not be attempted by strangers, who should anchor about one mile to the northward of the port, in 12 fathoms, with La Viuda bearing East. Great care must be taken after passing the Boca Grande of the extensive banks off Viciosa and Tumaco islands, as they are steep-to, having 14 fathoms within half a mile of their edge.

The chart, however, will be the best guide.

**CASCAJAL POINT,** a bold red cliff with two hills over it, forming a remarkable feature in this singularly flat country, is nearly 9 miles to the northward of El Morro, the coast between them forming a deep but shoal bay. Southward of this point lies the celebrated island of Gallo, where Pizarro drew the line on the sand, over which thirteen only of his followers crossed, and with these he remained, while Almagro returned to Panama for reinforcements. He afterwards retired to Gorgona as more secure, and equally adapted for his purpose. From this to Guascama point, a distance of 45 miles, the coast is a low and thickly wooded flat, forming the delta of the river Patia, which reaches the sea at this point, after a north-west course of 200 miles. The whole coast between points Mangles and Guascama should be approached with great caution, as with the exception of the island of El Morro and Point Cascajal it is a dead flat, the tree tops being the first points seen on the horizon ; the lead gives but little warning, a depth of 20 fathoms being found within one mile of the banks.

**OF GUASCAMA POINT,** the bank which fronts the low land extends 4 miles from the shore, and so runs parallel to the coast of the bay of Chocó the whole way to the river Buenaventura, a distance of 110 miles. About 5 miles from Guascama point is the mouth of the river Sanguianga, into which there is a passage through the breakers nearly a mile wide with a depth of $5\frac{1}{4}$ fathoms in it. The coast from this to the river Buenaventura is flat and monotonous, with the single exception of Tortuga peak, a small wooded hill 21 miles south of the river, and presents a most uninviting appearance, being low mangrove land converted into swamps by the overflowing of the numerous rivers ; in clear weather a distant range of mountains may be seen, clothed to their summits with trees.[*]

There are no less than 14 mouths of rivers on this length of the coast. These streams, although not large or deep for a continent, still draining as they do a country of some elevation, send a considerable volume of water into the sea ; and in the offing, freshes, rolling swells, and numerous trunks of trees are continually met with. The inhabitants, although not numerous, are yet frequently met with, and during the survey in 1846 a

---

[*] See Plan of the River Sanguianga.

house was generally in sight, especially at the entrance of the rivers. Inshore the flood sets N. by E. and the ebb S.S.W., about 1¼ miles an hour, but 40 miles from the coast there is generally a set to the north-east.

GORGONA ISLAND, a place famous in the annals of Pizarro, the Buccaneers, and earlier voyagers on this coast, is 24 miles to the north-east of Guascama point. It lies N.N.E. and S.S.W., is about 5 miles long by 1½ miles broad, and is remarkable from its three peaks, the highest and centre one being 1,296 feet above the sea. Gorgonilla, a rocky peninsula about one mile in length, lies off its south-west end, and 1½ miles to the westward of the southern point of the peninsula is La Roca, a singular sail rock, 60 feet high. This part of the island should not be approached by a ship, as it is foul and rocky. Gorgona is a beautiful island and forms a pleasant contrast to the low dense wood of the main-land; it is well watered and productive where it has been cultivated. After leaving Gallo island, Pizarro and his followers retreated to this island, where they lived five months, and having persuaded the vessel sent from Panama for their assistance to join them, finally sailed from Gorgona for Tumbez bay on the coast of Peru. The anchorage is off the watering bay on the east side of the island, in 30 fathoms about one-third of a mile from the shore. Water is good and easily obtained. It is high water, full and change, at Gorgona island at 4h. 10m.; the rise and fall being 10 feet. The current sets to the north-east, off the island.

BUENAVENTURA RIVER, in the bight of the bay of Chocó, 78 miles to the north-east of Gorgona island, is a broad deep stream, navigable by vessels drawing 24 feet as far as the town, a distance of 10 miles from the mouth. This is a port with great natural advantages, and promises to become a considerable emporium for the commerce of Nueva Granada. There being no land communication between Panama and Bogota, the capital, all traffic in the Pacific must go by way of Buenaventura; the roads, however, of the interior are a great bar to its prosperity, being rugged and difficult. The town, situated on the south bank, at present is a poor collection of houses, with a small barrack, battery, custom-house, and the residence of the governor, inhabited by negroes and mulattoes to the number of 1,000. It has, however, a considerable trade, importing salt, garlic, straw hats, and hammocks, and exporting rum, sugar, and tobacco; it is not considered healthy, and provisions are scarce and dear.*

Cule de Barca and Vigia de San Pablo are two off-lying islands on the north coast of the river, which is composed of red sandstone cliffs,

---

* See Plan of Buenaventura, No. 2,319; scale, m = 1·5 inches.

crowned with trees. The Vigia has the appearance of a gigantic wheat-sheaf, and is distant 4 miles from the entrance of the river, with Culo de Barca (boat's stern) 2 miles to the westward. There is a good stream of water in the sandy bay west of the Vigia, with a depth of 3½ fathoms within half a mile of the shore, and easy landing.

**Basan Point,** forming the north point of entrance to the river Buenaventura, is low and wooded with a few houses on it, at which a pilot may be obtained ; there is no regular establishment, but competent men are always to be had by sending a boat up to the town. Soldado point, on the southern shore, is a little more than a mile from Basan point, low and covered with mangroves. The river between these points is deep and clear, with the exception of a small shoal with 3 fathoms on it, lying 4 cables S.S.E. from the houses. The northern shore of the river is a low wooded bank, with an occasional cliff, but the southern is a mangrove swamp, intersected by small rivers, with the usual mud flat in front. Southward of Soldado point is a shoal bay, with breakers extending 5 miles from the land.

**NEGRILLAS ROCKS** are a low and dangerous reef, about 2 miles in circumference, nearly covered at high water, lying 8 miles W. by S. from Culo de Barca, consisting of one large and several detached rocks, with shoal patches ; they are dangerous to vessels bound to Buenaventura in the thick weather so frequently met with on this coast, and no vessel should approach nearer to them than 10 fathoms. In clear weather the Vigia de San Pablo kept open of the land bearing E. ½ S., will carry a vessel well to the southward of the reef.*

**TIDES.**—It is high water, full and change, off the entrance of Buenaventura river at 4h., but at the town, it is said to be at 6h., the rise and fall being 13 feet. Tides are regular, the ebb stream setting right out of the river, and running 2 knots ; the first of the flood comes from the northward, and runs little better than one knot.†

**DIRECTIONS.**—This river will easily be recognized by the red sandstone cliffs, to the north-west of the entrance, the more remarkable as they are the first met with northward of Point Cascajal. Large vessels, bound for Buenaventura, should make these cliffs, and thus avoid the shoal and dangerous bay south of Soldado point, taking care at the same time of the Negrillas rocks. They should not, however, approach the cliffs within 3 miles, or go into less than 5 fathoms, and anchor with Piedra point, the western extreme of the cliffs, bearing North and the

---

* *See* View on Plan No. 2,319.
† Remarks of Mr. V. G. Roberts, Master R.N., 1856.

entrance of the river E.N.E.   Small vessels may approach the land and
run along at a distance of half a mile from the cliffs, as far as Basan
point, from which a N.E. by E. ½ E. course should be steered for the
north point of entrance to the Rio Cayman, a stream on the south side of
the river, until the cliff north of the town opens of Arena point, on the
northern bank, then steer for the south end of the town, and anchor after
passing Arena point.

**MAGDALENA BAY** is a deep and spacious bay to the northward of
Piedra point, about 11 miles from the entrance of Buenaventura. Unlike
the rest of this coast its shores are cliffs of a moderate height, crowned
with trees ; it was not examined beyond the entrance in the survey of
1846.   This bay may be entered to the eastward or northward of Palmas
island, but the passage to the westward between the island and the
Negrillas rocks is rather shoal.   The northern is the best, between
Palmas island and Magdalena point, a good leading mark for which is the
northern rock off Palmas island on with the south cliff of the Mangrove
bay on the eastern coast of the harbour ; this will carry the vessel abreast
of Magdalena point, from which all is clear and a course may be steered
to any convenient anchorage.

**PALMAS ISLAND** is a small bold island one mile long by half a mile
broad, 3¼ miles from Point Piedra with detached rocks off each extreme ;
the eastern and northern sides appear clear, but the soundings are shoal
and irregular in the direction of the Negrillas reef.   The channel
between it and the coast to the eastward is 1½ miles wide, and forms a
good entrance to Magdalena bay for vessels drawing less than 20 feet
water.

**THE COAST.**—About 2 miles to the northward of Magdalena point the
cliffs cease and the low river mangrove coast recommences running in a
N.N.W. direction to Chirambirá point, a distance of 25 miles, forming the
delta of the river San Juan.   The water shoals quickly on approaching
the coast, 20 fathoms being found 8 miles from the shore.*

**POINT CHIRAMBIRA,** the southern point of entrance to the
Chirambirá mouth of the San Juan river, is remarkable as forming the
only harbour and convenient landing place between Magdalena bay and
Cape Corrientes ; the point has nothing to distinguish it, a rounding
series of low spits running one into the other being all that can be seen.
There is a considerable set from the river, and the bottom off the entrance,
like that of the Esmeraldas river, is singularly uneven, 3 fathoms and no

* *See* Chart of the West Coast of South America, Sheet 19, No. 2,257 ; scale, $m = 0.12$
of an inch.

bottom at 33, being within a cable of each other. Anchorage, however, may be found by vessels drawing less than 15 feet inside the point, but the passage should not be attempted by a stranger, the tides running with considerable force, and the channel not being 3 cables wide. The river is not deep inside the principal branch, only having a depth of 4 to 5 feet at a distance of 3 miles from the mouth; the water, however, was fresh at half that distance. There are a few houses on the north bank and a distillery rather on a large scale although rough in material. Bullocks, pigs, and vegetables can be obtained here, but are dear, and two or three days' notice is required to send up the river for them.*

**SAN JUAN RIVER** is a considerable stream, said to communicate, during the rainy season, with the Atrato (which flows into the Atlantic) by means of a canal in lat. 5° 10′ N., near the towns of Citera and Novita. Captain C. S. Cochrane, R.N., who visited this spot in 1824, found the distance between the streams to be about 400 yards, and the height of the ground necessary to be cut through in order to effect a navigable junction, 70 feet, composed chiefly of solid rock, in addition to which the streams would require deepening on each side for about a league, a considerable undertaking, if left to the natives, in a country where future reward is no temptation to present labour.

**TIDES.**—It is high water, full and change, at Point Chirambirá at 4h.; the rise and fall being 12 feet. The ebb stream is much stronger than the flood ; this, however, is local, and does not extend far from the mouth of the river.

**The COAST,** northward of Chirambirá point, trends in nearly a direct line North (true) to Cape Corrientes, a distance of 73 miles; it is somewhat similar to that of the bay of Chocó, being low land, intersected by numerous rivers. Apparently, it is less shoal and swampy, the breakers being found only off the entrances of the rivers, many of which could be entered by a boat, taking the ordinary precautions for crossing the bar. The water also shoals more gradually, 2 fathoms being generally found within one mile of the beach, and occasionally much nearer There is a low table land to the northward of the river Usaraga, about midway between Chirambirá point and Cape Corrientes ; two low peaks are also seen to the northward of the Usaraga, abreast of the rivers Baudo and Catripe, about half-way to the cape ; these are the only elevations, as the mountain range visible in the bay of Chocó was not seen on this part of the coast. Houses are frequently met with, especially at the mouths of the rivers ; the inhabitants, however, generally avoid communication.

---

* *See* Plan of Point Chirambirá ; scale, ᴤ = 2·0 inches.

**WINDS and WEATHER.**—Along the whole of the coast, from the river Guayaquil to Guascama point, the wind is mostly from South to West all the year round, following in some measure the trend of the shores ; thus, southward of the Equator, there is more southing than westing in the wind, and northward the reverse is found ; in both cases the wind veers to the southward as you leave the coast ; the exception to these rules are few, and generally occur in the fine season.   During the survey (1846 and 1847), in beating up this coast to the southward, and in running down it, the former in the months of May and June, the latter in those of October, November, and January, the wind was from S.S.E. to West (by the South); the only difference being that the winds were lighter and weather finer in the former months, as we got to the southward ; whilst the contrary took place during the latter season.   There is a constant current to the north-eastward, within 60 miles of the land, of great assistance to vessels bound to Panama ; the Pacific mail steamers generally pass between La Plata and the main, in their passage to Panama.*

After passing Guascama point and entering the Bay of Chocó, the winds become more variable and rains more frequent, and the following account by Dampier is perhaps as good as can be given :—" It is a very wet coast, and it rains abundantly here all the year long.   There are but few fair days ; for there is little difference in the seasons of the year between the wet and the dry ; only in that season which should be the dry time, the rains are less frequent and more moderate than in the wet season, for then it pours as out of a sieve."†   This kind of weather is found as far as Cape Corrientes, the prevailing wind being S.W. but N.E. winds were not uncommon.   Off-shore in this zone, between the parallels of 2° and 5° N., the winds are equally baffling, especially during the months of March, April, and May.   H.M.S. *Alarm*, in March 1859 only made 30 miles in 6 days.‡

**MALPELO or BALDHEAD**, a remarkable detached islet at the entrance of the Gulf of Panama, lying 203 miles W. by N. from Gorgona island, is barren, precipitous, and rises 1,200 feet above the sea, and may be seen 40 miles in clear weather.   The summit of the island bears a resemblance in several points of view to the crown of a head, and its being barren would warrant the name of Malpelo or Baldhead which the Spaniards have applied to it.   It is surrounded with islets, and the whole group may extend about 10 miles from north to south.   The strong currents in the vicinity cause the appearance of breakers, but it is believed to be steep to, 40 fathoms being found alongside, and 110 at the distance

---

* Remarks of Commander J. Wood, R.N., 1847.   † Dampier's Voyages, vol. i. p. 173.
‡ Remarks of Mr. J. Dathan, Master, R.N., 1859.

of a quarter of a mile. The island has never been surveyed, although visited by Commander Wood in the year 1847 ; but he could find no landing.

**RIVADENEYRA SHOAL.**—This very doubtful danger, said to lie 220 miles W. ¼ S. from Malpelo, was reported by M. Rivadeneyra in October 1842, to have a depth of 16 feet. By very imperfect observations it was said to be in lat. 4° 15′ N., long. 85° 10′ W. of Greenwich.[*] In November 1854 H.M.S. *Cockatrice*, and in August 1857 H.M.S. *Havannah*, each passed within 5 miles of this reported danger, without either of them perceiving any indications of shoal water.

**The GALAPAGOS** are a group of islands lying on the equator, and extending 90 miles on each side of it, at about 600 miles from the coast of Ecuador, to which, politically, they belong. They were first discovered by the Spaniards, and became in the end of the 17th century a great rendezvous of the Buccaneers who often resorted to them for refreshments, and as a place where they might refit, share out plunder, or plan new expeditions, without any risk of being molested.[†]

There are 6 principal islands, 9 smaller, and many islets, scarcely deserving to be distinguished from mere rocks. The largest island, Albemarle, is 60 miles in length and about 15 miles broad, the highest part being 4,700 feet above the sea. The constitution of the whole is volcanic. With the exception of some ejected fragments of granite, which have been curiously glazed by the heat, every part consists of lava or of sandstone. The higher islands generally have one or more principal craters towards their centre, with smaller ones on their flanks, and Darwin affirms that among the islands of this archipelago there are not less than 2,000 craters. There is a great similarity in the appearance of these islands, which consist of a tame rounded outline with peaks or extinct craters throughout, the lower parts are generally arid, but the summits, at an elevation of 1,000 feet, possess a tolerably luxuriant vegetation, especially on the windward side of the islands.[†]

**The CURRENTS** about these islands are remarkable for their velocity, which is from 2 to 5 miles an hour, and usually towards the north-west. Previous to making these islands, H.M.S. *Beagle*, was set 50 miles to the W.N.W., in September 1835. In addition to which there is a surprising difference in the temperature of streams of water moving

---

[*] Findlay's Pacific Directory, vol. ii., page 1046.  Bull. de la Soc. de Géogr. vol. ix. 1848.

[†] *See* chart of the Galápagos, No. 1,375 ; scale, *m* = 0·12 of an inch; and Voyages of H.M.S. *Adventure* and *Beagle*.

within a few miles of each other. On the northern side of Albemarle island the temperature of the sea a foot below the surface was 80° Fahr., but on the southern side it was less than 60°. These striking differences are probably owing to the cool current which comes from the southward, along the coasts of Chile and Peru (see page 368), and at the Galápagos encounters a far warmer body of water, moving from the Bay of Panama.

A curious instance of this meeting of the waters was observed on board H.M.S. *Havannah* on her passage from Callao to Central America, at noon of 29th April 1856, about 200 miles E.N.E. of the Galápagos, when she ran through a rippling extending N.W. and S.E. as far as could be seen, and most distinctly marked, the water to the southward being of a greener colour. Before entering it the temperature of the sea was 72° Fahr., a quarter of a mile farther inside 'or North of the rippling it rose to 78°, and 3 miles more to the northward it was 80°. Until this time the *Havannah* had experienced a set to the N.W. by W. of 30 miles in 24 hours, but on the 30th the current was only 9 miles in the same direction.*

CLIMATE.—Considering that these islands are placed directly on the equator, the climate is far from being excessively hot, a circumstance which, perhaps, is chiefly owing to the singularly low temperature of the surrounding sea. During the rainy season, from November to March, which is not, however, at all to be compared to a continental rainy season, there are calms, variable breezes, and sometimes westerly winds, though the latter are neither of long duration nor frequent. Dampier also states :—" The air of these islands is temperate enough, considering the clime. Here is constantly a fresh breeze all day and cooling refreshing winds during the night; therefore the heat is not so violent here as in most places near the equator. The time of the year for the rains is in November, December, and January; then there is oftentimes excessive hard tempestuous weather, mixed with much thunder and lightning. Sometimes before and after these months there are moderate refreshing showers ; but in May, June, July, and August the weather is always' very fair." Nothing need be added to this description except that heavy rollers occasionally break upon the northern shores of the Galápagos, during the rainy season, though no wind of any consequence accompanies them. They are probably caused by the Northers from Tehuantepec, and the Papagayos, or Northeast winds, which are so well known on the coast between Panama and Acapulco.

---

* Remarks of Captain T. Harvey, R.N., 1856.

**SUPPLIES.**—Formerly the principal production of these islands and from which they are named, was the terrapin or land tortoise (*Testudo indicus*). Dampier describes them as so numerous that 500 or 600 men might subsist on them alone for several months without any other sort of provision. Some of the larger weigh from three to four hundred pounds ; but their common size is between fifty and one hundred pounds. In shape they are similar to the small land tortoise, but have a very long neck which, together with their head, has a threatening appearance, resembling a large serpent ; they are, however, singularly timid and inoffensive. They form good live stock, being fat and sweet, no pullet eating more pleasantly, and keeping in good order for many days on board ship. This supply made the islands a favourite resort for Buccaneers in old times and the whalers in the present. The terrapin, however, is now said to be nearly exhausted, and an attempt has been made to compensate for the loss by the establishment of settlements on Charles and Chatham islands, where bullocks, pigs, goats and vegetables can generally be obtained after a few days' delay. Water is found at many of the islands in the rainy season, but the only watering place that can be depended on is on Chatham island, and that at times cannot be approached on account of the surf. Fish are plentiful and easily caught where the seine can be hauled.

**CHATHAM ISLAND,** the eastern of the group, is 19 miles long by 8 broad, the peaks at the south-western end being 1,650 feet high. This island is pronounced to be better adapted for a settlement than any of the others, having a fertile soil, several good anchorages on its western coast, and water being plentiful.[*]

Freshwater Bay, possessing the watering place before mentioned, is an open roadstead on the south side of the island. The anchorage, which is in 20 fathoms, and quite secure, being about 4 cables off the watering place, a fine stream falling from a lava cliff about 30 feet high ; ships well provided with ground-tackle may lie here and water without difficulty or danger. Wreck point, the south-east extreme of the island, has a small cove to the northward of it, in which is the settlement, a collection of huts ; about 4 miles inland are two plantations, at which most tropical productions are raised with ease. There is good landing in the cove, and Dalrymple rock, a singular islet 65 feet high, lies 2 miles to the northward of it.

Stephens bay having good anchorage in 10 to 12 fathoms half a mile from the shore, is 6 miles to the north-east of Wreck point. This bay may be easily known by the Kicker rock, a curious mass of stone 400 feet high, rising almost perpendicularly from the bottom of the sea,

---

[*] Remarks of Commander James Wood, R.N., 1847.

where it is 30 fathoms deep.   Finger point, on the north-east side of the bay, is a remarkable pinnacle 516 feet above the sea.   Terrapin road is an open anchorage on the north side of the island, with 12 and 14 fathoms half a mile from the shore, and Hobbs reef is a dangerous ledge extending over a mile from the north-west point.*

**HOOD ISLAND,** the southernmost of the group lying 27 miles from Chatham island, is small, round, and 640 feet high; it is rugged, covered with small sunburnt brushwood and bounded by a bold rocky shore. Gardner bay is an anchorage on its north-east shore, inside an island of the same name, 160 feet high, but care must be taken of a 14 feet patch lying near the centre of the bay, on which H.M.S. *Magicienne* struck in 1857.   It is a pin rock with 8 and 5 fathoms close to, lying in the centre of a line joining the horns of the bay, and may be avoided by keeping close to the shore of the small island.†

**CHARLES ISLAND,** 37 miles West of Hood island, is 24 miles in circumference, and peculiar in its aspect from a succession of round topped hills similar in shape though differing in size which show on every point of view.   The highest and largest of these hills rises 1,780 and the next about 1,500 feet.   On this island is Floriana, the largest settlement in the group, situated amongst the high hills and 4½ miles from Black beach road on the western side of the island.

**Supplies.**—On the plantations of Floriana, which are of considerable extent, every kind of tropical production is to be found in abundance, and the adjoining lands which had not been cleared (1849) appeared to be of the same character and to be available for increased subsistence if the island were peopled; at that time the number of settlers only amounted to 25.   There are said to be upwards of 2,000 head of wild cattle on this island; pigs and goats are also abundant.   Water is plentiful in all seasons of the year at the settlement, and could be easily led to the coast by means of pipes, although at present but little pains is taken to render it available for shipping.‡

Blackbeach road is an open anchorage, its only advantage being that it is the nearest landing place to the settlement.   Post Office Bay, on the north-west side of the island, is in every way superior, being a sheltered anchorage, with moderate depth of water, easy of access, and only wants fresh water to make it a desirable place for shipping.   The name of this

---

* *See* Plans on No. 1,375.

† Remarks of Mr. J. Harvey, Master R.N., 1857, and Plan on No. 1,375.

‡ Remarks of Admiral Sir George Seymour, in London Geographical Journal for 1849, page 22.

bay is the result of a custom established by the whalers. A letter box
was placed here, homeward bound ships examining the directions and
taking with them all which they might have the means of forwarding. It
is high water, full and change, in Post Office bay, at 2h. 10m.; the rise
being 6 feet.

Off the eastern side of Charles island are several outlying islets, Gardner,
the outer one, being over 4 miles from the shore, and 3 miles E. by S. ¼ S.
of this islet is a dangerous rock, awash.

MACGOWEN REEF, a dangerous shoal consisting of one rock, awash,
and another only a few feet under water, about half a mile to the eastward, lies in a line from Charles to Chatham island, and rather nearer the
latter, with the peak of Hood island bearing S.E. ¼ E. 22 miles. There
is no warning given by the lead of this shoal, 50 fathoms being found
within 1½ miles of it.

INDEFATIGABLE ISLAND, the third in magnitude of the group, consists of one large mountain lying 27 miles to the northward of Charles
island; Conway bay, on its northern side, appears to be the only anchorage.
It may be known by the Guy Fawkes islets, a straggling group to the
northward of the bay; there is good landing for boats, and terrapin were
abundant.  Duncan and Barrington are small islands, the former lying off
the western and the latter off the south-east side of Indefatigable island.

JAMES ISLAND, a favourite resort of the Buccaneers, being similar
to Chatham and Charles island, lies to the northwest of Indefatigable.
It has abundance of water on the higher lands, but the lower parts are so
broken and dry that little ever reaches the coast.  On this island is a salt
lake, a circular crater from which this useful article can be procured.
The best anchorage is in James bay on the west side of the island, to the
northward of the highest land of the island, a remarkable sugar loaf,
1,200 above the sea.  Vessels may anchor here in 14 fathoms within a
mile of a sandy beach with Albany island, an islet off a crater at the
north extreme of the bay bearing N. by W.  Sulivan bay at the east end
of James island is another anchorage, but it is open, with deep water. It
is high water, full and change, in James bay, at 3h. 10m.; the rise being
5 feet.*

ALBEMARLE ISLAND, the principal, and with Narborough, the western of the group, is a singular mass of volcanic ejections, and may be literally
described as consisting of six huge craters, whose bases are united from

---

* See Plans on No. 1,375.

their own overflowed lava.   The southern side which is exposed to the
trade wind, and completely intercepts it with the clouds it brings, is green
and thickly wooded, and doubtless has fresh water, but the heavy swell
prevented the examination of this portion of the island, which is so low
as not to be discernible until the surf is seen on the shore.   Four islets,
the remains of volcanoes, lie near the south-east extreme which, with
Brattle island, are extremely useful in warning vessels of their approach
to this dangerous piece of coast.   A long swell setting towards the land
and the generally light winds are additional reasons for avoiding this
shore of Albemarle island ; there is however anchorage in case of
necessity.

Point Essex, the south-west cape, is high and to the north of it at the
foot of the highest crater is Iguana cove, a wild looking anchorage
abounding in iguanas, which reptiles, although repulsive to the eye,
are good eating.   Northward of Iguana cove is Christopher point,
the south extreme of Elizabeth bay, the northern shore of which is
formed by Narborough island, a great volcano 3,720 feet high, utterly
barren and desolate.   Two craters were seen in action by H.M.S.
*Tagus,* in 1814, and a terrific eruption is described in Morrel's Voyages in
1825.

The passage between the two islands is in the north part of Elizabeth
bay, from 2 to 3 miles wide, and 55 fathoms deep.   On the Albemarle
side of this channel is Tagus cove, a snug anchorage formed by an old
crater, capable of containing 6 frigates, with from 6 to 14 fathoms water.
There are no dangers and the shores are so steep as to be almost inaccessible.   A few holes only could be found containing fresh water, yet
during the rainy season there must be considerable streams, judging by
the gullies worn in the rocks.*

Banks bay lies to the northward of Narborough island, between it and
Cape Berkeley the north-west point of Albemarle island ; there is no
anchorage in this bay, the water being very deep, no bottom with 150
fathoms being found half a mile from Narborough island.   The northern
point of Albemarle island bears the same name and has a reef off it
extending one mile from the shore ; 14 miles to the north-west is the
Reodondo rock, an islet 85 feet high.   It is high water, full and change,
in Iguana cove, at 2h. ; the rise being 6 feet.

**ABINGDON, BINDLOE,** and **TOWER ISLANDS,** respectively 1,950,
800, and 211 feet above the sea, form the north-eastern side of this archipelago.   The two former are of a similar nature to the other islands,
but the latter is different, being low and flat.   On the south-point of

---

* *See* Plans on No. 1,375.

Abingdon island Captain Basil Hall, R.N., landed in 1822, to obtain observations with the pendulum.

**CULPEPPER and WENMAN**, two high, rocky, and barren islets, lying N.W. and S.E. 20 miles from each other, complete the description of the group. Wenman, the southern, 830 feet high, is N. 29° W., 75 miles from Albemarle point, and, correctly speaking, consists of 3 islands and a large rock, but they are fragments of the same crater and appear as one from a distance. Culpepper, only 550 feet above the sea, is of a similar nature, with a reef off its south-east extreme.*

**VARIATION.**—The variation on this part of the coast of Ecuador, Nueva Granada, and the Galápagos islands, at the beginning of the year 1860, was as follows :—Off La Plata, 9° E. ; off Point Guascama, 8° E.; off Cape Corrientes, 7° E. ; and at Wenman island, Galápagos, 9° E. The curves of equal variation strike the shore in a S.E. direction, each degree being separated by an interval of about 270 miles. The variation is increasing about 1' annually.

---

* *See* Chart on No. 1,375.

# CHAPTER XVI.

### CAPE CORRIENTES TO PANAMA.

VARIATION 7° E. in 1860.   Annual change inappreciable.

---

**CAPE CORRIENTES,** easily known by the dome-like peaks of Anana, about 1,500 feet high, which rise directly over it, is the first high land north of Monte Christo, and generally makes like an island from the southward.   It is densely wooded from the summit of the peaks to the high-water mark, the almost constant rains giving a bright green colour to the peaks.   Alusea point, 8 miles north of the cape, forms the northern extreme of this remarkable promontory; the water off it is deep, 50 fathoms being found close to the rocks, and 100 at a distance of less than 3 miles; the cape is well named, as there appears to be a constant northerly set in the vicinity.

**CABITA BAY,** on the south-side of Cape Corrientes, lying to the eastward of a high rocky point, about 3 miles from the cape, although open to the southward, forms a good anchorage and capital watering place. Vessels may lie in 18 fathoms about 7 cables from the stream in the bight of the bay, with the western horn bearing S.W. by W.   On the eastern side of the bay the high bold land suddenly terminates, and a beach and low river land commences, extending, with the exception of the cliffs north of the river Buenaventura and that of Point Cascajal, as far south as the river Esmeraldas, a distance of 400 miles.   About 5 miles to the south-east of the watering place is the mouth of the river Jeya, to the southward of which is a remarkable perforated rock named the Iglesia (or Church) de Sevira.   There are a few houses in the bay, but the inhabitants are timid, and in 1847 always avoided the boats of H.M.S. *Herald.*[*]

**The COAST** from Alusea point trends to the eastward for 14 miles to the river Nuki, a small mountain stream, the shore between them being alternate bluffs and sandy beaches, with a few small streams similar to the Nuki.   At this river there were also houses, occupied in 1847 by an Englishman and some Indians.   About 8 miles N.N.W. from Nuki is a cluster of high rocks, 2 miles off the river Chiru, another of the mountain streams, and 3 miles to the northward of this group is Morro Mico,

---

[*] *See* Plan of Cabita Bay; scale, *m* = 2·0 inches.

a pinnacle of a similar nature; from thence the coast runs in the same direction, high, rugged, and woody, for nearly 8 miles to the entrance of Port Utria.

**PORT UTRIA,** unknown before the survey of the *Herald,* is a snug creek-like harbour about 3 miles long by half a mile broad, with an average depth of 12 fathoms. It is formed by a lofty but narrow peninsula, with two islets and some detached rocks off its south extreme. The entrance to the port is to the south-eastward of these, with no hidden dangers, the shores being steep-to. The eastern side is a sandy beach, running out to a spit 2¼ miles from the entrance. At this point the harbour is only 2½ cables across, but opens out after passing it, forming a commodious basin, in which a vessel may refit or heave down. This port will be easily known by Playa Baia, a beach about 4 miles long, fringed with cocoa-nut trees, situated to the northward of the peninsula, the land behind it being low. At its north extreme is the mouth of the little river Baia, with a rocky islet lying off it.*

**POINT FRANCISCO SOLANO,** 12 miles from the river Baia, is a long, rocky, tongue-like point, forming the western side of the deep bay of Solano. The coast southward is high and rocky, with occasional small beaches. There is a patch of rocks 3 cables from the shore, about 6 miles to the southward, and the water is deep, 40 fathoms being found within a mile of the coast. A reef, consisting of rocky patches, with deep water between, extends 1¼ miles from the point, and should be carefully avoided.

**SOLANO BAY,** a spacious but deep anchorage formed by Francisco Solano point on the south-west, and a lofty promontory, which juts out some 3 miles, on the north, abounds in fish, wood, water, and a great quantity of wild cocoa-nut palms; there are also considerable groves of vegetable ivory (*Phytelephas, sp.*), a beautiful palm-like plant found in low damp localities seldom intermixed with other trees or bushes, the ground beneath them appearing as if it had been swept. In 1847 a cargo of these nuts could have been collected with ease, the groves being close to the shore, in the vicinity of the sandy beaches. About 2 miles south-west of northern point is a small chain of rocky islets, nearly 1 mile long, the centre being a remarkable sugar-loaf. They are barren and frequented by large numbers of gannets.

**CUPICA BAY,** to the eastward of Point Cruces, nearly 22 miles from Francisco Solano point, is one of the best anchorages on the coast, and has

---

* *See* Plan of Port Utria.

some celebrity, in consequence of its being one of the points proposed for the junction of the Atlantic and Pacific oceans by means of the river Naipi, which rises in the hills some 5 miles to the eastward and flows into the Atrato after a course of about 17 miles. In 1847 the Alcalde at Cupica gave some information on this point, and conducted Captains Kellett and Wood to the stream by a rapid ascent, till an elevation of about 400 feet was gained, from whence it appeared level till they reached the Naipi; from this point to the place where loaded boats (bongos) stop the estimated distance was 12 miles. The Naipi, however, is only partly navigable, and unfitted for commerce.[*]

Major Alvarez, a Columbian officer, who travelled from the Atrato up the Naipi to Cupica bay, describes it as shallow, rapid, and rocky, passing over three sets of hills, and could perceive no possibility of making a canal between the Atrato and the Pacific. Captain Charles Friend, R.N., who explored the same ground in 1827, took five days to reach Cupica from the Atrato, and gives a similar account, the rapids being considerable and frequently met with. On the other hand it appears to be a well known fact that in 1820 a six-oared launch, belonging to the Chilian frigate *Andes*, was dragged across the land from Cupica,—an operation which occupied 10 hours,—launched into the Naipi, and conveyed Colonel Cancino and his suite without difficulty into the Atrato; and it must be remembered that all who have visited this district have been passing hastily across it, and have naturally sought the best path, i.e. the best travelling ground and not the lowest level.[†]

This project is by no means new. It was suggested to the Spanish Government by an intelligent Biscay pilot, Gogueneche by name, at an early date (when Spanish pilots were sailing masters); but so cautious were the Spaniards to prevent rather than encourage any scheme that might facilitate access to the west coast of America, or extend a knowledge of the mining localities near the Darien gulf, that it was prohibited, on pain of death, not only to navigate the Atrato, or pass by that river, but even to propose to take advantage of it in any way as a route.[‡]

Vessels may anchor in any part of the shores of Cupica bay in convenient depth of water. The nearest approach to the Naipi is in Limon bay, on the eastern shore, the land above it being about 500 feet high, over which is a waterfall named Quebrada del Mar. The bight of Cupica bay is a sandy beach 4½ miles long, at the west extreme of which is the mouth of river Cupica; on its banks there is a small

---

[*] Voyage of H.M.S. *Herald*, vol. i.

[†] *See* Travels in Columbia, vol. ii., and R.G.S. Journal, vol. xxxiii., No. 16.

[‡] *See* Rear Admiral FitzRoy on the American Isthmus, in London Geographical Journal, vols. xx. and xxiii., where is a full account of the several passes across.

village with some plantations from which vegetables can be obtained ; the houses, like those to the southward, are built on piles, in fact, this style of building extends from the river Tumbes to Garachiné point at the entrance of the Bay of Panama.  Point Cruces is a lofty straggling point, forming the west side of the bay, with outlying rocky islets extending 2 miles to the southward.  It is high water, full and change, in Cupica bay at 4h., the rise being about 13 feet.  Current in the offing sets to the northward.*

**CAPE MARZO.**—From Point Cruces the coast trends to the northwestward as far as this cape, which is of a similar nature, having detached islets, and also a bay to the eastward named Octavia, which, although smaller than Cupica, yet like it possesses convenient depth for anchoring.  In addition to the detached islets already mentioned there is another patch of high barren rocks of fantastic shapes lying about 1½ miles to the southward of the last, the passage between being deep and clear.  The western side of the cape is bold and rugged, but thickly wooded, running in a north direction for nearly 8 miles, and off its northern extreme are also detached islets, lying 1 mile from the coast with 16 fathoms inshore of them.

**The COAST** from these islets trends sharply to the eastward for 2 miles ; the cliffs ceasing at the mouth of the Corredo, a small stream easily entered by a boat, from which a continuous line of beach, with low land behind it, extends to the north-west for 14 miles, as far as Ardita bay.  On this coast are the mouths of two small streams, the Curachichi and the Ouredo, both of which are barred.  In Ardita bay, which may be known by a small islet lying off it, several canoes were seen ; and some little distance inland is a village called Jorado, the Alcalde of which, an Indian, visited the *Herald*, bringing a canoe full of plantains and sugar-cane for barter.  This whole coast, like that to the southward, is thinly inhabited, huts being generally found in the bays and the vicinity of the numerous small rivers.  From this bay to Point Piñas, a distance of 32 miles, the coast is high, rugged, and thickly wooded, having deep water close to the shore, with the exception of two small bays about 21 miles to the northward of Ardita.  The northern one, Gusgava, has convenient anchorage ; there is also a beach directly south of Point Piñas.†

**PIÑAS BAY,** about 3 miles to the northward of the point of the same name, is the best anchorage between Octavia bay and Garachiné point,

---

* *See* Plan of Cupica and Octavia Bays ; scale, m = 2·0 inches.

† *See* Chart of the West Coast of Central America, Sheet 1, No. 2,267 ; scale, m = 0·12 of an inch ; and Plan of Piñas Bay.

being about 2 miles long by 1¼ wide, with an average depth of 10 fathoms, but open to the S.W., from which quarter there are occasionally squalls in the wet season, throwing a considerable swell into the bay. The head of the bay is a beach, little more than 1 mile in length, with low land behind it; the sides are high and rocky. Good water is found in a stream running into the sea at the west extreme of the beach, protected from the swell by a small natural mole on its western side. Vessels may anchor about half a mile from the watering place in 8 fathoms. In the wet season they should be more on the west side of the bay in 12 fathoms, with the end of the mole bearing N. by W. Off the mouth of the harbour are the Centinelas, two high barren rocks.

GARACHINE POINT, the southern point of the bay of the same name, is 33 miles to the northward of Port Piñas, the coast between being high, bold, and wooded. About 3 miles south-west of the point is Cape Escarpado, off which lies the islet of Cajualo. The land over Garachiné is very lofty, Mount Zapo, a sharp conical peak, rising to an elevation of about 3,000 feet above the sea. Garachiné bay, lying to the north-east of the point, between it and Patino point, is shoal, the shore being low mangrove land, forming the mouths of the river Sambo, with mud banks extending 3 miles from the coast. Three mouths open into the bay. At the entrance of the western is the Pueblo of Garachiné, a small collection of huts. There is a bank 5½ miles long lying directly between Garachiné and Patino points, with patches of 15 feet water on it, having 4 and 5 fathoms inside; and 4 miles N.W. by W. from the former point, is a small patch of 4½ fathoms, with 6 and 8 fathoms close to. Vessels may anchor close off either of these points, the water being deep in their vicinity.*

The BAY of SAN MIGUEL, on the eastern side of Panama bay, was well-known to the Buccaneers, who used it as the entrance to the Pacific in their overland journeys from the Gulf of Darien, which they generally accomplished under 10 days. Dampier's account, in 1685, is singularly correct when compared with the recent surveys; since that time it has almost been forgotten until 1851, when the idea of a ship canal was suggested to connect it with Caledonia bay, in the Gulf of Darien. The entrance is 6½ miles wide, between Lorenzo point on the north and Patino point on the south.

Off Lorenzo point is Iguana island, and 3 miles to the northward the mouth of the river Congo, which Dampier wished to use in his journey

---

* *See* Chart of the Bay of Panama, No. 2,261; scale, m ─ 0·4 of an inch.

from the Pacific, and is thus described by him. "The river Congos (which is the river I would have persuaded our men to have gone up, as their nearest way in our journey over land,) comes directly out of the country, and swallows up many small streams that fall into it from both sides, and at last loses itself on the north side of the gulf, a league within Cape St. Lorenzo." From Iguana island the gulf opens, being nearly 11 miles across as far as Pierce point, a projecting rocky point on the north shore.*

**Buey Bank,** an extensive shoal drying, in patches at low water, on which a heavy sea breaks, is 6 miles in circumference, its inner edge lying nearly 1½ miles from Lorenzo point, but this passage should not be used, as there is only 10 feet water in it, and generally a heavy swell. A spit with 12 feet extends off its south-west end for 1½ mile, and 4½ fathoms only are found nearly 5 miles from it ; vessels should not stand within that depth. Colorado point, kept open of Patino point bearing N.E., is a good mark to clear this bank, and also for running up the gulf.

**The Western Side of the Gulf** is little known, but is reported to be shoal ; the eastern side has plenty of water along it. Off Patino point, which is just separated from the main enough to make it an island, there is no known danger ; 2¼ miles inside this, in a N.N.E. direction, is Colorado, a bold rocky point, with a conspicuous patch of reddish clay on its face, the coast between forming a bay ; 1¼ miles farther up the gulf the land gradually gets lower, and forms Point Hamilton, it then falls back to the eastward, and bends round again to the north-west, making a bay 3½ miles across, with low mangrove shores, having a village and anchorage in it, but there are some ledges of rocks that do not always show, so that great caution is necessary in using it. A boat should be first sent in to point out the deep water.

**Washington Island,** nearly 3½ cables N.N.E. from point Hamilton, is 3 cables long, the same broad, and densely covered with wood ; this island lays rather more than a mile from the inner point of the bay just spoken of, with several islets and rocks in the intermediate space. The channel up the gulf is to the northward of Washington island, between it and Jones island, a conspicuous little rock about 20 feet high, and covered with grass ; these two islands lie N.W. by W. and S.E. by E. 1¼ miles from each other.

The coast from abreast of Washington island takes a northerly direction for about 6 miles to Stanley island ; in this space are several little bays, lined with mangrove, the points generally being of small elevation rocky, and covered with bush, the channel lying between the coast and a group of islands on the west as far as Strain island, the western of

---

* *See* Plan of the Bay of San Miguel, and Darien Harbour, on No. 2,261 ; scale, m = 0·75 of an inch ; and Dampier's Voyages, vol. i.

the group. This little island is about 25 feet high, covered with trees and scrub, and surrounded by a ledge of rocks extending a short way off it towards the channel, but connected by mud banks with 2 islands westward of it.

At this point Barry rock, an islet 20 feet high, and covered with *cacti*, lying about 3 cables from the eastern shore, contracts the passage to 1 mile in width ; apparently there is deep water all round this rock, but the passage on its west side being by far the widest, most direct, and sounded, there can be no object in using the other. The channel continues of about the same breadth to Virago point, a distance of 2½ miles. In working through do not go within a line drawn from one island to the other, and avoid Bains bluff, one mile to the southward of Virago point, where there is a dangerous ledge of rocks at 3 cables from the shore.

**Stanley Island**, a low wooded island 1½ miles long by 1 broad, divides the channel into two passages, both leading into Darien harbour ; the principal one, or Boca Grande, being a continuation of the gulf in a northerly direction past the west and north sides of Stanley island, and the other, or Boca Chica, between its southern end and Virago point, on the south shore of the gulf. The latter channel, although much shorter, is too narrow for a sailing vessel to use with safety, on account of the rapid tide in it.

**BOCA CHICA** has two dangerous ledges of rocks at its outer entrance one on each side, the passage between them being barely 300 yards wide ; the southern one only shows at low-water spring tides, and lies one cable nearly west from Virago point. The Trevan rock on the north side is uncovered at half tide, and is about 2 cables from the shore of Stanley island. Mary island, the northern of the group before mentioned, kept just midway between the summit of the west part of Jorey island, the largest and western, and its west extreme, bearing S.W., offers an excellent mark for clearing these dangers, recollecting that if brought on with the summit, the vessel will get on the northern ledge, and if open to the westward she will be on the opposite one ; when past these rocks keep in mid-channel. Off the south-east point of Stanley island, there is a small ledge running out a short distance, having passed this the vessel will be in Darien harbour, and may anchor, as convenient, in from 5 to 10 fathoms sand and mud ; this channel is not, however, recommended, unless used at slack water, for during the strength of the tide it runs 6 or 7 knots ; the eddies making the steerage difficult.*

**BOCA GRANDE**, between the rocks outside the Boca Chica, and Milne island on the western shore, is one mile broad, and continues nearly the

---

* *See* Plan of the Boca Chica on No. 2,261 ; scale, *m* = 4·8 inches.

same width for 1½ miles between Stanley island and the shore. After
passing the Boca Chica steer to the N.E., so as to shut in Mary island
by Milne island, and do not approach Stanley island within 3 cables, as a
dangerous rock, showing only at about three quarters ebb and connected
by a ledge with the shore, lays off its north-west point. Milne island
just touching the eastern end of the islands connected with Strain
island, bearing S. ¼ E. is a good mark for running, and if working, when
north of Milne island tack directly Mary island opens of it on the one side,
and when Edith island is shut in on the other. The navigable channel at
this point is three-quarters of a mile wide, and begins to turn to the eastward
round the north end of Stanley island, narrowing to half a mile between
Ray and Jeannette islands on the north, and a large flat rock, nearly always
uncovered, and a little wooded island about a cable off Stanley island
on the south ; following the channel, it bends back to the south-east, and
continues of the same breadth between Ellen and Paley islands on the
west, and the main land on the east, into Darien Harbour.

**DARIEN HARBOUR**, a magnificent sheet of water extending for 11
or 12 miles, in a south-east direction as far as the village of Chapigana, is
formed by the junction of the Tuyra and Savannah rivers. The depth of
water from Paley island as far as the mouth of the Savannah, a distance
of 2 miles, is from 10 to 5 fathoms, beyond which there is not more than
12 to 18 feet at low-water springs. The best place for anchoring is off
the village of Palma, one mile to the southward of Price point on the
south side of the Boca Chica, in from 7 to 10 fathoms, at about 3 cables
from the shore. The Vaguila rock, showing at about half tide, lies off
the mouth of the Savannah, bearing East 1¾ miles from Palma point, with
a good channel between it and Graham point, the west point of entrance
to the river. The shores of Darien harbour are almost without exception one continuous line of mangrove, with densely wooded hills of from
100 to 300 feet high a short distance inland. Palma appears to be situated
on the best spot, and has an abundance of fresh water.

This harbour, with its extensive rivers, in the hands of an energetic
people who would cultivate the soil, would soon become valuable.
Every tropical produce of the western hemisphere could be grown,
mahogany can be had in abundance, the palm and india-rubber tree
also ; maize, rice, sugar, coffee, cocoa, yams, plantains, grow almost wild.

The climate, like most tropical ones, has its rainy and fine seasons,
the former commencing in May and lasting till November, accompanied
by the usual lightning, thunder, and winds of the West Indies ; the other
six months of the year fine ; and with common care the country is
healthy.*

---

* *See* Remark Book of Mr. G. H. Inskip, Master R.N., 1854.

The RIVER TUYRA, the Santa Maria of the Spaniards and Buccaneers, rises in lat. 7° 40′ N. and enters Darien harbour near the village of Chupigana. Twenty miles from this point, near the junction of the river Chucunaque, are the ruins of the old Spanish fort of Santa Maria, near which were the gold mines worked by the Spaniards in the 17th century. The river is described as being navigable 6 miles beyond the town, " abreast which it was reckoned to be twice as broad as the river Thames is at London. The rise and fall of the tide there was two fathoms and a half." The river Chucunaque rises in lat. 8° 50′ N., westward of Caledonia bay on the Atlantic ; its course appears to have been the favourite track of the Buccaneers from the Atlantic to the Pacific. Captains Coxon, Harris, and Sharp, with 330, men in April 1680, started from Golden island in Caledonia bay, and on the second day reached the head of the river, which they describe " as so serpentine that they had to cross it every half mile, sometimes up to their knees, sometimes to their middle, and running with a very swift current."

On the 5th day 70 of them embarked in canoes, but " the men in the canoes found that mode of travelling quite as wearisome as marching, for at almost every furlong they were constrained to quit their boats to launch them over rocks, or over trees that had fallen athwart the river, and sometimes over necks of land." Early on the 8th day they reached Yavisa, which is 15 miles from Santa Maria, at the junction of the river of same name, now the residence of the principal authorities of the province ; here they halted to prepare for the attack on the fort. " They also made paddles and oars to row with ; for thus far down the river, the canoes had been carried by the stream, and guided with poles ; but here the river was broad and deep." On the morning of the 10th they attacked and carried the fort without gaining the expected amount of plunder although a Buccaneer says, " we examined our prisoners severely."[*]

The SAVANNAH RIVER, the one generally preferred for the proposed junction of the oceans, rises in lat. 8° 44′ N., and a few miles from its source meets the river Lara, where the bottom is level with the half-tide. From this point the depth increases, 3 fathoms, low water being found 3¼ miles to the southward, and from thence to the mouth, a distance of 12 miles, the depth varies from 9 to 2 fathoms over a soft muddy bottom. The navigable entrance is nearly a mile wide between Graham point and Haydon bank, and the shores are low mangrove land, skirted with hills from two or three hundred feet high, within two miles of the banks. H.M.S. *Virago* anchored in 3½ fathoms, one mile to the north-east of Graham point. It is high water, full and change, in Darien harbour at 4h.

---

[*] Burney's Voyages, vol. ix., chapter ix.

and the spring rise is said to be 24 feet.  The tides in the narrows run proportionally strong, and great care should therefore be taken.

SAN JOSÉ BANK, a dangerous shoal, in the centre of which is the Trollope rock with only 2 feet water, lies in the fair way of ships bound to Panama from the southward, being 16 miles from Garachiné point and 9 from Galera island (see page 413), the south-eastern of the Pearl islands. The shoal is one mile long by about three-quarters broad, with 3½, 5, and 6 fathoms close to the rock, and 7 and 8 fathoms on its outer edge, vessels should not approach it within the depth of 10 fathoms.  It is easily avoided either by keeping on the main shore until Garachiné point bears to the southward of East, or by passing close to Galera island, which may be approached as near as 2 miles, taking care of the shoal patch and rocks off its southern side.

The PEARL ISLANDS, also known by the names of Islas del Rey, Islas del Istmo, and Islas de Colombia, form an archipelago on the eastern side of the Bay of Panama covering about 450 square miles, and consisting of 16 islands and several rocks.  Isla del Rey is the largest, San José, Gonzales, Casaya, Saboga, and Pacheca are of secondary, and the rest of minor importance.  There are from 30 to 40 fishing villages scattered about these islands, containing (1843) 1,941 inhabitants, chiefly engaged in the Pearl fishery, which is said to produce about two gallons of pearls a year; the shells also form a lucrative article of commerce much inquired after by French vessels.

These islands are low and wooded, the soil fertile, but not much cultivated; most of them belong to merchants at Panama, who employ negroes to plant and cultivate them.  The numerous cocoa nut groves, and bright sandy beaches, intersected by small rocky bluffs crowned with trees, give these islands a pleasant appearance.

PACHECA, SABOGA, and CONTADORA with the islets of Bartholomew and Chipre, are a group in the northern part of the archipelago, forming between them a good and capacious harbour, well suited as a depôt for steamers.  Saboga, the largest island, on the east side of which is a considerable village, has a reef extending 1½ miles to the northward, which, with Chipre to the southward, forms the western side of this harbour.  Pacheca and Bartholomew being on the northern, and Contadora on the south-eastern side; the latter island has 5 fathoms close to its north-east shore, which is low and well adapted for wharves.*

This harbour is about 2 miles long by nearly 1 broad, with an average

* See Plan of Pacheca, Saboga, and Contadora islands; scale, m = 2·0 inches.

depth of 9 fathoms ; it has three entrances, each possessing a 5 fathoms channel, which may be used as best suited to wind and tide.  The Pacheca channel lies southward of Pacheca, between it and the reef from Saboga; the Contadora is close round the northern end of Contadora islet ; and the Saboga channel lies between that island and Contadora.  It is high water, full and change, at Saboga island at 4h., the rise and fall being 14 feet.

**DIRECTIONS.**—Vessels using the Pacheca channel should pass within half a mile of the small island west of Pacheca, and stand to the southward until the centre of Bartholomew island bears E. by S. ½ S. ; steer for it on this bearing until the northern islets off Saboga open westward of Saboga, bearing S. by W. ½ W., when the vessel may haul to the southward for Contadora.  If entering by the Contadora channel, a ship should pass half a mile to the eastward of Bartholomew island and not stand to the westward before the nearest islets north of Saboga open northward of Saboga.  These islets kept just open bearing W. by S. ½ S., leads through this channel.  Care must be taken not to open them too much, as there is a 2 fathom patch to the north-east of Contadora to be avoided.

Vessels from the eastward using the Saboga channel should pass half a mile eastward of Contadora, and continue standing to the southward until the outer islet on the reef south of Saboga opens southward of the same island, bearing E. ½ S.  Steering on this course will clear the vessel of the sunk rock in this channel, which lies nearly one mile to the southward of Contadora, and when Pacheca is shut in by Saboga, bearing N. ¼ W., you may stand to the northward, and run through the channel which is steep-to on both sides.

If coming from the westward, close the islet of Catalina, which lies about 1½ miles south of Chipre, to avoid the shoal off the latter island.  From Catalina a N.E. by E. course should be steered for the north end of Chapera, the island next south of Contadora, until Pacheca touches Contadora, bearing N. by W. ; then steer N.N.W. ¼ W. for a small hill on Saboga, taking care *not to* shut in Pacheca with Saboga until the north point of Chipre bears southward of west, and that *it is* shut in before the south point of Saboga comes on the same bearing, then stand for mid-channel as before.  There is a channel northward of the sunk rock, off Contadora, but the southern one is the wider and better marked.  If this harbour were used, a few buoys would greatly assist the navigation.

**CHAPERA** and **PAJAROS** lie next to the southward, there is a 4 fathom channel between them, but the ground is foul and it should not be used ; a shoal with 13 feet lies one mile to the eastward of Pajaros, and to the southward the soundings are rocky and irregular.  No vessel should attempt the passages south of Chapera island, between it and Isla del Rey.

CASAYA, BAYONETA, and VIVEROS with several islets and rocks, are the largest islands on what may be termed an extensive reef stretching off from the north-west point of Isla del Rey, about 8 miles long by 5 broad, the passages between them being foul with occasional strong tides. A shoal 1½ miles long by three quarters of a mile wide, having only 9 feet water on its shoalest part lies nearly 4 miles to the eastward of the north point of Casaya. Caracoles and Cangrejo, small islets with foul ground around them, lie about 2 miles off the east point of Viveros. The whole of these islands should be avoided by vessels bound up the bay, by approaching no nearer on their western side than just to open San José eastward of Gonzales island (two large islands lying to the southward), bearing S. by E.; while on their eastern side, they should not open San Pablo, a small islet off the north-east point of Isla del Rey.

ISLA DEL REY, the main island of the group, is about 15 miles long by 7 broad, with several peaks on it, the highest being about 600 feet above the sea. Numerous islets, having deep water between them, lie off its western shore, extending 3 miles from the coast, but they should not be approached by strangers within the depth of 10 fathoms. Cocos point, its southern extreme, is a remarkable promontory jutting into the sea, 4 miles long by about one wide, its extreme cliff being crowned in 1859 by an umbrella-like tree, which makes it conspicuous. East of this point is the fine bay of St. Elmo, with convenient anchorage in all parts, and a good stream of water at Lemon point, in the bight of the bay.

The eastern shore has also islands off it, but they are steep-to, and may be approached within half a mile, with the exception of Canas island, at the eastern point, where there is a 3 fathom patch lying outside a sunk rock nearly 1½ miles from the shore; this may be easily avoided by not opening the islet of Mongé, eastward of St. Elmo, until Pablo island opens eastward of Muerta, a small barren island lying about one mile to the eastward of this patch, bearing N.W. by W.

SAN MIGUEL, the principal town of these islands, on the north side of Isla del Rey, is of some size, possessing a conspicuous church; it is, however, badly situated, landing being difficult at low water. Two hills—the Cerro Congo and Cerro Vali—lie to the southward of it, the former being 481 feet high. Supplies are uncertain and dear, being generally all sent to Panama. Vessels having to lie at San Miguel should run in between Caracoles and Cangrejo, and anchor in about 6 or 7 fathoms, when the church is shut in or behind Afuera, an islet lying off the town, bearing S.E. by S. Care must be taken, as the bottom is irregular and rocks abundant.

GALERA, a small island, generally the first land made by vessels bound to Panama, is 7½ miles to the south-east of Cocos point, like which it is remarkable for its umbrella tree. A cliff forms its southern side, sloping down to a beach on the north, and to the southward a reef runs off for nearly one mile. This island should not be approached within the depth of 10 fathoms, but there is a good passage between it and Cocos point, by using which the vessel will be clear of the San José bank.

GONZALES lies on the west side of Isla del Rey, with a broad deep channel between it and the islets before mentioned. It is about 12 miles in circumference, and has on its northern side two bays protected from the North by the islands of Señora and Señorita. These bays, called Perry and Magicienne, were re-examined in 1858 as to their capabilities for a depôt for steamers, and although not so good or so near to Panama as the harbour south of Pacheca, yet still have some advantages. They are divided by the little peninsula of Trapiche, off the east point of which is a rocky ledge, terminating in a shoal, with 14 feet water, at a distance of nearly 3 cables from the point ; inside this, to the southward, there is a small anchorage in 4½ fathoms.*

A large stream of water, found in full force in the month of April, at the end of what had been considered a remarkably dry season, runs into the sea on the western side of Magicienne bay. This bay, however, is small and shoal, without the advantages of Perry bay, which is one mile wide, and runs back for the same distance. Señora and Señorita, including the shoal off their eastern side, are about one mile long, and lie nearly the same distance northward of Trapiche, with a 7 fathoms channel between, steep-to on both sides.

TIDES.—It is high water, full and change, in Perry bay at 3h. 50m.; the rise being 16 feet. The tide stream is not felt in the anchorage, but there is a considerable set off the island, the flood setting to the northward, and ebb to the southward, the latter being generally the stronger.

DIRECTIONS.—Vessels may enter on either side of Señora and Señorita, taking care of the shoal to the eastward of them by keeping the eastern point of Gonzales island, a rocky peninsula, open of the point next north of it, bearing S.S.E., until the north end of Señora is shut in by Señorita, bearing N.W. by W. ¼ W., and a look-out must be kept if going into Perry bay for the shoal off Trapiche.

---

* *See* Plan of Perry and Magicienne Bays ; scale, m ~ 2·0 inches.

**SAN JOSÉ** lies 4 miles directly south of Gonzales, and is about 6½ miles long by 3 broad ; the summit forms a table land, said to be a considerable grazing ground.   Nearly 2 miles south-east from the Iguana point, the northern extreme of the island, is a large waterfall, running into the sea, and forming an excellent watering place.   On the south-east side of the island is a deep bay, but the swell sets in there with great violence.   Off the southern point are a number of high rocks of singular and fantastic shapes, also lashed by a heavy surf ; this part of the island should be avoided.   The western shore is bold and cliffy, with a small bay near the centre.

**PASSAGE ROCK** is a dangerous sunk rock, with 12 and 9 fathoms alongside of it, lying near the centre of the channel, between San José and Gonzales, which otherwise is deep and clear.   It is 1½ miles from Gonzales, and 2½ miles from San José.   The peak next south of the highest on Isla del Rey, just open southward of Coco island, one of the outlying islets before mentioned, bearing E. by N. ½ N., leads more than half a mile to the southward of the rock.   Vessels should keep between the San José shore and this mark.

**BRAVA POINT** forms with Lorenzo point, from which it is distant 2 miles, the northern point of entrance to the Gulf of San Miguel.   Both these points are edged with reefs and outlying rocks, on which the sea breaks with great violence ; this fact, together with the proximity of the Buey bank, makes this part of the coast dangerous, it should therefore be avoided, even by small vessels.

**FARALLON INGLES** is a small but high island, lying at the edge of the shoal off the river Buenaventura, about 5 miles to the northward of Brava point ; 12 and 15 feet water are found on its western side. It was in this river that Dampier and his party in 1681, being prevented by the Spaniards from going by way of the Santa Maria or Chepo rivers, sank their ship and commenced their journey to the Atlantic, which they reached in 23 days near the Concepcion cays, 60 miles westward of Golden island in Caledonia bay, having travelled 110 miles, crossing some high mountains ; but their common march was in the valleys among deep and dangerous rivers.   Gorda point, bold and woody with 4 fathoms close to, lies 4 miles northward of the Farallon ; there is less swell after passing this point.*

**The PAJAROS** are two small rocky islets 4 miles from Gorda point, with 4 and 5 fathoms to the westward, but only 12 feet between them

---

* Dampier's Voyages, vol. i.

and the shore.  At these islets the 5 fathoms shoal commences, which continues in front of the coast round the bay of Panama as far as Point Chamé, on its western shore.

**RIVER TRINIDAD,** 2½ miles from the northern islet, has a low rocky point forming its south-west point of entrance.  A 3 fathom channel was found into this river, extending 1½ miles from the point, beyond which it was not examined.  The northern bank of the river is composed of mangroves, which continue, with the exception of the bluffs of the rivers Chiman and Chepo, as far as Panama, a distance of nearly 70 miles.  Shag rock, a barren islet, frequented by birds, with shoal water round it, lies 2¼ miles from this entrance.

**MANGUE** and **MAJAGUAY,** 7 miles from the river Trinidad, are high, wooded islets at tide time, but not at low water, being situated on the south-west edge of a large mud flat, which extends from the north bank of the river Trinidad.  There is a depth from 10 to 12 feet water to the westward of them.

**RIVER CHIMAN,** to the northward of these islands, is wide at the mouth, but shoal, being nearly dry at low water, with small channels for canoes.  The entrance is well marked by the Mangue islets and the wooded bluffs on each side.  On the eastern side, under a hill, is the small town of Chiman.  This was the spot to which Pizarro retired in 1525, after beating about for 70 days with much danger and incessant fatigue, without being able to make any advance to the southward.  He was here joined by Almagro, and the following year they sailed again for Peru.

**PELADO ISLET,** W. by N. 4 miles from Mangue islet, directly off the mouth of the river Chiman, is a flat level islet of small extent and about 60 feet high ; it has no trees, but is covered with a coarse prickly shrub ; is steep-to on all sides, and forms a useful mark to vessels bound up the bay for Panama, who need not go inshore of it.

**CHEPILLO ISLAND,** 31 miles from Pelado, is described by Dampier as the most pleasant island in the Bay of Panama ; it lies off the mouth of the river Chepo, about 2 miles from the coast, and is one mile long by one-half broad, very fertile, being low on the north side, and rising by a gentle ascent towards the south, over which is a remarkable tree.  This tree also forms an excellent mark to vessels bound up the bay ; the southern point may be approached within a mile, but the other sides are shoal, a reef running off its northern point in the direction of the river.  The coast between this island and Pelado is low river land with mangrove bushes.

There are several small streams, the principal being the rivers Hondo and Corutu, but both are shoal at the entrance. The land north of these rivers is of some elevation. Column peak and Asses Ears, about 12 miles north of Chiman, and Thumb peak, at the west extreme of the range, are conspicuous.

Extensive mud banks, dry at low water, are found in front of this coast, stretching from 4 to 2 miles off the land ; outside these again the water is shoal for some distance, and vessels standing towards the main should tack in 9 fathoms.

**CHEPO RIVER** extends some distance into the interior of the isthmus, having its rise near the head of the Savannah river. The entrance is to the eastward of Chepillo island, through a 10 feet channel, about 3 cables broad ; there is a small hill, with a cliff under it, on the eastern bank, which, if brought to bear N.E. ¼ E., will lead through the deepest water. At the west bank of the river the mud flat recommences, and continues to Petillo point, just to northward of Panama ; a shoal bank lies in front of the flat, and vessels should stand no nearer than 6 fathoms between Chepillo and Panama.

**PANAMA,** a regular and formerly a well fortified city, standing on a rocky peninsula, has a noble appearance from the sea ; the churches, towers, and houses, showing above the line of the fortifications, stand out from the dark hills inland with an air of grandeur to which there is no equal on the west coast of South America. It is rendered still more conspicuous by mount Ancon, a beautiful hill, 540 feet high, lying nearly a mile to the westward of the city, to which it forms a pleasant background ; on each side of Ancon are flat hills, with copses of wood and savanas, grassy slopes and wild thickets, while to the southward the cultivated islets of Flamenco and Perico complete the scene, which, as Dampier says, " altogether " make one of the finest objects that I ever did see, in America " especially."*

The site of Panama has once been changed. The old city, built in 1518, which was taken and destroyed by the Buccaneers under Morgan in 1673, stood at the mouth of a creek, about 4 miles north-east of the present town. The spot is now deserted, but well marked by a tower, which, together with an arch, two or three piers of a bridge, and some fragments of wall, are the only remains of a once opulent city. The tower, in the afternoon, is still a conspicuous object from the anchorage.

The expectations formed of the modern city of Panama as seen from

---

* *See* Plan of Panama Road, No. 1,544 ; scale, m − 1·75 inches ; and Dampier's Voyages, vol. i, page 179.

the sea, are by no means realised on landing. The principal streets extend across the peninsula from sea to sea, intersected by the Calle Real or Royal street, which runs east and west, and has a quiet and stately, but comfortless, air. Heavy balconies in the upper stories are but little relieved by the unglazed grated windows, or any variety in the buildings.

The houses, mostly in the old Spanish style, are of stone, the larger having courts or patios; the public edifices comprise a cathedral, five convents, a nunnery, and a college, but most of these are in ruins. The cathedral is a large, lofty, building, on the west side of the Plaza, but the structure is hardly worthy of its situation, the towers alone redeeming it from insignificance, and forming, in the distance, an ornament to the city. The fortifications are well constructed, but, like the rest of the town, in many parts are in ruins, the north-east bastion having fallen in 1845; the south and west ramparts are still in good condition, forming a pleasant promenade. A great want is felt in Panama with regard to drainage. This is caused by neglect; for the elevation of the peninsula on which the city stands, together with the great rise of the tide, offers considerable advantages for cleansing the city, which duty at present is performed by the heavy rains of the wet season.*

The gold discoveries of California and British Columbia, by increasing the colonization and developing the great agricultural resources of those countries, have effected a change in the fallen fortunes and grass grown-streets of Panama; and comfortable inns and large well-stocked stores have been called into existence by the continuous transit of emigrants from Europe and the United States to San Francisco and Victoria.

The suburb of Santa Ana, situated on the isthmus which connects Panama with the mainland, is almost as extensive as the city, though not so well built. At its northern extreme is the terminus of the Panama railway to Colon or Aspinwall, on the Atlantic, a distance (by rail) of 47 miles. This railroad was only completed in January 1855, since which time the company has been constantly making improvements, until it is now one of the best appointed lines extant. Its gross revenue for 1857 amounted to 272,000*l.* or 5,725*l.* per mile, produced at the following proportions per cent., viz., passengers 53½, freight 27, treasure 9½, mails 8½, baggage 1¼, miscellaneous ¼.†

**TRADE.**—Some idea of the increasing trade of the isthmus of Panama may be formed from the fact that in the first week of June 1860 no less than 10 steamers arrived and sailed,—5 from Panama and 5 from Colon—all of them being sea-going vessels, of between 1,000 and 2,000 tons. These

---

* Voyage of H. M. S. *Herald*, vol. i.

† Remark Book of Captain Thomas Harvey, R.N., 1858; and Plan of Panama Railroad, No. 2,021; scale, m = 0·55 of an inch.

steamers, or others on the same lines, make regular semi-monthly trips direct from the isthmus to upwards of 50 different ports, in no less than 15 distinct countries. To carry on this trade a large fleet of first-class steamships is constantly employed, and few people, even among those engaged in the trade of the Pacific, have any idea of the amount of traffic these vessels bring to the isthmus. No less than 38 sea-going steamers, many of them registering from 2,000 to 3,000 tons, either arrive at or depart from the ports of Colon and Panama every month.

The merchants of the entire west coast of America, from British Columbia to Chiloe, receive their European mails and export all their specie *via* Panama. These facts will show the importance of this narrow neck of land. Every day the transit business increases, and promises soon to restore Panama to her old position, so well described by Dampier, when she was the highway between Spain and her colonies in the Pacific: " This is a flourishing city, by reason it is a thoroughfare, for all imported or exported goods and treasure, to and from all parts of Peru and Chile, whereof their store-houses are never empty. The road is seldom or never without ships. Besides, once in three years, when the Spanish armada comes to Portobel, then the Plate-fleet, also from Lima, comes hither with the king's treasure, and abundance of merchant-ships, full of goods and plate ; at that time the city is full of merchants and gentlemen; the seamen are busy in landing the treasure, and the carriers, or caravan masters, employed in carrying it overland on mules (in vast droves every day) to Portobel, and bringing back European goods from thence. Tho' the city be then so full, yet during this heat of business there is no hiring of an ordinary slave under a piece of eight a day ; houses, also chambers, beds, and victuals, are then extraordinary dear."*

SUPPLIES.—Panama affords the usual supplies which are to be obtained in tropical regions, but in 1857-8-9 they were generally dear ; provisions of excellent quality may, however, be obtained from the United States by ships requiring them ; and, when time will admit of it, getting such from the States is far preferable to purchasing in the markets of Colon or Panama, which in the above years were generally supplied with articles of an inferior quality ; biscuit especially will not keep in the hot climate of Panama. The United States squadron have all their stores and provisions sent across the isthmus. A store-ship as a depôt for that squadron was on her way from San Francisco to this port in 1859. The increasing importance of Panama as a central position from which to direct the

---

* Dampier's description, vol. i., page 179, of the arrival of the Plate fleet will, in a great measure answer for that of the Mail and passengers to or from California, recollecting that instead of " once in three years " it is once a fortnight.

movements of the Pacific Squadron, may render necessary a permanent depôt here, especially for receiving men and stores from England, and for sending invalids home.*

Water can be obtained at Panama from the tank of the U.S. mail steamers; but it is cheaper at Taboga (see page 424), where it may be purchased at two dollars a ton. Coal may be bought here at times from the mail companies, but it is generally dear; the cost of coal imported into Panama by way of Cape Horn being 16 dollars per ton, and by the railroad 15.† Consuls of all nations reside at Panama.

The home value on most of the products of this country has advanced very much since the completion of the railroad. Large quantities of Peru bark, balsam, cochineal, cocoa, coffee, hides, india-rubber, indigo, logwood, oil (whale and cocoanut), sarsaparilla, vanilla, gold, silver, and hundreds of other commodities of the Pacific, seek a market *viâ* this great central route of the globe. In 1858, 142 vessels, of 22,034 tons, entered inwards, and 136, of 94,912, cleared outwards; the value of imports being 11,373,424*l.*, and of exports 2,468,203*l.* The population of the isthmus in 1853 amounted to 144,108 persons. Population of the city may be 10,000 in 1860.

CLIMATE.—The geographical position of the Isthmus of Panama, the absence of high mountains, and the vast extent of forests and other uncultivated parts, tend to produce a hot and rainy climate, which nevertheless, with the exception of a few localities, as Chagres, Colon, and Portobello, is healthy and more favourable to the constitution of Europeans than that of most tropical countries. The most prevalent disease is intermittent fever, which makes its appearance during the change of the season; remittent fever is less frequent, but generally proves fatal. On board ship Panama is by far the most healthy place on the coast of Central America. Vessels of war have remained here many months at a times their crews continuing in a healthy state, excepting those men who had the will and opportunity to indulge in the vile spirit (*aguardiente*) of the country, which is cheap and easily procured. The yellow fever that existed at the Morro of Taboga in the early part of 1859 was confined to that spot, and, with few exceptions, the victims to it were men of drunken habits, and for this reason commanders should avoid giving liberty to their crews at Panama. H.M.S. *Herald,* when employed on the survey of the Bay in 1847-8, never gave liberty, and although the men were constantly in the boats the crew were always healthy.

The seasons are regularly divided into the wet and dry; the former commences in the latter end of May and lasts till November. Slight at

---

* Remarks of Capt. Harvey R.N., 1859.    † Naut. Mag. for March 1860, p. 164.

first, the rain gradually increases, and is fully established in June, when it falls occasionally in torrents, accompanied by thunder and lightning: the air is loaded with moisture, and calms or light variable winds prevail. The temperature varies from 75° to 87° Fahr.; still the atmosphere is oppressive, until cooled by the heavy rains and thunder storms before mentioned. About the end of June the rains are suspended for a short time; and the occurrence of this phenomenon is so regular that it is looked forward to by the inhabitants, who call it the veranito (little summer) de San Juan, probably from its taking place almost simultaneously with the feast of St. John (June 24th). In December the violent rains cease and the north-west wind sets in, producing an immediate change, and the climate now displays all its tropical beauties.*

Dampier's remarks on the climate of Panama are too true to be omitted :—" There are no woods nor marshes near Panama, but a brave, dry champaign land, not subject to fogs or mists. The wet season begins in the latter end of May, and continues till November. At that time the sea breezes are at S.S.W., and the land winds at North. The rains are not so excessive about Panama itself, as on either side of the bay; yet in the months of June, July, and August, they are severe enough. Gentlemen that come from Peru to Panama, especially in these months, cut their hair close, to preserve them from fevers; for the place is sickly to them, because they come out of a country which never hath any rains, but enjoys a constant serenity; but I am apt to believe this city is healthy enough to any other people."†

PETILLO POINT is a black rocky promontory with two small hills over it, between which is a rivulet admitting boats at high water; rocky ledges extend from this point for 1½ cables, and off their extreme a depth of 10 feet may be found. The coast between this point and Panama forms a bay nearly three-quarters of a mile deep, the bottom being mud edged with a sandy beach. A great portion of this bay is dry at low water springs, yet at its entrance there is a depth of 8 feet. It is termed El puerto or port of Panama, and it is here that most of the minor trade of the Gulf is carried on, by means of bongos, large canoes made from trees of such dimensions that some of them formed from a single trunk have measured 12 tons. These canoes, though clumsy in appearance, are well fitted for the navigation of the Gulf, and bring most of the tropical productions of the isthmus to Panama.‡

---

* Voyage of H. M. S. *Herald*, vol. i.          † Dampier's Voyages, vol. i, page 186.
‡ *See* enlarged Plan of Panama Road.

**BUEY POINT**, only seen after half-ebb, forms the southern horn of this bay, and the north-eastern point of the long rocky ledges that surround the eastern and southern shores of the peninsula of the city. They extend $3\frac{1}{2}$ cables from the N.E. bastion, 5 cables from the S.E. bastion in an easterly and $2\frac{1}{2}$ in a southerly direction, forming a bay southward of Buey point, in which is easy landing after half-flood, on a sandy beach in front of the Monk's gate, one of the principal entrances to the city. The general landing, however, is round Buey point, at the market place on the northern side of the town. From the commencement of the suburbs on the southern side, another ledge runs off for nearly three-quarters of a mile, east of which are Los Hermanos, 3 black rocks visible at first quarter‑ebb. Detached rocks, with 3 and 7 feet water between them, visible only at low water springs, lie off the south-east extreme of the rocks the outer one being 3 cables from the reef. These ledges, composed of rock with sand patches between, although now irksome and often dangerous to boats, afford every facility for erecting substantial piers and improving the port. As yet (1859) there is no attempt at works of this description, but the daily increasing trade must produce these necessary improvements.

**GUINEA POINT**, to the south-west of Panama, is the northern extreme of a large round hilly point, which forms the western side of Panama road. Between it and the town are the mouths of the Grande, Arena, and Falfan, small rivers, with cultivated banks. The water on this side of the roadstead is shoal as far as Tortola and Tortolita islands, which lie 2 miles to the southward of Batele point, the south extreme of the hilly point above mentioned. One mile W. by S. $\frac{1}{2}$ S. of the same point is Changarmi island, surrounded by the Pulperia reefs, while to the south west are Bruja and Venado points, rocky and projecting, with the outl,ing islets of Cocovi and Cocovicita. Although these dangers are mostly above water, yet this part of the bay of Panama should be avoided.

**PERICO and FLAMENCO**, with the outlying rock of San José, are a group of islands forming the south side of Panama road. Ilcñao and Culebra, the western and southern parts of Perico, are connected with it by an isthmus of beach and rocks; but at high water these present the appearance of three islands. Perico is the head quarters of United States mail steamers, the bay on its northern side forming a convenient anchorage, while on the isthmus, which is sandy on that side, steamers of 2,500 tons have been easily beached. Vessels using this anchorage after passing Flamenco should keep close round the north end of Perico, and anchor when the isthmus opens. Large vessels drawing over 20 feet may coal at Perico by passing west of the group at half tide, with Ancon hill, (which on that bearing makes like a cone) just open of Ileñao, N.N.W., pass about a cable's length from Ileñao, and anchor off its north-west end in 24 feet, when Perico opens. In both cases attention must be paid to the time of tide. The passage between Perico and Flamenco is shoal, and should not be used, but that between Flamenco and San José is deep, and the islands steep-to.

**DANAIDE ROCKS**, a patch of conical rocks, on the eastern ridge of the road, with only 12 and 15 feet on them, surrounded by 3½ and 4 fathoms, lie E. by S. 2¼ miles from the S.E. bastion. These rocks are awkwardly placed, lying in the track of vessels standing for the anchorage, keeping their luff with the land breeze. The Hermanos rocks, on with the hill between the rivers Grande and Falfan, bearing W. by S., will clear them, the ship passing to the northward, and the Cathedral towers kept open eastward of Ancon hill, bearing N.W. by W. ½ W., will lead the ship to the southward. It is a favourite fishing place, and vessels should avoid canoes seen in that vicinity, as they are probably fishing on the rocks.

**SULPHUR ROCKS**, a dangerous reef, lying one mile to the north-west of the Danaïde, 6 cables long by 3 broad, have a rock awash in their centre, with 6 and 9 feet around it, and outlying patches of 12 and 14 feet. The railroad flagstaff on with the centre of Ancon, bearing W. ½ S., will lead to the northward of them in 15 feet, but this passage should not be used at low water springs. Hermanos rocks, on with a round peak over the Rio Grande, clears them in passing to the southward in 18 feet.

**KNOCKER and TABOGA ROCKS** are 2 sunken rocks, with only 6 feet water on them; the former has a *red* buoy, with *staff and flag* on it, and lies nearly one mile E. ¼ N. from the S.E. bastion; the latter lies a little more than 2 cables to the south-west of the buoy, with 16 feet water between, and 12 feet inshore of them, but no stranger should attempt to pass west the Knocker buoy.

**TIDES.**—It is high water, full and change, in Panama at 3h. 23m. The springs range from 18 to 22 feet, and the neaps from 6 to 10 feet. The ebb sets South from 1 to 1½ miles an hour, and is stronger than the flood, which runs to the north-west. A long swell which occasionally sets into the road always ceases with the flowing tide. It has been remarked by the officers of the U.S. Pacific mail steamers that there is more rise in the small bay north of the town, and also in their own anchorage off Perico, than in the more open parts of the road.

**DIRECTIONS.**—Sailing vessels bound to Panama should endeavour to get within 3 or 4 miles of Chepillo island, especially between December and

and June, and so have all the advantages of the prevailing northerly wind.
From this point Ancon hill will be seen, and should be kept a little on the port
bow, as the wind hauls to the westward on approaching Panama. Vessels
drawing over 18 feet should pass south of the Danaïde rocks, by not
bringing San José on with the west point of Taboga (the largest of a
group of islands about 9 miles south of Panama), bearing S.S.W., until the
cathedral towers are open to the eastward of Ancon. Having passed the
Danaïde the ship is fairly in the road and may anchor according to her
draft ; if no more than 18 feet she may have Tortola just shut in by
Ileñao, bearing S.S.W. ¼ W., and San José open east of Taboguilla, the
eastern of the group above mentioned. Larger vessels drawing 24 feet
may come-to north of Perico, with the peak of Urava, the centre of the
Taboga group, on with the East point of Flamenco, bearing South, taking
care not to open Changarmi northward of Perico. If it is necessary
to work up the road to an inshore berth, tack on the western side just
before Perico and Flamenco touch, and in standing to the eastward do not
open San José of Taboga island.

Vessels drawing 14 feet may pass north of the Danaïde and south of the
Sulphur rocks, with the Hermanos rocks on with right side of the peak,
between the rivers Grande and Falfan, then San José on with the peak of
Taboguilla bearing S. ½ E. leads between the Sulphur and Knocker rocks,
and they may anchor north of the buoy in 16 feet, keeping it between
Perico and Flamenco, with Gabilan, a rocky peninsula west of the town,
just shut in by the S.E. bastion. During neap tides they may anchor
still farther to the N.W.*

Panama road, although shoal, may be considered secure ; the ground
being muddy holds well. A sailor, resident in Panama for five years,
remarks, that during that time there was no known case of a vessel being
driven from her anchor ; and with good ground tackle and common precaution a vessel might lie there all the year round with one anchor down.
Attention to the tides and soundings of the roadstead will enable a vessel
to lie close in at times for the discharge of cargo. The U.S.S. *St. Marys*,
drawing about 17 feet, lay at anchor during the neap tides inside the
Knocker buoy.†

TABOGA ISLAND, with those of Urava and Taboguilla form a
pleasant group of islands, about 4 miles long by 2 broad, lying 9 miles to
the southward of Panama. Taboga, the highest and largest, 930
feet above the sea, is well cultivated, with a considerable village on
its north-east side. To the northward of the village is the Morro of

---

* Remarks of Mr. Thomas A. Hull, Master R.N., 1858.  † Naut. Mag. for 1856, p. 333.

Taboga, a small hill, connected with the main island by a low, sandy isthmus, covered at high water. This place is the head-quarters of the Pacific Mail Company, who have here a steam factory and coal stores, also a gridiron, 300 feet long, on which H. M. S. *Magicienne*, a vessel of 1,255 tons, was repaired in 1858.

Vessels visit Taboga from Panama to procure water and supplies, both of which are more readily obtained than at the city. Water can be procured from the Company's tank at 2 dollars per ton. The anchorage formed by the Morro is convenient, being about 3 cables from the shore in 10 fathoms with the peak of Urava on with high cliff of Taboga and the church from S.W. ¼ S. to West. Vessels coaling at this island should avoid giving liberty to their crews. In 1859 H.M.S. *Alert* suffered severely from a fever contracted through the excesses of her men while at the Morro. Ships that have not given liberty have felt no bad effects after visiting this depôt.

Urava is a small, lofty island, separated from Taboga by a narrow and shoal channel; off its southern extreme is the small islet of Terapa. Taboguilla, 710 feet high, also well cultivated, with some islets off its south-west extreme, forms the north-east island of the group, with a wide and deep channel between it and Urava, in the centre of which is a sunk rock with 8 and 14 fathoms close to; it is easily avoided by closing either island, both being steep-to, or vessels may pass south of it by keeping the isthmus of the Morro open, bearing N.W. by W. ¼ W. Farallon, a small islet, also lies in this channel, but it is steep-to with 11 fathoms between it and Taboguilla.

**The COAST,** from Bruja point to Chamé point, a distance of 46 miles, forms a shoal bay, with several outlying banks and rocky islets, and vessels bound to Panama should keep near the islands of Taboga, and not approach this shore within the depth of 5 fathoms. Vique cove, in which is a small village, is 5 miles from Bruja point. About one mile to the north-east of Vique is a lofty treble-peaked hill, called Cerro de Cabra, forming a conspicuous object to vessels bound to Panama, and frequently mistaken for Taboga by those coming from the eastward. Vacamonto point, the western side of Vique cove, forms the only break in the mud flat which lies in front of this land, extending nearly 2 miles from the shore.

**CHAMÉ BAY,** at the head of which is a small river of the same name, is nearly filled up by large banks, of which the largest is the Cabra spit, lying in the middle, with Tabor isle on it. On the southern side is Chamé point, a singular, low, woody promontory jutting into the sea, 5½ miles long by half a mile broad. Between this and Cabra spit is a convenient harbour, 2 miles in length by about three-quarters of a mile in breadth,

with from 3 to 8 fathoms water in it, and from 16 to 18 feet, close to
the beach of Chamé point.    To the north-west of the river is a high range
called Sierra Capéro, and to the southward are the Cerro Chamé, a group
of wooded hills.*

**MELONES ISLAND** is a small rocky islet, 2½ miles to the north-west
of Taboga, with a rock above water, lying about half a mile to the northward of it.    Melones is steep-to, but vessels should be careful not to pass
to the westward of it.    Chamé island, with the Perique rock, are of a
similar nature, situated about the same distance southward of Taboga.
Valladolid is a large rock, nearly 2 miles to the south-west of Chamé
island, with 9 and 10 fathoms close to it.

**OTOQUE and BONA,** with Estiva island and the Redondo rock, lying
6 miles to the south-east of Chamé point, form a group similar but somewhat smaller than Taboga and Taboguilla, being cultivated, and having a
considerable village called La Goleta, in the bay on the western side of
Otoque.    Anchorage, in from 10 to 14 fathoms, may be found in any part
of this group, and all dangers are above water.    Both islands are high
and peaked, forming good land-marks to vessels using this side of the bay
of Panama.

**PARITA BAY,** large and open to the eastward, is nearly 20 miles
across, lying 45 miles to the south-west of Chamé point.    The coast between
is a continuous beach, called Playa Grande, in front of a low wooded
bank.    A depth of 4 and 5 fathoms is found about 2 miles off this beach,
except S.S.E. ½ E. of the Cerro Chamé where there is only 4 fathoms nearly
7 miles from the land, the bank extending from that to Chamé point.
Vessels from Parita bay should shape a course to pass about 2 miles to
the southward of Bona until Taboguilla is nearly touching Otoque, bearing
N. by E. ½ E., when they may steer up the bay inside, but nearer to the
islands.

The mud-flats are found again on the western side of Parita bay, the
coast being a low mangrove shore, intersected by the mouths of no less
than 5 small rivers ; the land to the westward is also low with several
hummocks.    At Liso point on the south side of the bay the hard bank
with sandy beach in front again commences and continues as far as Cape
Mala a distance of 38 miles, the coast trending to the south-east.

**IGUANA ISLAND,** a little higher than the adjacent coast, and thus
forming a conspicuous object, lies about 9 miles to the northward of Cape
Mala.    A ledge extends about 3 cables from its southern and also from
its eastern point, but otherwise the island is steep-to with 15 fathoms in

---

* See Plan of Chamé bay; scale m = 2·0 inches.

the channel between it and the main.  It is high water, full and change, at Iguana island at 4h.; the rise and fall being 15 feet.  The flood sets the northward and the ebb to the south-east, the latter being considerably the stronger, especially between December and June.

**CAPE MALA**, which forms the western point of entrance to the gulf of Panama, is a low but cliffy point with outlying rocky ledges, having deep water close to them.  The land from the north-west slopes gradually down to the sea at this point from a considerable distance, making the exact cape difficult to distinguish, unless the breakers are seen.  On opening the gulf round this a strong southerly set is generally experienced, especially in the dry season.

**NORTH and SOUTH FRAILES** are two low barren islets, lying N.N.W. ½ W. and S.S.E. ¼ E. about 2 miles from each other and 11 to the south-west of Cape Mala.  There is a reef about a cable off the north-west point of southern islet, but with that exception they are steep to, having from 20 to 30 fathoms within half a mile of the rocks ; this makes them dangerous in the thick squally weather of this coast to vessels keeping under the land of Cape Mala to avoid the current.

**MORRO PUERCOS.**—The coast from Cape Mala trends sharply to the westward, and continues low as far as Guanico point, a distance of 22 miles.  From this point it gradually rises for 7 miles to the Morro Puercos, a lofty headland which forms the commencement of a range of high coast land.  A 3-fathom patch lies about 3 miles to the north-east of the Morro, and 4 miles to the westward is a reef of rocks above water lying 1 mile from the shore.  To the north-east of Guanico point is an open bay, into which two small rivers, the Tomosi and Juera, empty themselves.  There is a patch of rocks close to the shore in the north-west part of this bay, and another off Raia point at its eastern end.

**MARIATO POINT**, a bold headland at the termination of the high land, of which Morro Puercos is the commencement, is 27 miles to the westward of the Morro.  The water off this coast is deep close to the rocks, no bottom being found with 100 fathoms within 3 miles of the shore.  Five miles to the north-west of Mariato point is Naranjas island, a rugged and rocky, but wooded, islet lying about half a mile to the westward of a bluff, to the northward of which the low land again commences, the coast trending to the northward towards the great bay of Montijo.  Mariato point is a good landfall for vessels bound to Panama from the westward, as by keeping under this land they will avoid the southerly set out of the Gulf.

**COCOS ISLAND**, a favourite rendezvous of the Buccaneers, and now much frequented by whalers, lies 380 miles W.S.W. from point Mariato. It is high on the western side, being visible from a distance of 20 leagues, and is about 13 miles in circumference. Its southern side, which has not been examined, consists of steep rugged cliffs, rising abruptly from the sea. The northern coast is indented into small bays with rocks and islets lying off them.*

**Supplies.**—Good water is plentiful and easily procured. Fish are abundant, and wild pigs numerous. The cocoa-nut trees, formerly so plentiful, have been cut down for fuel, and few remain that are accessible without some trouble.

**Chatham Bay** is on the north-east side of this island, having anchorage in about 14 fathoms half a mile from the shore, with Conic island off the eastern horn of the bay quite open of the land, bearing E.N.E., and the sandy beach in the bight of the bay, near which is the watering place, S. by E. This bay is open to the north, prevalent winds are south and S.W., with occasional squalls from the N.E., these, however, are of short duration.

**Wafer Bay**, about one mile to the westward of Chatham bay, is by no means such a good anchorage, having deeper water, and the heavy swell which occasionally rolls in makes the landing difficult. A small rugged rock lies off the mouth of the bay, about half a mile from Swain point, on the southern shore. Off Lionel head, the western part of the island, are the Wafer islands, 2 islets about a mile from the shore.

**The TIDES** at this island require attention. It is high water, full and change, in Chatham bay at 2h. 10m.; the rise and fall being 16 to 18 feet. The ebb sets to the East at the rate of 4 or 5 knots; the flood, which is weaker, sets to the West. The current off the island is strong and irregular, but generally setting to the north-eastward at the rate of 2 knots.

**GENERAL OBSERVATIONS.**—The navigation of the approaches to the Gulf of Panama, situated as they are in the region of the doldrums, with the land of Central America considerably affecting the northern trade; and between that port and the Galápagos islands, becomes to a vessel unaided by steam one of the most tedious, uncertain, and vexatious undertakings known to the sailor. Steam power will considerably simplify these difficulties, but the experiences of a sailing vessel may materially assist the navigation of the auxiliary screw steamer in this portion of the Pacific.

---

* *See* Plan of Cocos Island on No. 1,936; scale *m* = 2·0 inches.

**WINDS.**—The winds between Guayaquil and Cape Corrientes have already been described at page 393. Between Cape Corrientes and Panama, the prevalent winds are from the northward and westward, with frequent squalls and wet weather from the south-west between the months of June and December.

In the Gulf of Panama the winds are regulated by the seasons; the prevalent wind, however, is from the northward. In the fine season, commencing in December, these winds are regular and constant, bringing fine dry weather. To the southward of the gulf they blow much harder, and off the coast of Veragua a double reefed topsail breeze in January and February is not uncommon. In April and May the northerly winds are less regular, and have more westing in them, with calms, light sea, and land breezes, with occasional squalls from the south-westward. In June the rainy season sets in, and the southerly winds become stronger. Still the old north-west wind is mostly found after noon, and vessels sailing from Panama at all seasons will generally have a fair wind until south of Cape Mala.

Between the Galápagos islands and the coast, westward of the meridian of 80°, and southward of the parallel of 5° N., the winds are between south and west all the year round, and except between the months of February and June they are of sufficient strength and duration to make the navigation easy; but northward of lat. 5°, between 80° and 110° W., is a region of calm and doldrums, accompanied by rains and squalls of a most vexatious description. The weather met with can hardly be better illustrated than by the fact that in May 1848 H.M.S. *Herald*, in her passage towards the Sandwich Islands, although towed for 6 days as far west as 89° 20', still took 40 days from Panama to 110° W., owing to keeping between the parallels of 8° and 10° N., and in March of the following year, in the meridian of 87°, and the lat. of 8° N., only made 30 miles in 9 days.

**CURRENTS.**—The Gulf or great bay of Panama, formed by South America on the East, and Central America on the North, is also subject to varying currents, partly caused by the peculiar formation of the land, and apparently influenced in turns by the Peruvian or Mexican streams, according as the relative strength of each predominates. Thus Malpelo island is surrounded by a strong current, having much the appearance of breakers. Here Colnett found the current setting strongly into the gulf N.E. by E. at the rate of 2½ miles an hour, while other navigators describe them as running violently in the opposite direction. That these varying statements should be equally correct is not at all incompatible, considering the position of the island amidst conflicting winds. This uncertainty is another embarrassment to the navigation between Panama and the Galápagos. A steady current, however, has generally been found to set to the northward after passing Cape San Lorenzo, extending off shore for about 60 miles. This stream runs along the coast of the continent, round the Bay of Panama, and then sets with considerable force, especially in the dry season, to the southward down the western side of the bay. After passing Cape Mala it meets the Mexican current from the W.N.W. and thus causes the numerous ripplings and short uneasy sea so often met with at the entrance of the Gulf. This troubled water will be found more or less to the southward, according to the strength of the contending streams.

PASSAGES.—From the foregoing it will be seen that the passage from the southward into the Gulf of Panama is easily made during the greater part of the year, by keeping about 60 miles from the coast north of Guayaquil, and after crossing the line shaping a course for Galera island, at the same time taking care, especially in the dry season, to stand inshore with the first northerly winds. By so doing vessels will most probably have the current in their favour along the coast; whereas by keeping in the centre or on the western side of the gulf, a strong southerly set will be experienced.

After making Galera and clearing the San José bank, the navigation between the Pearl islands and the main is clear and easy, with the advantage of being able to anchor, should the wind fail and the tide be against the vessel. As a rule, this passage should be taken, but with a strong southerly wind, the navigator is tempted to run up the bay, in which case he should still keep on the western shore of the Pearl islands, where anchorage and less current will be found should the wind fail, an event always to be expected in these regions.

Vessels bound to Panama from the northward should make the island of Hicaron, which lies about 50 miles to the westward of Point Mariato, and from this endeavour to keep under the land as far as Cape Mala. If unable to do this, let them push across for the opposite coast of the continent, when the current will be found in their favour. On getting to the eastward of Cape Mala the safest plan is to shape a course for Galera island and to use the eastern passage. At the same time, if tempted up the gulf by a fair wind, vessels should endeavour to get on the western coasts of the Pearl islands, which have the advantages already explained.

The great difficulty, however, is the passage out of, or rather, from the Bay of Panama. Pizarro, the first man who attempted this, in November 1525, after beating about for 70 days was forced to return to the river Chiman, on the eastern side of the bay. The best plan for all sailing vessels, whether bound north or south from Panama, is to push to the southward and gain the S.E. trade ; by so doing they will not only avoid the doldrums and vexatious winds before described, but will have the additional advantage of salubrious weather, with the sea at a temperature of 75° instead of 83° Fahrenheit. The passage to the northward has been made by keeping close in shore after passing Cape Mala, and navigating by the sea and land breezes ; but this should only be attempted by vessels that are well found and manned, unless they are bound to the ports of Central America, when it is their only route.

The following directions, the best for sailing vessels, are chiefly by Lieutenant Maury, of the U.S. Navy.

From the Bay of Panama a vessel should make the best of her way south until she gets between lat. 5° N. and the equator ; on this course let her endeavour, if possible, to keep near the meridian of 80° W. From this make a S.W. course if the winds will allow. Should the wind be S.W. stand to the southward, but if S.S.W. stand to the West, if a good working breeze ; but if it be light and baffling, with rain, the vessel may know that she is in the doldrums, the quickest way to avoid which is by getting South.

From lat. 2° N., between June and January, vessels may stand off from the coast to the westward, and pass northward of the Galápagos islands taking care to keep to the southward of 5° N. As far as 95° they will have South and S.S.W. winds ; but after that meridian the wind will haul round to the southward, and vessels bound to the South Pacific may consider themselves fairly in the trade. Vessels bound northward, after passing the meridian of 100°, may edge away for the Clipperton rock ; after passing which they may push to the northward for the northern trade.

Between February and June it is better to cross the line before pushing to the westward. This will generally take a week, which outlay of time, however, is far preferable to encountering the vexatious weather met in that season north of the Galápagos. In this route it must be remembered that southward of lat. 1° N. the wind hauls to the eastward as you leave the coast, and in the meridian of 83° it is frequently found eastward of South ; but at the same time, vessels in standing off before crossing the equator, must take care to avoid being driven to the northward of that latitude. In fact, there are few passages in which so much depends on the skill and experience of the pilot as in leaving the Bay of Panama.

Vessels bound to the northward in the above season should keep south of the line until westward of 105° when a course may be shaped for 10° N. and 120° W. in which track they will probably find the northern trade.

The above difficulties will be easily avoided by auxiliary screw steamers, which vessels may at once proceed to the starting points above mentioned. The best plan will be to steam for the meridian of 85° W. on the equator, from which line a course may be shaped, according to their destination and season of the year. From that point their sails will be found to be as powerful as their engines.

The following facts will show the singular advantage of even small steam power in these regions: There was in 1859 an indifferent, old, screw steamer, the *Columbus*, belonging to the Panama Railway Company, that had been running with great regularity for upwards of a year between Panama and San José de Guatemala, a distance of about 1,020 miles, calling at Punta Arenas, Realejo, La Union, and Acajutla, both going and returning; at each place discharging and receiving cargo and mails. Sailing from Panama on the 17th of every month, and returning to that port on the 6th of the following; thus making the round in 19 days. It is estimated that it would take two months for a sharp sailing vessel under favourable circumstances to perform the same work.

## PACIFIC PASSAGE TABLE.*

| Place. | December, January, February. | | March, April, May. | | June, July, August. | | September, October, November. | |
|---|---|---|---|---|---|---|---|---|
| | No. of Passages noted. | Average in Days. | No. of Passages noted. | Average in Days. | No. of Passages noted. | Average in Days. | No. of Passages noted. | Average in Days. |
| **Acajutla to—** | | | | | | | | |
| Fonseca | | | 1 | 1† | | | 1 | 4 |
| Istapa | 1 | 2‡ | | | | | 2 | 2‡ |
| Realejo | | | 1 | 7 | | | | |
| **Acapulco to—** | | | | | | | | |
| Panama { | 4 | 27 | | | | | | |
| | 1 | 9† | 1 | 11† | | | | |
| Realejo | | | 1 | 10 | | | | |
| San Blas | 1 | 13 | 1 | 13 | | | | |
| Valparaiso | 1 | 38 | | | | | | |
| **Arica to—** | | | | | | | | |
| Chincha islands | | | 1 | 6 | | | | |
| Islay | 2 | 2 | 2 | 2 | | | | |
| **Bird Island to Honolulu** | 1 | 6 | | | | | | |
| **Buenaventura to Tahiti** | | | | | 1 | 48 | | |
| **Caldera to—** | | | | | | | | |
| Cobija | | | 3 | 4 | | | | |
| Valparaiso | 1 | 13 | 2 | 10 | | | | |
| **Callao to—** | | | | | | | | |
| Acapulco | | | 1 | 22 | | | | |
| Chincha islands | 2 | 3 | | | | | | |
| Coquimbo | | | 1 | 7† | | | 1 | 3 |
| Guanchaco | | | | | | | | |
| Guayaquil | | | 2 | 6 | 1 | 7 | 3 | 33 |
| Honolulu | 2 | 38 | 3 | 31 | 1 | 38 | | |
| Islay | | | 1 | 18 | | | | |
| Punta Arenas | | | 1 | 20 | | | | |
| Nukahiva | | | 1 | 29 | 2 | 22 | | |
| Panama { | | | 1 | 8† | | | | |
| | | | 2 | 16 | | | | |
| Payta | 2 | 5 | | | 1 | 5 | | |
| Petropaulski | | | 1 | 42 | | | | |
| Realejo | 1 | 17 | | | | | 2 | 32 |
| San Blas | | | 1 | 34 | | | | |
| San Francisco | | | 1 | 36 | | | | |
| San José de Guatemala | | | 1 | 21 | | | | |
| Valparaiso { | 1 | 8† | | | | | | |
| | 1 | 28 | 1 | 18 | 2 | 16 | | |
| Vancouver island | | | 1 | 46 | | | 1 | 48 |
| **Chincha islands to—** | | | | | | | | |
| Callao | 4 | 2 | | | | | | |
| Valparaiso | | | | | 1 | 18 | | |
| **Chiloe island (San Carlos) to Valparaiso** | | | | | 1 | 6 | | |
| **Concepcion to—** | | | | | | | | |
| Valparaiso | 2 | 2 | 1 | 2 | | | 3 | 2 |
| Valdivia | | | | | | | 1 | 6 |

* From Remark-book of H.M.S. *Havannah*, 1859.   † Passages made by steam only.
‡ Passages made by sail and steam.

[S.A.]

E E

| Place. | December, January, February. No. of Passages noted. | Average in Days. | March, April, May. No. of Passages noted. | Average in Days. | June, July, August. No. of Passages noted. | Average in Days. | September, October, November. No. of Passages noted. | Average in Days. |
|---|---|---|---|---|---|---|---|---|
| Falkland islands to— | | | | | | | | |
|   Valdivia - - | . | . | . | . | . | . | 1 | 40 |
|   Valparaiso - - | . | . | . | . | . | . | 2 | 33 |
| Fonseca to— | | | | | | | | |
|   Acajutla - - | . | . | . | . | . | . | 2 | 3 |
|   Callao - - | 1 | 11† | 1 | 8 † | . | . | . | . |
|   Realejo - - | . | . | 1 | 2 | . | . | . | . |
|   San Blas - - | 1 | 31 | 1 | . | . | . | 1 | 19 |
|   Vancouver - - | . | . | 1 | 49 | . | . | . | . |
|   Valparaiso - - | . | . | 1 | 53 | . | . | . | . |
| French Frigate shoal to Honolulu* - - | . | . | 1 | 7 | . | . | . | . |
| Guayaquil to— | | | | | | | | |
|   Fonseca - - | . | . | . | . | . | . | 1 | 16‡ |
|   Galápagos islands | 1 | 7 | . | . | . | . | 1 | . |
|   Panama - - | . | . | 1 | 12 | . | . | 1 | 8 |
| Galápagos islands to Atacames - - | 1 | 6 | . | . | . | . | . | . |
| Guaymas to— | | | | | | | | |
|   Acapulco - - | 1 | 13 | . | . | . | . | . | . |
|   Altata - - | . | . | 1 | 6 | . | . | . | . |
|   Mazatlan - - | 2 | 3 | 5 | 5 | . | . | . | . |
|   San Blas - - | . | . | . | . | . | . | 1 | 10 |
| Hengkong to San Francisco - - | . | . | . | . | 3 | 52 | . | . |
| Honolulu to— | | | | | | | | |
|   Behring strait - | . | . | 1 | 27 | 2 | 27 | . | . |
|   Hilo Hawaii - | . | . | 1 | 4 | . | . | . | . |
|   Hongkong - - | . | . | . | . | . | . | 1 | 30 |
|   Horn (Cape) - | . | . | 1 | 37 | . | . | . | . |
|   Tahiti - - | . | . | 1 | 27 | 1 | 24 | 1 | 22 |
|   Petropaulski - | . | . | 1 | 29 | 3 | 32 | . | . |
|   San Francisco - | . | . | 1 | 15 | 2 | 22 | 4 | 15 |
|   Valparaiso - | 1 | 59 | 2 | 58 | 1 | 50 | 2 | 67 |
|   Vancouver island | . | . | . | . | 2 | 30 | 1 | 17‡ |
| Cape Horn to— | | | | | | | | |
|   Concepcion - - | . | . | 1 | 22 | . | . | . | . |
|   Valdivia - - | . | . | . | . | . | . | 1 | 27 |
|   Valparaiso - | 8 | 21 | 1 | 20 | . | . | 1 | 21 |
| Huasco to Caldera - | . | . | 1 | 2 | . | . | . | . |
| †Islay to Chincha I. - | . | . | . | . | . | . | . | . |
| Istapa to— | | | | | | | | |
|   Acajutla - - | . | . | . | . | . | . | 1 | 2 |
|   Fonseca - - | 1 | 1† | . | . | . | . | 3 | 2 |
| Juan Fernandez to— | | | | | | | | |
|   Concepcion - - | 1 | 7 | . | . | . | . | . | . |
|   Valparaiso - . | . | . | . | . | . | . | 1 | 4 |
| Mazatlan to— | | | | | | | | |
|   Guaymas - - { | . | . | 5 | 9 | . | . | 2 | 11 |
| | . | . | 1 | 4† | . | . | . | . |
|   Honolulu - - | . | . | 1 | 31 | 1 | 18 | . | . |
|   La Paz - - | 1 | 6 | . | . | . | . | . | . |
|   Panama - - | . | . | . | . | 1 | 27 | . | . |
|   San Blas - - | 4 | 2 | 1 | 3 | 2 | 2 | 3 | 2 |
|   Cape San Lucas - | 1 | 5 | 1 | 3 | . | . | . | . |
|   Valparaiso - | . | . | 1 | 50 | . | . | . | . |
| Navita (Fiji islands) to Tonga Tábu - | . | . | . | . | 1 | 7 | . | . |
| New Caledonia to— | | | | | | | | |
|   Fiji islands - - | . | . | . | . | 1 | 7 | . | . |
|   Tahiti - - | . | . | 1 | 44 | . | . | 1 | 28 |
|   Valparaiso - - | . | . | . | . | . | . | 1 | 36 |

* This shoal is 450 miles from Honolulu.　　　† Passages made by steam only.
‡ Passages made by sail and steam.

| Place. | December, January, February. | | March, April, May. | | June, July, August. | | September, October, November. | |
|---|---|---|---|---|---|---|---|---|
| | No. of Passages noted. | Average in Days. | No. of Passages noted. | Average in Days. | No. of Passages noted. | Average in Days. | No. of Passages noted. | Average in Days. |
| **Nukahiva to—** | | | | | | | | |
| Honolulu | | | | | 4 | 13 | | |
| Tahiti | | | | | 2 | 7 | | |
| **Otaheite or Tahiti to—** | | | | | | | | |
| Honolulu | | | | | 1 | 16 | 1 | 17 |
| Hawaii | | | | | 1 | 25 | | |
| Panama | | | | | 1 | 20† | | |
| Pitcairn island | 1 | 19 | | | 1 | 17 | | |
| Raiatea | | | | | | | 1 | 1½ |
| San Francisco | | | | | 1 | 41 | | |
| Valparaiso | | | | | 1 | 34 | 1 | 33 |
| **Ovalau to San Francisco** | | | | | | | 1 | 47 |
| **Panama to—** | | | | | | | | |
| Acapulco | 1 | 29 | | | | | | |
| Buenaventura | | | 1 | 8 | 1 | 29 | | |
| Callao { | 1 | 10† | 1 | 9† | | | | |
| Cape Horn | | | | | 1 | 53 | | |
| Fonseca | | | 1 | 12 | | | | |
| Honolulu | | | | | 1 | 47 | | |
| Nukahiva | | | 1 | 25 | 1 | 31 | | |
| Payta | | | 1 | 25 | | | | |
| Petropaulski | | | | | 1 | 90 | | |
| Punta Arenas { | | | 1 | 9 | | | | |
| | | | 1 | 3† | 1 | 7½ | | |
| San Francisco | | | | | 2 | 54 | | |
| Valparaiso | 1 | 43 | 2 | 39 | 2 | 46 | 1 | 38 |
| Vancouver island | | | 1 | 72 | 1 | 70 | | |
| **Payta to—** | | | | | | | | |
| Callao | | | 1 | 9 | 1 | 23 | | |
| Guayaquil | 1 | 2 | | | 1 | 2 | 1 | 2 |
| Nukahiva | | | | | 1 | 19 | | |
| Panama | 1 | 11 | | | | | | |
| **Gulf of Peñas to Valparaiso** | | | | | | | 1 | 6½ |
| **Petropaulski to—** | | | | | | | | |
| Behring Strait | | | | | 2 | 17 | | |
| Mazatlan | | | | | | | 1 | 34 |
| Sitka | | | | | 2 | 19 | | |
| Vancouver island | | | | | | | 1 | 23 |
| **Cape Pillar to Valparaiso** | | | 1 | 19½ | | | | |
| **Pitcairn's island to—** | | | | | | | | |
| Chincha islands | | | | | 1 | 25 | | |
| Gambier island | | | 1 | 5 | | | | |
| Otaheite | | | 1 | 11 | | | 1 | 14 |
| Valparaiso | 1 | 24 | | | 2 | 21 | | |
| **Punta Arenas to—** | | | | | | | | |
| Fonseca | | | 1 | 5 | | | | |
| Honolulu | | | 1 | 48 | 1 | 50 | | |
| | | | 2 | 13 | 1 | 12 | | |
| Panama { | 1 | 3† | 2 | 3† | 1 | 3† | 1 | 3† |
| | | | | | | | 1 | 7 |
| **Raiatea to Tongatábu** | | | | | | | | |
| **Realejo to—** | | | | | | | | |
| Fonseca | 1 | 4 | 2 | 1 | | | | |
| Istapa | 1 | 2 | | | | | | |
| Panama | | | 1 | 16 | 1 | 7 | | |
| Punta Arenas { | | | 1 | 8 | | | | |
| | | | 2 | 2† | | | | |

† Passages made by steam only.    ‡ Passages made by sail and steam.

| Place. | December, January, February. | | March, April, May. | | June, July, August. | | September, October, November. | |
|---|---|---|---|---|---|---|---|---|
| | No. of Passages noted. | Average in Days. | No. of Passages noted. | Average in Days. | No. of Passages noted. | Average in Days. | No. of Passages noted. | Average in Days. |
| San Blas to— | | | | | | | | |
| Acapulco | 3 | 9 | 1 | 9 | . | . | 1 | . |
| Guaymas | 1 | 17 | . | . | . | . | 1 | 6 |
| Mazatlan | 1 | 6 | 3 | 4 | . | . | 3 | 5 |
| Manzanilla | 1 | 2 | . | . | . | . | . | . |
| Panama | 3 | 33 | . | . | 3 | 30 | . | . |
| Valparaiso | 1 | 77 | 5 | 55 | 2 | 46 | 1 | 55 |
| San Francisco to— | | | | | | | | |
| Acapulco | 1 | 16 | . | . | . | . | 1 | 18 |
| Guaymas | 1 | 20 | . | . | . | . | . | . |
| Honolulu | 4 | 16 | 1 | 17 | 3 | 16 | 9 | 15 |
| Cape Horn | 1 | 57 | . | . | . | . | . | . |
| Magdalena bay | . | . | . | . | . | . | 1 | 8 |
| Mazatlan | 1 | 14 | . | . | — | . | 1 | 18 |
| Monterey | . | . | . | . | . | . | 1 | 1 |
| Valparaiso | 2 | 47 | . | . | . | . | 3 | 55 |
| Sitka to— | | | | | | | | |
| Honolulu | 1 | 32 | . | . | . | . | . | . |
| San Francisco | . | . | . | . | 1 | 9 | . | . |
| Queen Charlotte sound | . | . | . | . | 1 | 4 | . | . |
| San José de Guatemala to— | | | | | | | | |
| Acajutla | . | . | 1 | 2 | . | . | . | . |
| Honolulu | . | . | . | . | 1 | 31 | . | . |
| Singapore to— | | | | | | | | |
| Vancouver island | 1 | 69 | . | . | . | . | . | . |
| Sydney to— | | | | | | | | |
| Cape Pillar | . | . | . | . | 1 | 30 | . | . |
| Tahiti | . | . | 1 | 14† | . | . | . | . |
| Valparaiso to— | | | | | | | | |
| Arica | . | . | 1 | 5 | . | . | . | . |
| Caldera | . | . | 3 | 4 | . | . | . | . |
| Callao | 7 | 10 | 2 | 10 | 2 | 10 | 2 | 10 |
| Concepcion | 3 | 7 | . | . | . | . | 2 | 8 |
| | . | . | . | . | . | . | 1 | 2† |
| Coquimbo | 1 | 2 | 4 | 3 | 1 | 1 | 2 | 2 |
| Honolulu | . | . | 3 | 40 | 1 | 43 | 1 | 39 |
| Cape Horn | 1 | 15 | . | . | . | . | 1 | 19 |
| Huasco | . | . | 1 | 2 | . | . | . | . |
| Islay | . | . | . | . | . | . | 1 | 8 |
| Juan Fernandez | 1 | 5 | . | . | . | . | . | . |
| Nukahiva | . | . | . | . | 1 | 31 | . | . |
| Pitcairns island | . | . | 2 | 26 | . | . | . | . |
| Punta Arenas | . | . | 1 | 19 | . | . | . | . |
| San Blas | 1 | 45 | 1 | 40 | . | . | . | . |
| San Francisco | . | . | . | . | . | . | 1 | 43 |
| Cape San Lucas | 1 | 39 | . | . | . | . | . | . |
| Vancouver island | . | . | 1 | 58 | . | . | . | . |
| Valdivia to Concepcion | . | . | . | . | . | . | 2 | 2 |
| Vancouver island to— | | | | | | | | |
| Honolulu | . | . | . | . | 1 | 18 | 3 | 21 |
| San Francisco | . | . | . | . | . | . | 4 | 6 |
| Valparaiso | 1 | 70 | . | . | 1 | 64 | . | . |

† Passages made by steam only.    ‡ Passages made by sail and steam.

# TABLE OF POSITIONS*

## ON THE

## COASTS OF SOUTH AMERICA.

| Place. | Observation Spot. | Latitude, South. | Longitude, West. | Tides. | |
|---|---|---|---|---|---|
| | | | | H. W. F. & C. | Rise on Springs. |

### EAST COAST.

| Place. | Observation Spot. | Latitude, South. | Longitude, West. | H. W. F. & C. | Rise on Springs. |
|---|---|---|---|---|---|
| | | ° ′ ″ | ° ′ ″ | h. m. | Feet. |
| †Rio de Janeiro - | Villegagnon islet well - | 22 54 40 | 43 08 45 | 3 00 | 4 |
| †Monte Video - - | Rat islet - - - - | 34 53 20 | 56 13 15 | Noon. | |
| Point Piedras - - | Extreme of grass - | 35 26 50 | 57 05 11 | | |
| Cape San Antonio - | North extreme - | 36 18 30 | 56 45 51 | | |
| Medano point - - | South-east summit - | 36 59 05 | 56 40 43 | 11 00 | 6 |
| Cape Corrientes - | Eastern summit - | 38 05 30 | 57 29 15 | 10 40 | 7 |
| Mogotes point - - | South-east summit - | 38 10 36 | 57 30 35 | | |
| Ventana mount - | Highest summit - | 38 11 45 | 61 56 18 | | |
| Gueguen river - | Mouth - - - | 38 36 00 | 58 40 00 | 9 55 | 8 |
| Bahia Blanco - - | Asuncion point - | 38 57 30 | 60 38 00 | 8 57 | 10 |

* In the above Table the longitudes, as found by Captain Robert Fitz Roy and the officers of the *Beagle*, with 15 chronometers, in the year 1832-6, have been followed as far as Malpelo point, at south entrance of Guayaquil gulf; from thence to Point Mariato, Panama Bay, they are by Captain H. Kellett, and depend upon the position of Panama. The former longitudes all depend upon the position of Rio de Janeiro (Villegagnon I.) which was assumed to be 43° 8′ 45″ W. of Greenwich; and whenever that position is altered the first part of this Table will change accordingly. Captain King, in his survey of a portion of this coast, assumed Rio de Janeiro to be in 43° ·5′ 0″, but although he differed in his starting point, his meridian distances by Monte Video, Port Desire, Port Famine, and San Carlos agreed with those of Captain Fitz Roy; all his longitudes therefore are 3′ 45″ farther east.

There is considerable discrepancy as to the longitude of Valparaiso :—

|  | ° ′ |
|---|---|
| Fitz Roy, as in the Table, finds it to be - - - - | 71 41 15 |
| Beechey, from 19 observations of the moon's transit (Naut. Mag. 1838) | 71 39 20 |
| Raper, in his discussion on longitudes (Naut. Mag. 1839) - - | 71 40 18 |
| Humboldt's long. of Callao, by transit of Mercury in 1802, and Fitz Roy's mer. dist. - - - - - - - | 71 39 15 |
| Kellett by meridian distance from Panama, N.E. bastion, 79° 31′ 9″ - | 71 35 54 |
| Don Carlos Moesta, Director of the National Observatory at Santiago | 71 37 13 |

This last position is the result of the mean of 70 observations of moon's culminations and of stars made at the end of 1852 with the great meridian circle of the observatory at Santiago. The meridian distance between the observatory and Valparaiso being obtained by electric telegraph.

Under these circumstances, and as all the Admiralty charts and plans of South America from the Rio de la Plata to Guayaquil are graduated to Fitz Roy's longitude, it seems desirable for the present to leave the Table unchanged. There is no error in it sufficient to affect navigation, and therefore no harm can arise. In the meantime it may be hoped that some more authoritative determination of the longitude of Rio de Janeiro will be made, and that a fresh meridian distance may be measured between Panama and Chagres.

It may be noticed that, taking King's longitude of Rio de Janeiro, Fitz Roy's meridian distances, Beechey's Valparaiso and Moesta's Santiago, they would give a mean of 71° 38′ 0″ as the longitude of Fort San Antonio, at Valparaiso; and it is not impossible that this longitude may hereafter prove to be very near the truth.

† Points from which meridian distances were measured.

| Place. | Observation Spot. | Latitude, South. | Longitude, West. | Tides. | |
|---|---|---|---|---|---|
| | | | | H. W. F. & C. | Rise on Springs. |

### EAST COAST—*cont.*

| Place. | Observation Spot. | Latitude, South. | Longitude, West. | H. W. F. & C. | Rise on Springs. |
|---|---|---|---|---|---|
| | | o ′ ″ | o ′ ″ | h. m. | Feet. |
| Bahia Blanco - - | Mount Hermoso summit - | 38 58 50 | 61 39 45 | 5 06 | 12 |
| Port Belgrano · - | Argentino fort - - | 38 43 50 | 62 14 41 | 6 00 | 12 |
| Labyrinth head - | Summit - - - - | 39 26 30 | 62 02 36 | 5 20 | 12 |
| Colorado river - | Mouth - - - - | 39 51 40 | 62 04 20 | 4 00 | 9 |
| Union bay - - | Indian head - - - | 39 57 30 | 62 07 00 | 3 10 | 12 |
| San Blas harbour - | Hog islet - - - | 40 32 52 | 62 09 00 | 2 0 | 12 |
| Rasa point - - | Summit over extreme - | 40 52 10 | 62 18 15 | 1 0 | |
| Rio Negro - - | Main point - - | 41 02 00 | 62 45 10 | 11 00 | 14 |
| Bermeja head - | Eastern summit - - | 41 11 00 | 63 07 30 | | |
| San Antonio port - | Villarino point - - | 40 49 00 | 64 53 55 | 10 40 | 23 |
| El Fuerte - - | Centre - - - | 41 06 30 | 65 10 30 | | |
| San Antonio Sierra - | Summit - - - - | 41 41 10 | 65 12 10 | | |
| San Jose port - | Point San Quiroga extreme | 42 14 15 | 64 27 10 | 10 00 | 30 |
| Valdes peninsula - | Norte point, Cliffy extreme | 42 03 00 | 63 47 40 | 10 00 | 20 |
| „ | Delgada point, S.E. cliff - | 42 46 15 | 63 36 30 | 8 30 | 12 |
| Nuevo Gulf - - | Point Ninfas, East cliff - | 42 58 00 | 64 19 30 | 7 00 | 10 |
| Chupat river - | Middle of entrance - - | 43 20 45 | 65 02 50 | 5 30 | 12 |
| Cape Raso - - | Eastern summit - - | 44 23 40 | 65 15 30 | | |
| Salaberria reef - | Extreme rock - - - | 44 25 00 | 65 07 20 | | |
| Santa Elena port | Spanish observatory - | 44 30 40 | 65 21 40 | 4 00 | 17 |
| Cape Dos Bahias - | Summit over extreme - | 44 56 30 | 65 32 00 | 3 20 | 12 |
| Leones island - | South-eastern summit - | 45 04 00 | 65 35 15 | 2 10 | 12 |
| Melo port - | Sugar loaf island - - | 45 04 10 | 65 47 40 | 3 40 | 15 |
| Medrano rocks - | Centre - - - - | 45 10 00 | 65 53 30 | | |
| Cape Aristazabal - | South-east pitch - - | 45 12 45 | 66 31 10 | | |
| Salamanca peak - | Peak - - - - | 45 34 00 | 67 19 30 | | |
| Murphy head - | Summit - - - - | 46 31 10 | 67 23 10 | 1 00 | |
| Cape Tres Puntas - | North-east pitch - - | 47 06 20 | 65 51 00 | 12 50 | |
| Cape Blanco - | North-east summit - - | 47 12 20 | 65 43 30 | 12 47 | 18 |
| †Desire port - | Spanish ruins - - - | 47 45 00 | 65 54 15 | 12 10 | 18½ |
| Sea Bear bay - | Observatory at south side | 47 15 15 | 65 45 40 | 12 45 | 20 |
| Shag rock - - | Centre - - - - | 48 08 30 | 65 53 30 | | |
| Cape Watchman - | Summit of Round Mount I. | 48 21 30 | 66 21 25 | Noon. | 24 |
| Belisco rock - | Summit - - - - | 48 29 20 | 66 12 15 | Noon. | 24 |
| Flat islet - - | Centre - - - - | 48 43 00 | 67 01 00 | Noon. | 25 |
| San Julian port - | Desengano head N.E. ext. | 49 14 30 | 67 36 10 | 10 45 | 30 |
| „ | Sholl monument - - | 49 15 20 | 67 42 00 | 10 30 | 30 |
| Francisco de Paula, c. | Extreme cliff - - | 49 41 10 | 67 36 00 | | |
| North point - - | Extreme - - - | 50 06 00 | 68 03 30 | | |
| Santa Cruz river - | Keel point - - - | 50 06 45 | 68 23 30 | 9 30 | 40 |
| Coy inlet - - - | Northern head - - | 50 54 10 | 69 04 20 | 9 39 | 40 |
| Cape Fairweather - | Extreme - - - | 51 32 05 | 68 55 20 | 9 00 | 30 |
| Gallegos river - | Observatory mound - | 51 33 20 | 68 59 10 | 8 50 | 46 |
| Cape Virgins - | South-east extreme - | 52 20 10 | 68 21 34 | 8 30 | 42 to 36 |

### FALKLAND ISLANDS.

| Place. | Observation Spot. | Latitude, South. | Longitude, West. | H. W. F. & C. | Rise on Springs. |
|---|---|---|---|---|---|
| West cay, Jason island | North-west extreme - | 50 59 47 | 61 27 30 | | |
| Grand Jason - - | Summit - - - - | 51 04 30 | 61 03 57 | | |
| Gibraltar reef - | White rock - - | 51 17 15 | 60 53 52 | | |
| Hope harbour - | Hope point - - | 51 20 51 | 60 40 14 | 8 10 | 7 |
| Sedge island - - | North-west extreme - | 51 10 30 | 60 27 20 | | |
| Port Egmont - - | Settlement cove, Saunders I. | 51 21 00 | 60 04 00 | 7 30 | 11 |

| Place. | Observation Spot. | Latitude, South. | Longitude, West. | Tides. | |
| --- | --- | --- | --- | --- | --- |
| | | | | H. W. F. & C. | Rise on Springs. |

### FALKLAND ISLANDS—*cont.*

| Place. | Observation Spot. | Latitude, South. | Longitude, West. | H. W. F. & C. | Rise on Springs. |
| --- | --- | --- | --- | --- | --- |
| | | ° ′ ″ | ° ′ ″ | h. m. | Feet. |
| Cape Tamar - - | North cliff summit - - | 51 16 50 | 59 29 50 | | |
| White rock point - | North-east extreme - - | 51 24 23 | 59 12 22 | | |
| Race point - - | Extreme cliff - - | 51 25 00 | 59 06 20 | | |
| Eddystone rock - | Centre - - - | 51 11 30 | 59 03 15 | | |
| Cape Bougainville - | North-east cliff - | 51 18 00 | 58 28 20 | | |
| Salvador port - - | Hut point - - | 51 24 00 | 58 16 30 | 8 10 | 8 |
| Macbride head - - | North cliff - - | 51 23 00 | 57 59 25 | | |
| Cape Carysfort - | North-east cliff - - | 51 25 40 | 57 51 00 | | |
| Volunteer point - | Eastern extreme - - | 51 31 15 | 57 43 40 | | |
| Uranie rock - - | Centre - - - | 51 31 45 | 57 41 00 | | |
| †Port Louis - - | Carenage - - - | 51 32 20 | 58 06 58 | 5 0 | 7 |
| Port William - - | Cape Pembroke lighthouse | 51 40 42 | 57 43 00 | 5 30 | 7 |
| Stanley harbour - | Observation spot - - | 51 41 10 | 57 51 30 | 5 30 | 7 |
| Fitzroy port - - | Extreme of East island - | 51 47 40 | 58 02 15 | 4 45 | 6 |
| Choiseul sound - | East cove Mare harbour - | 51 53 52 | 58 27 08 | 6 00 | 6 |
| Lively island - - | Prong point - - | 52 06 15 | 58 25 02 | | |
| Shag rock - - | Centre - - - | 52 14 30 | 58 39 42 | | |
| Sea Lion islands - | West extreme - - | 52 26 50 | 59 09 37 | | |
| Beauchene islands - | South extreme - - | 52 41 00 | 59 05 00 | | |
| Bull road - - - | Summit - - - | 52 20 50 | 59 19 57 | 6 00 | 6 |
| Barren island - - | South-east extreme - - | 52 24 36 | 59 42 22 | | |
| Elephant cays - - | West extreme of W. cay - | 52 09 00 | 59 52 52 | 7 00 | |
| Fox bay - - - | Summit of E. entrance - | 52 00 50 | 60 00 52 | 7 00 | 8 |
| Edgar port - - | Summit over south head - | 52 02 10 | 60 15 10 | 7 15 | 6 |
| Port Albemarle - | Albemarle rock - - | 52 14 20 | 60 23 27 | 7 15 | 7 |
| Cape Meredith - - | Southern cliff - - | 52 16 15 | 60 39 07 | | |
| Port Stephens - - | East ent. point summit - | 52 12 00 | 60 40 50 | 7 45 | 7½ |
| Cape Orford - - | West summit - - | 51 59 45 | 61 06 22 | | |
| New island - - | Landsend bluff - - | 51 42 00 | 61 20 30 | 10 30 | |
| Passage islands - | West extreme of Fourth I. | 51 33 20 | 60 54 20 | 9 30 | |

### OUTER COAST OF TIERRA DEL FUEGO.

| Place. | Observation Spot. | Latitude, South. | Longitude, West. | H. W. F. & C. | Rise on Springs. |
| --- | --- | --- | --- | --- | --- |
| Catherine point - | North-east extreme - - | 52 32 00 | 68 44 10 | | |
| Cape San Sebastian - | Northern height - - | 53 19 00 | 68 09 50 | 7 00 | 13 |
| Cape Peñas - - | South-east cliff - - | 53 51 30 | 67 33 20 | 6 42 | 12 |
| Cape San Paulo - | North-east cliff - - | 54 16 20 | 66 40 50 | | |
| Policarpo point - | Extreme - - - | 54 39 00 | 65 39 30 | | |
| Cape San Diego - | East extreme - - | 54 41 00 | 65 07 00 | 4 30 | 10 |
| Staten island - - | Cape St. Anthony - | 54 43 30 | 64 34 00 | | |
| ,, - - | Cape St. John - - | 54 42 50 | 63 43 45 | 4 30 | 9 |
| ,, - - | Cape San Bartholomew - | 54 53 45 | 64 45 30 | 4 45 | 9 |
| Good Success bay - | South head - - | 54 48 45 | 65 12 20 | 4 15 | 8 |
| Cape Good Success - | Southern extreme - - | 54 54 40 | 65 21 30 | | |
| New island - - | Waller point - - | 55 10 10 | 66 28 00 | | |
| Lennox cove - - | Summit of North head - | 55 17 00 | 66 49 00 | 4 40 | 8 |
| Goree road - - | Extreme of Guanaco point | 55 19 00 | 67 10 00 | 4 00 | 8 |
| Middle cove - - ⎫ Wollaston island - ⎭ | Centre beach - - | 55 35 30 | 67 19 00 | 3 30 | |
| Barnevelt island - | North-east extreme - | 55 48 25 | 66 44 40 | 3 40 | 8 |
| Cape Horn - - | Summit - - - | 55 58 40 | 67 16 00 | 3 40 | 9 |
| St. Martin cove - | Western bight - - | 55 51 20 | 67 34 00 | 3 50 | 7 |
| False Cape Horn - | South extreme - - | 55 43 15 | 68 05 40 | 3 28 | 6 |
| Orange bay - - | Forge cove, south shore - | 55 30 50 | 68 05 17 | 3 30 | 5 |

| Place. | Observation Spot. | Latitude, South. | Longitude, West. | Tides. H. W. F. & C. | Rise on Springs. |
|---|---|---|---|---|---|
| | | ° ′ ″ | ° ′ ″ | h. m. | Feet. |

**OUTER COAST OF TIERRA DEL FUEGO—*cont.***

| Place. | Observation Spot. | Latitude, South. | Longitude, West. | H. W. F. & C. | Rise on Springs. |
|---|---|---|---|---|---|
| Packsaddle island | Summit - - - - | 55 23 50 | 68 04 20 | 3 30 | 6 |
| Ramirez, Diego | North rock - - - | 56 25 00 | 68 43 00 | 4 00 | 6 |
| Ildefonso isles - | Southern rock - - | 56 53 30 | 69 17 00 | 3 20 | 6 |
| York Minster - | Summit - - - - | 55 24 50 | 70 02 30 | | |
| March harbour - | Observatory point - | 55 22 35 | 69 59 34 | 3 10 | 6 |
| Phillip rocks - | Largest - - - - | 55 14 10 | 70 57 00 | | |
| Stewart harbour - | Observatory pt., Shelter I. | 54 54 24 | 71 29 02 | 2 50 | 5 |
| Townshend harbour - | Observatory island - | 54 42 15 | 71 55 30 | 2 30 | 5 |
| Tower rocks - | Eastern rock - - | 54 36 40 | 73 02 50 | | |
| | | | | | |
| Cape Noir - | Extreme - - - | 54 30 00 | 73 05 30 | 2 25 | |
| Cape Gloucester | Summit - - - | 54 05 18 | 73 29 15 | 1 30 | 5 |
| Laura harbour - | Basin - - - | 54 07 00 | 73 18 45 | 1 00 | 4 |
| Cape Inman - | Cliff summit - - | 53 18 30 | 74 19 15 | 2 00 | 4 |
| Latitude bay - | Tent point - - | 53 18 40 | 74 15 44 | 2 05 | 4 |
| Dislocation harbour - | Tent island, south point - | 52 54 15 | 74 37 10 | 1 40 | 4 |

**MAGELLAN STRAIT.**

| Place. | Observation Spot. | Latitude, South. | Longitude, West. | H. W. F. & C. | Rise on Springs. |
|---|---|---|---|---|---|
| Cape Virgins - | South-east extreme - | 52 20 10 | 68 21 34 | 8 30 | 36 to 42 |
| Catherine point | North-east extreme - | 52 32 00 | 68 44 10 | | |
| Cape Possession | Middle of cliff - | 52 17 00 | 68 56 20 | 8 40 | 40 |
| Cape Orange - | North extreme - | 52 27 10 | 69 28 00 | 9 00 | |
| | | | | | |
| Cape Gregory - | Extreme - - - | 52 39 00 | 70 13 40 | 10 00 | 20 to 25 |
| Oazy harbour - | Entrance, west head - | 52 42 00 | 70 36 35 | | |
| Peckett harbour - | Peckett point - - | 52 46 45 | 70 44 00 | Noon | 6 |
| Santa Marta island | Summit - - - | 52 50 00 | 70 34 45 | Noon. | 10 |
| Cape Negro - | South-west extreme cliff - | 52 56 40 | 70 49 00 | | |
| Cape Valentyn - | Summit at extreme - | 53 33 30 | 70 33 45 | | |
| †Famine port - | Tent - - - | 53 38 15 | 70 57 45 | Noon. | 7 |
| Cape San Isidro - | Extreme - - - | 53 47 00 | 70 57 50 | | |
| San Antonio port - | Humming-bird cove | 53 54 00 | 70 52 55 | Noon. | 7 |
| Vernal mount - | Summit - - - | 54 06 28 | 71 01 24 | | |
| | | | | | |
| Buckland mount - | Summit - - - | 54 26 00 | 70 22 30 | | |
| Sarmiento mount - | North-east peak - | 54 27 15 | 70 51 15 | | |
| Cape Froward - | Summit of bluff - | 53 53 43 | 71 18 15 | 1 00 | 6 |
| Cape Holland - | South-east extreme - | 53 48 33 | 71 39 25 | 10 40 | 6 |
| Pond mount - | Summit - - - | 53 51 45 | 71 56 30 | | |
| Port Gallant - | Wigwam point - | 53 41 45 | 72 00 41 | 9 00 | 6 |
| Charles island - | Wallis mark - - | 53 43 57 | 72 05 45 | | |
| Bachelor river - | Entrance - - - | 53 33 00 | 72 19 15 | 1 40 | 5 |
| Cape Quod - | Extreme - - | 53 32 10 | 72 33 25 | | |
| Cape Notch - | Extreme - - - | 53 25 00 | 72 48 55 | | |
| | | | | | |
| Playa Parda - | Summit of Shelter island - | 53 18 45 | 73 01 30 | 1 08 | 6 |
| Half-port bay - | Point - - - | 53 11 40 | 73 18 45 | 2 00 | 6 |
| Havannah point - | Extreme - - - | 53 07 30 | 73 16 00 | | |
| Cape Upright - | North extreme - | 53 04 08 | 73 36 00 | 1 30 | |
| Cape Tamar - | South extreme - | 53 55 30 | 73 48 10 | 3 05 | 6 |
| Cape Phillip - | Summit - - - | 52 44 20 | 73 56 44 | | |
| Valentine harbour - | Mount island - - | 52 55 00 | 74 18 45 | 2 00 | 6 |
| Cape Parker - | Western summit - | 52 42 00 | 74 14 30 | | |
| Mercy harbour - | Summit of Observation islet | 52 44 58 | 74 39 14 | 1 20 | 4 |
| Westminster hall - | Eastern summit - | 52 37 18 | 74 24 10 | | |
| | | | | | |
| Cape Pillar - | Northern cliff - | 52 42 50 | 74 43 20 | 1 00 | 6 |
| Los Evangelistas - | Sugar loaf islet - | 52 24 18 | 75 06 40 | 1 00 | 5 |
| Cape Victory - | Extreme - - - | 52 16 10 | 74 54 39 | | |

| Place. | Observation Spot. | Latitude, South. | Longitude, West. | Tides. | |
|---|---|---|---|---|---|
| | | | | H. W. F. & C. | Rise on Springs. |

## WEST COAST.—INNER CHANNEL.

| Place. | Observation Spot. | Latitude, South. | Longitude, West. | H. W. F. & C. | Rise on Springs. |
|---|---|---|---|---|---|
| | | ° ′ ″ | ° ′ ″ | h. m. | Feet. |
| Deep harbour - - | South point of entrance - | 52 41 10 | 73 48 20 | | |
| Goods bay - - | Enterprise rock - | 52 34 16 | 73 46 30 | 12 30 | 7 |
| Fortune bay - - | Summit of low land - | 52 15 48 | 73 44 26 | 12 50 | 7 |
| Bessel point - - | Extreme - - - | 52 00 40 | 73 46 40 | | |
| Mount Trafalgar - | Summit - - - | 51 38 00 | 74 24 45 | | |
| Cape Flamsteed - | Extreme rock - - | 51 46 25 | 73 51 45 | | |
| Cape Kendall - - | Extreme - - - | 51 27 15 | 74 10 04 | | |
| Puerto Bueno - - | N. point of Schooner cove | 50 58 40 | 74 11 00 | 1 40 | |
| Guia narrows - - | Passage point - - | 50 44 36 | 74 21 45 | 2 10 | |
| Cape San Andres - | Extreme - - - | 50 18 00 | 74 40 45 | | |
| Sandy bay - - | East point - - | 49 45 40 | 74 16 45 | | |
| Saumarez island - | Bold head - - | 49 32 48 | 74 06 15 | | |
| Eden harbour - - | Eden island - - | 49 09 00 | 74 18 00 | 12 30 | 5 |
| Halt bay - - - | Centre of islet - | 48 54 15 | 74 14 20 | 12 50 | 5 |
| Middle island - - | North point extreme - | 48 27 35 | 74 21 40 | | |
| Island harbour - - | Centre of island - | 48 07 00 | 74 28 00 | | |
| Fatal bay - - - | South islet - - | 47 55 00 | 74 43 30 | | |
| Cape San Roman - | North extreme - - | 47 44 30 | 74 52 30 | | |
| Ayantau islands - | Summit of largest - | 47 34 15 | 74 40 20 | | |

## WEST COAST.

| Place. | Observation Spot. | Latitude, South. | Longitude, West. | H. W. F. & C. | Rise on Springs. |
|---|---|---|---|---|---|
| Cape Isabel - - | Summit - - - | 51 52 00 | 75 10 00 | | |
| Cape Santa Lucia - | Summit - - - | 51 30 00 | 75 29 00 | | |
| Cape Santiago - | Summit - - - | 50 42 00 | 75 28 00 | | |
| Port Henry - - | Observatory - - | 50 00 18 | 75 18 55 | 11 45 | 5 |
| Cape Pimero - - | Extreme - - - | 49 50 05 | 75 35 30 | | |
| Off-shore islet - | Centre - - - | 49 25 10 | 75 36 00 | | |
| Western rock - - | Centre - - - | 49 01 00 | 75 48 40 | | |
| Rock of Dundee - | Summit - - - | 48 06 15 | 75 42 00 | | |
| Port Barbara - - | Wreck point - - | 48 02 20 | 75 29 20 | 11 45 | 6 |
| Guaianeco islands - | Northern islet summit - | 47 38 10 | 75 14 00 | | |
| Xavier island - - | North-east extreme - | 47 03 15 | 74 16 00 | | |
| Cape Tres Montes - | Extreme - - - | 46 59 57 | 75 27 50 | | |
| Port Otway - - | Tent - - - | 46 49 32 | 75 19 20 | 11 37 | 6 |
| Cape Raper - - | Rock close to - - | 46 49 10 | 75 40 55 | | |
| San Andres bay - | Christmas cove - | 46 35 00 | 75 34 05 | 12 45 | 5 |
| Hellyer rocks - - | Middle - - - | 46 04 00 | 75 14 00 | | |
| Cape Taytao - - | Western extreme - | 45 53 20 | 75 08 00 | | |
| Anna Pink bay - | Summit of Puentes island | 45 51 36 | 74 51 25 | 12 25 | 6 |
| Menchuan island - | Summit - - - | 45 36 00 | 74 56 00 | | |
| Vallenar road - | Three-finger I., S.E. ext. - | 45 18 30 | 74 36 15 | 12 18 | 5 |
| Huamblin island - | West head - - | 44 49 30 | 75 14 45 | | |
| Narborough island - | Extreme of John point - | 44 40 40 | 74 48 30 | | |
| Guaytecas islands - | Observatory islet, Port Low | 43 48 30 | 74 03 05 | 12 40 | 7 |
| Huafo island - | Weather point summit - | 43 35 30 | 74 48 40 | | |
| Cape Quilan - | South-west extreme - | 43 17 10 | 74 26 00 | | |
| Cape Matalqui - | West extreme - - | 42 10 40 | 74 14 00 | | |
| Cocotue head - | Summit - - - | 41 56 40 | 74 05 35 | | |
| Huspilacuy point - | Lighthouse - - | 41 46 45 | 73 55 45 | | |
| †Port San Carlos - | Point Arena - - | 41 51 20 | 73 56 00 | 12 14 | 6 |
| Tres Cruces point - | Extreme - - - | 41 49 30 | 73 31 40 | 1 15 | 16 |
| Fort Calbuco - | East end of island - | 41 46 10 | 73 10 45 | 1 18 | 22 |
| Reloncavi sound - | Port Montt - - | 41 30 30 | 72 58 00 | | |

| Place. | Observation Spot. | Latitude, South. | Longitude, West. | Tides. | |
|---|---|---|---|---|---|
| | | | | H. W. F. & C. | Rise on Springs. |

### WEST COAST—cont.

| Place. | Observation Spot. | Latitude, South. | Longitude, West. | H. W. F. & C. | Rise on Springs. |
|---|---|---|---|---|---|
| | | ° ′ ″ | ° ′ ″ | h. m. | Feet. |
| Ancud gulf - - | Changues I., N. summit - | 42 15 00 | 73 18 00 | | |
| „ - - | Oscuro cove - - - | 42 04 00 | 73 29 00 | 1 00 | 20 |
| Castro town - - | Eastern part - - - | 42 27 45 | 73 49 20 | 12 11 | 18 |
| Corcovado gulf - | Talcan harbour - - | 42 47 00 | 72 58 00 | | |
| „ - | Centenela point - - | 42 59 00 | 73 22 30 | | |
| „ - | Port San Pedro - - | 43 19 35 | 73 45 20 | 12 30 | 9 |
| Godoy point - - | South-west extreme - | 41 34 15 | 73 50 20 | | |
| Cape Quedal - - | Summit - - - | 41 03 00 | 73 59 50 | | |
| Galera point - - | West extreme - - | 40 02 00 | 73 46 40 | | |
| Valdivia - - | Fort Corral - - | 39 52 53 | 73 29 00 | 10 35 | 5 |
| Cauten head - - | Cliff summit - - | 38 40 40 | 73 30 20 | | |
| Mocha island - - | South summit - | 38 24 10 | 73 56 50 | | |
| Tucapel point - - | Extreme - - - | 37 42 00 | 73 43 00 | | |
| Cape Rumena - - | Summit of N.W. cliff - | 37 12 45 | 73 42 00 | | |
| Lavapie point - - | Extreme - - - | 37 06 50 | 73 38 20 | | |
| Arauco fort - - | Centre - - - | 37 15 00 | 73 23 00 | | |
| Santa Maria island - | Aguada point - - | 37 02 50 | 73 34 00 | 10 20 | 6 |
| Tumbes point - - | North-west cliff - | 36 37 15 | 73 10 20 | | |
| Talcahuano - - | Fort Galvez - - | 36 42 00 | 73 10 00 | 10 14 | 5 |
| Coliumo head - - | North extreme - - | 36 31 30 | 73 01 15 | | |
| Cape Carranza - - | South-west extreme - | 35 37 20 | 72 42 20 | | |
| Maule river - - | Church rock - - | 35 19 40 | 72 29 20 | | |
| Topocalmo point - | Summit on extreme - | 34 00 50 | 72 05 00 | | |
| Rapel shoal - - | - - - - | 33 51 00 | 71 56 30 | 9 45 | |
| White rock point - | White rock - - | 33 29 00 | 71 46 50 | | |
| Curaumilla point - | Rock off - - - | 33 06 00 | 71 48 00 | | |
| Valparaiso - - | Lighthouse - - - | 33 01 10 | 71 41 30 | | |
| † „ - | Fort San Antonio - | 33 01 53 | 71 41 15 | 9 32 | 5 |
| Quintero rocks - - | Centre - - - | 32 52 20 | 70 37 00 | | |
| Papudo bay - - | Landing place - - | 32 30 09 | 71 30 45 | | |
| Pichidanque - - | South-east point of island | 32 07 55 | 71 36 00 | 9 20 | 5 |
| Maytencillo cove - | North head - - | 31 17 05 | 71 42 05 | | |
| Mount Talinay - - | Summit - - - | 30 50 45 | 71 41 45 | | |
| Lengua de Vaca - | Extreme - - - | 30 13 40 | 71 41 30 | | |
| Coquimbo bay - | Pajaros niños, N. rock - | 29 55 10 | 71 25 10 | 9 08 | 5 |
| Pajaros islets - - | Summit of southern - | 29 35 00 | 71 36 25 | | |
| Chañeral island - | South-west summit - | 29 01 15 | 71 39 05 | | |
| Port Huasco - - | Outer rock of inner port - | 28 27 15 | 71 19 00 | 8 30 | 6 |
| Herradura de Carrisal | Landing-place - - | 28 05 45 | 71 15 45 | | |
| Salado bay - - | Summit of Cachos point - | 27 39 20 | 71 06 25 | | |
| Morro de Copiapo - | Summit - - - | 27 09 30 | 71 01 45 | | |
| Port Caldera - - | Summit of island - | 27 02 56 | 70 56 10 | | |
| Port Flamenco - - | South-east corner - | 26 34 30 | 70 47 30 | 9 10 | 5 |
| Pan de Azucar - - | Summit - - - | 26 09 15 | 70 47 05 | | |
| Lavata bay - - | Outer cove - - | 25 39 30 | 70 47 15 | 9 20 | 5 |
| Grande point - - | Outer summit - - | 25 07 00 | 70 33 30 | | |
| Jara head - - | Summit - - - | 23 53 00 | 70 35 45 | | |
| Mount Moreno - - | Summit - - - | 23 26 30 | 70 38 15 | | |
| Leading bluff - - | Extreme - - - | 23 01 20 | 70 34 40 | | |
| Cobija bay - - | Landing place - - | 22 34 00 | 70 21 05 | 9 54 | 4 |
| Algodon bay - - | Extreme point - - | 22 06 00 | 70 17 05 | | |
| San Francisco head - | West pitch - - | 21 55 50 | 70 14 45 | | |
| River of Loa - - | Mouth - - - | 21 28 00 | 70 06 15 | | |
| Patache point - - | Extreme - - - | 20 51 05 | 70 18 15 | | |
| Grueso point - - | Extreme - - - | 20 23 00 | 70 16 00 | | |

| Place. | Observation Spot. | Latitude, South. | Longitude, West. | Tides. | |
|---|---|---|---|---|---|
| | | | | H. W. F. & C. | Rise on Springs. |
| | WEST COAST—*cont.* | | | | |
| | | ° ′ ″ | ° ′ ″ | h. m. | Feet. |
| Port Iquique - - | Centre of island - - | 20 12 30 | 70 14 30 | 8 45 | 5 |
| Pichalo point - - | Extreme - - - - | 19 36 30 | 70 19 00 | | |
| Gully of Camarones - | Centre - - - | 19 12 30 | 70 20 00 | | |
| Cape Lobos - - | Summit - - - | 18 45 40 | 70 25 30 | | |
| Arica bay - - - | Mole - - - - | 18 28 05 | 70 23 45 | 8 00 | 5 |
| Morro de Sama - | Highest summit - | 17 58 35 | 70 56 15 | | |
| Coles point - - | Extreme - - | 17 42 00 | 71 26 15 | | |
| Ylo town - - - | Mouth of rivulet - | 17 37 00 | 71 23 45 | 8 20 | 6 |
| Tambo valley - - | Mexico point - - | 17 10 50 | 71 52 00 | | |
| Port Islay - - - | Custom house - - | 17 00 00 | 72 10 15 | 8 53 | 7 |
| | | | | | |
| Cornejo point - - | Extreme - - - | 16 52 00 | 72 22 00 | | |
| Mount Camana - | Summit - - - | 16 37 00 | 72 25 00 | | |
| Pescadores point - | South-west extreme - | 16 23 50 | 73 20 25 | | |
| Atico - - - | East cove - - | 16 13 30 | 73 45 15 | 8 53 | 5 |
| Chala point - - | Extreme - - - | 15 48 00 | 74 31 00 | | |
| Cape Lomas - - | Summit - - - | 15 33 15 | 74 54 45 | | |
| Port San Juan - - | Needle hummock - | 15 20 56 | 75 13 20 | 5 10 | 3 |
| Nasca point - - | Summit - - - | 14 57 00 | 75 34 30 | | |
| Infiernillo rock - | Summit - - - | 14 40 00 | 75 59 00 | | |
| Independencia bay - | Santa Rosa I., S. point | 14 18 15 | 76 13 30 | 4 50 | 4 |
| | | | | | |
| Carretas head - | South extreme - - | 14 11 00 | 76 20 00 | | |
| San Gallan island - | Northern summit - - | 13 50 00 | 76 31 15 | | |
| Pisco town - - | Centre - - - | 13 43 00 | 76 16 30 | 4 50 | 4 |
| Chincha islands - | Boat slip on northern - | 13 38 20 | 76 27 30 | | |
| Frayles point - - | Extreme - - - | 13 01 00 | 76 34 50 | | |
| Asia rock - - | Summit - - - | 12 48 00 | 76 41 55 | | |
| Chilca point - - | South-west pitch - | 12 31 00 | 76 52 40 | | |
| Morro Solar - - | Summit - - - | 12 11 30 | 77 06 15 | | |
| Cape San Lorenzo - | Lighthouse - - | 12 04 00 | 77 19 30 | | |
| †Callao - - - | Arsenal Flagstaff - | 12 04 00 | 77 13 30 | 5 47 | 4 |
| | | | | | |
| Hormigas rocks - | Southern - - | 11 58 00 | 77 50 00 | | |
| Pelado islet - - | Summit - - | 11 27 10 | 77 53 00 | | |
| Salinas hill - - | Summit - - - | 11 15 30 | 77 39 55 | | |
| Huacho point - - | Extreme - - - | 11 08 45 | 77 40 15 | 4 44 | 3 |
| Supé bay - - - | Head of bay - - | 10 49 45 | 77 47 00 | | |
| Callejones point - | Extreme - - - | 10 30 40 | 78 00 30 | | |
| Legarto head - - | Summit - - - | 10 07 00 | 78 14 00 | 6 10 | 2 |
| Mount Mongon - | West summit - - | 9 38 15 | 78 21 15 | | |
| Casma bay - | Inner south point - | 9 28 00 | 78 25 35 | | |
| Samanco bay - - | Cross point - - | 9 15 30 | 78 32 45 | 6 30 | 2 |
| | | | | | |
| Santa bay - - | Santa head - - - | 9 00 00 | 78 41 30 | | |
| Chao islet - - | Centre - - - | 8 46 30 | 78 49 00 | | |
| Guañape islands - | Summit of highest - | 8 34 50 | 78 59 15 | | |
| Truxillo town - - | Church - - - | 8 07 30 | 79 04 00 | | |
| Macabi islet - - | Summit - - - | 7 49 15 | 79 30 55 | | |
| Malabrigo hill - | Summit - - - | 7 43 30 | 79 28 30 | | |
| Pacasmayo point - | North-west extreme - | 7 25 15 | 79 37 25 | | |
| Sana point - - | Extreme - - - | 7 10 35 | 79 43 30 | | |
| Eten hill - - - | Summit - - - | 6 55 00 | 79 54 00 | | |
| Lambayeque road - | Centre of beach - - | 6 46 00 | 79 59 30 | 4 00 | 3 |
| | | | | | |
| Lobos de Afuera - | Cove on east side - | 6 56 45 | 80 43 55 | | |
| Lobos de Tierra - | Central summit - - | 6 26 45 | 80 52 50 | | |
| Aguja point - - | Western cliff - - | 5 55 30 | 81 10 00 | | |
| Payta point - - | North extreme - - | 5 05 00 | 81 10 00 | 3 20 | 3 |
| Pariña point - - | Extreme - - - - | 4 40 50 | 81 20 45 | | |

| Place. | Observation Spot. | Latitude, South. | Longitude, West. | Tide. H. W. F. & C. | Rise on Springs. |
|---|---|---|---|---|---|
| | | ° ′ ″ | ° ′ ″ | h. m. | Feet. |

### WEST COAST—cont.

| Place. | Observation Spot. | Latitude, South. | Longitude, West. | H. W. F. & C. | Rise on Springs. |
|---|---|---|---|---|---|
| Cape Blanco - - | Centre cliff - - - | 4 16 40 | 81 15 45 | | |
| Picos point - - | Extreme cliff - - - | 3 45 10 | 80 47 30 | | |
| Malpelo point - - | Mouth of Rio Tumbez - | 3 30 40 | 80 30 30 | | |
| *Santa Clara island - | Lighthouse - - - | 3 10 40 | 80 24 34 | 4 00 | 11 |
| †Puna island - - | Puna Church cross - - | 2 44 06 | 79 53 11 | 6 00 | 11 |
| Piedras point - - | Extreme - - - - | 2 26 18 | 79 49 49 | | |
| †Guayaquil - - | Arsenal, South end of city | 2 12 24 | 79 51 24 | 7 00 | 11 |
| Point Santa Elena - | West extreme - - - | 2 11 30 | 80 59 47 | 1 18 | 8 |
| Pelado island - - | Summit - - - - | 1 56 00 | 80 48 35 | | |
| Salango bay - - | Watering place - - | 1 35 14 | 80 50 37 | | |
| Callo point - - | Extreme cliff - - - | 1 23 15 | 80 45 20 | | |
| La Plata island - - | East point - - - | 1 16 55 | 81 03 00 | | |
| Cape San Lorenzo - | Marlingspike rock - - | 1 08 30 | 80 55 00 | | |
| Monte Christo - - | Summit - - - - | 1 03 40 | 80 40 00 | | |
| †Port Manta - - | North west point of town - | 0 56 46 | 80 42 47 | 3 4 | 6 |
| Caracas bay - - | Punta Playa - - - | 0 35 25 | 80 24 29 | | |
| Cape Pasado - - | Extreme - - - - | 0 21 30 | 80 29 42 | | |
| Jama point - - | Extreme - - - - | 0 09 40 | 80 20 35 | | |
| | | North. | | | |
| Pedernales point - | Outer rock - - - | 0 04 15 | 80 06 45 | | |
| Cape San Francisco - | South-west extreme - - | 0 40 00 | 80 07 00 | | |
| Point Galera - - | North extreme - - - | 0 50 10 | 80 04 45 | | |
| †Atacames - - | Entrance of river Sua - | 0 52 30 | 79 51 57 | 3 37 | 13 |
| Esmeraldas river - | West point of entrance - | 0 59 52 | 79 41 13 | | |
| Point Verde - - | Extreme - - - - | 1 05 38 | 79 26 42 | | |
| River Santiago - - | Tola village - - - | 1 12 20 | 79 05 45 | 3 30 | 13 |
| Point Mangles - - | South pt. of ent. to creek | 1 36 00 | 79 02 35 | | |
| †Tumaco - - - | South-west pt. of El Morro I. | 1 49 36 | 78 44 34 | 2 33 | 12 |
| Point Cascajal - - | Gallo island - - - | 1 59 00 | 78 38 44 | | |
| Guascama point - | Extreme - - - - | 2 37 10 | 78 23 29 | | |
| Gorgona island - - | Watering bay - - - | 2 58 10 | 78 10 20 | 4 10 | 10 |
| †Buenaventura river : | Basan point - - - | 3 49 27 | 77 10 50 | 6 00 | 13 |
| ”          ” | Vigia de San Pablo - | 3 49 00 | 77 15 00 | | |
| Negrillas rocks - | Centre of largest - - | 3 56 00 | 77 23 30 | 4 00 | 13 |
| Chirambirá point - | North extreme - - - | 4 17 06 | 77 28 49 | 4 00 | 12 |
| C. Corrientes - - | South-west extreme - | 5 28 46 | 77 32 33 | | |
| Alúsea point - - | North extreme - - - | 5 36 20 | 77 29 25 | | |
| Port Utria - - | Centre of southern islet - | 5 58 30 | 77 20 20 | | |
| Solano point - - | North extreme - - - | 6 17 55 | 77 27 30 | | |
| †Cupica bay - - | Entrance of Cupica river - | 6 41 19 | 77 29 36 | 4 00 | 12 |
| Cape Marzo - - | South-east extreme - - | 6 49 45 | 77 40 00 | | |
| †Port Piñas - - | North-east bight - - | 7 34 37 | 78 09 50 | | |
| Garachine point - | North-east extreme - - | 8 06 00 | 78 21 15 | 4 00 | 15 |
| Patino point - - | Centre of islet - - - | 8 16 20 | 78 17 10 | | |
| Darien harbour - - | Graham point - - - | 8 28 50 | 78 04 40 | 4 00 | 24 |

### BAY OF PANAMA.

| Place. | Observation Spot. | Latitude, South. | Longitude, West. | H. W. F. & C. | Rise on Springs. |
|---|---|---|---|---|---|
| Galera island - - | Centre - - - - - | 8 11 20 | 78 45 45 | | |
| San José bank - - | Trollope rock - - - | 8 06 40 | 78 37 40 | | |
| Isla del Rey - - | Extreme of Cocos point - | 8 12 30 | 78 53 45 | | |
| †    ” - - | Church of S. Miguel - | 8 27 00 | 78 55 35 | | |
| Saboga island - - | Church - - - - | 8 37 10 | 79 03 10 | 4 00 | 14 |

* From Santa Clara island to Point Mariato the longitudes are reckoned from the N.E. bastion, Panama.

| Place. | Observation Spot. | Latitude, South. | Longitude, West. | Tides. | |
|---|---|---|---|---|---|
| | | | | H. W. F.& C. | Rise on Springs. |

BAY OF PANAMA—*cont.*

| | | o ' " | o ' " | h. m. | Feet. |
|---|---|---|---|---|---|
| Gonzales island - - | Havannah head - - | 8 25 00 | 79 05 50 | 3 50 | 16 |
| San José island - | Iguana point - - - | 8 18 25 | 79 06 30 | | |
| Brava point - - | West extreme - - - | 8 20 36 | 78 24 30 | | |
| Pajaros islands - - | North-west island - - | 8 32 20 | 78 32 10 | | |
| Pelado island - - | Centre - - - - | 8 37 35 | 78 41 40 | | |
| Chepillo island - | The tree - - - - | 8 56 32 | 79 07 00 | 4 00 | 16 |
| †Panama - - - | North-east bastion - - | 8 56 56 | 79 31 09 | 3 23 | 15 to 22 |
| †Flamenco island - | North point - - - | 8 54 30 | 79 30 20 | | |
| Bona island - - | Peak - - - - | 8 33 35 | 79 34 05 | | |
| Point Chamé - - | Extreme - - - - | 8 39 00 | 79 40 50 | | |
| Parita bay - - | Liso point - - - | 7 58 10 | 80 20 40 | | |
| Iguana island - - | Centre - - - - | 7 37 05 | 79 59 00 | | |
| Cape Mala - - | Extreme - - - - | 7 27 40 | 79 58 30 | 4 00 | 16 |
| South Fraile - - | Centre - - - - | 7 19 30 | 80 07 10 | | |
| †Morro Puercos - | South extreme - - - | 7 13 45 | 80 25 10 | | |
| Point Mariato - - | Extreme - - - - | 7 12 00 | 80 51 30 | | |

ISLANDS OFF THE COAST.*

| | | South. | | | |
|---|---|---|---|---|---|
| Juan Fernandez - | Fort in Cumberland bay - | 33 37 36 | 78 53 00 | | |
| Mas a fuera island - | West point - - - | 33 49 00 | 80 56 00 | | |
| St. Ambrose and St. } Felix islands - } | Peterborough Cathedral - | 26 16 12 | 80 11 43 | | |
| Galapágos islands. | — | — | — | | |
| Chatham island - - | Freshwater bay - - | 0 56 25 | 89 33 25 | | |
| Hood island - - | Gardner bay - - - | 1 22 10 | 89 44 00 | | |
| †Charles island - - | Post-office bay - - - | 1 15 25 | 90 31 30 | | |
| Macgowan reef - | Centre - - - - | 1 08 30 | 89 59 30 | 2 10 | 6 |
| Barrington island - | Summit - - - - | 0 50 30 | 90 10 00 | | |
| Indefatigable island - | Eden island, Conway bay - | 0 33 25 | 90 37 45 | | |
| James island - - | Sugar loaf near west end - | 0 15 20 | 90 56 48 | 3 10 | 5 |
| Narborough island - | North-west extreme - - | 0 20 00 | 91 44 45 | | |
| Albemarle island - | Iguana cove - - - | 0 59 00 | 91 32 15 | 2 00 | 6 |
| " | Tagus cove - - - | 0 15 55 | 91 26 45 | | |
| | | North. | | | |
| " | Point Albemarle - - | 0 10 00 | 91 27 10 | | |
| Bindloes island - - | Southern summit - - | 0 18 50 | 90 33 55 | | |
| Towers island - - | Western cliff - - - | 0 20 00 | 90 02 30 | | |
| Abingdon island - | Summit - - - - | 0 34 25 | 90 48 10 | | |
| Wenman island - | North-west summit - - | 1 22 55 | 91 53 30 | | |
| Culpepper island - | Summit - - - - | 1 39 30 | 92 04 30 | 2 10 | |
| Malpelo island - - | Summit - - - - | 4 00 00 | 81 32 00 | | |
| †Cocos island - - | Chatham bay - - - | 5 32 57 | 86 58 22 | 2 10 | ?7 |

* These longitudes are reckoned from Valparaiso, excepting Malpelo and Cocos.

# INDEX.

| | Page |
|---|---|
| Abingdon island | 399 |
| Abra channel | 198 |
| —— inlet | 151, 198, 199 |
| Abtao island | 269 |
| Acari morro | 338 |
| Achilles bank | 254 |
| Aconcagua mount | 286 |
| Acuy island | 266 |
| Adelaide passage | 179 |
| Admiralty sound | 173, 175, 176 |
| Adventure bay | 250 |
| —— bridge | 164 |
| —— cove | 143, 144 |
| —— harbour | 94 |
| —— passage | 144 |
| —— sound | 93, 94 |
| Afuera islands | 363, 364, 369 |
| Agnes islands | 147, 148 |
| Agnada point | 384 |
| Aguirre bay | 135 |
| Aguja point | 364–366 |
| Aguy point | 254, 255 |
| Ahoni point | 265 |
| Aborcados islands | 320 |
| Ainsworth harbour | 176 |
| Alacran reef | 330 |
| Alan island | 261 |
| Albany island | 398 |
| Albermarle island | 394, 398 |
| —— rock and port | 102, 120 |
| Alcalde point | 301 |
| Aldunate inlet | 243 |
| Alexander mount | 248 |
| Algarroba cove | 285 |
| Algodon bay | 325 |
| Alikhoolip island | 144 |
| Ali rocks | 48 |
| Alquilqua bay | 205 |
| Alusea point | 401 |
| Almargos point | 275 |

| | Page |
|---|---|
| Amatape mountains | 366 |
| Ambrose, St., island | 319 |
| Amortajada, or Santa Clara island | 371 |
| —— light | 371 |
| —— shoals | 372 |
| Anachachi rock | 308, 309 |
| Analao islet | 249 |
| Anana peaks | 401 |
| Anchorage bay | 105 |
| Anchor bay | 229 |
| —— inlet | 119 |
| Anchorstock hill | 11, 14 |
| Ancon bay | 198, 351 |
| —— hill | 416 |
| —— port | 351 |
| Ancon Sin Salida | 22 |
| Ancud gulf | 252, 258, 271, 272 |
| Andres head | 6 |
| Andrews bay | 186 |
| Andrew sound | 229, 230 |
| Anegada bay | 20 |
| Anegadiza point | 277 |
| Angeles point | 286, 288 |
| Angosto port | 201, 202 |
| Anna Pink bay | 248 |
| Anne Shoal | 49 |
| Ann, St., island | 200 |
| Ano-nuevo cape | 227 |
| Anthony, St., cape | 127 |
| Antonio, San, cape | 1, 2, 3, 274 |
| —— cove | 285 |
| —— fort | 288 |
| Antony creek | 114 |
| Anxious pass | 106 |
| —— point | 177 |
| Apabon point and reef | 264, 265 |
| Apiau island | 261 |
| Apostolos rocks | 152, 220, 222 |
| April peak | 235 |
| Arauco bay and fort | 279, 280 |

| | Page |
|---|---|
| Arauz bay - - - | 189, 190 |
| Arcana mount - - | 361 |
| Arce bay - - - - | 195 |
| —— islet - - - | 44 |
| Arch islands and anchorage - | 120 |
| Arena point, 254, 255, 281, 282, 326, 373, | |
| | 375 |
| Arenal valley - - - | 292 |
| Arenas Gordas bank - - | 2 |
| Arenas point - - - | 132 |
| Arequipa - 331, 333, 334, 335, 336 | |
| Argentino Fuerte - - - | 11 |
| Ariadne island - - - | 15 |
| —— point - - | 177 |
| Arica - - 287, 330, 331, 332, 369 | |
| Ariel rocks - - - | 7 |
| Aristazabal cape - - 46, 47 | |
| Arrayan bight - - - | 299 |
| Arrow harbour - - - | 91 |
| Asaurituan - - - | 240 |
| Asia island - - - | 345 |
| Asuncion point - - 7, 8, 10 | |
| Atacames - - - | 384 |
| —— bay - - - | 384 |
| —— ledge - - - | 384 |
| —— river - - - | 384 |
| Atahuanqui point - - | 354 |
| Atequipa valley - - - | 337 |
| Atico point - - - | 336, 337 |
| Atlantic to Pacific, round Cape | |
| Horn - - - | 209–212 |
| —— by Magellan | |
| Strait - - - | 212–216 |
| Atlas point - - - | 42 |
| Atrato river - - - | 403 |
| Augusta island - - | 235 |
| Awash rock - - - | 103 |
| Ayangui point - - | 380 |
| Ayautau islands - - | 240 |
| Aymond mount - - | 160 |
| Aytay cape - - - | 266 |
| Azua point - - - | 339 |
| Azucar Pan de - - | 46 |
| | |
| Bacalao point - - | 318 |
| Bachelor bay and river - | 188 |
| Back harbour - - - | 128 |
| Bad bay - - - | 243, 244 |
| Bahia Blanco - - | 8, 10 |
| Baja rock - - - | 288, 378 |

| | Page |
|---|---|
| Bajas point - - - | 352 |
| Bald island and road - | 117 |
| Ballena point - 290, 291, 314 | |
| Ballenita port - - | 314 |
| Ballista islands - - | 342 |
| Balsas - - - | 296, 361 |
| Baia, playa - - - | 402 |
| Banks bay - - - | 399 |
| Banks off Rio Negro - - | 26 |
| Baracura head - - | 255 |
| Barbara channel, 130, 145, 146, 147, 175, | |
| 179, 180, 182, 183, 184, 190, 210 | |
| ——, south entrance - | 147 |
| ——, tides - - | 149 |
| ——, port - - 238, 239, 246 | |
| Barcelo bay - - | 194 |
| Barnevelt isles - - | 141 |
| Bar of Rio Negro - 26, 28, 29 | |
| Barranca hills - - | 27, 29 |
| —— ledge - 161, 214, 219 | |
| Barrancas - - - | 327 |
| Barranquilla bay - - | 307 |
| Barranquilla de Copiapó - | 307 |
| Barren reefs - - | 97 |
| Barrington island - - | 398 |
| Barrister bay - - | 151 |
| Barrow harbour - - | 95 |
| —— head - - 178, 179 | |
| Barry rocks - - - | 407 |
| Bartholomew island - 410, 411 | |
| Basan point - - | 390 |
| Basil Hall port - 123, 124, 125 | |
| —— volcano - | 176 |
| Batele point - - | 421 |
| Bay of harbours - - | 96 |
| —— islands - - | 205 |
| Bayoneta island - | 412 |
| Beacon point - - | 117 |
| Beagle channel - 131, 136, 141, 144 | |
| —— island - - | 235 |
| —— mountains - - | 353 |
| —— rock - - | 52 |
| Beaubasin port - - | 183 |
| Beauchêne island - | 83, 94 |
| Beaver harbour - - | 116 |
| —— island - - | 116 |
| Bedford bay - - | 179, 180 |
| Begueta bay - - | 353, 354 |
| Belen bank - - | 282 |
| —— bluff - - | 31, 32 |
| Belgrano port - - | 9, 11–15 |

| | Page |
|---|---|
| Bellacas point - - - 382 |
| Bellaco rock - - - 54, 55 |
| Bell bay - - - - 184 |
| —— mount - - - 358 |
| —— mountain - - - 135 |
| Bending cove - - - 189, 190 |
| Bense harbour and islands - - 111 |
| Berkeley cape - - - - 399 |
| —————— sound - 70, 78, 79, 82, 84 |
| Bermeja head - - - 26, 31 |
| Bessel point - - - 225, 226 |
| Beware point - - - 339 |
| Billy rock - - - - 80, 83 |
| Bindloe island - - - 399 |
| Bio Bio river - - - 280 |
| Bird island - - - 119 |
| ——, islets - - - 55 |
| Blackbeach road - - - 397 |
| Black point - - - 7, 10 |
| —— river - - 243, 244 |
| —— spot, or Medano mark - 14 |
| Blanca island and rocks - - 343, 358 |
| ———— or Lobo point - - 327 |
| Blanco cape and shoals - - 49, 209 |
| ———— cape, Peru - - - 366 |
| ————, Bahia - - 8, 10 |
| Blas San, banks and harbour - 21-25, 31, 63 |
| Bleaker island - - - 93, 94 |
| Boca chica island - - 407 |
| —— grande - - - 387, 407 |
| —— jambeli - - - 372, 373 |
| Bocas de Canales - - 240, 241 |
| Bodega cove - - - 285 |
| Bodie creek - - - 91 |
| Bold head - - - - 231 |
| —— point - - - 86, 88, 99 |
| Bolivia, boundary line of - - 326 |
| Bona island - - - 425 |
| Bonet bay - - - 191 |
| Bonifacia head - - - 275 |
| Boqueron cape - - 161, 164, 172 |
| ———— mount - - - 177 |
| ———— of Callao - - 350, 351 |
| ———— of Pisco - - - 341 |
| Boquita point - - - 283 |
| Borja bay - 189, 192, 193, 213, 215 |
| Borrachos point - - - 382 |
| Bouchage bay - - 170 |
| Bougainville cape - - - 77 |
| ———————— cove - - 170, 171 |

| | Page |
|---|---|
| Bouganiville creek - - 79 |
| Bournand bay - - - 170 |
| Bradley cove - - - 184 |
| Brattle island - - - 399 |
| Brava point - - 230, 414 |
| Brazo Ancho - - - 230 |
| Brazo de Norte - - - 237 |
| Breaker coast - - - 150 |
| Break-pot rock - - - 281 |
| Brea mountains - - 365 |
| Breaksea island - - - 238 |
| Brecknock passage - - 145 |
| Brenton cape - - - 237 |
| —— loch - - - 100 |
| —— sound - - - 174 |
| Brett harbour - - - 107 |
| Brightman inlet - - 15, 16 |
| Broad channel - - 25 |
| —— road - - - 135 |
| Broderip bay - - - 180 |
| Brookes harbour - - - 176 |
| Brothers rocks - - 332 |
| Browns bay - - - 179 |
| Bruja point - - - 421 |
| Brunswick peninsula - - 189 |
| Bucalemo head - - 284, 285 |
| Buckland mount - 125, 175, 17 |
| Buenaventura river - 388, 389, 414 |
| —————— town - 390 |
| Bueno river - - - 274 |
| Buey bank - - 406, 414 |
| —— point - - - 421 |
| Bufadero cliff - - - 355 |
| Bull road - - 92, 93, 96 |
| Bunche river - - - 383 |
| Burgess island - - - 184 |
| Burney mount - - - 224 |
| Burns inlet - - - 248 |
| Burnt harbour and island - - 107 |
| Bustamente bay - - 47 |
| Bynoe islands - - 147, 239 |
| Byron island - - - 239 |
| —— shoal - - - 49 |
| —— sound - - 107, 108 |
| | |
| | |
| Caballos road - - - 339 |
| Cabeza de Vaca point - - 311, 312 |
| Cabita bay - - - 401 |

| | Page |
|---|---|
| Cabra bspit | 424 |
| Cachos point | 306 |
| Caduhuapi rocks | 252 |
| Cahuache island | 260, 261 |
| Cajualo islet | 405 |
| Calbuco or El Fuerte | 269, 270 |
| Caldera point and port | 287, 308, 311, 312, 314 |
| ———, light | 311 |
| Caldereta peninsula | 310 |
| Calderillo, The | 311 |
| Calista islands | 101, 104 |
| Callao | 287, 346, 347, 348, 349, 368 |
| —— bay and point | 345, 346, 347, 348, 350, 351 |
| Callo island | 381 |
| —— point | 380 |
| Calms | 393, 428 |
| Calvario point | 357 |
| Calvary cove | 275 |
| Camana valley and Monte | 336 |
| Camarones bay | 44 |
| ———— gully | 330 |
| Cambridge island | 235 |
| Camden islands | 145, 146, 178 |
| Campana de Quillota | 286 |
| ———— mount | 360 |
| Cañada creek | 18 |
| Canal of the mountains | 227, 228 |
| Canas island | 412 |
| Cañaveral cove | 248 |
| Candelaria cape | 236 |
| Cañete river | 345 |
| Cangrejo island | 412 |
| Canning isles | 230 |
| Canoas rocks | 191 |
| Canoitad rocks | 252 |
| Cantor point | 37 |
| Capa point | 337 |
| Cape de Ros | 138 |
| Capstan rocks | 144 |
| Caracas river | 382 |
| Caracoles island | 412 |
| Carampangue river | 280 |
| Carcass island | 108, 109, 110 |
| —— reef | 108 |
| Carelmapu islets | 255, 273 |
| Carew harbour | 114 |
| Carlos III. island | 191, 222 |
| —— San, fort | 254, 275 |

| | Page |
|---|---|
| Carlos port | 253–256, 273 |
| Carmen San, town | 22, 26, 30 |
| Carnero bay and head | 278 |
| ——— point | 379 |
| Carquin bay and head | 353 |
| Carranza cape | 363 |
| Carrasco mount | 327 |
| Carretas head and mount | 340, 341, 360 |
| Carrisal bay and cape | 300, 301 |
| —— cove | 305 |
| Cartagena beach | 285 |
| Carva island | 269 |
| Carysfort cape | 76, 78, 84 |
| Casa de Josefina | 378 |
| Casamayor point | 48 |
| Casaya island | 412 |
| Cascade harbour | 184 |
| Cascajal point | 378, 388 |
| Casma bay | 287, 357 |
| Castellano isles | 185 |
| Castillos point | 46 |
| Castlereagh cape | 144 |
| Castle rock | 119 |
| Castro isles | 185 |
| —— point | 41, 263 |
| —— town and churches | 263, 270, 272 |
| Casualidad rock | 291 |
| Catalina bay | 165 |
| Cathedral mount | 237 |
| Catherine point, Tierra del Fuego | 132, 157, 158, 161 |
| Catripe river | 392 |
| Caucahuapi head | 253 |
| Caucahue island | 258, 259 |
| Cauten head and river | 276 |
| Caxa Chica rock | 307–310 |
| —— Grande | 308, 309 |
| Cayetano islands | 45, 180, 184, 185 |
| Caylin island | 268 |
| Cayman river | 391 |
| Centinela point | 266, 375 |
| Centinelas islands | 405 |
| Ceres island | 225 |
| Cerra de Cabra | 424 |
| Cerro Asul | 345 |
| —— chamé | 425 |
| —— de St. Ynez | 291 |
| —— Mongon | 357 |
| Chacao head and bay | 257 |
| Cerros de la Cruz, Los | 376 |

| | Page |
|---|---|
| Chacao narrows - | 252, 254–257, 269 |
| Chaffers gullet - - | - 120 |
| Chala morro and point | - - 387 |
| Chamé bay - - - | - 424 |
| —— island - - | - 425 |
| —— point - - | 415, 424 |
| —— river - - | - 424 |
| Chancay bay - - | - 352 |
| Chanchan cove - - | - 275 |
| Chanduy heights - - | - 379 |
| —— shoals - - | - 379 |
| Chañeral bay, island and port | - 301 |
| —— bay - - | - 313, 314 |
| Changarmi island - | - 421 |
| Changues islands - - | - 259, 260 |
| Chanticleer island - - | 138, 139 |
| —— rocks - - | - - 138 |
| Chao islands - - - | - 359 |
| Chapera island - - | - 411 |
| Charapota point - - | - 382 |
| Charles island - - | - 397, 398 |
| ——' islands - | 149, 179, 190, 191 |
| Chartres river - - | - - 112 |
| Chasco cove - - | - - 306 |
| Chatham bay - - | - 427 |
| —— harbour - - | - 117 |
| —— island - - | 229, 230, 396 |
| Chaulin island - - | - - 267 |
| Chaulinec island - - | 261, 265 |
| Chavini point - - | - - 337 |
| Cheape channel - - | - 241 |
| Chelin island - - | - 262, 265 |
| Chepillo island - | - 415, 416 |
| Chepo river - | 414, 415, 416 |
| Chidhuapi island - - | - - 269 |
| Chigua Loco cove - - | - 292 |
| Chilca point and port - | - - 345 |
| Child bluff - - - | 189 |
| Chile, boundary of - - | - 316 |
| ——, currents on the coast of - | 316 |
| ——, passages - - | - 318 |
| —— tides - - | - 316 |
| —— winds - - | - 316–318 |
| Chilen bluff - - - | 258 |
| Chileno point - - | - 326 |
| Chiloe island - | 249, 251–253, 262 |
| —— tides - | - 262, 267, 272 |
| Chiman river - | 414, 415 |
| —— town - | - - 415 |
| Chimborazo mountain - | - 377 |

| | Page |
|---|---|
| Chimbote village - - | - 352 |
| Chimu valley - - | - 361 |
| Chincha islands - | 341–343, 345 |
| —— river - - | - - 345 |
| Chipana bay - - | - - 326 |
| Chipre island - - | - 410 |
| Chirambira point - - | - 391 |
| —— river - - | - 391 |
| Chiut island - - | - 261 |
| Chivilingo river - - | - 280 |
| Choco bay - - | - 388, 389 |
| Chocoy head - - | 254, 255 |
| Chogon point - - | - - 259 |
| Choiseul bay - - | 70, 191 |
| —— sound - - | 90, 91, 100 |
| —— tides - - | - 92 |
| Chomache village - - | - 327 |
| Chonos archipelago - | 249–252, 271 |
| Chorillos bay and town - | 346, 351 |
| Choros bank - - | - 282 |
| —— cape and islands - | - 299 |
| Christmas cove - - | - 247 |
| —— harbour - | 111, 112 |
| —— sound - | 143, 144, 153 |
| Christopher point - - | - 399 |
| Chuapa river - - | - 292 |
| Chulin island - - | - 261 |
| Chungunga island - - | - 300 |
| Chupat river - - | - 39, 40 |
| —— bar - - | - 40 |
| Church rock - - | - 283, 284 |
| Churruca bay - - | - 205 |
| Cirujano island - - | - 242, 243 |
| Clapperton inlet - - | - 224 |
| Clarence island - | 177–179, 183, 184 |
| Clearbottom bay - - | - 143 |
| Clements island - - | - 249 |
| Clerke port - - | - 144 |
| Cliff cove - - | - 247 |
| —— end - - | - 33 |
| —— island - - | - 115 |
| Climate, Coquimbo - - | - 297 |
| —— Falkland islands - | - 71 |
| Cloyne reefs - - | - 226 |
| Coal - | 273, 280, 282, 295 |
| Cobija bay - | - 287, 323–325 |
| Cobre hill - - | - 299 |
| Cocale head - - | - 275 |
| Cochinos islet - - | - 254, 255 |
| Cockatrice rock - - | - 278 |

Page

Cockburn channel 146, 177-179, 183, 216
Coco island - - - - 414
Cocos island - - - 427
——— point - - - - 412
Cosotue head - - - 253
Cocovi islet - - - - 421
Cocovicita island - - - 421
Codo point - - - - 346
Cogimies shoals - - - 383
Coles point - - - - 332
Colina Redonda - - - 356
Colita island - - - 268, 269
Coliumo bay and head - - 283
College rocks - - - 150
Colliers rocks - - - - 116
Colorado point - - - - 406
——— river - - 8, 16, 18
Colworth cape - - - 224
Committee bay - - - 106
Compu inlet - - - 266, 267
Concepcion bay - - 280, 281
——— channel - 219, 229, 230
——— strait - - 223, 235
Conchali bay - - - 291
Concon rocks - - - 289
Condesa bay - - - 196, 197
Cone inlet - - - - 247
Congo river - - - 405
Constitucion harbour - - 322
——— town - - - 284
Contadora island - - 410, 411
Conway bay - - - 398
Cook bay - - - 136, 144
——— port - 123, 124, 126-128
Cooke port - - - - 176
Copaipó bay - - - - 307
——— directions for - 309, 310
——— Morro de - - 307, 310
Copper cove - - - 325
Coquimbo - 287, 294, 296-298
——— climate at - - 297
Corcovado gulf - 250, 252, 258, 271
——— island - - 359
Cordes bay - - - 186
Cordillera - - - 227, 252
Cordova cove - - 47, 48
——— islet - - - 191
——— peninsula - - 194, 200
Coredo river - - - 404
Cornejo point - - - 334, 335

Page

Cornejos islet - - - 357
Cornish opening - - 246
Corona head - - 253-255
Coronel point - 256, 269, 280
Coronilla cove - - - 189
Corral fort - - - 275
Corrientes cape - 5, 75, 391, 401
Corso mount island - - 237
Cortado cape - - - 206
Corutu river - - - 416
Courtenay sound - - 145, 178
Coventry cape - - 186, 215
Cow bay - - - - 78
Coy inlet - - - - 60, 61
Creek hill - - - 19, 20
——— island - - - 21
——— pass - - - 106
Crooked reach - 184, 192, 194, 213, 215, 217, 221, 222
Crossley bay - - - 127
Cruces point - - - 404
——— river - - 274, 275
Cruz bay - - - - 43
Cuaviguilgua bay - - 205
Cucao bay - - - 253
Cuevas cape - - - 206
Culebra island - - 421
Culebras cove and point - 356
Culla-calla river - 274, 275
Cullin island - - - 270
——— point - - - 283
Culo de Barca - - 389
Culpepper island - - 399
Cumba river - - - 332
Cumberland bay - - 319
Cupica bay - - 402-404
Cupola Mount - - 208
Curachichi river - - 404
Curauma head - - 286
Curaunilla point - 286, 288
Curioso cape - - - 56
Currents of bay of Panama - 428
Currents:—Coast of Chile - 363
——— Coast of Peru - 368
——— near Falkland islands - 69, 155, 156
——— near Staten island and Cape Horn - - 66
Cutter cove - - 189, 190
Cygnet harbour - - - 100

| | Page |
|---|---|
| Dædalus island - - - | 138 |
| ——— rock - - - | 138 |
| Dalcahue channel and village - | 262 |
| Dallas point - - - | 307–309 |
| Dalrymple rock - - - | 396 |
| Danaide rocks - - | 422, 423 |
| Danoso reef - - - - | 55 |
| Danson harbour - - | 101, 105 |
| Darby cove - - - - | 206 |
| Dardo head - - - - | 340 |
| Darien harbour - - | 408, 409 |
| Dark hill - - - - | 248 |
| Darwin bay and channel - - | 136, 249 |
| ——— harbour - - - | 91 |
| ——— mount - - - | 355 |
| David, St., head - - | 192, 194 |
| ——— sound - - | 191, 192, 222 |
| Dawson island - - | 167, 173–176 |
| Deadtree island - - - | 242 |
| Dean harbour - - - | 180 |
| Deceit cape - - - | 140, 141 |
| ——— island - - - | 140 |
| Deep harbour - - - | 223 |
| Deepwater sound - - - | 150 |
| Deer island - - - - | 21 |
| De Gennes river - - - | 171 |
| Delfin point - - - | 41 |
| Delgada point - - - | 38 |
| Deseado cape - - - | 152 |
| Desecho - - - - | 244 |
| Desengaño head - - - | 56 |
| Desertores islands - - | 261, 262 |
| Desire port - - - | 50–52 |
| Desolate bay - - - | 136, 145 |
| Desolation cape - - | 145 |
| ——— island - - | 151, 204 |
| Desvelos bay - - - | 55 |
| Detif head and point - - | 264, 265 |
| Diana islands - - - | 226 |
| ——— peak - - - | 234 |
| Dick point - - - - | 113 |
| Diego Ramirez islands - - | 142, 143 |
| Diego, San, cape - - - | 133, 134 |
| Difficulties of navigation - - | 429 |
| Dighton bay - - - | 182 |
| Dinero mount - - | 157, 159 |
| Direction bluff - - - | 338 |
| ——— hills - 33, 35, 160, 161, 219 |
| ——— islets - - - | 233 |
| Disappointment bay - - | 228 |
| Dislocation harbour - - | 152 |
| | Page |
|---|---|
| Division mount - - - | 359 |
| Divis island - -- - | 236 |
| Dog bank - - - | 19, 20 |
| Dolphin cape - - | 76, 99 |
| ——— rock - - - | 221 |
| Dome of St. Paul's mount - | 243, 244 |
| Doña Maria table-land - - | 339 |
| Don Martin island - - | 353 |
| Dorah peak - - - | 238 |
| Doris cove - - - | 145 |
| Dormido rock - - | 278, 279 |
| Dos Bahias cape - - | 44 |
| —— Hermanas - - | 31, 183 |
| ——— islands - - | 183 |
| Double creek - - - | 114 |
| Driftwood point - - | 93 |
| Duck harbour - - - | 235 |
| Duende island - - | 248 |
| Duff bay - - - | 143 |
| Duke of York island - | 235 |
| Duncan harbour and rock - | 235 |
| ——— island - - | 398 |
| Dundee rock - - - | 238 |
| Dungeness point - - 132, 157–159 |
| Dunnose head - - - | 112 |
| Dyer cape - - | 237, 238 |
| Dyke island - - - | 118 |
| Dyneley bay - - | 237–239 |
| ——— sound - - | 179 |
| Dynevor Castle mountain - | 190 |
| ——— inlet - - | 161 |
| ——— sound - - | 199 |
| Eagle bay - - | 169, 170 |
| —— passage - - | 75, 97 |
| Earnest cape - - | 227 |
| East cove - - - | 91 |
| —— gate post - - | 14 |
| —— point - - | 310 |
| Easter bay - - - | 228 |
| Echenique point - - | 205 |
| Eddystone rock - - | 54 |
| ——— East Falkland - 76, 103 |
| Eden harbour - - | 231, 232 |
| —— island - - | 231, 232 |
| Edgar port - 102, 114, 120, 121 |
| Edgeworth cape and shoal - | 181, 182 |
| Edith island - - | 408 |
| Edye creek - - - | 114 |
| Egg harbour - - | 45 |

Page
Egmont cays and port - - 105–107
Elephant cays - - - 98
———— cove - - 106, 117
———— point - - - 107
El Frayle rock - - - 280
— Fuerte, Port San Antonio - - 33
———— or Calbuco town - 269, 270
— Gobernador - - - 290
— Morion - - 192–194, 221
— Morro - - - - 387
— Rincorn - - - 8
— Roca - - - - 389
— Yunque mountain - - 318
Elisa bay - - - - 179
Elizabeth bay - - 187, 215, 399
———— island - 159, 162, 163, 169, 175,
189, 214, 215, 218
Ellen bay - - - - 229
—— island - - - - 408
Elmo, St., bay - - - 412
———— island - - - 412
Elvira point - - - - 185
Engaño bay - - - 40
English cove - - - 333
———— creek - - - 277
———— narrows - - 231–233
———— reach - 147, 188, 191, 192, 215, 221
Entrance isles - - - 245
———— mount - - 57, 59, 60
Ercules point - - - 37, 38
Erizos rocks - - - 356
Escape bay - - - - 229
Escondido creek - - - 34
Esmeraldas - - - - 385
———— river - - - 385
Española house - - - 373
———— patch - - - 373
———— point - - - 373
Esperanza island - - 228, 229
Espinosa heights - - 48, 282
Espiritu Santo cape - 132, 157, 158, 161, 164
Essex point - - - - 399
Estaguillas point - - - 273
Estero Balsa - - - 380
—— Salado - - - - 379
Estevan channel - 219, 223, 228, 229
————, San, port - - 247, 248
Estiva island - - - 425
Eten point - - - - 362
Euston bay - - - 149, 150
———— opening - - - 190

Page
Evangelistas islets - - 208, 220, 222
Evouts isles - - - 141
Exeaquil cove - - - 196
Exmouth promontory - - - 231
Expectation bay - - - 230
Eyre sound - - - - 231

Fairway isles - - 204, 223
Fairweather cape - - 61, 62, 157
Falfan river - - - 421
Falkland islands - 68–121, 210
————, appearance of - 69
————, approaching - 74
————, climate - - 71
————, currents and tides - 69
————, productions, - 72–74
————, remarks on - - 68
————, soil - - 72
———— sound - - 68, 99–105
————, directions for - 103–105
————, north entrance - 103, 104
————, south entrance - 105
————, tides - 102, 103
————, West, tides, north coast of, 108, 109
————, winds - - - 70
Fallos channel - - - 239
Falsa bay - - - 15
—— cape - - - - 274
False Cape Horn - 137, 141, 143
—— Corona isles - - - 189
—— cove - - - - 134
—— Maule - - - - 284
—— passage - - - - 112
—— Sisters rocks - - - 26, 31
Famine port 165–169, 174, 175, 199, 211–
213, 215, 217, 220
———— reach - - 164, 167, 215
Fanning head - - - 103
Fanny road - - - 96
Farallon de Castillo rock - - 387
Farallon ingles - - - 414
Fatal bay - - - 234
Felipe, St., mount - - 167–169
Felix point - - - - 206
——, St., island - - - 319
Ferrol bay - - - 358
Field bay - - - 146, 180
Fincham islands - - - 150
Findlay harbour - - - 100

Page

Finger point - - - 397
First Narrows - 159–162, 165, 213, 214, 218, 219, 221
Fitton harbour - - 175, 176
Fitz basin - - - - 87
Fitz-Roy channel - - 189, 190
———— island - - - 178
———— port - - - 86–88
Flamenco bay and port - 312, 313
———— island - - 416, 421
Flat islet - - - - 55
—— point - - - 27, 29
—— rock point - - 333, 335
Flat-top hill - - - 17, 20
Flinders bay - - - 127
Flinn sound - - - - 239
Flores bay - - - - 195
Florido island - - - - 43
Foca island - - 364, 365
Fog bay - - - - 228
Forelius peninsula - - 243, 244
Forsyth island - - - 322
Fortescue bay - 182, 186, 187, 190, 215
Fortune bay - - 224, 225
Forty-five cape - - - 189
Foul bay - - - - 99
Fox harbour - - - - 95
———— West Falkland 75, 102, 120
———— Magellan strait- 174, 175
Frailes rocks - - 46, 426
Francis, St., bay - - - 140
Francisco, San, cape - 325, 383
————, island - - 346
————, river - - - 384
———— Solano, point - - 402
Franklin sound - - 127, 137, 138
Frayle point - - - 345
French harbour - - - 117
Freshwater bay 166, 167, 173, 213, 396
———— cove - - 184
Fronton bank - - - 350
———— island - 349, 350
Froward cape 171, 172, 183–185, 215, 218
———— reach - 164, 183, 215
Fuerte Argentino - - - 11
Fuerto Viejo cove - - - 280
Furies rocks - - 145, 146, 147
Fury cove - - - - 231
—— harbour - - - 147
—— island - - 147, 148

Page

Fury rocks - - - - 145

Gabriel channel - - 175, 176
Galapagos islands - - 394–396
Galera island - 273, 274, 349, 383
———— point - - 410, 413
Galiano isles - - - 47
Gallan, San, island - 256, 257
———— point - 341, 342, 369
Gallant port - 175, 186, 187, 200, 213
—— cape - - - 221
Gallegos cape - - - 247
———— port - - - 62
———— river - - - 157
Gallo island - - - 388
—— point - - 285, 286
Gamboa river - - - 263
Garachiné point - 405, 410
———— town - - 405
Gardner bay - - - 397
———— island - - - 398
Garita hill - - - 360
Garrido island - - - 249
Gate Post, east - - - 14
Gateway channel - - 14
Gente Grande bay - 164, 165
Gente point - - - 164
George cape - - - 235
————, St., gulf - - 47, 48
Gibraltar reef - - - 108
———— rock - - 108, 146
Gidley cove - - - 175
Gilbert island - - 144, 145
Gill bay - - - 45
Glacier bay - - 197, 198
———— sound - - 204
Glasscott point - - 169, 172
Gloucester cape - - 149, 150
Godoy point - - - 273
Goleta, la - - - 425
Golding island - - - 106
Gonzales head - - 274, 275
———— island - 412, 413
———— narrows - - 184
Good harbour - - - 239
Goodluck bay - - - 195
Goods bay - - - 224
Good Success bay and cape - 134–136, 210

Page

Gorda point - 329, 330, 365, 414
Goree road - - 136, 137
Gorgona island - - 389, 393
Governor channel - - - 116
—— island - - - - 117
Gracia point - - - 162, 163
Grafton islands - - - 149
Gramadel bay - - - - 355
Grande isla - - - 308, 309
—— point - - 315, 321
Grantham bay - - 99, 100, 104
Grave cove - - - 108
Graves island - - - 151
—— mount - - - 173
Great Black rock - - 143, 144
—— Gat - - - - 23
—— island - 95, 101, 104, 105
Green bay and island - - 15, 16
—— island - - 377, 378
——— and spit - - 16, 113
—— point - - - 378
Greenough peninsula - - 183
Gregorio cove - - - 44
Gregory bay 162, 165, 213, 214, 218, 221
——— cape - - 161, 162, 218
Gretton bay - - - - 138
Grey cape - - - - 228
—— channel - - 115, 116
Grueso point - - - 327, 328
Guaban head - - - 253
Guaianeco islands - - - 239
Guambacho bay and town - - 357
Guanaco point - - 136, 137
Guañape hill and islands - 359, 360
Guanico point - - - 426
Guano - - 325, 327, 343, 362
Guard bay - - - - 230
Guarmey bay and town - 355, 356
Guascama point - - - 388
——— river - - - 388
Guata cove - - - 336
Guayaquil gulf - - - 371
——— river - - 371, 376
——— town - - - 376
Guayteca grande - - - 250
Guaytecas islands - - - 250
Gueguen river - - - 7
Guia narrows - - 228, 229, 235
Guilmen heights - - 255
Guinea point - - - 421

Page

Guirior bay - - - 195
Gulf of Peñas - 201, 204, 233, 234, 241
—— San Estevan - - 241, 242
—— San Matias - - - 31, 33
—— San Rafael - - - 245
—— Tres Montes - 242, 244, 245
Gull harbour - - - 118
Gun bay - - - - 170
Guy Fawkes islets - - 398

Half port bay - - 199, 200, 215
—— Way cove - - 98, 114, 115
Halt bay - - - - 233
—— island - - - 93
Hamilton point - - - 406
Hamper bay - - - 225
Hanover island - 204, 223, 229, 230, 235
Harbour islet - - - 44, 357
Hardy peninsula - - 141, 143
Harmless point - - - 339
Harriet port - - 82, 84, 86
Harris bay - - - 175
Harvey bay - - - 239
Hawkins bay - - - 183
Hazard isles - - - 240
Hector rock - - - 278
Helgat bank - - - 23, 24
Hellyer rocks - - - 248
Henderson island - - - 143
Henry port - - 235-237, 246
Hermanas Dos - - - 31, 183
Hermeneg point - - - 6
Hermite, or Cape Horn islands - 138, 139, 141
Hermoso mount - - 7, 9, 14
Herradura de Carrisal - - 304, 305
——— Coquimbo - 290, 294, 296
——— Mexillones - - 323
——— point - - 304
Herschel island - - - 138
Hewett bay - - 147, 179
Heywood's pass - - - 226
Hidden harbour - - - 184
—— islet - - - 41
High Cliff island - - - 104
Hill cove - - - 108
—— Gap port - - 99, 104
—— islands - - - 180
Hind island - - - 143
Hobbs reef - - - 397

| | Page |
|---|---|
| Hog island beacon | 21, 24, 83 |
| Holland cape | 175, 185, 186, 221 |
| Holloway sound | 242, 245 |
| Hondo river | 416 |
| Hood island | 397 |
| Hope harbour | 108–110, 149, 177 |
| —— mount | 176 |
| —— point | 109 |
| —— reef | 110 |
| Hoppner port | 123, 127 |
| —— sound | 244 |
| Horace peaks | 145 |
| Horadada rocks | 350, 351 |
| Horca hill | 355 |
| Horcon bay and head | 289, 290 |
| Hormigas de Afuera | 351 |
| —— Tierra, or Ants | 352 |
| Horn, Cape | 122, 130, 140, 142, 154–156, 210–213, 216 |
| —— appearance of | 140 |
| —— currents off | 66, 140, 155, 156 |
| —— rocks off | 140 |
| —— time for rounding | 154, 155, 220 |
| —— spit | 12, 14, 15 |
| Hornby hills | 99 |
| Horse Block islet | 117 |
| Hose harbour | 224 |
| House cove | 118 |
| Howard port | 99, 101 |
| Huacas point | 341 |
| Huacho bay | 287, 353 |
| Huafo island | 251 |
| Huamblin island | 249, 250 |
| Huamlad passage | 252 |
| Huanaquero hill and point | 294 |
| Huanchaco road and peak | 287, 360, 361 |
| Huano cove | 332 |
| Huantacayhua mines | 327 |
| Huaspacho shoal | 253, 255 |
| Huapilacuy light | 254 |
| —— point | 254, 255 |
| Huapilinao head | 258 |
| Huar island | 271 |
| Huasco point and port | 287, 302–304 |
| —— tides | 316 |
| Huaura islands | 352, 353 |
| Huechucucuy head | 253, 254 |
| Hueso Parado | 315, 321 |
| Huildad inlet | 267, 268, 272 |
| —— shoal | 268 |

| | Page |
|---|---|
| Humming Bird cove | 174 |
| Hummock island | 111, 116 |
| Humos cape | 283 |
| Hunter island | 225, 226 |
| Icy sound | 182, 204 |
| Ignacio bay | 241 |
| Iguana cove | 399 |
| —— island | 405, 425 |
| —— point | 414 |
| Ilay bay and point | 333, 334 |
| Ildefonso islands | 142, 143 |
| Ileñao island | 421 |
| Imel bank | 265 |
| Inchin islands | 248, 249 |
| Indefatigable island | 398 |
| Independencia bay | 340, 341 |
| Indian bay | 170 |
| —— cove | 143 |
| —— head | 19, 20 |
| —— reach | 231 |
| —— sound | 184 |
| Infiernillo rock | 339 |
| Inglefield island | 189, 190 |
| Inglesia de Sevira | 401 |
| Ingles port | 308 |
| Inlet bay | 225 |
| Inman bay | 183 |
| —— cape | 150, 151 |
| Inner point | 366 |
| Inocentes channel | 223, 228–230 |
| —— islands | 229, 230, 235 |
| Interior sounds | 227 |
| Ipswich island | 149 |
| Iquique port and reef | 287, 327–329 |
| Isabel cape | 234, 235 |
| Isabella island | 149 |
| Isidro point | 162 |
| Isla del Rey | 412 |
| Isla Grande | 308, 309 |
| —— pass | 308, 309 |
| Isla Nueva | 379 |
| Island harbour | 89, 233, 234 |
| Islay bay | 333, 334 |
| Isquiliac island and mount | 249 |
| Isthmus bay | 225 |
| Jack harbour | 170 |
| Jago, St., bay | 161 |
| Jaguey point | 355 |
| Jama point | 382 |

|  | Page |
|---|---|
| Jama river | 383 |
| James bay | 398 |
| —— island | 149, 191, 250, 398 |
| Jampa point | 380 |
| Jannette island | 408 |
| Jara head | 322 |
| Jaron mount | 322 |
| Jason islands | 109, 110, 116 |
| Jerome channel | 185, 187–190, 200 |
| —— tides and variation | 190, 222 |
| Jesuit sound | 241 |
| Jeya river | 401 |
| John, Cape St., tide race | 66, 123, 156, 210 |
| ——, St., harbour | 123, 124 |
| Johnson harbour | 70, 79 |
| Jones island | 406 |
| Jorey island | 407 |
| Jorge bay | 322 |
| —— mount | 322 |
| Jorgino mount | 322 |
| José, San, shoal | 270 |
| —— —— bank | 410 |
| —— —— island | 414 |
| Juan Dias river | 331 |
| —— Fernandez island | 318, 319 |
| —— Soldado mount | 299 |
| —— San river | 392 |
| Judge rocks | 152, 220, 222 |
| Jupiter rocks | 148 |
| | |
| Kater peak | 138 |
| Keats sound | 177 |
| Keel point | 57 |
| Kelly harbour | 241, 242 |
| Kelp lagoon | 90 |
| —— remarks on | 74, 86, 130 |
| Kempe harbour | 183 |
| —— island | 147, 148 |
| Kentish isles | 230 |
| Kent mount | 86, 88 |
| Keppel sound | 105, 106 |
| Kicker rock | 396 |
| Kidney island | 78, 82 |
| King George bay | 111, 112, 115, 116 |
| —— directions for | 115, 116 |
| King harbour | 100 |
| —— island | 178 |
| Kirke narrows | 228 |
| —— rocks | 178 |
| Knocker rock | 422 |

|  | Page |
|---|---|
| Laborde bushes | 14 |
| Labyrinth islands | 117 |
| —— shoals and head | 15, 16 |
| Lacao bay | 257 |
| Lacuy peninsula | 253, 254 |
| Lago de Botella | 184, 180 |
| La Goleta | 425 |
| Lambayeque road | 287, 362 |
| Lami bank | 269 |
| Landfall islands | 150, 151 |
| Langara bay | 194 |
| —— port | 185 |
| La Playa Brava bay | 322 |
| Laraquete beach and river | 280 |
| Laredo bay | 164, 165, 214, 218, 221 |
| Las Animas | 313 |
| —— Bajas point | 352 |
| Last harbour | 192 |
| —— Hope inlet | 228 |
| Latitude bay | 151 |
| Laura harbour | 149, 150 |
| Lavapie point | 278, 279 |
| Lavata bay | 315 |
| Law peak | 152 |
| Laytec island | 268 |
| Leading bluff | 323, 325, 356 |
| —— hill | 27, 29, 143 |
| —— island | 149 |
| Lechuza mount | 343 |
| Lee bay | 164 |
| —— rocks | 152 |
| Leeward bay | 227 |
| Legarto head | 355, 356 |
| Lelbun point | 265 |
| Le Maire strait | 122, 130, 134, 135, 209, 210 |
| —— tide race | 127, 210 |
| —— tides | 66, 122, 127, 135, 210 |
| Lemu island | 262, 264, 265 |
| Lengua de Vaca | 293, 294 |
| Lennox island | 136 |
| Leones cape | 301 |
| —— cove and isle | 44 |
| L'Etoile cape | 199 |
| Leübu river | 277, 278 |
| Level bay | 231, 232 |
| —— islands | 231 |
| Liebre island | 250 |
| Lights, Caldera point | 311 |
| ——, Huapilacuy | 254 |
| ——, Pembroke cape | 80, 83, 84 |

|  | Page |
| --- | --- |
| Lights, San Lorenzo cape - - 349 |
| ———, Santa Clara - - 371 |
| ———, Valparaiso - - 286 |
| Ligua bay and river - - 290 |
| Liles point - - - 289 |
| Lima - - - - 348, 349 |
| Limari river - - 292 |
| Limon bay - - - 403 |
| Linao cove - - - 258 |
| Linlin island - - 260 |
| Linna islands - - 260 |
| Lintinao island - - 263, 264 |
| Lion bay - - - 233 |
| —— cove - - - 195 |
| Lirquen point and rock - 282 |
| Liso point - - - 425 |
| Litis isles - - - 290 |
| Little black rock - - 143 |
| —— gat - - - 23 |
| —— island - - - 94 |
| Lively island and sound - 83, 90, 92, 93 |
| Lliuco village - - 258 |
| Loa gully and river - - 324, 326 |
| Loberia head - - - 281, 282 |
| Lobo peak - - - 38 |
| —— point - 281, 304, 326, 327 |
| Lobos bank - - - 15, 33, 34 |
| —— cape - - - 330 |
| —— head - - 41, 259 |
| —— island - - 3, 290 |
| —— isles - - - 47 |
| —— de Afuera islands - 362, 363, 368 |
| —— de Tierra island - 363 |
| Locos island - - 291 |
| Logan rock - - - 245 |
| Lomas bay 132, 159-161, 173, 174 |
| —— point and road - 337 |
| —— valley - - 337, 338 |
| London islands - - 145, 146 |
| Londonderry islands - 144 |
| Long island - - 79, 224 |
| Long reach - 192, 194, 197, 200, 201, 215, 220, 221 |
| ————— appearance of coast in - 194 |
| ————— tides - - 221 |
| ————— weather in - 194 |
| Lookout point - - 55 |
| Lorenzo point - - 405 |
| Lorenzo, San, island - 347, 349, 350, 351 |
| Lort bay - - - 142 |
| Los Hermanos - - 421, 423 |

|  | Page |
| --- | --- |
| Lota cove - - - 286 |
| Louis port - - - 79 |
| Low bay - - - 92 |
| —— mount - - 78, 81 |
| —— port - - 250, 251 |
| Lower Toro bank - 11, 14, 15 |
| Lucas bay and reef - - 120 |
| Lucky ledge - - 191 |
| Luco bay - - - 279 |
| Lunes cape - - - 201 |
| Lurin river - - - 345 |
| Lyell sound - - 183, 184 |

| Macabi island - - 362 |
| Macbride head - - 76, 78 |
| Machado cape - - 240 |
| Machala creek - - 372 |
| Macgowen reef - - 398 |
| McKinnon bay - - 406 |
| Madre islands - 223, 230, 235-237 |
| Magdalen sound 175, 176-178, 183 |
| Magdalena bay - - 391 |
| ————— point - 391 |
| ————, Santa, island and dangers 163, 164, 165, 215, 218 |
| Magellan cove - - 79 |
| ————— strait of - 152, 157-234 |
| ————— tides - 134, 221, 222 |
| ————— passage through from the Pacific to the Atlantic 216, 222 |
| ————— sounds and channels between, and Gulf of Penas 223, 234 |
| Magill isles - 146, 178, 179 |
| Majaqual - - 386 |
| Main point - 25 |
| —— passage - 104 |
| —— or Redonda point - 26, 28 |
| Maipu river - - 285 |
| Majaguay islands - 415 |
| Mala bank - 373, 374 |
| —— cape 425, 426 |
| —— hill - 373 |
| Malabrigo hill - 361 |
| —— road - 361 |
| Malaspina port - 47 |
| Malpelo island - 393 |
| ————— point - 366, 372, 375 |

| | Page |
|---|---|
| Mamilla water | 325 |
| Manao bay | 258 |
| Mancora valley | 366 |
| Mandinga point | 373 |
| Mangles island | 387 |
| —— point | 387 |
| Mangue island | 415 |
| Manning bay | 200 |
| Manoel point | 276 |
| Manta port and village | 381 |
| Manybranch harbour | 99, 101 |
| Mansano bank | 282 |
| —— cove | 274 |
| Manzera bank and island | 275 |
| Mar Chiquito | 4, 5 |
| March harbour | 144 |
| Mare harbour | 90–92 |
| Maria bay | 149 |
| —— Doña, table-land | 339 |
| —— Santa, island | 278 |
| —— point | 339 |
| —— road | 279 |
| Marian cove | 199, 215 |
| Mariato point | 426 |
| Marine islands | 244 |
| Martin, Don, island | 353 |
| —— St., cove | 137–140 |
| Mary island | 407, 408 |
| —— St., point | 166, 167 |
| Marzo cape | 404 |
| Masa island | 378 |
| Masafuera island | 318–320 |
| Masatierra island | 318 |
| Matalqui cape and paps | 252, 253 |
| Matamores | 305 |
| Matorillos island | 377 |
| Matias, San, gulf of | 31, 33 |
| Maule head and river | 283, 284 |
| —— False valley | 284 |
| Maullin inlet | 273 |
| Maury's remarks | 430 |
| Maxwell island | 137 |
| —— port | 139 |
| Mayllen island | 271 |
| May reef | 53 |
| Maytencillo cove | 292 |
| Mazaredo bay | 184 |
| Mazorque island | 352, 353 |
| Medal bay | 200 |
| Medano bank | 4 |
| —— black spot | 14 |
| Medano point | 3, 4 |
| Medio cape | 133 |
| —— island | 237 |
| —— point | 27, 310 |
| Medrano shoal | 47 |
| Mehuin river | 275 |
| Mellersh cove | 185 |
| Melo port | 46 |
| Melones island | 425 |
| —— rock | 425 |
| Melville sound | 146 |
| Menchuan island | 248 |
| Mercury sound | 178 |
| Mercy port | 207, 216, 127 |
| Messier channel | 223, 233, 234, 240 |
| —— anchorages | 234 |
| Meulin island | 260 |
| Mexico point | 334 |
| Mexillones bay and mount | 323 |
| Mexillon island | 329 |
| Michael bay and point | 184, 230 |
| Michael's, St., channel | 181, 184 |
| Mid channel island | 232 |
| Middle bank | 23 |
| —— bay | 99, 306 |
| —— cove | 137 |
| —— island | 90, 106, 117, 233 |
| —— point | 198 |
| —— rock and shoal | 96 |
| Milagro cove | 274 |
| Milky Way rocks | 148 |
| Mill point | 274 |
| Millar cove | 185 |
| —— island | 234 |
| Millon point | 277 |
| Milne island | 407, 406 |
| Mocha channel | 277 |
| —— island | 275–277 |
| Moffat harbour | 101 |
| Mogotes point | 5 |
| Molguilla point | 277 |
| Mollendito cove | 334 |
| Mollendo cove | 333 |
| Monday cape | 199–201, 217 |
| Mondragon channel | 372 |
| —— island | 374, 377 |
| —— point | 374, 377 |
| Mongé island | 412 |
| Mongoncilla point | 356 |
| Mongon mount | 357 |
| Monmouth cape | 164 |

| | Page |
|---|---|
| Monmouth island | 191 |
| Montague bay | 226 |
| Montanita point | 380 |
| Monte Christo | 381 |
| Monte Camana | 336 |
| —— Gordo | 331 |
| —— Hurtado | 9 |
| —— Jorgino | 322 |
| —— Video hill | 55 |
| Montijo bay | 426 |
| Montt port | 271 |
| Moreno bay and mount | 322 |
| —— islet | 44 |
| Morro de Acari | 338 |
| —— Sama | 331 |
| —— channel | 378 |
| —— island | 387 |
| —— Mico | 491 |
| —— point | 310, 379 |
| ——, port del | 236 |
| —— Santa Agueda | 172 |
| —— Solar | 346, 348 |
| Morton island | 143 |
| Motley island | 92, 93 |
| Muelles bay and point | 290 |
| Muisme river | 383 |
| Mulatas cape | 351, 352 |
| Murphy cape | 48 |
| Murray cove | 184 |
| Murrell river | 82 |
| Mussel bay | 191 |
| —— island | 186 |
| Mutico point | 255 |
| | |
| Nahuelhuapi lake | 271 |
| Naipi river | 403 |
| Naranjal river | 377 |
| Narborough island | 398 |
| —— islands | 208, 221, 250 |
| Narrow bank | 160 |
| —— creek | 226 |
| Nasca cape | 339 |
| Nash bay | 192 |
| Nassau bay | 136, 137, 141, 210 |
| —— channel | 171 |
| —— island | 170, 171 |
| Natividad bay | 284 |
| Navarin islands | 141, 143 |
| Nayahue islet | 261, 262 |
| Needle rocks | 106–110 |

| | Page |
|---|---|
| Neesham bay | 237, 239 |
| Negrillos rocks | 390 |
| Negro cape | 163–165, 213, 218 |
| —— Rio | 8, 25, 26–30 |
| Nelson strait | 223–225, 230, 235 |
| Nena cove | 276 |
| Neptune rocks | 148 |
| Neuke mount | 282 |
| Neuman inlet | 245 |
| New cove | 43 |
| —— island | 110, 115–117, 136 |
| —— Year harbour | 118, 123, 124 |
| —— islands | 123, 125, 210 |
| —— sound | 142, 143 |
| Newhaven | 100 |
| Niebla castle | 275 |
| Nihuel island | 261 |
| Ninfas point | 38, 39 |
| —— reef | 39 |
| Nipple hill | 35 |
| Nodales peak | 172 |
| Noir island and road | 148, 210 |
| Nombre head | 132 |
| Non-entry bay | 174 |
| Nonura point | 363 |
| Noratos cove | 336 |
| Nort bay | 179 |
| Norte point | 32, 36 |
| North bank | 11 |
| —— Barranca | 26 |
| —— basin | 86 |
| —— bay | 224 |
| —— cove | 147 |
| —— island | 96, 115, 174 |
| —— port | 111 |
| —— road | 137 |
| —— rock | 142 |
| North-west islets | 104 |
| —— pass | 105, 106 |
| Nose peak | 173 |
| Notch cape | 194–197, 200 |
| Novales shoal | 47 |
| Nuestra Señora bay | 321 |
| Nuevo gulf | 38 |
| —— head | 38 |
| Nuki river | 401 |
| Nuñez creek | 189 |
| Nutland bay | 180 |
| | |
| Oake bay | 224 |

|  | Page |
|---|---|
| Oasy harbour | 163 |
| Obispito cove | 312 |
| Obispo cove | 312 |
| Observation islet | 207 |
| —————— mount | 206 |
| Obstruction sound | 190, 226 |
| Ocoña valley | 336 |
| Octavia bay | 404 |
| Ofqui isthmus | 244 |
| Old Settlement cove | 107 |
| Open bay | 230 |
| Oracion bay | 225 |
| Orange bank | 160, 161, 221 |
| —————— bay | 137, 141, 142 |
| —————— cape | 132, 160, 161, 219 |
| Orford cape | 70, 118 |
| Orozco table | 133 |
| Ortiz islets | 189, 192, 193, 221 |
| Oscuro cove | 258, 259, 272 |
| Osorno bay | 194, 299 |
| Otoque island | 425 |
| Otter islands | 224 |
| Otway bay | 151, 199 |
| —————— port | 245, 246 |
| —————— water | 164, 188, 189, 190 |
| Our Lady's bay | 239 |
| Oven cove | 45, 63 |
| Ouredo river | 404 |
| Owen port | 174 |
| —————— road | 97, 98 |
| Oyarvide heights | 327 |
| Oyster fishery | 372 |
| Pabellon de Pica | 327 |
| Pablo island | 412 |
| Pacasmayo road | 361, 362 |
| Pachacamac islands | 345, 346 |
| Pacheca island | 410 |
| Packsaddle bay | 141 |
| Pacocha road and village | 332 |
| Painter's muller | 123 |
| Paiton | 386 |
| Pajaros islets | 299 |
| —————— Niños | 281 |
| —————— islets | 297, 298 |
| —————— point | 301 |
| —————— islands | 411, 414 |
| Pajonal cove | 306 |
| Paley island | 408 |
| Palmas island | 391 |

|  | Page |
|---|---|
| Palmer point | 225 |
| Palominos rocks | 349 |
| Panama bay | 416, 430 |
| —————— gulf | 416, 430 |
| —————— N.E. bastion | 420 |
| —————— railroad | 417 |
| —————— road | 423 |
| —————— tides | 422 |
| —————— town | 416, 430 |
| Pancha point | 351 |
| Pan de Azucar | 46 |
| Panguapi | 387 |
| Paposo village | 315, 321 |
| Papudo port | 289, 290 |
| Papuya cove | 284 |
| Paquiqui cape | 325 |
| Paraca cape and village | 341, 342 |
| Paracas bay | 343 |
| —————— peninsula | 342 |
| —————— shoal | 341 |
| Pareja creek | 15 |
| Pariña point | 366 |
| Parita bay | 425 |
| Park bay | 178 |
| Parker bay | 207 |
| —————— cape | 204, 207, 208 |
| Parrot cliff | 9 |
| Parry harbour | 176 |
| —————— port | 123, 125, 127 |
| Pasado cape | 382 |
| Passage islands | 106, 111, 112, 113 |
| —————— point and shoal | 187 |
| —————— rock | 414 |
| —————— table | 432 |
| Passages—Atlantic to Pacific, round | |
| Cape Horn | 209–212 |
| —————— by Magellan strait | 212–222 |
| —————— coast of Chile | 316 |
| —————— coast of Peru | 368–370 |
| —————— Pacific to Atlantic, by Magellan strait | 216–222 |
| —————— to and from Bay of Panama | 429–431 |
| Patache point | 327 |
| Patagonia, west coast, Var. and Tides | 65 |
| Patch cove | 246 |
| Patino point | 405 |
| Patillo point | 354 |
| Paul, Mount St. | 147 |

|  | Page |
|---|---|
| Payana point | - 371 |
| ——— shoals | - 371, 372 |
| Paysan | - 361 |
| Payta | - 287, 364, 365, 368 |
| —— point | - 365 |
| Paz bank | - 16 |
| —— island | - 250 |
| Pea point | - 119 |
| Pearl islands | - 410 |
| Pebble island and sound | - 105, 106 |
| Pecheura point | - 255 |
| Peckett barbour | - 163 |
| Pedernales point | - 383 |
| Pedro, San, harbour and passage | 252, 268, 287 |
| ——— bay | - 274 |
| ——— island | - 252 |
| Peel inlet | - 228, 229, 230 |
| Pelado island | - 353, 415 |
| ——————— Panama bay | - 380 |
| Pelepelgua | - 198 |
| Pelican rock | - 297, 298 |
| Pembroke cape | - 78, 80, 82, 84 |
| ——————— light | - 80, 83, 84 |
| Penamarca rock | - 421 |
| Peñas cape | - 133 |
| —— gulf | - 201, 204, 233, 234, 241 |
| Penco | - 281 |
| Penguin cove | - 118 |
| ——————— island | - 52, 53, 248 |
| Penitente point and rock | - 291, 292 |
| Periagua islet | - 183 |
| ——————— rocks | - 256 |
| Perico island | - 416, 421 |
| Perique rock | - 425 |
| Perry bay | - 413 |
| Pescador islands | - 352 |
| Pescadores point | - 336 |
| Peterborough Cathedral | - 319 |
| Petillo point | - 416, 420 |
| Petncura rock | - 257 |
| Philip bay | - 161, 165 |
| ———, St., Mount | - 167, 168, 169 |
| Phillip cape | - 201, 203, 204, 207, 223 |
| Phillips rocks | - 144 |
| Philomel road and port | - 113-115 |
| Piazzi island | - 225, 226, 228, 235 |
| Pica, Pabellon of | - 327 |
| Pichalo point | - 329 |
| Pichidanque bay | - 289-291 |
| Pickthorn point | - 111 |
| Picos point | - 366, 375 |
| Picton island | - 136 |
| —— opening | - 237, 239 |
| Piedras bank | - 1 |
| —— point | - 1, 328, 329, 378 |
| Pillar cape | - 130, 140, 152, 207, 208, 211, 217, 220, 222 |
| Piñas bay and point | - 404 |
| Pinero rock | - 341 |
| Piojo point | - 275 |
| Pirulil head | - 253 |
| Pisagua bay and river | - 327, 329 |
| Pisco bay and town | - 287, 341-343, 345, 369 |
| Pisura point | - 363, 364 |
| Pitt channel | - 229, 230 |
| Piura river | - 364 |
| Plaid island | - 163 |
| Plata island | - 381 |
| —— point | - 321 |
| Playa Baia | - 402 |
| —— grande | - 425 |
| —— Parda cove | - 151, 197, 198, 199, 215, 217 |
| Pleasant port | - 86, 88 |
| ——————— road | - 90 |
| Policarpo cove | - 134 |
| Polillao cove | - 301 |
| Pond bay and mount | - 184 |
| Ponsonby sound | - 136, 141 |
| Poqueldon village | - 264 |
| Poroto point | - 299 |
| Portete point | - 383 |
| Portland bay | - 230 |
| Posadas bay | - 195 |
| Posa harbour | - 386 |
| Possession bank | - 221 |
| ——————— bay | - 159, 214, 219 |
| ——————— cape | - 159-161, 219 |
| Post-office bay | - 397 |
| Pratt passage | - 145 |
| Preservation cove | - 173 |
| Primero cape | - 236, 237 |
| Pringle cape | - 247 |
| Providence cape | - 201, 202, 205, 217 |
| Prowse islands | - 178, 179 |
| Pucari shoal | - 271 |
| Puchachailgua | - 205 |
| Puercos, Morro de | - 426 |

| | Page |
|---|---|
| Puerto Bueno | 229 |
| Pulluche inlet | 249 |
| Pulmun reef | 260 |
| Pulperia reefs | 421 |
| Puluqui channel and island | 270 |
| Puna island | 371 |
| —— Patch | 374 |
| —— town | 374 |
| —— vieja creek | 373 |
| Punoun point | 256, 257 |
| Punta Arenas shoal | 374 |
| Punta Mas-oeste | 226 |
| —— Oeste | 226 |
| Purcell island | 242, 244 |
| Puyo islands | 248 |
| Pyramid hill | 178 |
| —— road and rock | 39 |
| | |
| Quarter-Master island | 162, 164 |
| Quebrada de Camarones | 330 |
| Quedal cape | 273, 274 |
| Queen Adelaide archipelago | 201, 223, 234 |
| —— island | 204 |
| —— Charlotte bay | 112–115, 118 |
| Quehuy island | 265 |
| Quelan bay, channel, and cove | 266, 267 |
| —— point | 266 |
| Quemado point | 340, 341 |
| Quenac island | 260 |
| Queniao point | 258 |
| Quenu island | 269 |
| Quicavi bluff, laguna, and race | 259, 260 |
| Quiebra olla, or Break-pot rock | 281 |
| Quilan island | 252 |
| —— cape and road | 253 |
| Quilca river and valley | 334, 336 |
| Quilimari village | 291 |
| Quillahua point | 273 |
| Quillota, Campana de | 286 |
| Quinchao channel and island | 260, 262 |
| Quinched harbour and village | 264 |
| Quintano isles | 47 |
| Quintay cove | 286 |
| Quintergen point | 259 |
| Quintero bay and rocks | 289 |
| Quintraquin point | 257 |
| Quiriquina channel and island | 281, 282 |
| Quod cape | 189, 192–194, 196, 200, 221 |
| Rabbit island | 111 |

| | Page |
|---|---|
| Race point | 99, 101, 103 |
| —— rocks | 107, 109 |
| Rapel shoal | 284, 285 |
| Raper cape | 246, 247 |
| Rare cove | 283 |
| Rasa islet | 44 |
| —— point | 3, 22 |
| Raso cape and cove | 42 |
| —— reef off | 42 |
| Ray island | 408 |
| Red cape | 231, 351 |
| Redonda point | 26, 28 |
| Redondo rock | 399, 425 |
| Rees islets | 177 |
| Refuge port | 248 |
| Rejoice harbour | 229 |
| Relan, cove, reef, and village | 262 |
| Relief harbour | 226, 228 |
| Reloncavi sound | 270, 271 |
| Remarquable cape | 170 |
| Remolinos point | 257 |
| Rennell island | 225 |
| Reparo bank | 34 |
| Rescue point | 247, 248 |
| Retford cape | 228 |
| Retreat bay | 224 |
| Rey island | 412 |
| Richards port | 114 |
| Richmond road | 136 |
| Rimac river | 348 |
| Rincon or Bahia Blanca | 8, 10 |
| —— point | 321 |
| Rio Negro | 8, 25, 26–30 |
| —— tides | 28 |
| Rivadeneyra shoal | 393 |
| Robledo rocks | 47 |
| Roca, El | 389 |
| Rocky bight | 231 |
| —— cove | 226 |
| —— nook and point | 167, 365 |
| Rodney cove | 118 |
| Rosario port | 236 |
| —— shoal | 271 |
| Rosas bay | 33 |
| —— mount | 2 |
| Round head | 201 |
| —— island, Magellan strait | 203 |
| —— West Falkland | 113 |
| Rous creek | 113 |
| —— sound | 143 |

| | Page |
|---|---|
| Rowlett cape | 250 |
| Roy cove | 111 |
| Royal road | 163, 214 |
| Rubia head | 20, 23, 24 |
| —— point | 23, 24 |
| Ruggles bay | 101, 105 |
| Rumena cape | 279 |
| Rundle pass | 239 |
| Rupert island | 187, 191, 222 |
| Ryan bushes | 13 |
| | |
| Saboga islands | 410, 411 |
| Saddle island | 115 |
| Sail rock | 113 |
| Sal point | 366 |
| Salaberria reef | 42, 43 |
| Salado bay and point | 306, 307 |
| —— river | 2 |
| Salamanca peak | 47 |
| Salango island | 380 |
| Salcedo rock | 342 |
| Salinas bay | 352 |
| —— point | 290, 352, 371, 373 |
| —— rocks | 352 |
| Salvador port | 77 |
| Sama Morro and point | 331 |
| Samanco or Guambacho bay | 357, 358 |
| —— head | 357, 358 |
| San Andres bay | 229, 230, 247 |
| ——Antonio cape | 1, 2, 3, 274 |
| —— cove | 285 |
| —— fort | 288 |
| —— port, Magellan strait | 173, 174 |
| —— Patagonia | 33, 34, 63 |
| —— Blas | 21–25, 31, 63 |
| —— channel | 235 |
| —— Carlos fort | 254, 275 |
| —— port, Chiloe island | 253–256, 273 |
| —— East Falkland | 100, 105 |
| —— Carmen | 22, 26, 30 |
| —— Diego cape | 133, 134 |
| —— Estevan gulf | 241, 242 |
| —— port | 247, 248 |
| —— shoal | 55 |
| —— Fernando islands | 248 |
| —— Francisco cape | 325, 383 |
| —— island | 346 |
| San Gallan island | 341, 342, 369 |
| —— point | 256, 257 |
| —— Ildefonso cape | 201, 202 |
| —— Isidro cape | 167, 169, 171, 176 |
| —— Jose bank | 410 |
| —— rock | 410 |
| —— shoal | 270 |
| —— Josef port | 36 |
| —— Juan island | 174 |
| —— port | 338 |
| —— river | 392 |
| —— Julian port | 56 |
| —— Lorenzo cape | 381 |
| —— island | 347, 349, 350, 351 |
| —— light | 349 |
| —— Matheo point | 381 |
| —— Matias gulf | 31, 33 |
| —— Miguel bay | 405–407 |
| —— port | 186, 412 |
| —— Nicolas bay | 171–173, 187, 215 |
| —— port | 338 |
| —— Pablo cape | 133 |
| —— Pasqual reef | 46 |
| —— Pedro bay | 274 |
| —— harbour and passage | 152, 268, 287 |
| —— island | 252 |
| —— point | 315 |
| —— sound | 184 |
| —— Policarpo | 240 |
| —— Quentin sound | 243, 246 |
| —— Rafael gulf | 245 |
| —— Roman cape | 240 |
| —— Roque point | 45 |
| —— Sebastian bay and cape | 132, 133, 172 |
| —— channel | 172 |
| —— Tadeo river | 242–244 |
| —— Vicente cape | 134 |
| —— port | 280, 281 |
| Sanborombon bay and river | 2 |
| Sandbar island | 104 |
| Sandy bay | 225, 231 |
| —— point | 165–167, 173, 221 |
| Santa Anna point | 167–169 |
| —— bay and island | 358, 359 |
| —— Casilda cape | 205 |
| —— Clara, or Amortajada island | 371 |
| —— light | 371 |
| —— tides | 374 |

| | Page |
|---|---|
| Santa Cruz river | 57 |
| —— Elena | 380 |
| ———— bay | 380 |
| ———— point | 379 |
| ———— port | 43 |
| —— Ines cape | 133 |
| —— Lucia cape | 235 |
| —— Magdalena island | 163–165, 215, 218 |
| —— Maria island | 278 |
| ———— point | 339 |
| ———— road | 279 |
| —— Marta island and reef | 162–164, 215, 218 |
| ———— bank | 382 |
| —— Monica port | 205 |
| —— Rosa island | 340 |
| ———— river | 372 |
| Santay island | 376 |
| Santiago cape | 235 |
| Sarco bay | 301 |
| Sarmiento bank | 158, 159 |
| ———— channel | 219, 225, 226, 228, 235 |
| ———— mount | 130, 131, 146, 175, 177 |
| Saturday harbour | 151 |
| Saumarez island | 230, 231 |
| Saunders island | 107–109 |
| Savanna river | 408, 409 |
| Schapenham bay | 142 |
| Schetky cape | 150 |
| Schomberg cape | 146, 148 |
| Schooner cove | 229 |
| Scotchwell harbour | 250 |
| Scourfield bay | 137 |
| Sea Bear bay | 53 |
| —— Dog island | 118 |
| —— Lion islands | 57, 59, 60, 83, 93, 94 |
| ———— rocks | 79 |
| —— reach | 201, 215, 216, 221 |
| Seal cove | 92, 189 |
| —— point | 85 |
| —— rocks, Falkland islands | 80–83, 115, 116 |
| ———— Patagonia | 236, 237 |
| ———— Peru | 336, 362 |
| Seasons, coast of Chile | 316, 317 |
| Sebastiana bank | 254, 273 |
| ———— island | 254, 273 |
| Sechura bay | 364 |
| Second Barranca, or Piedras point | 22 |
| | Page |
|---|---|
| Second narrows | 162–165, 214, 215, 218 |
| Secretary Wren island | 191 |
| Sedge island | 106, 107, 109, 110 |
| Sedger river | 168, 169 |
| Selaian rock | 257 |
| Señora island | 413 |
| Señorita island | 413 |
| Serena, la | 296 |
| Serpent bank | 19, 20 |
| Serrate channel | 340 |
| Sesambre | 137 |
| Sevira island | 401 |
| Shag harbour | 99, 104 |
| —— island | 54, 56, 144 |
| —— narrows | 180, 181, 184 |
| —— rock | 414 |
| ———— East Falkland | 83, 93 |
| ———— Patagonia | 54 |
| Shallow harbour | 113, 114, 116 |
| Shelter island | 198 |
| Shingle point | 343 |
| Ship gat | 23 |
| —— harbour | 115–117 |
| —— island | 44 |
| Shipton cove | 185 |
| Shoal island | 224 |
| Sholl bay | 177, 204 |
| —— point | 56 |
| —— port | 183 |
| Shoulder peak | 152 |
| Sierra Tandil | 5 |
| —— Ventana | 8, 10 |
| —— Vulcan | 5 |
| Sierras de San Antonio | 35 |
| Silla de Payta | 364 |
| Silvester point | 162–164 |
| Simon bay | 181, 184, 185 |
| Skyring harbour | 207 |
| ———— island | 146, 249 |
| ———— mount | 146, 147, 175 |
| ———— water | 184, 190, 228 |
| Small islet | 96, 115 |
| ———— craft bight | 230 |
| ———— sheep coves | 251 |
| Smylie channel | 118 |
| Smyth channel | 204, 219, 223, 225, 226, 228, 235 |
| ———— harbour | 180–182 |
| Snow sound | 196 |
| Snowy channel | 197 |

|  | Page |
|---|---|
| Snug bay - - - | 185, 215 |
| Socorro island - - | - 249 |
| Solano bay and point - - | - 402 |
| Solar bay - - - | - 350 |
| —— point - - | - 346, 349 |
| Soldado point - - | 270, 390 |
| Solitaria - - - | - 262 |
| Sorrell ledge - - | - 50 |
| South Barrancas - - | - 27 |
| —— cape - - | - 45, 127, 135 |
| —— point - - | - 27, 99 |
| Spaniard harbour - - | - 135 |
| Sparrow cove - - | - 81 |
| Spartan pass - - | - 237 |
| Speedwell bay - - | - 239 |
| —— island - - | - 97, 98 |
| Spiring bay - - | - 54 |
| Split island - - | - 111 |
| Squally point - - | - 177 |
| St. Anthony cape - - | - 127 |
| — Elmo island - - | - 412 |
| Staats island - - | - 116 |
| Stag road - - | - 79 |
| Staines peninsula - - | - 228 |
| Stanley harbour - - | 72, 73, 81, 84 |
| —— island - | 406, 407, 408 |
| Staples inlet - - | - 183 |
| Starve island - - | - 20 |
| Staten island - - | 122–129, 209–211 |
| ——, geology of - - | - 128 |
| ——, natural history of - | - 128 |
| —— tides - - | - 66, 122 |
| —— vegetation - | 128, 219 |
| Steep point and reef - | - 338 |
| Stephens bay - - | - 396 |
| —— port and bluff - | 118–120 |
| Stewart bay, Magellan strait - | 194, 197 |
| ——, Patagonia - | - 247 |
| —— harbour - | - 145 |
| —— island - | - 144, 145 |
| Stokes creek - - | - 184 |
| —— inlet - - | - 177 |
| —— point - - | - 189 |
| Stop cove - - | - 118 |
| Stormy bay - - | - 178 |
| Strain island - - | - 406 |
| Stragglers rocks - - | - 204 |
| Sea head and village - | - 384 |
| Sugar loaf, James island - | - 398 |
| —— Holloway sound - | 242, 244 |
| —— island, Chile | - 314, 316 |

|  | Page |
|---|---|
| Sugar loaf island, Magellan strait | 189, 208 |
| Sulivan bay - - - | - 398 |
| —— harbour -- - | - 95 |
| —— sound - - | - 189 |
| Sulphur rocks - - | - 422 |
| Summer isles - - | - 224 |
| Sunday cape - - | - 133, 151 |
| Sunk rock, East Falkland - | - 103 |
| —— Magellan strait - | - 203 |
| Supé bay and town - | - 354, 355 |
| Sur, cape del - - | - 45 |
| Susanna shoal - - | - 50 |
| Sussex port - - | - 100 |
| Swallow bay - | - 196, 197, 220 |
| —— harbour - | - 215 |
| Swan inlet - - | - 91 |
| —— islands - | - 69, 102–104 |
| —— passage - | - 104 |
| Sweepstakes foreland - | - 162 |
| Symonds harbour - | - 114 |
| Tabla cape - - | - 291, 292 |
| Table of Orozco - | - 133 |
| Taboga island - - | - 423 |
| —— rock - - | - 422 |
| Taboguilla island - | - 423 |
| Tabon island - - | - 269 |
| Tabor island - - | - 424 |
| Tacna - - | - 330 |
| Tagus cove - - | - 399 |
| Talara point - - | - 365 |
| Talcahuana port - - | 281, 282 |
| Talcan inlet and island - | 261, 262 |
| —— rocks - - | - 261 |
| Talinay mount - - | - 292 |
| Taltal point - - | - 315, 321 |
| Tamar cape - | - 201–205, 217, 219 |
| —— harbour - | - 105 |
| —— island - | - 203, 204 |
| —— pass - | - 105, 106 |
| —— port - | - 105, 203, 216 |
| Tambo valley - - | - 333, 334 |
| Tanque, Playa de, and village - | 293, 294 |
| Tantil island - - | - 270 |
| Tarn bay - - | - 233, 234, 240 |
| —— mount - - | - 169 |
| Tate cape - - | - 150 |
| Taura river - - | - 377 |
| Taylor point - - | - 184 |

| | Page |
|---|---|
| Taytao cape | 248, 249 |
| Tea channel | 116, 117 |
| —— island | 117 |
| Teatinos point | 299 |
| Temblador cove | 300 |
| Tembleque islands | 372 |
| Tenglo island | 271 |
| Tenoun point | 260, 262 |
| Tenquehuen islands | 248, 249 |
| Tenquelil island | 261 |
| Tenuy point | 254 |
| Teran isles | 184 |
| Terapa island | 424 |
| Terhiten island | 137 |
| Terrapin road | 397 |
| Tetas point | 322 |
| Texada point | 14 |
| Thetis bay | 134 |
| Thieves sound | 145, 178 |
| Thomas cape | 354 |
| Three brothers | 133 |
| —— crowns hill | 119 |
| —— fingers island | 249 |
| —— island bay | 189 |
| —— peaks cape | 235 |
| Tianitau | 240 |
| Tickle pass | 104 |
| Tide creek | 33 |
| —— races 37, 66, 76, 102, 118, 127, 134, | 210, 259 |
| —— rock | 103 |
| —— table, Ancud gulf | 272 |
| Tides are under their respective headings. | |
| Tierra del Fuego | 129–156 |
| —— currents | 66, 155, 166 |
| —— general observations upon the sea coast of | 130, 131, 153–156 |
| —— S. coast, soundings | 131 |
| —— winds and weather | 153–155 |
| Tilgo islet | 299 |
| Tilli road | 48 |
| Tilly bay | 191 |
| Tiquia reef | 261 |
| Tirua cape | 276, 277 |
| Todos Santos lake | 271 |
| Tolten river | 276 |
| Tom bay | 230 |
| Tom port or harbour | 146, 147 |
| Tombo point | 41 |
| Toms narrows | 181–184 |
| Tongoy bay | 293 |
| —— Playa de | 293, 294 |
| Topar island | 230, 237 |
| Topocalma point | 284, 285 |
| Toro bank | 11, 15 |
| —— point | 285 |
| —— reef | 300 |
| Tortola island | 421 |
| Tortolas islets | 315 |
| Tortolita island | 421 |
| Tortoral baxo | 305 |
| —— cove | 293, 305 |
| Tortoralillo bay | 299, 300, 312 |
| Tortuga island | 357 |
| —— mount | 358 |
| —— peak | 388 |
| —— rocks | 289 |
| Tosca | 7 |
| Tova island and cove | 46 |
| Tower island | 399 |
| —— rock | 50, 52, 235 |
| —— rocks | 148 |
| Town point | 111, 112 |
| Townshend harbour | 145, 146 |
| Tranque island | 266, 267 |
| Transition bay | 177 |
| Trapiche peninsula | 413 |
| Treble island | 144 |
| Tree bluff | 175 |
| Trefusis bay | 143 |
| Tres Cruces point | 257, 285 |
| —— Montes cape | 244–247 |
| —— gulf | 242, 244, 245 |
| —— Puntas cape | 48, 50 |
| —— shoals | 48 |
| Trevan rock | 407 |
| Trinidad channel | 230, 236, 237 |
| —— gulf | 219, 223, 235–237, 239 |
| —— river | 415 |
| Trinchera point | 379 |
| Triton bank | 161 |
| Trollope rock | 410 |
| Trujillana entrance | 340 |
| Tumaco island and town | 387 |
| Truxillo | 361 |
| —— bay | 206 |
| Tubul river | 279 |
| Tucapel head and point | 277 |
| Tuesday bay | 206 |
| —— cove | 206, 216 |
| Tumbes heights | 281 |

| | Page |
|---|---|
| Tumbez bay | 372 |
| —— river | 366, 372 |
| Tungo village | 340 |
| Tunquen bight | 286 |
| Turn cape | 177, 178 |
| Tussac island | 87 |
| —— rocks | 146, 147 |
| Tuyra river | 409 |
| Tuyu river and bank | 2, 3 |
| Twins rocks | 110 |
| Two Sisters cape | 31 |
| Tyssen islands | 100, 104 |
| —— island passage | 104 |
| —— patch | 104 |
| Ulloa peninsula | 191, 192, 196 |
| Union bay | 19, 20 |
| —— point | 41 |
| —— sound | 226, 227 |
| Upright bay and cape | 201, 202, 204 |
| —— cape anchorages | 204, 205 |
| Uranie rock | 78 |
| Urava island | 423, 424 |
| Uriarte port | 205 |
| Usborne islands | 248 |
| —— mount | 354, 355 |
| Useless bay | 172, 173, 176 |
| —— cove | 247 |
| Utria, port | 402 |
| Vacamonte cove and point | 424 |
| Vaca point, Lengua de | 293, 294 |
| Vaguila rock | 408 |
| Valao chico | 372 |
| Valdes creek | 32, 37 |
| Valdez port | 174 |
| Valdivia | 273, 274, 380 |
| Valentine harbour | 204, 206, 216 |
| Valentyn bay | 135 |
| —— cape | 167, 169, 173 |
| Valladolid island | 425 |
| Vallenar islands and road | 249, 250 |
| Valle point | 161 |
| Valparaiso bay | 286–289 |
| —— light | 286 |
| Vancouver island | 226, 228, 229, 235 |

| | Page |
|---|---|
| Vancouver port | 122–124, 127, 128 |
| Van isles | 237 |
| Vano points | 292 |
| Variation 30, 67, 190, 235, 298, 320, 370, 400 | |
| Vascuñan cape | 302 |
| Venado island and point | 421 |
| Ventanilla point | 289 |
| Vera bay | 42 |
| Verde point | 386 |
| Vernal, the | 177, 183 |
| Viana isles | 47 |
| Vicente, San, port | 280, 281 |
| Viciosa island | 387 |
| Victor Gully | 330, 331 |
| Victoria harbour | 91 |
| Victory cape | 208, 216, 220, 234 |
| —— pass | 223, 225, 226, 228 |
| Video, Monte | 55 |
| Viejas island and spit | 340 |
| Vigia de San Pablo | 389 |
| Vilcun mount | 262 |
| Villarino point | 33 |
| Villena cove | 195 |
| Villiers point | 189 |
| Viper bank | 21 |
| Vique cove and town | 424 |
| Virago point | 407 |
| Virgins cape | 63, 157–159, 221 |
| —— reef | 158 |
| Viuda island | 357 |
| —— rock | 387 |
| Viveros island | 412 |
| Vivian island | 189 |
| Voces bay | 169 |
| Vogelborg rock | 279 |
| Volcano, Capt. B. Hall's | 176 |
| Volunteer point | 75, 78, 82, 84 |
| Wafer bay | 427 |
| Wager island | 239 |
| Wales point | 124 |
| Walker bay | 230 |
| —— shoal | 163, 164 |
| Wallis mark | 191 |
| Wallis shoal | 159, 215 |
| Warp bay | 178 |
| Warrah river | 106 |
| Warrington cove | 182 |
| Washington island | 406 |

| | Page |
|---|---|
| Watchman cape - - 54, 55 |
| Water cove - - - 141 |
| Waterfall bay - - - 233 |
| Watering - 11, 16, 19, 22, 42, 47, 55, 61, |
| 79, 81, 82, 91, 107, 112, 115, |
| 127, 145, 148, 151, 152, 160, |
| 168, 185, 204, 213, 236, 245, |
| 283, 287, 318, 342, 397, 401, |
| 405, 419, 424. |
| Waterman island - - 143, 144 |
| Weather on the coast of Chile - 316 |
| ———— Peru - 367 |
| ———— rocks - - - 152 |
| Weddell bluff - - - 57 |
| ———— island - - 116–118 |
| Week islands - - 131, 151, 152 |
| Weir creek - - - - 82 |
| Welcome bay - - - 225 |
| Wellington island 204, 223, 233–237, 239 |
| Wells point - - - 53 |
| Wenman island - - - 399 |
| West bay - - - 319 |
| —— cape - - - 138, 141 |
| —— channel - - - 30, 235 |
| —— cove, East Falkland - - 91 |
| ——, Patagonia - - 43 |
| —— island - - - 104 |
| —— passage - - - 112 |
| West point island and passage - 108–110 |
| Westminster Hall - - - 208 |
| Wet island - - - 181 |
| Whale's back shoal - - 347, 349 |
| Whale-boat bay - - - 227 |
| ———— sound - - 136, 145 |
| Whale passage - - - 113 |
| ——— point - - - 191 |
| ——— sound - 182, 192, 196 |
| Whaler bay - - - 111 |
| ——— passage - - - 105 |
| Wharton harbour - - - 100 |
| White horse island - - - 235 |
| ——— islets - - - 335 |
| ——— kelp cove - - 233 |
| ——— narrows - - - 228 |
| ——— patch - - - 383 |
| ——— point - - 87 |
| ——— rock bay - 101, 103–105 |
| ———, Chile - 305, 321 |
| ———, East Falkland - 99, 103 |
| ——— harbour - 75, 101 |
| ——— point - 101, 285 |

| | Page |
|---|---|
| White Stone Cobija - - 334 |
| Wickham island - - 174, 249 |
| Wide Channel - 223, 230, 231 |
| Widow rock - - - 387 |
| Wigwam point - - - 187 |
| Willes bay - - - 175 |
| William port - - 80, 82–84 |
| Williwaws - - - 140 |
| Winds, Falkland islands - - 70 |
| ——— on the coast of Chile - 316–318 |
| ——— Peru - 367 |
| Winds and weather, Patagonia, East coast - - - 63 |
| ———, West coast - - - 246 |
| ——— in bay of Panama 393, 428 |
| Windward bay - - - 237 |
| Wolf island - - - 101 |
| —— rock - - 81–83 |
| Wollaston islands - - 137, 138 |
| Woodcock mount - - - 181 |
| Wood cove - - - 189 |
| —— islands - - - 143 |
| —— mount - - 55, 56 |
| —— shoal - - 105, 121 |
| Woods bay - - 186, 215 |
| Worsley bay and sound - - 228 |
| Wreck island - 106, 107, 109 |
| ——— point - - 238, 396 |
| Xavier island and port - - 241, 242 |
| Xaultegua gulf - 184, 194, 199, 200, 202, 217 |
| Yal bay, cove, and point - - 265 |
| Yannas Cove - - - 278 |
| Yca River - - - 339 |
| —— village - - - 340 |
| Yemcouma isle - - - 253 |
| Yerba Buena hamlet - - 299 |
| Ylo road and village - - 332 |
| Ymerquiña island - - - 261 |
| Ynche-mo island - - 248, 249 |
| Yngles bank - - 255, 256 |
| Yngles bay and port - - 310 |
| York Minster - - 143, 144, 153 |
| —— road - 187, 215, 222 |

| | Page | | | | Page |
|---|---|---|---|---|---|
| Ypun or Narborough island | - 249, 250 | Zapallar point | - | - | - 290 |
| Ysla Blanca bay | - 315 | Zapo peak | - | - | - 405 |
| Zampo Palo | 371, 373 | Zarate island | - | - | - 341 |
| Zach Peninsula - | - 225, 226 | Zuraita island - | - | - | - 11 |

LONDON :

Printed by GEORGE E. EYRE and WILLIAM SPOTTISWOODE,
Printers to the Queen's most Excellent Majesty.
For Her Majesty's Stationery Office.

SD - #0023 - 290724 - C0 - 229/152/27 - PB - 9780282657420 - Gloss Lamination

# Forgotten Books

*Forgotten Books' Classic Reprint Series utilizes the latest technology to regenerate facsimiles of historically important writings.*

*Careful attention has been made to accurately preserve the original format of each page whilst digitally enhancing the quality of the aged text.*

---

*Philosophy ~ Classics ~ Science ~ Religion History ~ Folklore ~ Mythology*

---

Forgotten Books